Topics in Functional Analysis

ADVANCES IN Mathematics
SUPPLEMENTARY STUDIES

Topics in Functional Analysis

Essays Dedicated to M. G. Krein on the Occasion of His 70th Birthday

ADVANCES IN MATHEMATICS
SUPPLEMENTARY STUDIES, VOLUME 3

EDITED BY

I. Gohberg

*Department of Mathematical Sciences
Tel Aviv University
Tel Aviv, Israel
and
Department of Pure Mathematics
The Weizmann Institute of Science
Rehovot, Israel*

M. Kac

*Department of Mathematics
The Rockefeller University
New York, New York*

With the Editorial Board
of *Advances in Mathematics*

ACADEMIC PRESS New York San Francisco London 1978
A Subsidiary of Harcourt Brace Jovanovich, Publishers

ACADEMIC PRESS, INC.
111 Fifth Avenue, New York, New York 10003

United Kingdom Edition published by
ACADEMIC PRESS, INC. (LONDON) LTD.
24/28 Oval Road, London NW1 7DX

Library of Congress Cataloging in Publication Data
Main entry under title:

Topics in functional analysis.

Includes bibliographical references.
CONTENTS: DeBranges, L. and Trutt, D. Quantum Cesaro operators.--Case, K. M. Inverse scattering, orthogonal polynomials, and linear estimations.--Dym, H. and Kravitsky, N. On recovering the mass distribution of a string from its spectral function. [etc.]
1. Functional analysis--Addresses, essays, lectures.
2. Krein, Mark Grigor'evich. I. -Krein, Mark Grigor'evich, Date II. Gokhberg, Izrail' Tsudikovich. III. Kac, Mark.
QA320.T66 515'.7 78-13184
ISBN 0-12-287150-2

PRINTED IN THE UNITED STATES OF AMERICA

M. G. KREIN

Contents

List of Contributors xiii
Biography of M. G. Krein xv
Preface xix

Quantum Cesàro Operators 1

Louis de Branges and David Trutt

References 24

Inverse Scattering, Orthogonal Polynomials, and Linear Estimation

K. M. Case

I. Introduction 25
II. Orthogonal Polynomials 26
III. The Discrete Inverse Scattering Problem 33
IV. Relations between Inverse Scattering and Orthogonal Polynomials 38
V. Linear Estimation 40
VI. Conclusion 42
References 43

On Recovering the Mass Distribution of a String from its Spectral Function

Harry Dym and Naftali Kravitsky

Notation 45
1. Introduction 46
2. Factorization Identities 50
3. Auxiliary Identities 57
4. The Main Theorem 66
5. The Gelfand–Levitan Construction 70
6. Ramifications 73
7. Examples 77
8. Extensions of the Construction 84
References 89

Spectral Analysis of Families of Operator Polynomials and a Generalized Vandermonde Matrix, 1. The Finite-Dimensional Case

I. Gohberg, M. A. Kaashoek, and L. Rodman

I. Introduction and Preliminaries 91
II. The Generalized Vandermonde Matrix 96
III. Systems of Equations 112
References 128

Orbit Structure of the Möbius Transformation Semigroup Acting on H^∞ (Broadband Matching)

J. William Helton

1. Introduction 129
2. Reduction to a Linear Problem 134
3. The Boundary Case 140
4. Orbit Structure 143
5. Simplifying the Scalar Case and Sensitivity of Orbit 146
6. Explicit Computation 148
7. Motivation and History 153
References 155

On the Asymptotic Number of Bound States for Certain Attractive Potentials

M. Kac

1. 159
2. 159
3. 162
4. 164
5. 165
References 167

Generalized Krein–Levinson Equations for Efficient Calculation of Fredholm Resolvents of Nondisplacement Kernels

Thomas Kailath, Lennart Ljung, and Martin Morf

I. Introduction 169
II. Generalized Sobolev Identities 174

III. Generalized Krein–Levinson Equations 177
IV. Further Results and Extensions 181
References 183

Trotter's Product Formula for an Arbitrary Pair of Self-Adjoint Contraction Semigroups

Tosio Kato

1. Introduction 185
2. Proof of the Theorem, Part 1 186
3. Proof of the Theorem, Part 2 189
4. Applications 191
5. Generalizations to Improper Operators 192
Addendum 194
References 195

The Time Delay Operator and a Related Trace Formula

Peter D. Lax and Ralph S. Phillips

1. Introduction 197
2. The Time Delay Operator 201
3. A Related Trace Formula 207
References 215

Boussinesq's Equation as a Hamiltonian System

H. P. McKean

1. Introduction 217
2. Eigenfunctions of L 218
3. Hamiltonian Flows 220
4. Estimating the Periodic Spectrum of L 222
5. The Solitary Wave 224
References 226

On the Set of Minimal Extensions of a Subadditive Functional

D. P. Milman

1. 227
2. 228

3. 228
4. Sketch of the Proof of Theorem B 229
5. The Special Cases 229
6. Details of Proof of Theorem B: Case (a)—*E* Is a Vector Space 230
7. Details of Proof of Theorem B: Case (b)—*E* Is a Fréchet Space 235
8. Addition and Conclusion 238
References 240

Symmetric Singular Integral Operators with Arbitrary Deficiency

Joel D. Pincus

Introduction 241
I. Riemann–Hilbert Problems 243
II. 253
III. 260
IV. 262
V. The Deficiency Spaces Reexamined 266
VI. The Domain of *L* and Its Extensions 269
VII. Remarks 271
VIII. Symmetric Toeplitz Operators 278
References 281

Change of Variables Formulas with Cayley Inner Functions

Marvin Rosenblum and James Rovnyak

1. Introduction 283
2. Composition Formulas for Harmonic Functions 285
3. Composition with Inner and Cayley Inner Functions 292
4. Composition Formulas Involving Singular Integrals 300
5. Applications. Some Distribution Function Results 302
6. Applications. Ahiezer Transforms 306
7. Applications. Generalized Chebychev Expansions 311
References 318

The Complex-Wave Representation of the Free Boson Field

I. E. Segal

1. Introduction 321
2. The Free Boson Field 325
3. The Partial Fourier–Wiener Transform 335
4. Standardization of the Real-Wave Representation 338
5. The Intertwining Operator between the Complex and Real Representations 340

References 343

Families of Pseudodifferential Operators

Harold Widom

I. Introduction 345
II. Pseudodifferential Families on R^n 350
III. Pseudodifferential Families on Manifolds 359
IV. The Symbol 360
V. Compact Manifolds 376
VI. Applications 386

References 394

List of Contributors

Numbers in parentheses indicate the pages on which the authors' contributions begin.

LOUIS DE BRANGES (1), Department of Mathematics, Purdue University, Lafayette, Indiana 47907

K. M. CASE (25), The Rockefeller University, New York, New York 10021

HARRY DYM (45), Department of Mathematics, The Weizmann Institute of Science, Rehovot, Israel

I. GOHBERG (91), Department of Mathematical Sciences, Tel Aviv University, Tel Aviv, Israel, and Department of Pure Mathematics, The Weizmann Institute of Science, Rehovot, Israel

J. WILLIAM HELTON (129), Department of Mathematics, University of California, San Diego, La Jolla, California 92093

M. A. KAASHOEK (91), Wiskundig Seminarium der Vrije Universiteit, Amsterdam, The Netherlands

M. KAC (159), Department of Mathematics, The Rockefeller University, New York, New York 10021

THOMAS KAILATH (169), Department of Electrical Engineering, Stanford University, Stanford, California 94305

TOSIO KATO (185), Department of Mathematics, University of California, Berkeley, Berkeley, California 94720

NAFTALI KRAVITSKY (45), Department of Mathematics, The Weizmann Institute of Science, Rehovot, Israel

PETER D. LAX (197), Courant Institute of Mathematical Sciences, New York University, New York, New York 10012

LENNART LJUNG† (169), Department of Electrical Engineering, Stanford University, Stanford, California 94305

H. P. McKEAN (217), Courant Institute of Mathematical Sciences, New York University, New York, New York 10012

D. P. MILMAN (227), Department of Mathematical Sciences, Tel Aviv University, Tel Aviv, Israel

MARTIN MORF (169), Department of Electrical Engineering, Stanford University, Stanford, California 94305

RALPH S. PHILLIPS (197), Department of Mathematics, Stanford University, Stanford, California 94305

† Present address: Linköping University, Sweden.

JOEL D. PINCUS (241), Department of Mathematics, State University of New York, Stony Brook, Stony Brook, New York 11794

L. RODMAN (91), Department of Mathematical Sciences, Tel Aviv University, Tel Aviv, Israel

MARVIN ROSENBLUM (283), Department of Mathematics, University of Virginia, Charlottesville, Virginia 22903

JAMES ROVNYAK (283), Department of Mathematics, University of Virginia, Charlottesville, Virginia 22903

I. E. SEGAL (321), Department of Mathematics, Massachusetts Institute of Technology, Cambridge, Massachusetts 02139

DAVID TRUTT (1), Department of Mathematics, Lehigh University, Bethlehem, Pennsylvania 18015

HAROLD WIDOM (345), Natural Science Division, University of California, Santa Cruz, Santa Cruz, California 95064

Biography of M. G. Krein

Mark Grigorevich Krein, one of the most eminent mathematicians of our time, celebrated his seventieth birthday on April 3, 1977, thus marking more than fifty years of creative work of unsurpassed breadth and quality.

Krein is the author of more than 250 papers and monographs in algebra, function theory, probability theory, differential and integral equations, geometry of Banach spaces, operator theory, and, last but not least, mathematical physics and mechanics. His work opened up new areas of mathematics while greatly enriching the more traditional ones. In addition to having educated dozens of brilliant students in his native land, he has inspired the work of many more in all corners of the earth.

Krein was born in Kiev into a Jewish family of modest means, and he was one of four brothers and two sisters. His father was a lumber merchant.

Having completed his secondary school education at the age of fourteen, he at once started to attend evening lectures of Professor Delone at the Kiev Polytechnic Institute. These lectures were on calculus and descriptive geometry, which were part of the traditional engineering curriculum. In addition Krein also participated in Delone's seminar on number theory and Galois theory, and he gave his first scientific talk in this seminar when still fourteen years of age. For the next three years, though too young to register as a regular student, he attended lectures in the Mathematical Section of the Kiev Institute for People's Education and took part in a number of seminars organized by D. A. Grave.

In the spring of 1924, at the age of seventeen, Krein ran away from home and went to Odessa. During the summer, despite difficult living conditions, Krein wrote his first scientific paper, which was published two years later under the title "Le système dérivé et les contours dérivés." When in the autumn of 1924 he gave a talk on this work at the Odessa Institute for People's Education (later to become the Odessa State University) he caught the attention of Professor N. G. Chebotarev, who was in attendance. Chebotarev was so impressed that in 1926 and with the help of Professor S. O. Shatounovsky he arranged for Krein to be admitted as an aspirant (equivalent of a Ph.D. student in the West). In the intervening two years Krein supported himself by tutoring in mathematics; and one of his students, R. L. Romen, later became his wife. During this period

Krein also attempted a career as a sailor and then even that of an acrobat. Fortunately for mathematics he failed in both.

Krein completed his graduate studies in 1929. No advanced degrees were awarded in those days in the Soviet Union (the awards were reestablished in the late thirties), but Krein presented to the mathematics faculty seven papers (three in algebra and four in analysis, including one in function theory), of which one (in function theory) was chosen as a dissertation for defense. All seven papers were subsequently published.

After working for about four years in various institutes in Odessa, Krein was appointed to the chair of Function Theory and later to the chair of Analysis in the newly reestablished Odessa State University. During this period (1934–1940) Krein worked on the moment problem, geometry of Banach spaces, quadratures, integral equations, and spectral theory of operators. Much of this work (some of it in collaboration with N. I. Achieser and F. R. Grantmacher) is today justly considered classical.

Krein, an inspiring and enthusiastic teacher, attracted also a group of highly gifted students, among whom were A. B. Artemenko, M. S. Livschits, D. P. Milman, M. A. Naimark, V. P. Potapov, M. A. Rutman, and V. L. Shmulyan.

In 1939 the Moscow State University took the almost unprecedented step of nominating Krein for the Doctor's degree (the highest academic degree in the Soviet Union) without the usual requirement of a defense of a dissertation. In the same year the Academy of Sciences of the Ukrainian Soviet Republic elected Krein a corresponding member.

Krein spent the war years (1941–1944) in Kuibyshev on the Volga where he held the chair of Theoretical Mechanics at the local industrial institute, and he took an active part in organizing the Kuibyshev Aviation Institute.

Not long after his return to Odessa in 1944 Krein was dismissed from the university, and his students had to leave. Thus the excellent and famous school of functional analysis at this university came to an end.

From the university Krein moved to the Odessa Institute of Naval Engineering and then (in 1954) to the Odessa Institute of Civil Engineering. In both institutes he held the chair of Theoretical Mechanics.

From 1944 to 1952 Krein also held a part-time position as head of the department of Functional Analysis and Algebra at the Mathematical Institute of the Ukrainian Academy of Sciences in Kiev, but then he as well as Achieser were dismissed on the grounds that they were not permanent residents of Kiev. A number of other mathematicians, who also were not residents of Kiev, retained their posts.

Clearly the postwar years were not easy for Krein. He nevertheless continued to work; many of his best known and most striking results, in particular, those concerned with the spectral theory of the vibrating string and the inverse Sturm–Liouville problem, date from this period.

In the same period he developed the method of directive functionals, ways of continuing helical arcs in Hilbert spaces, and obtained fundamental results in the stability theory of Hamiltonian systems.

Deprived of a university base, it is remarkable that he nevertheless assembled around him a group of highly gifted young mathematicians, many of whom though officially affiliated with other institutions, received their principal guidance and inspiration from him. In this group we find among others V. M. Adamyan, D. Z. Arov, Y. M. Berezansky, I. S. Iokhidov, M. A. Krasnoselski, I. S. Kac, S. C. Krein (M. G.'s youngest brother), H. K. Langer, A. A. Nudelman, C. Y. Popov, and Y. L. Shmulyan.

With few exceptions the mathematical community in the Soviet Union, in full realization of Krein's great achievements, had tried to gain for him a measure of the official recognition that he so richly deserves. He was repeatedly nominated by the Moscow Mathematical Society, the Mathematics Faculty of the Moscow State University, and by other distinguished institutions for the Lenin and the Stalin prizes, for membership in the Soviet Academy of Sciences, and for full membership in the Ukrainian Academy of Sciences, all to no avail. Worse yet, recently some of Krein's best known students and collaborators have also encountered difficulties.

Krein was retired recently from his academic post (there is no obligatory retirement age in the Soviet Union, and many mathematicians older than Krein continue to hold their positions). He continues to live in Odessa, a city which he has always regarded with love and affection. A grandson has recently entered the University as a student in mathematics.

Krein remains active in research, and his standards remain as high as ever. A perfectionist in the extreme he sometimes exasperates his collaborators. When one of them was asked how the manuscript of a book he was writing with Krein was coming along he replied with sadness that it was 90% complete. "Why then do you sound so sad?" came the next question. "Because," replied the junior author, "if you had asked me a week ago I would have told you that the manuscript was 95% complete."

In dedicating this volume to M. G. Krein on the occasion of his seventieth birthday and in recognition of his fundamental and brilliant contributions to mathematics, the authors and editors join his other friends and admirers in wishing him many happy and productive years to come.

Preface

The first paper in this volume, "Quantum Cesàro Operators," by Louis de Branges and David Trutt, is related to eigenfunction expansions for the Cesàro operator of summability theory. A striking aspect of the expansion is the existence of plane measures with respect to which the Newton polynomials form an orthogonal set. In this work a generalization is obtained in which Newton polynomials are replaced by Heine polynomials.

K. M. Case, in "Inverse Scattering, Orthogonal Polynomials, and Linear Estimation," discusses the relations between these fields. He shows that all three can be reduced to a generalized form of the discrete Gelfand–Levitan equation for which an explicit solution is readily obtained.

Harry Dym and Naftali Kravitsky, in their paper "On Recovering the Mass Distribution of a String from Its Spectral Function," use the Gohberg–Krein theory of factorization to study the influence of small perturbations in the spectral function on the mass distribution of a vibrating string. An algorithm for computing the mass distribution of the perturbed string is derived. The well-known procedure of Gelfand–Levitan emerges as a special case.

"Spectral Analysis of Families of Operator Polynomials and a Generalized Vandermonde Matrix," by I. Gohberg, M. A. Kaashoek, and L. Rodman, develops a new theory with a spectral flavor that makes transparent many operations on noncommutative polynomials with matrix coefficients. It sheds new light on the old problem of finding common multiples for a family of matrix polynomials. The classical Vandermonde matrix is modernized and used as the main tool. Although here everything is carried out for the finite-dimensional case, the origin of the problems is the theory of nonself-adjoint operators.

Linear multiplication operators on H^p spaces are heavily studied. The paper by J. William Helton, "Orbit Structure of the Möbius Transformation Semigroup Acting on H^∞ (Broadband Matching)," uses these linear techniques to settle questions about certain nonlinear (linear fractional) multiplication operators. The results give a theoretical solution to the engineering problem: What is the best matching circuit for transferring power to a given n-port passive circuit from n decoupled power sources?

In "On the Asymptotic Number of Bound States for Certain Attractive

Potentials,'' M. Kac derives by two different methods A. Martin's asymptotic formula for the number of bound states of the Schrödinger equation

$$\frac{1}{2}\nabla^2\phi + \mathrm{AV}(\mathbf{r})\phi = -\mathrm{E}\phi$$

in the limit $A \to \infty$ for a nonnegative integrable $V(\mathbf{r})$, $\mathbf{r} \in R^3$.

Thomas Kailath, Lennart Ljung, and Martin Morf show, in "Generalized Krein–Levinson Equations for Efficient Calculation of Fredholm Resolvents of Nondisplacement Kernels," how the computational advantages of having *Toeplitz* (or *displacement*) kernels can be extended in a natural way to arbitrary kernels by introducing the notion of *displacement ranks* of kernels. The ideas here seem to have several interesting ramifications.

Tosio Kato, in "Trotter's Product Formula for an Arbitrary Pair of Self-Adjoint Contraction Semigroups," settles the convergence question for the famous product formula in the case of two semigroups in Hilbert space with arbitrary nonpositive self-adjoint generators by showing that it always converges strongly and by determining the limit. An addendum, due to an idea of B. Simon, gives a generalization to the case of two semigroups with (negative) sectorial generators.

Peter D. Lax and Ralph S. Phillips, in "The Time Delay Operator and a Related Trace Formula," study the time delay operator for the classical wave equation in the framework of their scattering theory; the spectral representation of this operator is the logarithmic derivative of the scattering operator. They also study a related trace formula.

"Boussinesq's Equation as a Hamiltonian System," by H. P. McKean, discusses the existence of infinitely many conserved quantities for the equation of Boussinesq. The results, due for the most part to Zakharov, are here derived in a new and more explicit way paralleling the known development for the equation of Korteweg–de Vries.

In "On the Set of Minimal Extensions of a Subadditive Functional," D. P. Milman presents a theorem that includes as special cases the three well-known principles of linear functional analysis, the Krien–Milman theorem, and a few others. It also contains a case in which a convex set (of linear functionals–minimal extensions) in a space without topology has "sufficiently many" extreme points.

Joel Pincus, in "Symmetric Singular Integral Operators with Arbitrary Deficiency," finds an explicit description of the deficiency spaces of a class of integral operators in terms of the so-called principal function invariant associated with a generator of an almost commuting algebra.

Marvin Rosenblum and James Rovnyak, in "Change of Variables Formulas with Cayley Inner Functions," develop a calculus for substitution operations in definite and singular integrals of functions on a circle or line.

They give applications to distribution function formulas, inversion formulas for singular integrals, and orthogonal expansions.

I. E. Segal's "The Complex-Wave Representation of the Free Boson Field" treats holomorphic functions on Hilbert space from the standpoint of functional integration. A simply defined Hilbert space of such functions with natural additional structure is shown to be canonically equivalent to the Fock–Cook representation for the boson field, whose real-wave representation is also treated.

The last paper included in this volume, "Families of Pseudodifferential Operators," by Harold Widom, introduces an invariant complete symbol, and develops a corresponding symbolic calculus, for pseudodifferential operators on manifolds and certain families of such. As an application it is shown how the coefficients in various asymptotic expansions (e.g., the heat expansion associated with a strongly elliptic differential operator on a compact Riemannian manifold) may be mechanically computed.

TOPICS IN FUNCTIONAL ANALYSIS
ADVANCES IN MATHEMATICS SUPPLEMENTARY STUDIES, VOL. 3

Quantum Cesàro Operators

LOUIS DE BRANGES

Department of Mathematics
Purdue University
Lafayette, Indiana

AND

DAVID TRUTT

Department of Mathematics
Lehigh University
Bethlehem, Pennyslvania

TO MARK KREIN ON HIS SEVENTIETH BIRTHDAY

Eigenfunction expansions are found for quantum generalizations of the Cesàro operator. Certain Hilbert spaces of entire functions, which are the closed span of polynomials, appear in the expansion. General problems for Hilbert spaces of analytic functions and polynomial approximation are also considered.

Some choice of quantum q, $0 < q < 1$, is assumed in what follows. The quantum Cesàro operator takes the power series $a_0 + a_1 z + a_2 z^2 + \cdots$ into the power series $b_0 + b_1 z + b_2 z^2 + \cdots$, where

$$b_n = (a_0 + a_1 q + \cdots + a_n q^n)/(1 + q + \cdots + q^n)$$

for every nonnegative integer n. The transformation is considered in various Hilbert space norms. Let k be a given number, $k < 1$. Consider the unique Hilbert space of power series which has the powers of z as an orthogonal basis and in which the square of the norm of z^n is

$$\frac{(1-k)(1-qk)\cdots(1-q^{n-1}k)}{(1-q)(1-q^2)\cdots(1-q^n)}\, q^n.$$

The quantum Cesàro operator is a closed linear transformation in the space. Its inverse is the transformation which takes $f(z)$ into $g(z)$ whenever $f(z)$ and $g(z)$ are elements of the space such that

$$g(z) = (1 - z/q)[f(z/q) - qf(z)]/(1 - q).$$

ISBN 0-12-287150-2

Operators L_+, L_-, and D are defined in the space as follows: D takes $f(z)$ into $g(z)$ whenever $f(z)$ and $g(z)$ are elements of the space such that

$$g(z) = f(qz);$$

L_- takes $f(z)$ into $g(z)$ whenever $f(z)$ and $g(z)$ are elements of the space such that

$$g(z) = [f(z) - f(0)]/z;$$

and L_+ takes $f(z)$ into $g(z)$ whenever $f(z)$ and $g(z)$ are elements of the space such that

$$g(z) = zf(z) - qzf(qz).$$

The identities

$$\begin{aligned} qDL_- &= L_-D, \\ DL_+ &= qL_+D, \\ L_-L_+ &= 1 - qD, \end{aligned}$$

and

$$L_+L_- = 1 - D$$

are satisfied formally. The operator D is self-adjoint. The operators $(1 - kD)qL_-$ and L_+ are adjoints.

Heine's generalization of the hypergeometric series is written

$$\begin{aligned} &\varphi(a, b; c; z) \\ &\quad = 1 + \frac{(1-a)(1-b)}{(1-q)(1-c)} z + \frac{(1-a)(1-qa)(1-b)(1-qb)}{(1-q)(1-q^2)(1-c)(1-qc)} z^2 + \cdots. \end{aligned}$$

The confluent series is

$$\begin{aligned} \varphi(a; c; z) &= \lim_{b \to \infty} \varphi(a, b; c; -z/b) \\ &= 1 + \frac{1-a}{(1-q)(1-c)} z + \frac{(1-a)(1-qa)}{(1-q)(1-q^2)(1-c)(1-qc)} qz^2 + \cdots, \end{aligned}$$

where $q^{n(n-1)/2}$ appears as a factor in the coefficient of z^n. Note that

$$\varphi(0; 0; z) = \prod_{n=0}^{\infty} (1 + q^n z).$$

The quantum gamma function $\gamma(z)$ is defined by

$$\gamma(z) = \frac{\varphi(0;0; -q)\varphi(0; 0; z)\varphi(0; 0; q/z)}{\varphi(0; 0; 1)\varphi(0; 0; q)\varphi(0; 0; -z)}.$$

The representation

$$\gamma(z) = \sum_{n=-\infty}^{+\infty} z^n/\varphi(0; 0; q^n)$$

holds when $z\bar{z} < 1$.

A unique Hilbert space $\mathcal{Q}(k)$ exists, whose elements are entire functions, such that the expression $\varphi(z, \bar{w}; k; q)$ belongs to the space as a function of z for every complex number w and such that the identity

$$F(w) = \langle F(z), \varphi(z, \bar{w}; k; q)\rangle_{\mathcal{Q}(k)}$$

holds for every element $F(z)$ of the space. The space admits an orthogonal basis consisting of the polynomials

$$\frac{(1 - z)(1 - qz) \cdots (1 - q^{n-1}z)}{(1 - q)(1 - q^2) \cdots (1 - q^n)} q^n$$

for nonnegative integers n. The square of the norm of the nth polynomial is

$$\frac{(1 - k)(1 - qk) \cdots (1 - q^{n-1}k)}{(1 - q)(1 - q^2) \cdots (1 - q^n)} q^n.$$

An isometric transformation exists of the starting Hilbert space onto the space $\mathcal{Q}(k)$ for which the above function corresponds to z^n for every nonnegative integer n. Let $f(z)$ and $g(z)$ be elements of the starting Hilbert space and let $F(z)$ and $G(z)$ be the corresponding elements of $\mathcal{Q}(k)$. Then D takes $f(z)$ into $g(z)$ if, and only if,

$$G(z) = F(qz) + [F(z) - F(qz)]/z,$$

L_- takes $f(z)$ into $g(z)$ if, and only if,

$$G(z) = [F(z) - F(z/q)]/z,$$

and L_+ takes $f(z)$ into $g(z)$ if, and only if,

$$G(z) = q(1 - z)F(qz).$$

The condition

$$g(z) = (1 - z/q)f(z/q) + zf(z)$$

holds if, and only if,

$$G(z) = zF(z).$$

The identity

$$\langle zF(z), G(z)\rangle_{\mathcal{Q}(k)} = \langle F(z), G(z/q) - k[G(z/q) - G(z)]/z\rangle_{\mathcal{Q}(k)}$$

holds for all elements $F(z)$ and $G(z)$ of the space such that $zF(z)$ and $G(z/q) - k[G(z/q) - G(z)]/z$ belong to the space. The transformation $F(z)$ into $(z-k)F(qz)$ takes $\mathcal{Q}(q^2k)$ onto the set of elements of $\mathcal{Q}(k)$ which vanish at k, and the identity

$$\|(z-k)F(qz)\|^2_{\mathcal{Q}(k)} = \|F(z)\|^2_{\mathcal{Q}(q^2k)}(1-k)(1-qk)/q$$

holds for every element $F(z)$ of $\mathcal{Q}(q^2k)$.

A computation of inner products is the reason for interest in the spaces.

THEOREM 1. *The identity*

$$\begin{aligned} &2\pi\gamma(k)\|F(z)\|^2_{\mathcal{Q}(k)} \\ &= \sum_{n=0}^{\infty} \frac{q^{n(n-1)/2}}{(1-q)(1-q^2)\cdots(1-q^n)} \\ &\times \int_0^{2\pi} |\gamma(k^{1/2}q^{n/2}e^{-i\theta})F(k^{1/2}q^{-n/2}e^{i\theta}|^2\, d\theta \end{aligned}$$

holds for every element $F(z)$ of $\mathcal{Q}(k)$ when $0 < k < 1$. The identity

$$\begin{aligned} &2\pi\varphi(0;0;1)\varphi(0;0;q)\varphi(0;0;s)\varphi(0;0;q/s)\|F(z)\|^2_{\mathcal{Q}(0)} \\ &= \sum_{n=-\infty}^{+\infty} q^{n(n-1)/2}\varphi(0;0;s^{1/2}q^{n/2}e^{i\theta}) \\ &\times \varphi(0;0;s^{-1/2}q^{1-n/2}e^{-i\theta})F(s^{1/2}q^{-n/2}e^{i\theta})|^2\, d\theta \end{aligned}$$

holds for every element $F(z)$ of $\mathcal{Q}(0)$ when $s > 0$.

The spaces are related to a quantum generalization of Laguerre polynomials due to Stieltjes [1]. If k is a given number, $0 < k < 1$, a unique inner product on polynomials exists such that the indentities

$$\langle tf(t), g(t)\rangle_{\mathcal{S}(k)} = \langle f(t), tg(t)\rangle_{\mathcal{S}(k)}$$

and

$$\langle f(t), g(t)\rangle_{\mathcal{S}(k)} = \langle (t+k)f(qt), g(qt)\rangle_{\mathcal{S}(k)}$$

hold for all polynomials $f(z)$ and $g(z)$ and such that the constant polynomial one has norm one. The polynomials $\varphi(q^{-n}; k; q^n z)$ form an orthogonal set with respect to the inner product. The square of the norm of the nth polynomial is

$$\frac{(1-q)(1-q^2)\cdots(1-q^n)}{(1-k)(1-qk)\cdots(1-q^{n-1}k)}\, q^{-n}.$$

The inner product is given explicitly by

$$\gamma(k)\langle f(t), g(t)\rangle_{\mathscr{S}(k)} = \sum_{n=-\infty}^{+\infty} f(kq^n)\overline{g}(kq^n)k^n/\varphi(0; 0; q^n)$$

when $0 < k < 1$. The identity

$$\varphi(0; 0; -q)\varphi(0; 0; s)\varphi(0; 0; q/s)\langle f(t), g(t)\rangle_{\mathscr{S}(0)}$$

$$= \sum_{n=-\infty}^{+\infty} f(sq^n)\overline{g}(sq^n)q^{n(n-1)/2}s^n$$

holds for every positive number s when $k = 0$. An isometry follows.

THEOREM 2. *A unique linear transformation $f(z)$ into $F(z)$ of polynomials into polynomials exists such that*

$$F(z) = \frac{(1 - z)(1 - qz)\cdots(1 - q^{n-1}z)}{(1 - k)(1 - qk)\cdots(1 - q^{n-1}k)}$$

whenever

$$f(z) = \varphi(q^{-n}; k; q^n z),$$

and the identity

$$\|f(z)\|_{\mathscr{S}(k)} = \|F(z)\|_{\mathscr{Q}(k)}$$

is satisfied. Let $f(z)$ and $g(z)$ be polynomials and let $F(z)$ and $G(z)$ be the corresponding elements of $\mathscr{Q}(k)$. Then the condition

$$G(z) = zF(z)$$

is equivalent to the condition

$$g(z) = (z + k)f(qz).$$

The condition

$$G(z) = (z - k)F(z/q)$$

is equivalent to the condition

$$g(z) = zf(z).$$

The condition

$$G(z) = F(qz) + [F(z) - F(qz)]/z$$

is equivalent to the condition

$$g(z) = f(qz) - k[f(z) - f(qz)]/z + q[f(z/q) - f(z)]/z.$$

The condition

$$G(z) = [F(z) - F(z/q)]/z$$

is equivalent to the condition

$$g(z) = k[f(z/q) - f(z)]/z - q[f(z/q^2) - f(z/q)]/z.$$

The condition

$$G(z) = q(1 - z)F(qz)$$

is equivalent to the condition

$$\begin{aligned} g(z) = &-qzf(q^2z) + (k + q)f(qz) - k(1 + q)f(q^2z) \\ &- kq[f(z) - f(qz)]/z + k^2[f(qz) - f(q^2z)]/z. \end{aligned}$$

If $0 < k < 1$, *then*

$$\gamma(k/z)F(z) = \sum_{n=-\infty}^{+\infty} f(kq^n)(k/z)^n/\varphi(0; 0; q^n)$$

for $z\bar{z} > k^2$. *If* $k = 0$, *then*

$$\varphi(0; 0; -q)\varphi(0; 0; s/z)\varphi(0; 0; qz/s)F(z) = \sum_{n=-\infty}^{+\infty} f(sq^n)(s/z)^n q^{n(n-1)/2}$$

for every positive number s.

Define $\lambda(r, s; n)$ to be the function of nonnegative integers r, s, and n which is zero unless $\max(r, s) \leqslant n \leqslant r + s$, in which case it is

$$(-1)^{r+s-n} q^{r(r+1)/2} q^{s(s+1)/2} q^{-n(r+s)} q^{n(n-1)/2} \times \frac{(1-q)\cdots(1-q^{\max(r,s)})(1-q)\cdots(1-q^n)}{(1-q)\cdots(1-q^{n-r})(1-q)\cdots(1-q^{n-s})(1-q)\cdots(1-q^{r+s-n})}.$$

These numbers appear as coefficients in the expansion

$$\begin{aligned} &\frac{(1-z)(1-qz)\cdots(1-q^{r-1}z)}{(1-q)(1-q^2)\cdots(1-q^r)} q^r \frac{(1-z)(1-qz)\cdots(1-q^{s-1}z)}{(1-q)(1-q^2)\cdots(1-q^s)} q^s \\ &= \sum_{n=0}^{\infty} \lambda(r, s; n) \frac{(1-z)(1-qz)\cdots(1-q^{n-1}z)}{(1-q)(1-q^2)\cdots(1-q^n)} q^n, \end{aligned}$$

which holds for all nonnegative integers r and s.

Assume that an inner product on polynomials is given such that the transformation which takes $F(z)$ into $F(qz) + [F(z) - F(qz)]/z$ is formally self-adjoint. If a nonnegative measure μ on the Borel subsets of the complex plane exists such that the indentity

$$\|F(z)\|^2 = \int |F(z)|^2 \, d\mu(z)$$

holds for every polynomial $F(z)$, then the inequality

$$\sum f(m, n)\bar{f}(r, s)\lambda(m, s; k)\lambda(n, r; k) \left\| \frac{(1 - z)(1 - qz)\cdots(1 - q^{k-1}z)}{(1 - q)(1 - q^2)\cdots(1 - q^k)} q^k \right\|^2 \geqslant 0$$

holds for every function $f(m, n)$ of nonnegative integers m and n which vanishes outside of a finite set.

An unsolved problem is to determine all inner products on polynomials which satisfy these conditions.

Specialized problems for Hilbert spaces of analytic functions are difficult because of unsolved problems in the general theory. A fundamental one is to construct all nonnegative measures μ on the Borel subsets of the complex plane, which are supported in a given compact set, such that the identity

$$f(0) = \int f(z) \, d\mu(z)$$

holds for every polynomial $f(z)$. The set is convex and is also compact in the weakest topology for which $\int f(z) \, d\mu(z)$ is continuous in μ for every continuous function $f(z)$. By the Krein–Milman convexity theorem [2], the set is the closed convex span of its extreme points. A characterization of extreme points is an underlying principle in connection with the Stone–Weierstrass theorem [3]. An element μ of the set is an extreme point of it if, and only if, the functions of the form $f(z) + \bar{g}(z)$ for polynomials $f(z)$ and $g(z)$ are dense in $L^1(\mu)$. This condition has found application in the analogous extreme point problem for norm determining measures associated with Hilbert spaces of entire functions [4].

Notable among extremal measures are some related to the theory [5] of the space $\mathscr{C}(z)$ of square summable power series $f(z) = \sum a_n z^n$,

$$\|f(z)\|^2_{\mathscr{C}(z)} = \sum |a_n|^2.$$

THEOREM 3. *Let μ be a nonnegative measure on the Borel subsets of the complex plane, with compact support but not a point mass concentrated at the origin, such that the identity*

$$f(0) = \int f(z) \, d\mu(z)$$

holds for every polynomial $f(z)$. Assume that the functions of the form $f(z) + \bar{g}(z)$ for polynomials $f(z)$ and $g(z)$ are dense in $L^\infty(\mu)$ taken in its weak topology induced by $L^1(\mu)$. Then a function ψ exists which is bounded and analytic in the *unit disk, is zero at the origin, and has distinct values at distinct points of the disk, such that the identity*

$$\int |f(z)|^2 \, d\mu(z) = \|f(\psi(z))\|^2_{\mathscr{C}(z)}$$

holds for every polynomial $f(z)$. The support of μ is the boundary of the region onto which ψ maps the unit disk. The measurable boundary value function of ψ on the unit circle takes its values in the support of μ. The μ-measure of every Borel set S is the Lebesgue measure of the set of real numbers t modulo one such that $\psi(\exp(2\pi it))$ belongs to S.

An inequality is derived from the maximum principle.

THEOREM 4. *Let ψ_1 and ψ_2 be functions which are bounded and analytic in the unit disk, are zero at the origin, and have distinct values at distinct points of the disk. Let μ_1 and μ_2 be the nonnegative measures on the Borel subsets of the complex plane such that the μ_k-measure of every Borel set S is the Lebesgue measure of the set of real numbers t modulo one such that $\psi_k(\exp(2\pi it))$ belongs to S. Then the identity*

$$f(0) = \int f(z) \, d\mu_k(z)$$

holds for every polynomial $f(z)$. The inequality

$$\int \log|f(z)| \, d\mu_1(z) \leqslant \int \log|f(z)| \, d\mu_2(z)$$

holds for every polynomial $f(z)$ if, and only if, the region onto which ψ_1 maps the unit disk is contained in the region onto which ψ_2 maps the unit disk.

The same inequality is obtained more generally using the balayage principle.

THEOREM 5. *Let μ be a nonnegative measure on the Borel subsets of the complex plane, with compact support, such that the identity*

$$f(0) = \int f(z) \, d\mu(z)$$

holds for every polynomial $f(z)$. Let ψ be a function which is bounded and analytic in the unit disk, is zero at the origin, and has distinct values at distinct points of the disk. Then a unique nonnegative measure ν on the Borel subsets of the complex plane, with compact support, exists such that the identity

$$f(0) = \int f(z) \, d\nu(z)$$

holds for every polynomial $f(z)$, such that ν assigns zero measure to the region onto which ψ maps the unit disk, and such that a nonnegative measure σ on the Borel subsets of the reals modulo one exists for which, for every Borel set S contained in the complement of the region onto which ψ maps the unit disk, $\nu(S) - \mu(S)$ is the σ-measure of the set of real numbers t modulo one such that $\psi(\exp(2\pi it))$ belongs to S. The inequality

$$\int \log|f(z)|\, d\mu(z) \leqslant \int \log|f(z)|\, d\nu(z)$$

holds for every polynomial $f(z)$.

The study of such inequalities is suggested as being of value in determining the structure of the set of nonnegative measures μ on the Borel subsets of the complex plane, with support in a given compact set, such that the identity

$$f(0) = \int f(z)\, d\mu(z)$$

holds for every polynomial $f(z)$. A compact convex subset of this set is the set of such measures for which the inequality

$$\log|f(0)| \leqslant \int \log|f(z)|\, d\mu(z)$$

holds for every polynomial $f(z)$. The measures constructed in Theorem 3, together with the point mass at the origin, are conjectured to be the only extreme points of the set.

The expansion theory for Cesàro operators [6] is obtained from the quantum theory in the limit $q = 1$. The Cesàro operator takes the power series $a_0 + a_1 z + a_2 z^2 + \cdots$ into the power series $b_0 + b_1 z + b_2 z^2 + \cdots$, where

$$b_n = (a_0 + a_1 + \cdots + a_n)/(n + 1)$$

for every nonnegative integer n. The transformation can also be considered in various norms. Let h be a given positive number. Consider the unique Hilbert space which has the powers of z as an orthogonal basis and in which the square of the norm of z^n is

$$\frac{h(h+1)\cdots(h+n-1)}{1 \cdot 2 \cdots n}.$$

The Cesàro operator is a closed linear transformation in the space. Its inverse is the transformation which takes $f(z)$ into $g(z)$ whenever $f(z)$ and $g(z)$ are elements of the space such that

$$g(z) = (1 - z)[zf'(z) + f(z)].$$

Operators L_+, L_-, and D in the space are defined as follows: D takes $f(z)$ into $g(z)$ whenever $f(z)$ and $g(z)$ are elements of the space such that

$$g(z) = zf'(z);$$

L_- takes $f(z)$ into $g(z)$ whenever $f(z)$ and $g(z)$ are elements of the space such that

$$g(z) = [f(z) - f(0)]/z;$$

and L_+ takes $f(z)$ into $g(z)$ whenever $f(z)$ and $g(z)$ are elements of the space such that

$$g(z) = z^2 f'(z) + zf(z).$$

The identities

$$\begin{aligned} DL_- - L_-D &= -L_-, \\ DL_+ - L_+D &= L_+, \\ L_-L_+ &= 1 + D, \\ L_+L_- &= D \end{aligned}$$

are satisfied formally. The operator D is self-adjoint. The operators L_+ and $(D + h)L_-$ are adjoints.

The hypergeometric series is written

$$F(a, b; c; z) = 1 + \frac{ab}{1c} z + \frac{a(a+1)b(b+1)}{1 \cdot 2c(c+1)} z^2 + \cdots.$$

The corresponding confluent series is

$$\begin{aligned} F(a; c; z) &= \lim_{b \to \infty} F(a, b; c; z/b) \\ &= 1 + \frac{a}{1c} z + \frac{a(a+1)}{1 \cdot 2c(c+1)} z^2 + \cdots. \end{aligned}$$

The hypergeometric series converges when $z = 1$ if $\operatorname{Re}(c - a - b) > 0$. An identity due to Gauss,

$$F(a, b; c; 1) = \frac{\Gamma(c)\Gamma(c - a - b)}{\Gamma(c - a)\Gamma(c - b)}$$

expresses it in terms of the gamma function.

Hilbert spaces of analytic functions in the eigenfunction expansions of Cesàro operators have previously been studied [7]. Let $\mathcal{N}(h)$ be the unique Hilbert space, whose elements are functions analytic in the half-plane $z + \bar{z} > -h$, such that the expression $F(-z, -\bar{w}; h; 1)$ belongs to the space as a function of z for every point w in the half-plane and such that the identity

$$F(w) = \langle F(z), F(-z, -\bar{w}; h; 1)\rangle_{\mathcal{N}(h)}$$

holds for every element $F(z)$ of the space. The space admits an orthogonal basis consisting of the polynomials

$$(-1)^n \frac{z(z-1)\cdots(z-n+1)}{1\cdot 2\cdots n}$$

for nonnegative integers n. The square of the norm of the nth polynomial is

$$\frac{h(h+1)\cdots(h+n-1)}{1\cdot 2\cdots n}.$$

An isometric transformation exists of the starting Hilbert space onto the space $\mathcal{N}(h)$ for which the above polynomial corresponds to z^n for every nonnegative integer n. Let $f(z)$ and $g(z)$ be elements of the starting Hilbert space and let $F(z)$ and $G(z)$ be the corresponding elements of $\mathcal{N}(h)$. Then D takes $f(z)$ into $g(z)$ if, and only if,

$$G(z) = z[F(z) - F(z-1)],$$

L_- takes $f(z)$ into $g(z)$ if, and only if,

$$G(z) = -[F(z+1) - F(z)],$$

and L_+ takes $f(z)$ into $g(z)$ if, and only if,

$$G(z) = -zF(z-1).$$

The condition

$$g(z) = (z-1)[zf'(z) + f(z)] + f(z)$$

holds if, and only if,

$$G(z) = -zF(z).$$

The identity

$$\langle zF(z), G(z)\rangle_{\mathcal{N}(h)} = \langle F(z), (z+h)[G(z+1) - G(z)]\rangle_{\mathcal{N}(h)}$$

holds for all elements $F(z)$ and $G(z)$ of $\mathcal{N}(h)$ such that

$$zF(z) \qquad \text{and} \quad (z+h)[G(z+1) - G(z)]$$

belong to the space. The transformation $F(z)$ into $(z+h)F(z-1)$ takes $\mathcal{N}(h+2)$ onto $\mathcal{N}(h)$ and the identity

$$\|(z+h)F(z-1)\|^2_{\mathcal{N}(h)} = h(h+1)\|F(z)\|^2_{\mathcal{N}(h+2)}$$

holds for every element $F(z)$ of $\mathcal{N}(h+2)$.

THEOREM 6. *Every element $F(z)$ of $\mathcal{N}(h)$ has a boundary value function on the line $z + \bar{z} = -h$ and the identity*

$$2\pi\Gamma(h)\|F(z)\|^2_{\mathcal{N}(h)}$$
$$= \sum_{n=0}^{\infty} \frac{1}{n!} \int_{-\infty}^{+\infty} |\Gamma(n/2 + h/2 + it)F(n/2 - h/2 + it)|^2 \, dt$$

is satisfied.

A relation exists between these spaces and the theory of Laguerre polynomials [9]. If h is a given positive number, a unique inner product on polynomials exists such that the identities

$$\langle tf(t), g(t)\rangle_{\mathcal{L}(h)} = \langle f(t), tg(t)\rangle_{\mathcal{L}(h)}$$

and

$$\langle tf'(t), g(t)\rangle_{\mathcal{L}(h)} + \langle f(t), tg'(t)\rangle_{\mathcal{L}(h)} = \langle (t-h)f(t), g(t)\rangle_{\mathcal{L}(h)}$$

hold for all polynomials $f(z)$ and $g(z)$ and such that the constant polynomial one has norm one. The inner product is given explicitly by

$$\Gamma(h)\langle f(t), g(t)\rangle_{\mathcal{L}(h)} = \int_0^{\infty} f(t)\bar{g}(t)\exp(-t)t^{h-1} \, dt.$$

The polynomials $F(-n; h; z)$ form an orthogonal set with respect to the inner product. The square of the norm of the nth polynomial is

$$\frac{1 \cdot 2 \cdots n}{h(h+1)\cdots(h+n-1)}.$$

An isometry results.

THEOREM 7. *For every measurable function $f(t)$ of $t > 0$ such that*

$$\int_0^{\infty} |f(t)|^2 \exp(-t)t^{h-1} \, dt < \infty,$$

a unique element $F(z)$ of $\mathcal{N}(h)$ exists such that

$$\Gamma(z+h)F(z) = \int_0^{\infty} f(t)\exp(-t)t^{z+h-1} \, dt$$

when $z + \bar{z} > -h$, and the identity

$$\|F(z)\|^2_{\mathcal{N}(h)} = \int_0^{\infty} |f(t)|^2 \exp(-t)t^{h-1} \, dt$$

is satisfied. If $f(z) = F(-n; h; z)$, then

$$F(z) = (-1)^n \frac{z(z-1)\cdots(z-n+1)}{h(h+1)\cdots(h+n-1)}.$$

If $f(z)$ and $g(z)$ are polynomials and if $F(z)$ and $G(z)$ are the corresponding elements of $\mathcal{N}(h)$, then the condition

$$G(z) = -zF(z)$$

is equivalent to the condition

$$g(z) = zf'(z) - (z-h)f(z).$$

The condition

$$G(z) = (z+h)F(z+1)$$

is equivalent to the condition

$$g(z) = zf(z).$$

The condition

$$G(z) = F(z) - F(z-1)$$

is equivalent to the condition

$$g(z) = f'(z).$$

Define $\Lambda(r, s; n)$ to be the function of nonnegative integers r, s, and n which is zero unless $\max(r, s) \leqslant n \leqslant r + s$, in which case it is

$$\frac{(-1)^{r+s-n} 1 \cdot 2 \cdots \max(r, s) 1 \cdot 2 \cdots n}{1 \cdot 2 \cdots (n-r) 1 \cdot 2 \cdots (n-s) 1 \cdot 2 \cdots (r+s-n)}.$$

These numbers appear in the identity

$$(-1)^r \frac{z(z-1)\cdots(z-r+1)}{1 \cdot 2 \cdots r} (-1)^s \frac{z(z-1)\cdots(z-s+1)}{1 \cdot 2 \cdots s}$$
$$= \sum_{n=0}^{\infty} \Lambda(r, s; n)(-1)^n \frac{z(z-1)\cdots(z-n+1)}{1 \cdot 2 \cdots n},$$

which holds for all nonnegative integers r and s.

Assume that an inner product on polynomials is given such that the transformation which takes $F(z)$ into $z[F(z) - F(z-1)]$ is formally self-adjoint. If a nonnegative measure μ on the Borel subsets of the complex plane exists such that the identity

$$\|F(z)\|^2 = \int |F(z)|^2 \, d\mu(z)$$

holds for every polynomial $F(z)$, then the inequality

$$\sum f(m, n)\bar{f}(r, s)\Lambda(m, s; k)\Lambda(n, r; k) \left\| (-1)^k \frac{z(z-1)\cdots(z-k+1)}{1 \cdot 2 \cdots k} \right\|^2 \geqslant 0$$

holds for every function $f(m, n)$ of nonnegative integers m and n which vanishes outside of a finite set. An unsolved problem is to determine all inner products on polynomials which satisfy these conditions.

Proof of Theorem 1. Consider first the case $0 < k < 1$. For each nonnegative integer n, define an inner product on polynomials by

$$2\pi\langle F, G\rangle_n = \int_0^{2\pi} F(k^{1/2}q^{-n/2}e^{i\theta})G^*(k^{1/2}q^{-n/2}e^{-i\theta}) \times \gamma(k^{1/2}q^{n/2}e^{-i\theta})\gamma(k^{1/2}q^{n/2}e^{i\theta})\,d\theta,$$

where $G^*(z) = \bar{G}(\bar{z})$. It is obvious that the identity

$$\langle zF(z), zG(z)\rangle_n = kq^{-n}\langle F(z), G(z)\rangle_n$$

holds for all polynomials $F(z)$ and $G(z)$. The identity

$$\langle (z/k - 1)F(z/q), G(z)\rangle_n = \langle F(z), G(z)\rangle_{n+1}$$

is verified by Cauchy's formula using the identity

$$\gamma(qz) = (1/z - 1)\gamma(z).$$

Define an inner product on polynomials by

$$\gamma(k)\langle F(z), G(z)\rangle = \sum_{n=0}^{\infty} \frac{q^{n(n-1)/2}}{(1-q)(1-q^2)\cdots(1-q^n)} \langle F(z), G(z)\rangle_n.$$

Convergence is easily verified. The identity

$$\langle zF(z), G(z)\rangle = \langle F(z), G(z/q) - k[G(z/q) - G(z)]/z\rangle$$

is obtained for all polynomials $F(z)$ and $G(z)$ by using the previous identities. It follows that the inner product is a constant multiple of the inner product of $\mathcal{Q}(k)$. Since the polynomials are dense in $\mathcal{Q}(k)$, the theorem follows once it is shown that the square $W(k)$ of the norm of one is equal to one. The identity $W(k) = W(q^2k)$ holds because the transformation $F(z)$ into $(z - k)F(qz)$ takes $\mathcal{Q}(q^2k)$ into $\mathcal{Q}(k)$ in such a way that

$$\|(z - k)F(qz)\|^2_{\mathcal{Q}(k)} = \|F(z)\|^2_{\mathcal{Q}(q^2k)}(1 - k)(1 - qk)/q.$$

Since $W(k)$ is the restriction of a function analytic in the unit disk, it is a constant. The theorem follows on showing that $W(k) = 1$ in the limit $k = 1$. This is true because

$$2\pi = \lim(1 - k)\int_0^{2\pi} |\gamma(k^{1/2}e^{i\theta})|^2\,d\theta$$

since $\gamma(q) = 1$.

A similar argument applies when $k = 0$. For each integer n, define an inner product on polynomials by

$$2\pi\langle F(z), G(z)\rangle_n = \int_0^{2\pi} F(s^{1/2}q^{-n/2}e^{i\theta})G^*(s^{1/2}q^{-n/2}e^{-i\theta}) \times |\varphi(0;0;s^{1/2}q^{n/2}e^{i\theta})\varphi(0;0;s^{-1/2}q^{1-n/2}e^{-i\theta})|^2\,d\theta.$$

Then the identity

$$\langle zF(z), zG(z)\rangle_n = sq^{-n}\langle F(z), G(z)\rangle_n$$

clearly holds for all polynomials $F(z)$ and $G(z)$. The identity

$$\langle zF(z), G(z)\rangle_n = s\langle F(z), G(z)\rangle_{n+1}$$

is obtained using Cauchy's formula and the identity

$$\varphi(0;0;z)\varphi(0;0;q/z) = z\varphi(0;0;qz)\varphi(0;0;1/z).$$

Define an inner product on polynomials by

$$2\pi\varphi(0;0;1)\varphi(0;0;q)\varphi(0;0;s)\varphi(0;0;q/s)\langle F(z), G(z)\rangle = \sum_{n=-\infty}^{+\infty} q^{n(n-1)/2}\langle F(z), G(z)\rangle_n.$$

Convergence is easily verified. The identity

$$\langle zF(z), G(z)\rangle = \langle F(z), G(z/q)\rangle$$

is obtained for all polynomials $F(z)$ and $G(z)$ by applying the previous identities. It follows that the inner product is a constant multiple of the inner product of $\mathcal{Q}(0)$. The theorem follows once it is shown that the norm of the constant polynomial one is equal to one. This result is obtained by a straightforward calculation from Jacobi's identity

$$\varphi(0;0;-q)\varphi(0;0;z)\varphi(0;0;qz) = \sum_{n=-\infty}^{+\infty} q^{n(n-1)/2}z^n.$$

An alternative proof of Theorem 1 can be given using the results of Theorem 2, which is proved independently of Theorem 1. The present proof, which is longer and more difficult to justify, is preferred because it proceeds directly from axiomatic properties of the spaces whose inner products are computed.

Proof of Theorem 2. The stated results are obtained by straightforward calculations using known properties [1] of Stieltjes polynomials.

Proof of Theorem 3. Argue by contradiction, assuming that the origin belongs to the support of μ. Then a sequence of measurable real valued functions $e_n(z)$ of z exists such that $e_n(z)$ vanishes outside of the disk of radius $1/n$ about the origin, where

$$\int e_n(z)\, d\mu(z) = 1$$

for every n, and where the sequence of numbers

$$\int |e_n(z)|\, d\mu(z)$$

is bounded. Since the identity

$$\int h(z)\, d\mu(z) = \lim_{n\to\infty} \int h(z) e_n(z)\, d\mu(z)$$

holds for every polynomial $h(z)$, it follows for all functions of the form $h(z) = f(z) + \bar{g}(z)$ for polynomials $f(z)$ and $g(z)$. Since such functions are weakly dense in $L^\infty(\mu)$ by hypothesis, the same conclusion holds for every element of $L^\infty(\mu)$. In particular, $h(z)$ can be chosen to be zero in some disk of radius ϵ about the origin and to be one outside of the disk. Then the identity states that μ has zero mass outside of the disk of radius ϵ. By the arbitrariness of ϵ, μ is a point mass concentrated at the origin, which is contrary to hypothesis. It follows that the origin does not belong to the support of μ.

If $f(z)$ is any polynomial, $[f(z) - f(0)]/z$ is a polynomial which has value $f'(0)$ at the origin. By hypothesis the identity

$$f'(0) = \int [f(z) - f(0)]/z\, d\mu(z)$$

is satisfied. Since the origin does not belong to the support of μ, the transformation which takes $f(z)$ into $[f(z) - f(0)]/z$ is bounded in the metric of $L^2(\mu)$. It follows that the linear functional on polynomials which takes $f(z)$ into $f'(0)$ is continuous in the metric of $L^2(\mu)$. A unique element $\varphi(z)$ of $L^2(\mu)$ exists which belongs to the closure of the polynomials, has norm one in $L^2(\mu)$, is orthogonal to every polynomial whose derivative vanishes at the origin, and whose inner product with any polynomial having a nonzero derivative at the origin is a positive multiple of that derivative.

The identity

$$f(0) = \int f(z)\varphi(z)\bar{\varphi}(z)\, d\mu(z)$$

then holds for every polynomial $f(z)$. So $\varphi(z)\bar{\varphi}(z)$ is a function of z which belongs to $L^1(\mu)$ and which satisfies the identity

$$\int h(z)\, d\mu(z) = \int h(z)\varphi(z)\bar{\varphi}(z)\, d\mu(z)$$

for every function $h(z)$ of the form $h(z) = f(z) + \bar{g}(z)$ for polynomials $f(z)$ and $g(z)$. Since such functions are dense in $L^\infty(\mu)$ taken in the weak topology induced by $L^1(\mu)$, $\varphi(z)\bar{\varphi}(z) = 1$ almost everywhere with respect to μ.

If w is a point of the complex plane such that the linear functional on polynomials defined by taking $f(z)$ into $f(w)$ is continuous in the metric of $L^2(\mu)$, define $K(w, z)$ to be the unique element of $L^2(\mu)$ which belongs to the closed span of the polynomials such that the identity

$$f(w) = \int f(z)\bar{K}(w, z)\, d\mu(z)$$

holds for every polynomial $f(z)$. An argument earlier in the proof will show that w cannot belong to the support of μ when $K(w, z)$ is defined. If $f(z)$ is any element of $L^2(\mu)$ which belongs to the closed span of the polynomials, the same identity is used to define the value of $f(z)$ at w. These conditions are satisfied by hypothesis when $w = 0$ and $K(w, z) = 1$.

If α is a given point of the complex plane such that $K(\alpha, z)$ is defined, then the identity

$$f(w) \int (z - \alpha)/(z - w)\bar{K}(\alpha, z)\, d\mu(z) = \int f(z)(z - \alpha)/(z - w)\bar{K}(\alpha, z)\, d\mu(z)$$

holds whenever w is not in the support of μ. An obvious estimate shows that $K(w, z)$ is defined if the integral

$$\int (z - \alpha)/(z - w)\bar{K}(\alpha, z)\, d\mu(z)$$

is nonzero. Since the integral is a continuous function of w which has value one at α, it is nonzero in some neighborhood of α. It follows that the set of numbers w for which $K(w, z)$ is defined is open. Since the integral is an analytic function of w, it is nonzero with isolated exceptions in the connected component of the open set which is the complement of the support of μ. So the set of numbers w for which $K(w, z)$ is defined is also closed in the subspace topology of this open set. It is clear that $K(w, z)$ is not defined in the unbounded component of the open set.

The hypotheses imply that the functions of the form $f(z) + \bar{g}(z)$ for polynomials $f(z)$ and $g(z)$, with $g(0) = 0$, are dense in $L^2(\mu)$. They also imply that such functions $f(z)$ and $\bar{g}(z)$ are orthogonal in $L^2(\mu)$. So the orthogonal complement in $L^2(\mu)$ of the polynomials is the closed span of the functions of the form $\bar{g}(z)$ for a polynomial $g(z)$ which vanishes at the origin. But multiplication by $\varphi(z)$ is an isometry of $L^2(\mu)$ onto itself. It follows that the orthogonal complement of the functions of the form $\varphi(z)f(z)$ for polynomials $f(z)$ is the set of functions of the form $\varphi(z)\bar{g}(z)$ where $g(z)$ belongs to the closed span of the polynomials which vanish at the origin. Assume that some such product $h(z) = \varphi(z)\bar{g}(z)$ belongs to the closed span of the polynomials. Then for every polynomial $f(z)$,

$$\int f(z)\bar{h}(z)\,d\mu(z) = \int f(z)g(z)\bar{\varphi}(z)\,d\mu(z),$$

where

$$f(0)\int g(z)\bar{\varphi}(z)\,d\mu(z) = \int f(z)g(z)\bar{\varphi}(z)\,d\mu(z).$$

It follows that the identity

$$f(0)\int \bar{h}(z)\,d\mu(z) = \int f(z)\bar{h}(z)\,d\mu(z)$$

holds for every polynomial $f(z)$. The condition implies that $h(z)$ is a constant and that $g(z)$ is a constant multiple of $\varphi(z)$. So every element $g(z)$ of $L^2(\mu)$ which belongs to the closed span of the polynomials and which vanishes at the origin is of the form $g(z) = \varphi(z)f(z)$ for an element $f(z)$ of $L^2(\mu)$ which belongs to the closed span of the polynomials.

Let w be any nonzero number for which $K(w, z)$ is defined. Since multiplication by $\varphi(z)$ is isometric in $L^2(\mu)$, $\varphi(z)K(w, z)\bar{\varphi}(w)$ belongs to $L^2(\mu)$ and the identity

$$f(w) = \int f(z)\bar{\varphi}(z)\bar{K}(w, z)\varphi(w)\,d\mu(z)$$

holds for every element $f(z)$ of $L^2(\mu)$ of the form $f(z) = \varphi(z)g(z)$, where $g(z)$ belongs to the closed span of the polynomials in $L^2(\mu)$. But an element $f(z)$ of $L^2(\mu)$ in the closed span of the polynomials is of this form if, and only if, $f(0) = 0$. So the expression

$$K(w, z) - \varphi(z)K(w, z)\bar{\varphi}(w)$$

is an element of $L^2(\mu)$ which is in the closed span of the polynomials and which is orthogonal to every element of the closed span of the polynomials that vanishes at the origin. It follows that the expression is a constant, equal to one by its value at the origin. A derivation of the identity

$$K(w, z) = 1/[1 - \varphi(z)\bar{\varphi}(w)]$$

has been given.

The identity implies that the integral powers of $\varphi(z)$ form an orthogonal basis for $L^2(\mu)$. The nonnegative integral powers of $\varphi(z)$ form an orthogonal basis for the closed span of the polynomials in $L^2(\mu)$. The nonpositive integral powers of $\varphi(z)$ form an orthogonal basis for the closed span of the conjugates of the polynomials in $L^2(\mu)$. The identity also implies that $\varphi(w)\bar{\varphi}(w) < 1$ whenever $K(w, z)$ is defined.

If α and β are distinct points for which $K(\alpha, z)$ and $K(\beta, z)$ are defined, then a polynomial exists which has different values at α and β. It follows that $K(\alpha, z)$ and $K(\beta, z)$ are unequal. This condition means that $\varphi(\alpha)$ and $\varphi(\beta)$ are unequal.

Since $\varphi(z)$, considered in the set where $K(w, z)$ is defined, is a nonconstant analytic function with values in the unit disk, its range is a nonempty open subset of the disk. The range is also closed in the subspace topology of the disk. For if (w_n) is a sequence of points such that $K(w_n, z)$ is defined for every n and such that $\lim \varphi(w_n)$ exists and belongs to the disk, then the expression $K(w_n, w_n)$ has a finite limit. Since the set of points w for which $K(w, z)$ is defined is bounded, the sequence can be chosen with no loss of generality so that $w = \lim w_n$ exists. Then $f(w) = \lim f(w_n)$ for every polynomial $f(z)$. Since the inequality

$$|f(w_n)|^2 \leqslant K(w_n, w_n) \int |f(z)|^2 \, d\mu(z)$$

holds for every n, the inequality

$$|f(w)|^2 \leqslant \lim K(w_n, w_n) \int |f(z)|^2 \, d\mu(z)$$

holds for every polynomial $f(z)$. The condition means that $K(w, z)$ is defined. Since $\varphi(z)$ is continuous, $\lim \varphi(w_n) = \varphi(w)$.

Since the unit disk is connected, $\varphi(z)$ maps the set of points w for which $K(w, z)$ is defined onto the unit disk, and this set of points w is a connected set. Let ψ be the unique function, defined and analytic in the unit disk, such that $\psi(\varphi(w)) = w$ whenever $K(w, z)$ is defined. Then $\psi(z)$ is a bounded analytic function which has distinct values at distinct points of the disk, has value zero at the origin, and maps the unit disk onto the set of points w for which $K(w, z)$ is defined.

The transformation which takes $f(z)$ into $f(\varphi(z))$ is an isometry of $\mathscr{C}(z)$ onto the closed span of the polynomials in $L^2(\mu)$. If $f(z) = \sum a_n z^n$, the corresponding element of $L^2(\mu)$ is the sum $g(z)$ of the orthogonal series $\sum a_n \varphi^n(z)$. The expansion is also valid in the sense that $g(w) = \sum a_n \varphi^n(w)$ whenever $K(w, z)$ is defined. The identity $f(w) = g(\psi(w))$ holds at all points of the unit disk.

Since $\psi(z)$ is represented by a square summable power series, it has a boundary value function $\psi(e^{i\theta})$ on the unit circle. It is a bounded measurable function of θ whose values are taken in the closure of the region onto which ψ maps the unit disk. If $f(z) = \sum a_n z^n$ is a polynomial, then the identity

$$f(\psi(e^{i\theta})) = \sum a_n \psi^n(e^{i\theta})$$

holds for almost all real θ.

Define a nonnegative measure ν on the Borel sets of the complex plane so that, for every Borel set S, $\nu(S)$ is the Lebesgue measure of the set of real numbers t modulo one such that $\psi(\exp(2\pi it))$ belongs to S. Then the support of ν is contained in the closure of the region onto which ψ maps the unit disk, and the identity

$$\int |f(z)|^2 \, d\nu(z) = \int_0^1 |f(\psi(\exp(2\pi it)))|^2 \, dt$$

holds for every polynomial $f(z)$. But if $f(\psi(z)) = \sum b_n z^n$, then

$$\int |f(z)|^2 \, d\mu(z) = \sum |b_n|^2 = \int |f(z)|^2 \, d\nu(z).$$

It follows that the identity

$$\int f(z)\bar{g}(z) \, d\mu(z) = \int f(z)\bar{g}(z) \, d\nu(z)$$

holds for all polynomials $f(z)$ and $g(z)$.

By the Stone—Weierstrass theorem, every continuous function on the union of the supports of μ and ν can be approximated uniformly by polynomials in the two variables z and $\bar{z}$. Since the identity

$$\int h(z) \, d\mu(z) = \int h(z) \, d\nu(z)$$

holds for every such polynomial $h(z)$, it holds for every continuous function $h(z)$. It follows that $\mu(S) = \nu(S)$ for every Borel set S.

By the construction of ν, the support of μ is contained in the closure of the region onto which ψ maps the unit disk. Since the support of μ contains no point of the region, it is contained in the boundary of the region. It remains to show that the support of the measure contains all points of the boundary.

Argue by contradiction, assuming that some point of the boundary is not in the support of μ. Then the reciprocal $f(z)$ of a linear function exists which is bounded by one on the support of μ, is bounded in the region into which ψ maps the unit disk, but is not bounded by one in the region. For every positive integer n, $f^n(\psi(z))$ is represented by a square summable power series because it is analytic and bounded in the unit disk. It follows that $f^n(z)$ belongs to the closure of the polynomials in $L^2(\mu)$. Since the function is bounded by one on the support of μ, the inequality

$$|f^n(w)|^2 \leqslant K(w, w)$$

holds at all points of the region onto which ψ maps the unit disk. A contradiction is obtained because some such choice of w exists for which the modulus of $f(w)$ is greater than one.

Proof of Theorem 4. Since ψ_k is bounded and analytic in the unit disk, it has a measurable boundary value function on the unit circle. By the definition of μ_k, the identity

$$\int f(z)\,d\mu_k(z) = \int f(\psi_k(\exp(2\pi it)))\,dt$$

holds for every polynomial $f(z)$. The identity

$$f(0) = \int f(z)\,d\mu_k(z)$$

follows from the identity

$$g(0) = \int g(\exp(2\pi it))\,dt,$$

which holds for every function $g(z)$ which is bounded and analytic in the unit disk, since $\psi_k(z)$ has value zero at the origin. The identity

$$\int \log|f(z)|\,d\mu_k(z) = \int \log|f(\psi_k(\exp(2\pi it)))|\,dt$$

holds for every polynomial $f(z)$.

Assume that the region onto which ψ_1 maps the unit disk is contained in the region onto which ψ_2 maps the unit disk. Then $\psi_1(z) = \psi_2(B(z))$ for a function $B(z)$ which is analytic and bounded by one in the disk and which has value zero at the origin. The desired inequality

$$\int \log|f(z)|\,d\mu_1(z) \leqslant \int \log|f(z)|\,d\mu_2(z)$$

is obtained by showing that the inequality

$$\int \log|g(B(\exp(2\pi it)))|\,dt \leqslant \int \log|g(\exp(2\pi it))|\,dt$$

holds for every function $g(z)$ which is bounded and analytic in the unit disk. It is clearly sufficient to consider only the case in which $g(z)$ is bounded by one. The inequality holds trivially if the boundary value function of $g(z)$ has absolute value one on the unit circle. By Nevanlinna's factorization [5], it is sufficient to obtain the inequality in the case that $\log g(z)$ can be defined analytically in the disk. By [5, Theorem 20], the inequality

$$\|g(B(z))^{1/r}\|_{\mathscr{C}(z)} \leqslant \|g(z)^{1/r}\|_{\mathscr{C}(z)}$$

holds for every positive integer r. It follows that the inequality

$$\int |g(B(\exp(2\pi it)))|^{1/r}\,dt \leqslant \int |g(\exp(2\pi it))|^{1/r}\,dt$$

holds for every positive integer r. The desired inequality follows from the identity

$$\log(x) = \lim_{h\searrow 0}(x^h - 1)/h$$

using the Lebesgue monotone convergence theorem.

Assume that the region onto which ψ_1 maps the unit disk is not contained in the region onto which ψ_2 maps the unit disk. Then the region onto which ψ_1 maps the unit disk is not contained in the closure of the region onto which ψ_2 maps the unit disk. A point w exists which belongs to the region onto which ψ_1 maps the unit disk but which does not belong to the closure of the region onto which ψ_2 maps the unit disk. It follows that

$$\log|w| = \int \log|z - w|\, d\mu_1(z)$$

and that

$$\log|w| > \int \log|z - w|\, d\mu_2(z).$$

Proof of Theorem 5. Since ψ has distinct values at distinct points of the unit disk, an inverse function φ exists. It is defined and analytic in the region onto which ψ maps the unit disk and its values are taken in the unit disk.

Consider any nonnegative measure α on the Borel subsets of the complex plane whose support is contained in the closure of the region onto which ψ maps the unit disk, which has finite total mass, and which assigns mass zero to the boundary of the region. Then

$$\int \frac{1 + w\bar{\varphi}(z)}{1 - w\bar{\varphi}(z)}\, d\alpha(z)$$

is an analytic function of w whose real part is nonnegative in the unit disk. By the Poisson representation [5], the identity

$$\int \operatorname{Re} \frac{1 + w\bar{\varphi}(z)}{1 - w\bar{\varphi}(z)}\, d\alpha(z) = \int \operatorname{Re} \frac{1 + w\exp(-2\pi it)}{1 - w\exp(-2\pi it)}\, d\beta(t)$$

holds for a unique nonnegative measure β on the Borel subsets of the reals modulo one. The theorem is proved by showing that the inequality

$$\int \log|g(\psi(z))|\, d\alpha(z) \leqslant \int \log|g(\exp(2\pi it))|\, d\beta(t)$$

holds for every function $g(z)$ which is bounded and analytic in the unit disk.

By linearity it is sufficient to obtain the inequality in the case that α has total mass one. The set of all such measures α is then the weakly closed convex span of its extreme points, which are point masses. By linearity and continuity, it is sufficient to obtain the inequality when α is the measure with mass one concentrated at some point w of the region onto which ψ maps the unit disk. In this case β is absolutely continuous with respect to Lebesgue measure and has density

$$\beta'(t) = \frac{1 - |\varphi(w)|^2}{|1 - \varphi(w)\exp(-2\pi it)|^2}.$$

The desired inequality states that

$$\log|g(\varphi(w))| \leqslant (1 - |\varphi(w)|^2) \int \frac{\log|g(\exp(2\pi it))|\, dt}{|1 - \varphi(w)\exp(-2\pi it)|^2}.$$

The inequality follows from Nevanlinna's estimate [5] since $g(\varphi(z))$ is bounded and analytic in the unit disk.

When α is chosen to agree with μ on the subsets of the region onto which ψ maps the unit disk, then α is a nonnegative measure σ on the Borel subsets of the reals modulo one. Let v be the nonnegative measure on the Borel subsets of the complex plane which is zero on the region onto which ψ maps the unit disk and for which, for every Borel subset S of the complement of the region, $v(S) - \mu(S)$ is the σ-measure of the set of real numbers t modulo one such that $\psi(\exp(2\pi it))$ belongs to S. If $f(z)$ is a polynomial, then $g(z) = f(\psi(z))$ is bounded and analytic in the unit disk. Since $f(z) = g(\varphi(z))$, the inequality

$$\int \log|f(z)|\, d\mu(z) \leqslant \int \log|g(\exp(2\pi it))|\, d\sigma(t)$$

holds with integration on the left over the region onto which ψ maps the unit disk. By the definition of σ,

$$\int \log|g(\exp(2\pi it))|\, d\sigma(t) = \int \log|f(z)|\, dv(z) - \int \log|f(z)|\, d\mu(z),$$

where the integrals on the right are taken over the complement of the region onto which ψ maps the unit disk. Since v assigns zero measure to the region,

$$\int \log|f(z)|\, d\mu(z) \leqslant \int \log|f(z)|\, dv(z)$$

with integration over the complex plane.

Proof of Theorem 6. Define an inner product on polynomials so that

$$4\pi\|F(z)\|^2 = \sum_{n=0}^{\infty} \frac{1}{n!} \int_{-\infty}^{+\infty} |\Gamma(n/2 + h/2 + it)F(n/2 - h/2 + it)|^2\, dt$$

for every polynomial $F(z)$. By Cauchy's formula, the identity

$$\begin{aligned} &\int_{-\infty}^{+\infty} F(n/2 - h/2 + it)G^*(n/2 - h/2 - it + 1) \\ &\qquad \times \Gamma(n/2 + h/2 - it + 1)\Gamma(n/2 + h/2 + it)\, dt \\ &\quad = \int_{-\infty}^{+\infty} F(n/2 - h/2 + it + \tfrac{1}{2})G^*(n/2 - h/2 - it + \tfrac{1}{2}) \\ &\qquad \times \Gamma(n/2 + h/2 - it + \tfrac{1}{2})\Gamma(n/2 + h/2 - it + \tfrac{1}{2})\, dt \end{aligned}$$

holds for all polynomials $F(z)$ and $G(z)$ if n is a nonnegative integer. Since $z\Gamma(z) = \Gamma(z+1)$, the identity

$$\langle zF(z), G(z)\rangle = \langle F(z), (z+h)[G(z+1) - G(z)]\rangle$$

holds for all polynomials $F(z)$ and $G(z)$. It follows that the inner product is a constant multiple of the inner product of $\mathcal{N}(h)$. Since the polynomials are dense in $\mathcal{N}(h)$, the theorem follows once it is shown that the polynomial one has norm one. This is true because

$$2\pi\Gamma(h) = \sum_{n=0}^{\infty} \frac{1}{n!} \int_{-\infty}^{+\infty} |\Gamma(n/2 + h/2 + it)|^2 \, dt$$

by the binomial theorem since

$$2\pi 2^{-n-h}\Gamma(n+h) = \int_{-\infty}^{+\infty} |\Gamma(n/2 + h/2 + it)|^2 \, dt$$

for every nonnegative integer n by the theory of Pollaczek polynomials [10].

Proof of Theorem 7. The stated results are obtained by straightforward calculations.

References

1. L. de Branges, "Espaces hilbertiens de fonctions entières," Masson, Paris, 1972.
2. M. Krein and D. Milman, On extreme points of regular convex sets, *Studia Math.* **19** (1940), 133–138.
3. L. de Branges, The Stone–Weierstrass theorem, *Proc. Amer. Math. Soc.* **10** (1959), 822–824.
4. D. Trutt, Extremal norm-determining measures for Hilbert spaces of entire functions, *J. Math. Anal. Appl.* **20** (1967), 74–89.
5. L. de Branges and J. Rovnyak, "Square Summable Power Series," Holt, New York, 1966.
6. T. L. Kriete and D. Trutt, The Cesàro operator in l^2 is subnormal, *Amer. J. Math.* **93** (1971), 215–225.
7. E. Kay and D. Trutt, Newton spaces of analytic functions, *J. Math. Anal. Appl.* **60** (1977), 325–328.
8. E. Kay, H. Soul, and D. Trutt, Some subnormal operators and hypergeometric kernel functions, *J. Math. Anal. Appl.* **53** (1976), 237–242.
9. K. F. Klopfenstein, A note on Hilbert spaces of factorial series, *Indiana Univ. Math. J.* **25** (1976), 1073–1081.
10. L. de Branges and D. Trutt, Meixner and Pollaczek spaces of entire functions, *J. Math. Anal. Appl.* **12** (1968), 12–24.

AMS (MOS) subject classification: 47A15.

TOPICS IN FUNCTIONAL ANALYSIS
ADVANCES IN MATHEMATICS SUPPLEMENTARY STUDIES, VOL. 3

Inverse Scattering, Orthogonal Polynomials, and Linear Estimation†

K. M. CASE

The Rockefeller University
New York, New York

DEDICATED TO M. G. KREIN ON THE OCCASION OF HIS SEVENTIETH BIRTHDAY

The relations between inverse scattering, orthogonal polynomials, and linear estimation are discussed. It is found that all three can be reduced to a generalized form of the discrete Gelfand–Levitan equation. Explicit solutions are readily obtained. However, specific properties of these solutions seem to be particularly easy to obtain when viewed from the viewpoint of inverse scattering theory.

I. INTRODUCTION

In various places [1, 2, 3] in the literature it has been shown that there are relations between the topics mentioned in the title of this article. Here we would like to spell out these relations explicitly. In particular, we shall see how the results of inverse scattering theory have implications for the other two.

Since this article is somewhat pedagogical in nature we restrict ourselves to discrete problems in one dimension. Corresponding results for the continuous case are readily obtained (heuristically) by appropriate limiting procedures. Also, in contrast to [2], we consider polynomials defined on a finite segment of the real line instead of on the unit circle. Since we know that there are close relations between these two kinds of polynomials this is not a significant distinction. However, the relations seem somewhat clearer when stated in terms of polynomials on the line.

The main result is that all three topics are described by the same equation, which we shall call the generalized Gelfand–Levitan equation (G.G.L.E.) [1, 4]. Remarkably, an explicit general solution can be given. While numerical evaluation may be tedious, in some cases the connections that we shall find will enable us to find efficient approximations.

† This work was supported in part by the Air Force Office of Scientific Research, Grant 722187.

ISBN 0-12-287150-2

II. Orthogonal Polynomials

The classical approach is the following. Suppose we are given a nondecreasing function $\rho(\lambda)$ which is constant for all but a finite part of the real λ axis. We want to construct polynomials $\phi(\lambda, n)$ of degree n such that

$$\int \phi(\lambda, n)\phi(\lambda, m)\, d\rho(\lambda) = \delta(n, m). \tag{1}$$

Let

$$\phi_0(\lambda, i), \qquad i = 0, 1, 2, \ldots n, \ldots, \tag{2}$$

be any linearly independent polynomials of degree i. Then we set

$$\phi(\lambda, n) = \sum_{m=0}^{n} K(n, m)\phi_0(\lambda, m). \tag{3}$$

Since $\phi(\lambda, n)$ is orthogonal to all $\phi(\lambda, r)$ for $r \leqslant n-1$, it is orthogonal to all polynomials of degree less than $n-1$. Therefore,

$$\begin{aligned}\int \phi(\lambda, n)\phi_0(\lambda, r)\, d\rho(\lambda) &= \sum_{m=0}^{n} K(n, m) \int \phi_0(\lambda, m)\phi_0(\lambda, r)\, d\rho \\ &= 0, \qquad r \leqslant n-1.\end{aligned} \tag{4}$$

Thus if we define $\mu(m, r)$ by

$$\mu(m, r) = \int \phi_0(\lambda, m)\phi_0(\lambda, r)\, d\rho(\lambda) \tag{5}$$

and $\kappa(n, m)$ by

$$\kappa(n, m) = K(n, m)/K(n, n) \tag{6}$$

we obtain

$$\sum_{m=0}^{n-1} \kappa(n, m)\mu(m, r) = -\mu(n, r), \qquad 0 \leqslant r \leqslant n-1. \tag{7}$$

A very important property of $\mu(m, r)$ is its symmetry,

$$\mu(m, r) = \mu(r, m), \tag{8}$$

which is obvious from equation (5). Equation (7) may quite fairly be called a discrete version of the Gelfand–Levitan equation.[1] Actually, it will be

[1] Here we are developing a result for polynomials on a line analogous to that for polynomials on a unit circle [2].

convenient to consider a slight generalization of equation (7), namely

$$\sum_{m=0}^{n-1} \bar{\kappa}(n, m)\mu(m, r) = -\bar{\mu}(n, r), \qquad 0 \leqslant r \leqslant n-1, \tag{9}$$

where the μ and $\bar{\mu}$ may or may not be the same. We call equation (9) the generalized Gelfand–Levitan equation (G.G.L.E.). Using Cramer's rule we can readily write down the solution of equation (9). Thus, let

$$[\det \mu]_0^{n-1} = \begin{vmatrix} \mu(0,0) & \mu(0,1) & \cdots & \mu(0, n-1) \\ \mu(1,0) & \mu(1,1) & \cdots & \mu(1, n-1) \\ \vdots & \vdots & \cdots & \vdots \\ \mu(n-1,0) & \mu(n-1,1) & \cdots & \mu(n-1, n-1) \end{vmatrix} \tag{10}$$

and let $|\bar{\mu}_{n-1}|r$ denote the determinant obtained by replacing the rth column in the determinant above by $-\bar{\mu}(0, n), -\bar{\mu}(1, n), \ldots, -\bar{\mu}(n-1, n)$. Then we have

$$\bar{\kappa}(n, r) = |\mu_{n-1}|r/[\det \mu]_0^{n-1}. \tag{11}$$

Returning to our orthogonal polynomials (equation (7)) we see that

$$\kappa(n, m) = |\mu_{n-1}|m/[\det \mu]_0^{n-1}. \tag{12}$$

To obtain the $K(n, m)$ we still need $K(n, n)$. This is readily obtained from the normalization condition

$$\int \phi^2(\lambda, n)\, d\rho(\lambda) = 1. \tag{13}$$

Using the expansion of equation (3) for one of the $\phi(\lambda, n)$ gives

$$\begin{aligned} 1 &= \sum_{m=0}^{n} K(n, m) \int \phi_0(\lambda, m)\phi(\lambda, n)\, d\rho \\ &= K(n, n) \int \phi_0(\lambda, n)\phi(\lambda, n)\, d\rho \end{aligned} \tag{14}$$

since $\phi(\lambda, n)$ is orthogonal to all polynomials of degree less than n. Inserting the expansion again for $\phi(\lambda, n)$ and dividing by $K^2(n, n)$ yields

$$\frac{1}{K^2(n, n)} = \mu(n, n) + \sum_{m=0}^{n-1} \kappa(n, m)\mu(m, n). \tag{15}$$

If we now insert the $\kappa(n, m)$ given by equation (12) we readily obtain

$$\frac{1}{K^2(n, n)} = \frac{[\det \mu]_0^n}{[\det \mu]_0^{n-1}},$$

or

$$K(n, n) = ([\det \mu]_0^{n-1}/[\det \mu]_0^n)^{1/2}. \tag{16}$$

Therefore

$$K(n, m) = |\mu_{n-1}|m/([\det \mu]_0^n[\det \mu]_0^{n-1})^{1/2}. \tag{17}$$

If we define $[\det(\mu, \phi)]_0^n$ by

$$[\det(\mu, \phi)]_0^n = \begin{vmatrix} \mu(0, 0) & \mu(0, 1) & \cdots & \mu(0, n) \\ \mu(1, 0) & \mu(1, 1) & \cdots & \mu(1, n) \\ \vdots & \vdots & \vdots\vdots\vdots & \vdots \\ \mu(n-1, 0) & \mu(n-1, 1) & \cdots & \mu(n-1)n \\ \phi_0(\lambda, 0) & \phi_0(\lambda, 1) & \cdots & \phi_0(\lambda, n) \end{vmatrix} \tag{18}$$

then combining equations (3) and (17) yields the well-known result

$$\phi(\lambda, n) = [\det(\mu, \phi)]_0^n/([\det \mu]_0^{n-1}[\det \mu]_0^n)^{1/2}. \tag{19}$$

For small n this is a simple formula from which to compute the $\phi(\lambda, n)$. We will develop more efficient formulas to compute the determinants when n is not small.

First, though, we give some indications as to why orthogonal polynomials are related to the inverse scattering and linear estimation problems.

The connection to the inverse scattering problem arises from the fact that our orthogonal polynomials satisfy a particularly simple three term recursion formula. Indeed, since $\lambda\phi(\lambda, n)$ is a polynomial of degree $n + 1$ we can expand so that

$$\lambda\phi(\lambda, n) = \sum_{m=0}^{n+1} \gamma(n, m)\phi(\lambda, m). \tag{20}$$

Multiplying by $\phi(\lambda, r)\rho$ and integrating gives

$$\begin{aligned} \int \phi(\lambda, r)\lambda\phi(\lambda, n)\, d\rho &= \sum_{m=0}^{n+1} \gamma(n, m) \int \phi(\lambda, r)\phi(\lambda, m)\, d\rho \\ &= \gamma(n, r). \end{aligned} \tag{21}$$

Now if $r < n - 1$, $\lambda\phi(\lambda, r)$ is a polynomial of degree less than n and the left side of equation (21) is zero. Thus the only nonvanishing coefficients are

$$\begin{aligned} \gamma(n, n+1) &= \int \phi(\lambda, n+1)\lambda\phi(\lambda, n)\, d\rho \equiv a(n+1), \\ \gamma(n, n) &= \int \phi(\lambda, n)\lambda\phi(\lambda, n)\, d\rho \equiv b(n), \\ \gamma(n, n-1) &= \int \phi(\lambda, n)\lambda\phi(\lambda, n-1)\, d\rho \equiv a(n). \end{aligned} \tag{22}$$

Therefore, we have the recursion relation

$$\lambda\phi(\lambda, n) = a(n+1)\phi(\lambda, n+1) + b(n)\phi(\lambda, n) + a(n)\phi(\lambda, n-1). \quad (23)$$

It is natural to ask for the relations of the $a(n)$ and the $b(n)$ to the determinants involved in the solution of the Gelfand–Levitan equation. To find these we introduce the expansions of equation (3) into the recursion relation and obtain

$$a(n+1)\sum_{m=0}^{n+1} K(n+1, m)\phi_0(\lambda, m) + b(n)\sum_{m=0}^{n} K(n, m)\phi_0(\lambda, m)$$
$$+ a(n)\sum_{m=0}^{n-1} K(n-1, m)\phi_0(\lambda, m) = \lambda\sum_{m=0}^{n} K(n, m)\phi_0(\lambda, m). \quad (24)$$

The desired relations can be obtained by equating the coefficients of λ^{n+1} and λ^n on both sides. The result is somewhat inelegant except for two important special cases.

(a) *The* $\phi_0(\lambda, n)$ *are also orthogonal polynomials.* Thus there is a weight function σ such that

$$\int \phi_0(\lambda, n)\phi_0(\lambda, m)\, d\sigma(\lambda) = \delta(n, m), \quad (25)$$

and the $\phi_0(\lambda, n)$ satisfy a recursion formula

$$\lambda\phi_0(\lambda, n) = a_0(n+1)\phi_0(\lambda, n+1) + b_0(n)\phi_0(\lambda, n) + a_0(n)\phi_0(\lambda, n-1) \quad (26)$$

Then equation (24) becomes

$$a(n+1)\sum_{m=0}^{n+1} K(n+1, m)\phi_0(\lambda, m) + b(n)\sum_{m=0}^{n} K(n, m)\phi_0(\lambda, m)$$
$$+ a(n)\sum_{m=0}^{n-1} K(n-1, m)\phi_0(\lambda, m)$$
$$= \sum_{m=0}^{n} K(n, m)[a_0(m+1)\phi_0(\lambda, m+1) + b_0(m)\phi_0(\lambda, m)$$
$$+ a_0(m)\phi_0(\lambda, m-1)] \quad (27)$$

Multiplying this by $\phi_0(\lambda, n+1)\, d\sigma$ and integrating gives

$$a(n+1)/a_0(n+1) = K(n, n)/K(n+1, n+1), \quad (28)$$

while doing the same with $\phi_0(\lambda, n)$ gives

$$b(n) = b_0(n) + a_0(n)\kappa(n, n-1) - a_0(n+1)\kappa(n+1, n). \quad (29)$$

(b) *The $\phi_0(\lambda, n)$ are the powers, i.e., $\phi_0(\lambda, n) \equiv \lambda^n$.* Then equating powers of λ^{n+1} and λ^n in equation (24) is very simple and gives the results

$$a(n+1) = K(n, n)/K(n+1, n+1) \tag{30}$$

and

$$b(n) = \kappa(n, n-1) - \kappa(n+1, n). \tag{31}$$

We give some applications of these relations.

(1) Suppose we have found the $\phi(\lambda, i)$, $i = 0, 1, 2, \ldots, n$. Then we also have $K(i, j)$, $0 \leqslant j \leqslant i \leqslant n$. To find $\phi(\lambda, n+1)$ we can use the recursion relation in the form

$$\phi(\lambda, n+1) = \frac{\lambda\phi(\lambda, n)}{a(n+1)} - \frac{b(n)\phi(\lambda, n)}{a(n+1)} - \frac{a(n)\phi(\lambda, n-1)}{a(n+1)}. \tag{32}$$

The only new nontrivial quantities then to be calculated are $[\det \mu]_0^{n+1}$ and $|\mu_{n+1}|n$. (If, as is frequently the case, $b(n) \equiv 0$, there is only the one new determinant to be found.)

(2) As we shall see the $K(n, n)$ usually play very significant roles. A natural question, then, is the following: Suppose we have two different sets of ϕ_0's—call them $\phi_1(\lambda, n)$ and $\phi_2(\lambda, n)$. What is the relation between $K^{(1)}(n, n)$ and $K^{(2)}(n, n)$?

From equations (22) we see the $a(n+1)$ depends only the $\phi(\lambda, n)$ and $\rho(\lambda)$. Accordingly, the two different expressions for $a(n+1)$ must be equal. Thus, from equation (28) we obtain

$$\begin{aligned} a^1(n+1)K^{(1)}(n, n)/K^{(1)}(n+1, n+1) \\ = a^2(n+1)K^{(2)}(n, n)/K^{(2)}(n+1, n+1)]. \end{aligned} \tag{33}$$

If we define $\Lambda(n)$ by

$$\Lambda(n) = K^{(2)}(n, n)/K^{(1)}(n, n), \tag{34}$$

equation (33) is the recursion relation

$$\Lambda(n+1)/\Lambda(n) = a_2(n+1)/a_1(n+1), \tag{35}$$

whose solution is

$$\Lambda(n) = \Lambda(0) \prod_{i=1}^{n} \frac{a_2(i)}{a_1(i)}.$$

But

$$\Lambda(0) = K^{(2)}(0, 0)/K^{(1)}(0, 0) = \phi_1{}^2(\lambda, 0)/\phi_2{}^2(\lambda, 0).$$

Hence

$$\frac{K^{(2)}(n, n)}{K^{(1)}(n, n)} = \frac{\phi_1{}^2(\lambda, 0)}{\phi_2{}^2(\lambda, 0)} \prod_{i=1}^{n} \frac{a_2(i)}{a_1(i)}. \tag{36}$$

Some Remarks. (1) If $\phi_2(\lambda, n)$ are the powers, we see from equation (30) that in (36) we must put $a_2(i) \equiv 1$, $\phi_2(\lambda, 0) = 1$. Thus equation (36) becomes

$$K^{(1)}(n, n) = \frac{K^{(2)}(n, n)}{[\phi_1(\lambda, 0)]^2} \prod_{i=1}^{n} a_1(i). \tag{37}$$

(2) By a quite similar argument we can also determine the relations of the determinants associated with polynomials $\phi_1(\lambda, n)$ and $\phi_2(\lambda, n)$. We find

$$[\det \mu^{(2)}]_0^0/[\det \mu^{(1)}]_0^0 \equiv \mu^{(2)}(0, 0)/\mu^{(1)}(0, 0) = [\phi_2(\lambda, 0)/\phi_1(\lambda, 0)]^2 \tag{38}$$

and

$$\frac{[\det \mu^{(2)}]_0^n}{[\det \mu^{(1)}]_0^n} = \left[\frac{\phi_1(\lambda, 0)}{\phi_2(\lambda, 0)}\right]^{4n+2} \prod_{i=1}^{n} \left[\frac{a_2(i)}{a_1(i)}\right]^{2(n+1-i)}, \qquad n = 1, 2, \ldots . \tag{39}$$

(3) A similar question is the following. If we are given some orthogonal polynomials we know that our $\phi(\lambda, n)$ can be constructed in the form

$$\phi(\lambda, n) = \sum_{m=0}^{n} K^{(2)}(n, m)\phi_0(\lambda, m). \tag{40}$$

However, both the $\phi(\lambda, n)$ and $\phi_0(\lambda, m)$ can be constructed directly in terms of $(\lambda)^i$, $i = 0, 1, 2, \ldots$, i.e.

$$\phi(\lambda, n) = \sum_{m=0}^{n} K^{(1)}(n, m)\lambda^m$$

and

$$\phi_0(\lambda, n) = \sum_{m=0}^{n} K^{(0)}(n, m)\lambda^m. \tag{41}$$

What is the relationship between the three K's? (Alternatively, what is the relationship between the solutions of the corresponding three generalized Gelfand–Levitan equations?) This is found by replacing the ϕ and ϕ_0 in equation (40) by their expressions from (41) and then comparing coefficients of powers of λ. The result can be expressed as follows: Let $K^{(i)}$ be triangular matrices such that

$$K^{(i)}{}_{nm} \begin{cases} = K^{(i)}(n, m), & m \leqslant n, \\ = 0, & m > n, \end{cases}$$

Then

$$K^{(2)} = K^{(1)}[K^{(0)}]^{-1}. \tag{42}$$

In particular,

$$K^{(2)}(n, n) = K^{(1)}(n, n)/K^{(0)}(n, n). \tag{43}$$

Now let us give some indication as to why orthogonal polynomials are related to linear estimation problems. Such problems are essentically least squares fits to some data. Therefore, let us consider the following question.[2]

Let $\phi_0(\lambda, m)$, $m = 0, 1, 2, \ldots, n$, be a set of linearly independent polynomials of degree m. We want to choose constants $C(n, m)$, $m < n$, such that

$$Q_n = \int g_n{}^2(\lambda)\, d\rho(\lambda) \tag{44}$$

is a minimum. Here $\rho(\lambda)$ is some weight function and

$$g_n(\lambda) = \phi_0(\lambda, n) + \sum_{m'=0}^{n-1} C(n, m')\phi_0(\lambda, m'). \tag{45}$$

Explicitly we have

$$Q_n = \int \Bigg[\phi_0{}^2(\lambda, n) + 2 \sum_{m'=0}^{n-1} C(n, m')\phi_0(\lambda, m')\phi_0(\lambda, n) + \sum_{m'=0}^{n-1} \sum_{m''=0}^{n-1} C(n, m')C(n, m'')\phi_0(\lambda, m')\phi_0(\lambda, m'') \Bigg] d\rho(\lambda). \tag{46}$$

For Q_n to be a minimum we demand that

$$\delta Q_n/\delta C(n, m) = 0, \tag{47}$$

which yields

$$\sum_{m'=0}^{n-1} C(n, m')\mu(m', m) = -\mu(m, n),\ 0 \leqslant m \leqslant n-1. \tag{48}$$

Here $\mu(m', m)$ is as given by equation (5). Comparing with equation (7) we see that this is just the generalized Gelfand–Levitan equation associated with the weight function $\rho(\lambda)$ and polynomials $\phi_0(\lambda, m)$. Thus to minimize Q_n we choose

$$C(n, m) = \kappa(n, m) = K(n, m)/K(n, n).$$

The corresponding g_n is

$$g_n = \phi_0(\lambda, n) + \sum_{m=0}^{n-1} \frac{K(n, m)}{K(n, n)} \phi_0(\lambda, m) \equiv \frac{\phi(\lambda, n)}{K(n, n)}. \tag{49}$$

[2] See footnote 1.

The minimum Q_n is then

$$Q_n(\min) = \int [\phi^2(\lambda, n)/K^2(n, n)]\, d\rho(\lambda) \equiv 1/K^2(n, n). \tag{50}$$

III. The Discrete Inverse Scattering Problem

Consider the following eigenvalue problem:[3]

$$a(n+1)\phi(\lambda, n+1) + b(n)\phi(\lambda, n) + a(n)\phi(\lambda, n-1) = \lambda\phi(\lambda, n), \tag{51}$$

where $a(0)\phi(\lambda, -1) = 0$, $\phi(\lambda, 0) = C$. Here we assume that $a(n)$, $b(n)$ approach finite limiting values $a(\infty)$, $b(\infty)$ (sufficiently rapidly).

With these assumptions it is readily shown that the spectrum consists of

(i) a finite number of real discrete eigenvalues corresponding to square summable solutions of equation (51) and the boundary conditions;

(ii) a continuous spectrum (corresponding to bounded solutions), where

$$b(\infty) - 2a(\infty) \leqslant \lambda \leqslant b(\infty) + 2a(\infty). \tag{52}$$

These solutions are conveniently described by Z such that

$$\lambda = b(\infty) + a(\infty)[Z + Z^{-1}]. \tag{53}$$

Then the continuous spectrum corresponds to Z on the unit circle, i.e. $Z = e^{i\theta}$.

It is useful to introduce two auxiliary solutions, $\phi_{\pm}(\lambda, n)$, of equation (51) which satisfy the boundary conditions

$$\lim_{n\to\infty} |\phi_{\pm} - Z^{\pm n}| \to 0. \tag{54}$$

Further, we define functions $f_{\pm}(Z)$ by equation (51) as

$$f_{\pm}(Z) = a(0)\phi_{\pm}(Z, -1). \tag{55}$$

(The function $f_{+}(Z)$ is called the Jost function. It is fundamental for all that follows.)

In terms of these auxiliary functions the continuum functions are

$$\phi(\lambda, n) = [C/a(\infty)2i\sin\theta][f_{-}(Z)\phi_{+}(Z, n) - f_{+}(Z)\phi_{-}(Z, n)], \qquad Z = e_1^{i\theta}. \tag{56}$$

(Note that the linearly independent functions are obtained for Z running over the upper half of the unit circle, i.e., $0 \leqslant \theta \leqslant \pi$).

From equation (56) we see that the asymptotic behavior of our continuium functions as $n \to \infty$ is

$$\phi(\lambda, n) \sim [C|f_{+}(Z)|/a(\infty)\sin\theta]\sin(n\theta + \delta), \tag{57}$$

[3] More details are given in [1].

where

$$\delta(\theta) = -\arg f_+(Z). \tag{58}$$

The inverse scattering problem can now be formulated. Suppose we are given $\delta(\theta)$ for $0 \leqslant \theta \leqslant \pi$, the positions of the discrete eigenvalues λ_i, and certain constants associated with the normalization of the discrete eigen functions. What are $a(n)$ and $b(n)$? We shall sketch two solutions. But first we need to list some properties that are derived elsewhere [1].

(1) $f_+(Z)$ is analytic within the unit circle except for a simple pole at $Z = 0$.

(2) The zeros of $f_+(Z)$ on or within the unit circle are the discrete eigenvalues. For simplicity we will *from now on* assume there are no such eigenvalues.

(3) On the unit circle $f_+{}^*(Z) = f_-(Z)$.

(4) One can prove the completeness of the eigenfunctions in the form

$$\int_{-\infty}^{\infty} \phi(\lambda, n)\phi(\lambda, m)\, d\rho(\lambda) = \delta(n, m), \tag{59}$$

where

$$d\rho(\lambda) = \begin{cases} 0 & \text{unless} \quad b(\infty) - 2a(\infty) \leqslant \lambda \leqslant b(\infty) + 2a(\infty) \\ \rho'(\lambda)\, d\lambda & \text{if} \quad b(\infty) - 2a(\infty) \leqslant \lambda \leqslant b(\infty) + 2a(\infty). \end{cases} \tag{60}$$

Here

$$\rho'(\lambda) = a(\infty) \sin\theta / \prod C^2 |f_+|^2. \tag{61}$$

Solution 1. This proceeds via the Gelfand–Levitan equation. Thus from equation (59) we recognize that the $\phi(\lambda, n)$ are orthogonal polynomials with weight function $\rho(\lambda)$. Hence, if we can construct $\rho'(\lambda)$ we can apply the results of the proceding section. In particular, from equations (28) and (29) we find $a(n)$, $b(n)$.

It is, however, easy to find $\rho'(\lambda)$. Since $Zf_+(Z)$ is analytic and nonzero (within the unit circle) $\ln Zf_+(Z)$ is analytic there. Its imaginary part *on* the unit circle is known. Hence the problem is reduced to the well-known one of finding a function analytic within the unit circle given its imaginary part on the boundary. Thus we readily construct $f_+(Z)$. From this we read off $|f_+|$, and then ρ' is given by equation (61).

Let us recapitulate. The procedure is to go from $\delta(\theta)$ to f_+ and then to ρ'. Then apply the results of the theory of orthogonal polynomials.

We refer the reader elsewhere for the explicit construction of f_+ from $\delta(\theta)$ [1]. However, for what follows we note that the process is invertible. Given ρ' we can construct δ. Thus given ρ' we know the real part of $\ln Zf_+(Z)$ on the unit circle.

An application of the Poisson–Jensen theorem then gives

$$Zf_+(Z) = \exp\{(1/4\pi)\int_{-\pi}^{\pi}[\ln|a(\infty)\sin\theta'/\pi C^2\rho'|][\exp(i\theta'+Z)/\exp(i\theta'-Z)]\,d\theta'\}. \tag{62}$$

Then if we let Z approach the unit circle from equation (62) we obtain

$$\exp(i\delta) = \exp\left\{i\theta - \frac{P}{4\pi}\int_{-\pi}^{\pi}\left[\ln\left|\frac{a(\infty)\sin\theta'}{\pi C^2\rho'}\right|\right]\left[\frac{\exp(i\theta')+\exp(i\theta)}{\exp(i\theta')-\exp(i\theta)}\right]d\theta'\right\}. \tag{63}$$

Solution 2. A second approach is via the Marchenko equation [5]. Thus consider a comparison equation

$$a_0(n+1)\phi_0(\lambda, n+1) + b_0(n)\phi_0(\lambda, n) + a(n)\phi_0(\lambda, n-1) = \lambda\phi_0(\lambda, n). \tag{64}$$

Let $\phi_0{}^{\pm}$ be solutions subject to the conditions

$$\lim_{n\to\infty}|\phi_0{}^{\pm} - Z^{\pm n}| \to 0. \tag{65}$$

Then expand $\phi^+(Z, n)$ in the form

$$\phi^+(Z, n) = \sum_{m=n}^{\infty} A(n, m)\phi_0{}^+(Z, m). \tag{66}$$

The simplest example (and the only one which we shall consider here) is when $a_0(n) \equiv a(\infty)$, $b_0(n) \equiv b(\infty)$. In this case

$$\phi_0{}^{\pm}(Z, n) \equiv Z^{\pm n}. \tag{67}$$

First note that if we have the $A(n, m)$, we have solved our inverse problem. Indeed, if we insert the expansion of equation (66) into (64) and equate coefficients of Z^n and Z^{n+1} we find

$$a(n) = a(\infty)A(n, n)/A(n-1, n-1) \tag{68}$$

and

$$b(n) = b(\infty) + a(\infty)[\alpha(n, n+1) - \alpha(n-1, n)], \tag{69}$$

where

$$\alpha(n, m) = A(n, m)/A(n, n).$$

The similarity to the solution via the first method (equations (28) and (29)) is striking.

To find equations to determine the $A(n, m)$ let us rewrite equation (56) in the form

$$[a(\infty)(Z - Z^{-1})/Cf_+(Z)]\phi(\lambda, n) = -\phi^-(Z, n) + S\phi^+(Z, n), \tag{70}$$

where $S \equiv f_-/f_+ = e^{2i\delta}$. Multiply this by $Z^{m-1}/2\pi i$ and integrate around the unit circle. The left-hand side is evaluated using analytic properties of ϕ and f_+ and their readily obtained behavior for small Z. On the right hand we introduce the expansion of equation (66). The net result is the equations

$$\frac{\delta(m, n)}{A(n, n)} = A(n, m) + \sum_{m'=n}^{\infty} A(n, m')F(m', m), \tag{71}$$

where

$$F(m', m) = F(m, m') = -(1/2\pi i)\oint SZ^{m+m'-1}\, dZ.$$

If $t(m', m) = \delta(m', m) + F(m', m)$ we can rewrite these as

$$\frac{1}{A^2(n, n)} = t(n, n) + \sum_{m'=n+1}^{\infty} \alpha(n, m')t(m', n) \tag{72}$$

and

$$\sum_{m'=n+1}^{\infty} \alpha(n, m')t(m', m) = -t(n, m), \qquad m > n. \tag{73}$$

The similarity to G.G.L.E. (9) is again very close. Indeed, the only differences are the limits. Accordingly it is not surprising that can write down the solution immediately using Cramer's rule. Thus, if

$$[\det t]_n^\infty = \begin{vmatrix} t(n, n) & t(n, n+1) & t(n, n+2) & \cdots \\ t(n+1, n) & t(n+1, n+1) & \cdot & \cdots \end{vmatrix}, \tag{74}$$

then

$$A^2(n, n) = [\det t]_{n+1}^\infty / [\det t]_n^\infty, \tag{75}$$

and if $|n + 1|_m$ is the same as $[\det t]_{n+1}^\infty$ with the mth column replaced by $-t(n, n+1), -t(n, n+2), -t(n, n+3), \ldots,$

$$\alpha(n, m) = |n + 1|_m / [\det t]_{n+1}^\infty. \tag{76}$$

Formally, this present solution seems more complicated than Solution (1). Here we are dealing with infinite rather than finite determinants. However, we note:

(1) For continuous problems they are equally difficult Fredholm determinants.

(2) The $K(n, m)$ and $A(n, m)$ are simply related. For practical calculations one or the other is easier to evaluate, depending on whether n is small or large.

To see that there must be some relations we note that the ratio of two A's or that of two K's gives the same $a(n)$. Similarly, a combination of $\kappa(n, n-1)$ and $\kappa(n+1, n)$ gives the same $b(n)$ as a combination of $\alpha(n, n+1)$ and $\alpha(n-1, n)$. Of course, the exact form of the relations will depend on the $\phi_0(\lambda, m)$ and $\phi_0{}^+(Z, m)$ used in equations (3) and (66). Clearly, we expect this to be simplest when the $\phi_0(\lambda, m)$ satisfy the same recursion relation as the $\phi_0{}^+(Z, m)$. Accordingly, we consider the simple case when both satisfy

$$a(\infty)\phi(\lambda, n+1) + b(\infty)\phi(\lambda, n) + a(\infty)\phi(\lambda, n-1) = \lambda\phi(\lambda, n). \tag{77}$$

Then the $\phi_0{}^+(Z, n) \equiv Z^n$, and the $\phi_0(\lambda, n)$ are Tchebycheff polynomials.
The relations follow from the orthogonality relations expressed by equation (59) [6]. We state the result for the present case as a

THEOREM. *Let $K'(n, m)$ be the inverse of $K(n, m)$, i.e.,*

$$\phi_0(\lambda, n) = \sum_{m=0}^{n} K'(n, m)\phi(\lambda, m). \tag{78}$$

Then

$$K'(n, m) = \sum_{l=m}^{n} A(m, l)p(l-n), \qquad m \leqslant n, \tag{79}$$

where

$$p(l-n) = \frac{a(\infty)}{2\pi i}\oint \frac{Z^{l-n-1}\,dZ}{Zf_+(Z)}. \tag{80}$$

We remark that the $p(m)$ are readily expressed in terms of the Fourier coefficients of $\ln \rho'$. Indeed, since $a(\infty)/Zf_+(Z)$ is analytic within the unit circle, we see from equation (80) that $p(m)$ is just the coefficient of Z^m in the Taylor series expansion of $a(\infty)/Zf_+(Z)$ around the origin. However, we see from equation (62) that if we make the expansion

$$\frac{1}{4\pi}\int_{-\pi}^{\pi}\left[\ln\left|\frac{a(\infty)\sin\theta'}{\pi C^2\rho}\right|\right]\left[\frac{e^{i\theta'}+Z}{e^{i\theta'}-Z}\right]d\theta' = -\sum_{m=0}^{\infty}\gamma(m)Z^m, \tag{81}$$

then

$$a(\infty)/Zf_+(Z) = a(\infty)\exp\left[+\sum_{m=0}^{\infty}\gamma(m)Z^m\right], \tag{82}$$

where explicitly we have

$$\gamma(0) = (-1/4\pi) \int_{-\pi}^{\pi} \ln|a(\infty) \sin\theta'/\pi C^2 \rho'|\, d\theta'$$
$$\equiv (-1/2\pi) \int_{0}^{\pi} \ln|a(\infty) \sin\theta'/\pi C^2 \rho'|\, d\theta' \tag{83}$$

and

$$\gamma(n) = (-1/2\pi) \int_{-\pi}^{\pi} [\ln|a(\infty) \sin\theta'/\pi C^2 \rho'|]\, e^{-in\theta'}\, d\theta'$$
$$\equiv (-2/\pi) \int_{0}^{\pi} [\ln|a(\infty) \sin\theta'/\pi C^2 \rho'|] \cos n\theta'\, d\theta', \qquad n > 0. \tag{84}$$

Expanding the exponential in equation (82) and collecting powers of Z we then directly compute the $p(m)$. The first few are:

$$\begin{aligned} p(0) &= a(\infty) \exp\gamma(0); \\ p(1) &= p(0)\gamma(1); \\ p(2) &= p(0)\{\gamma(2) + \tfrac{1}{2}[\gamma(1)]^2\}; \\ p(3) &= p(0)\{\gamma(3) + \gamma(1)\gamma(2) + \tfrac{1}{6}[\gamma(1)]^3\}. \end{aligned} \tag{85}$$

Given the $p(m)$ it is straightforward to invert equation (79) so as to obtain the $K(n, n)$ and $\kappa(n, m)$ in terms of $A(n, n)$ and $\alpha(n, m)$. For example, we find

$$K(n, n) = p(0)/A(n, n), \tag{86}$$

$$\kappa(n, n-1) = -[\gamma(1) + \alpha(n-1, n)], \tag{87}$$

$$\kappa(n, n-2) = \frac{[\gamma(1)]^2}{2} - \gamma(2) + \alpha(n-1, n)[\gamma(1) + \alpha(n-2, n)] - \alpha(n-2, n). \tag{88}$$

Notice that equation (86) implies that when $a(n)$ is given by (68) it is also given by (28). Similarly, equations (87) and (69) imply (29).

IV. Relations between Inverse Scattering and Orthogonal Polynomials

In the previous section we saw that the discrete inverse scattering problem can be reduced to one in the theory of orthogonal polynomials. Does this mean that inverse scattering theory throws no light on orthogonal polynomials? Not really. Thus:

(1) Inverse scattering immediately calls attention to the basic importance of the Jost function $f_+(Z)$. That the weight function ρ' has a factorization of the form $\rho' \approx 1/|f_+|^2$ is obvious. Further, the fact that f_+ is a solution of the

difference equations evaluated at $n = -1$ enables us to obtain many of its properties by very elementary means.

(2) The inverse scattering approach also suggests many new relations.

(a) Elsewhere [1] we have shown that there are an infinite number of sum rules. These relate sums of powers of the $a(n)$ and $b(n)$ to the Fourier coefficients of equations (83) and (84).

(b) The results of scattering theory can be used to materially simplify the calculation of various quantities of interest. Here we make use of an idea introduced by Dyson [7] in a somewhat different context. This uses the fact that we have both a Gelfand–Levitan and a Marchenko representation. They have complementary regions of usefulness. Thus, while our expression of the $K(n, m)$ as ratios of determinants is simple enough to calculate for small n, it is horrendous if n is very large. On the contrary, the explicit expressions for $A(n, m)$ are horrendous for small n but are simple as $n \to \infty$. Thus, for our polynomials $\lim_{n\to\infty} A(n, m) = \delta(n, m)$. Using the relations obtained in Section III between the $K(n, m)$ and $A(n, m)$ we can calculate each in the region where the explicit expression is difficult to compute. We give two examples.

EXAMPLE 1. In the minimization problem of Section II we saw that the minimum was $1/K(n, n)$. A natural question is: What is the limit as $n \to \infty$? Direct computation from the determinants would be ridiculous. However, from equation (86) we see that

$$\lim_{n\to\infty} K(n, n) = \lim_{n\to\infty} [p(0)/A(n, n)] = p(0). \tag{89}$$

EXAMPLE 2. The asymptotic behavior of the polynomials as $n \to \infty$ is of frequent interest. Let us see how the inverse scattering approach treats this. We rewrite equation (56) in the form

$$\phi(\lambda, n) = [C|f_+|/a(\infty)2i\sin\theta][e^{i\delta}\phi_+(Z, n) - e^{-i\delta}\phi_-(Z, n)] \tag{90}$$

For $\phi_\pm$ we introduce the Marchenko representation

$$\phi_\pm(Z, n) = \sum_{m=n}^{\infty} A(n, m)Z^{\pm n}. \tag{91}$$

Equation (90) is then

$$\phi(\lambda, n) = \frac{C|f_+|}{a(\infty)\sin\theta} \sum_{m=n}^{\infty} A(n, m)\sin(m\theta + \delta). \tag{92}$$

Thus to find $\phi(\lambda, n)$ for large n we only need $A(n, m)$ for large n. But then $A(n, m) \approx \delta(n, m)$ and improved values can be obtained from the Marchenko equations by perturbation theory. Thus, let us write equations (72) and (73) in the forms

$$\frac{1}{A^2(n, n)} = 1 + F(n, n) + \sum_{m \leqslant n+1}^{\infty} \alpha(n, m')F(m', n), \tag{93}$$

and

$$0 = \alpha(n, m) + F(n, m) + \sum_{m'=n+1}^{\infty} \alpha(n, m')F(m', m), \qquad m > n. \tag{94}$$

Here now we want n large, and therefore the $F(n, m)$ are small.

In zeroth approximation we have

$$A(n, n) = 1, \qquad \alpha(n, m) = A(n, m) = 0. \tag{95}$$

The next approximation is

$$A(n, n) = 1 - \tfrac{1}{2}F(n, n), \qquad \alpha(n, m) = -F(n, m). \tag{96}$$

Then our improved asymptotic form for $\phi(\lambda, n)$ is

$$\phi(\lambda, n) \approx \frac{C|f_+|}{a(\infty)\sin\theta}\left\{\left[1 - \frac{F(n, n)}{2}\right]\sin(n\theta + \delta) + \sum_{m=n+1}^{\infty} F(n, m)\sin(m\theta + \delta)\right\}. \tag{97}$$

Further iteration of equations (93) and (94) will give successively improved forms.

V. Linear Estimation

Consider the following simple, but rather general, problem. We are given an input $Z(t)$ which is the sum of a signal $X(t)$ and noise $y(t)$, i.e.,

$$Z(t) = X(t) + y(t) \tag{98}$$

for $-T \leqslant t \leqslant 0$. Further, the correlation functions $\langle Z(t)Z(t')\rangle$, $\langle X(t)X(t')\rangle$, and $\langle X(t)Z(t')\rangle$ are presumed known. It is desired to find a linear combination of the $Z(t)$ such that a least squares best estimate is obtained for $X(\tau)$. (If $\tau > 0$, this is prediction, $\tau = 0$ is filtering, and $\tau < 0$ is smoothing).

Let us consider the discrete case, where we are given Z at $t = 0, -1, -2, \ldots, -(n-1)$. We want $X(s)$. As our estimate we choose

$$X^*(s) = \sum_{m=0}^{n-1} \kappa(n, m)Z(m + 1 - n). \tag{99}$$

Require that the "error" ϵ_n be a minimum. Here we choose

$$\epsilon_n{}^2 = \langle[X(x) - X^*(s)]^2\rangle. \tag{100}$$

Thus

$$\epsilon_n^2 = \mu_x(S+n-1, S+n-1) - 2\sum_{m'=0}^{n-1} \kappa(n, m')\mu_{xZ}(S+n-1, m')$$
$$+ \sum_{m'm''=0}^{n} \kappa(n, m')\kappa(n, m'')\mu_Z(m', m''), \tag{101}$$

where we have defined the μ's as

$$\begin{aligned} \mu_x(S+n-1, S+n-1) &= \langle X(S)X(S)\rangle, \\ \mu_{xZ}(S+n-1, m') &= \langle X(S)Z(m'+1-n)\rangle, \end{aligned} \tag{102}$$

and

$$\mu_Z(m', m'') = \langle Z(m'+1-n)Z(m''+1-n)\rangle.$$

We require

$$\partial\epsilon_n^2/\partial\kappa(n, m) = 0, \qquad 0 \leqslant m < n. \tag{103}$$

This yields

$$\sum_{m'=0}^{n-1} \kappa(n, m')\mu_Z(m', m) = \mu_{xZ}(S+n-1, m), \qquad 0 \leqslant m < n. \tag{104}$$

We see that this is precisely our generalized Gelfand–Levitan equation (9) with the replacements

$$\begin{aligned} \mu(m', m) &\to \mu_Z(m', m) \\ \bar{\mu}(n, m) &= -\mu_{xZ}(S+n-1, m). \end{aligned} \tag{105}$$

It is now hardly surprising that linear estimation and orthogonal polynomials are related. They are both governed by the same fundamental equation.

For the minimum ϵ_n^2 we have, since $\kappa(n, m)$ satisfies equation (104),

$$\epsilon_n^2(\text{min}) = \mu_x(S+n-1, S+n-1) - \sum_{m'=0}^{n-1} \kappa(n, m')\mu_{xZ}(S+n-1, m'). \tag{106}$$

It has been seen that the G.G.L.E. is readily solved using Cramer's rule. Accordingly, we merely state the results.

Let

$$[\det \mu_Z]_0^{n-1} = \begin{vmatrix} \mu_Z(0) & \mu_Z(0, 1) & \cdots & \mu_Z(0, n-1) \\ \mu_Z(1, 0) & \mu_Z(1, 1) & \cdots & \mu_Z(1, n-1) \\ \vdots & \vdots & & \vdots \\ \mu_Z(n-1, 0) & \mu_Z(n-1, 1) & \cdots & \mu_Z(n-1, n-1) \end{vmatrix}, \tag{107}$$

$$[\det(\mu_Z, \mu_{xZ}, \mu_x]_0^{n-1}$$

$$= \begin{vmatrix} \mu_Z(0,0) & \mu_Z(0,1) & \cdots & \mu_Z(0,n-1) & \mu_{xZ}(S+n-1,0) \\ \mu_Z(1,0) & \mu_Z(1,1) & \cdots & \mu_Z(1,n-1) & \mu_{xZ}(S+n-1,1) \\ \vdots & \vdots & \cdots & \vdots & \vdots \\ \mu_Z(n-1,0) & \mu_Z(n-1,1) & \cdots & \mu_Z(n-1,n-1) & \mu_{xZ}(S+n-1,n-1) \\ \mu_{xZ}(S+n-1,0) & \mu_{xZ}(S+n-1,1) & \cdots & \mu_{xZ}(S+n-1,n-1) & \mu_x(S+n-1,S+n-1) \end{vmatrix} \tag{108}$$

and

$$[\det(\mu_Z, \mu_{xZ}, Z)]_0^{N-1}$$

$$= \begin{vmatrix} \mu_{xZ}(S+n-1,0) & \mu_Z(0,0) & \mu_Z(0,1) & \cdots & \mu_Z(0,n-1) \\ \mu_{xZ}(S+n-1,1) & \mu_Z(1,0) & \mu_Z(1,1) & \cdots & \mu_Z(1,n-1) \\ \mu_{xZ}(S+n-1,n-1) & \mu_Z(n-1,0) & & \cdots & \mu_Z(n-1,n-1) \\ 1 & \dfrac{Z(1-n)}{[\mu_x(S+n-1,S+n-1)]^{1/2}} & \dfrac{(Z-n)}{[\mu_x(S+n-1,S+n-1)]^{1/2}} & \cdots & \dfrac{Z(0)}{[\mu_x(S+n-1,S+n-1)]^{1/2}} \end{vmatrix} \tag{109}$$

Then

$$\epsilon_n{}^2(\min) = [\mathrm{Det}(\mu_Z, \mu_{xZ}, \mu_x)]_0^{n-1}/[\det \mu_Z]_0^{n-1}, \tag{110}$$

and

$$\frac{X^*(S)}{[\mu_x(S+n-1, S+n-1)]^{1/2}} = 1 - \frac{[\mathrm{Det}(\mu_Z, \mu_{xZ}, Z)]_0^{n-1}}{[\det \mu_Z]_0^{n-1}}. \tag{111}$$

One simple case may be noted. If there is no noise and we are estimating only one unit of time ahead we have $S = 1$, $\mu_x = \mu_Z = \mu_{xZ}$. Then equation (110) becomes

$$\epsilon^2(\min) = [\det \mu_Z]_0^n/[\det \mu_Z]_0^{n-1}, \tag{112}$$

which is just our earlier $1/K^2(n, n)$. Also

$$1 - \frac{X^*(1)}{[\mu_Z(n, n)]^{1/2}}$$

is just our "orthogonal polynomial" formed with the "linearly independent polynomials" $1, Z(m + 1 - n)/[\mu_Z(n, n)]^{1/2}, m = 0, 1, \ldots n - 1$.

VI. Conclusion

We have seen that the three topics in the title are closely related. They all are governed by the discrete generalized Gelfand–Levitan equation. Since this admits an explicit solution the subject would seem to be closed. However, the explicit solution is usually quite cumbersome. In obtaining specific properties the insight given by inverse scattering theory seems particularly helpful.

References

1. K. M. Case and M. Kac, A discrete version of the inverse scattering problem, *J. Math. Phys.* **14** (1973), 594; K. M. Case, Orthogonal polynomials from the viewpoint of scattering theory, *J. Math. Phys.* **15** (1974), 2166; K. M. Case, Orthogonal polynomials II, *J. Math. Phys.* **16** (1975), 1435.
2. U. Grenander and G. Szego, "Toeplitz Forms and Their Applications," Univ. of California Press, Berkeley, California 1958.
3. F. J. Dyson, Photon noise and atmospheric noise in active optical systems, J. Opt. Soc. Amer. **65** (1975), 551.
4. I. M. Gelfand and B. M. Levitan, On the determination of a differential equation from its spectral function, *Izv. Akad. Nauk SSSR* **15** (1951), 30a; *Amer. Math. Soc. Transl.* **1** (1956) 253.
5. Z. S. Agranovic and V. A. Marchenko, "The Inverse Problem in Scattering Theory," Gordon and Breach, New York, 1963.
6. K. M. Case and S. C. Chui, The discrete version of the Marchenko equation in the inverse scattering problem, *J. Math. Phys.* **14** (1973), 1643.
7. F. J. Dyson, Fredholm determinants and inverse scattering problems, *Commun. Math. Phys.* **47** (1976), 171.

TOPICS IN FUNCTIONAL ANALYSIS
ADVANCES IN MATHEMATICS SUPPLEMENTARY STUDIES, VOL. 3

On Recovering the Mass Distribution of a String from Its Spectral Function†

HARRY DYM AND NAFTALI KRAVITSKY

Department of Mathematics
The Weizmann Institute of Science
Rehovot, Israel

DEDICATED TO M. G. KREIN ON THE OCCASION OF HIS SEVENTIETH BIRTHDAY

Contents

Notation
1. Introduction
2. Factorization Identities
3. Auxiliary Identities
4. The Main Theorem
5. The Gelfand–Levitan Construction
6. Ramifications
7. Examples
8. Extensions of the Construction
References

NOTATION

The superscript * stands for complex conjugate when applied to numbers: $(a + ib)^* = a - ib$, for real a and b, and designates the adjoint when applied to operators. Unspecified limits of integration, as in $\int g(x)\,dx$ are always understood to be $\pm\infty$. The convention for the Stieltjes integral

$$\int_a^b g(x)\,dm(x)$$

with right continuous monotone m, is to count the jump in m at the point b, if any, but not the jump in a:

$$\int_a^b g(x)\,dm(x) = \int_a^{b-} g(x)\,dm(x) + g(b)\{m(b) - m(b-)\},$$

$$\int_{a-}^a g(x)\,dm(x) = g(a)\{m(a) - m(a-)\}.$$

$m[x] = m(x) - m(x-)$ stands for the jump in m at the point x.

† Communicated by Mark Kac.

ISBN 0-12-599102-9

1. Introduction

The objective of this paper is to study the effect of small perturbations in the *principal spectral functions* Δ of the vibrating string equation

$$A(x, \gamma) = 1 - \gamma^2 \int_0^x d\xi \int_{0-}^{\xi} A(\eta, \gamma)\, dm(\eta), \qquad 0 \leqslant x < l \tag{1.1}$$

upon the *mass distribution* m of the string. The latter is always taken right continuous and nondecreasing on the interval $0 \leqslant x \leqslant l$, with $m(0-) = 0$ and $0 < m(x) < \infty$ for $0 < x < l$; $l \leqslant \infty$ denotes the *length* of the string, and γ is a complex parameter. The point $x = l$ is assumed to be a point of growth of m. However, a jump in m is permitted at $x = l$ only if $l + m(l-) < \infty$.

Equation (1.1) is the integrated version of a generalized second-order differential equation. Indeed, if m increases smoothly, i.e., if $dm(\eta) = m'(\eta)\, d\eta$ and $m'(\eta) > 0$, then A is readily seen to be the solution of the ordinary differential equation

$$\frac{1}{m'(x)} \frac{d^2 A}{dx^2} = -\gamma^2 A, \qquad 0 < x < l,$$

subject to the initial data

$$A(0, \gamma) = 1, \qquad A'(0, \gamma) = 0.$$

The advantage of format (1.1) is that it permits more general mass distributions. The spectral theory of such generalized second-order differential equations was first developed by M. G. Krein in a series of Doklady notes (1952a,b, 1953a,b, 1954a,b), largely devoid of proof; the recently translated expository paper by Kac and Krein (1974) helps to fill that gap. An independent exposition of Krein's work and a number of related ideas due to L. de Branges, appears in Dym and McKean (1976). The latter contains a proof of every fact which is stated below without proof.

An odd nondecreasing function Δ subject to the constraint

$$\int_{-\infty}^{\infty} \frac{d\Delta(\gamma)}{\gamma^2 + 1} < \infty$$

is said to be a *spectral function* for the string *lm* if the transform based upon the solution A of (1.1),

$$\hat{f}(\gamma) = \int_{0-}^{l} f(x) A(x, \gamma)\, dm(x),$$

is a norm preserving map (apart from a factor of $\sqrt{\pi}$) of $\mathbf{L}^2([0, l], dm)$ into $\mathbf{L}^2(R^1, d\Delta)$:

$$(1/\pi) \int_{-\infty}^{\infty} \left| \int_{0-}^{l} f(x) A(x, \gamma)\, dm(x) \right|^2 d\Delta(\gamma) = \int_{0-}^{l} |f(x)|^2\, dm(x). \tag{1.2}$$

The range of the A transform is actually contained in the closed subspace of *even* members of $\mathbf{L}^2(R^1, d\Delta)$, since $A(x, \gamma)$ is an even function of γ. If the range of the A transform fills out this subspace, i.e., if the mapping $f \to \hat{f}$ is *onto* this subspace and not just into, then Δ is said to be a *principal spectral function* for the string. For example, $d\Delta(\gamma) = d\gamma$ is a spectral function for the string with mass function $m(x) = x$ and *any* length $l \leqslant \infty$. However, Δ is a principal spectral function for this string if and only if $l = \infty$. In either case, l finite or not, $A(x, \gamma) = \cos \gamma x$ for $x \leqslant l$ and (1.2) just expresses the well-known norm preserving properties of the cosine transform. There is also an analog of the sine transform based upon the function

$$B(x, \gamma) = \gamma \int_{0-}^{x} A(y, \gamma)\, dm(y).$$

The B transform is (apart from a factor $\sqrt{\pi}$) a norm preserving map of the class of functions $f \in \mathbf{L}^2([0, l], dx)$ which are constant on those intervals across which m (or, equivalently, B) is constant onto $\mathbf{L}^2(R^1, d\Delta)$:

$$(1/\pi) \int \left| \int_0^l f(x) B(x, \gamma)\, dx \right|^2 d\Delta(\gamma) = \int_0^l |f(x)|^2\, dx. \tag{1.3}$$

Notice that the range of the B transform is contained in the closed subspace of *odd* members of $\mathbf{L}^2(R^1, d\Delta)$, since $B(x, \gamma)$ is an odd function of γ.

To every string there corresponds at least one spectral function. If the string is *long*, i.e., if $l + m(l-) = \infty$, then there is exactly one and it is a principal spectral function. If the string is *short*, i.e., if $l + m(l-) < \infty$, then there are uncountably many nonprincipal spectral functions. In addition, for each short string, there is a one parameter family of *principal spectral functions* Δ_k, $k \geqslant 0$, corresponding to the boundary condition

$$k\{A^-(l, \gamma) - \gamma^2 A(l, \gamma) m[l]\} + A(l, \gamma) = 0, \qquad 0 \leqslant k \leqslant \infty, \tag{1.4}$$

which is imposed at the right-hand end point of the string, subject to the proviso that k be strictly positive if

$$m[l] = m(l) - m(l-)$$

is.

Amplification. The left- and right-hand derivatives of A may be evaluated explicitly from (1.1) as

$$A^-(x, \gamma) = -\gamma^2 \int_{0-}^{x-} A(\eta, \gamma)\, dm(\eta)$$

and

$$A^+(x, \gamma) = -\gamma^2 \int_{0-}^{x} A(\eta, \gamma)\, dm(\eta),$$

respectively. The difference

$$A^+(x, \gamma) - A^-(x, \gamma) = -\gamma^2 A(x, \gamma)m[x],$$

where $m[x]$ denotes the jump in m at x, if any:

$$m[x] = m(x) - m(x-).$$

If $k = \infty$, then (1.4) is interpreted to mean

$$A^+(l, \gamma) \equiv A^-(l, \gamma) - \gamma^2 A(l, \gamma)m[l] = 0.$$

The restriction $k \geqslant 0$ ensures that the spectrum γ^2 of the underlying operator $[-d^2/dm\,dx]$ is nonnegative and so permits the convenience of working with odd spectral functions Δ in the variable γ.

The principal spectral function Δ_k for a short string with k fixed as in (1.4) is just a step function with jumps of height

$$\Delta[\gamma_n] = (\pi/2)\left\{\int_{0-}^{l} [A(x, \gamma_n)]^2\, dm(x)\right\}^{-1}$$

placed at the nonzero roots[1] $\pm\gamma_n$, $n \geqslant 1$, of (1.4) and a jump of height

$$\Delta[0] = \pi\left\{\int_{0-}^{l} [A(x, 0)]^2\, dm(x)\right\}^{-1} = \pi/m(l)$$

at $\gamma_0 = 0$ if the latter is a root of (1.4). Therefore,

$$\begin{aligned}\frac{1}{\pi}&\int |\hat{f}(\gamma)|^2\, d\Delta_k(\gamma)\\ &= \sum_{\gamma_n \geqslant 0} |\hat{f}(\gamma_n)|^2 \left\{\int_{0-}^{l} [A(x, \gamma_n)]^2\, dm(x)\right\}^{-1}\\ &= \sum_{\gamma_n \geqslant 0} \left[\left|\int_{0-}^{l} f(x)A(x, \gamma_n)\, dm(x)\right|^2 \Big/ \int_{0-}^{l} [A(x, \gamma_n)]^2\, dm(x)\right]\end{aligned}$$

and the isometry (1.2) is seen to be just a restatement of the fact that the eigenfunctions $A(x, \gamma_n)$, $\gamma_n \geqslant 0$, of (the Sturm–Liouville type problem) (1.1) subject to (1.4) are complete in $\mathbf{L}^2([0, l], dm)$. The fact that the eigenfunctions $A(x, \gamma_n)$, $\gamma_n \geqslant 0$, form an orthogonal system in $\mathbf{L}^2([0, l], dm)$ is deduced in the usual way from the identity

$$(\gamma_n{}^2 - \gamma_j{}^2) \int_{0-}^{l} A(x, \gamma_n)A(x, \gamma_j)\, dm(x) = 0.$$

[1] The roots of (1.4) are symmetrically displaced about 0 since A and $A^{\pm}$ are even functions of γ. The roots are labeled in increasing order: $0 < \gamma_1 < \gamma_2 < \cdots$ with $\gamma_{-n} = -\gamma_n$.

It follows easily from the orthogonality that Δ_k is indeed a principal spectral function: the A transform maps $\mathbf{L}^2([0, l], dm)$ *onto* the class of *even* members of $\mathbf{L}^2(R^1, d\Delta_k)$.

Every odd nondecreasing function Δ which meets the summability constraint

$$\int \frac{d\Delta(\gamma)}{\gamma^2 + 1} < \infty$$

is *either* the principal spectral function of exactly one short string *lmk* *or* the principal spectral function of exactly one long string *lm*. In both these instances Δ will, from now on, be referred to as the principal spectral function of the string *lmk* with the understanding that if the string is long, then the k is superfluous and should be disregarded.

Now suppose that Δ is the principal spectral function of a string *lmk* and that

$$\Delta^{\cdot} = \Delta + \tilde{\Delta}$$

is the principal spectral function of a string $l^{\cdot}m^{\cdot}k^{\cdot}$. The problem under study is to deduce the structure of the $^{\cdot}$-string from the string *lmk* corresponding to Δ under the assumption that the perturbation $\tilde{\Delta}$ is suitably small. The main result is a prescription for computing the mass distribution of at least the initial segment of the perturbed string. That is given in Section 4, and a variation thereof is discussed in Section 6. The verification of the prescription rests upon an elaborate interplay between the solutions $K_i(x, \cdot)$, $i = 1, 2$, of the integral equations

$$K_1(x, y) + F_1(x, y) + \int_{0-}^{x} K_1(x, s)F_1(s, y)\, dm(s) = 0,$$

$$K_2(x, y) + F_2(x, y) + \int_{0}^{x} K_2(x, s)F_2(s, y)\, ds = 0,$$

in which the kernels

$$F_1(x, y) = \lim_{R\uparrow\infty} \frac{1}{\pi} \int_{-R}^{R} A(x, \lambda)A(y, \lambda)\, d\tilde{\Delta}(\lambda)$$

and

$$F_2(x, y) = \lim_{R\uparrow\infty} \frac{1}{\pi} \int_{-R}^{R} B(x, \lambda)B(y, \lambda)\, d\tilde{\Delta}(\lambda).$$

A basic tool is the factorization principle of Gohberg and Krein (1970) applied to (suitable restrictions of) the operators $(I + F_i)^{-1}$. A special case of that factorization is derived in Section 2 and then utilized, in Sections 2 and 3, to deduce a number of auxiliary identities.

The procedure for recovering the perturbed string equation involves assumptions on two kernels, $F_1(x, y)$ and $F_2(x, y)$. In the special case ($m(x) = x$, $l = \infty$, $A(x, \gamma) = \cos \gamma x$, $B(x, \gamma) = \sin \gamma x$, $d\Delta = d\gamma$) considered by Gelfand and Levitan (1951) it turns out that the assumptions imposed on F_2 are fulfilled automatically and so need not be itemized separately. Moreover, the Gelfand–Levitan procedure for reconstructing the potential of the Schrödinger equation emerges as an easy corollary of Theorem 4.1. All that is discussed in Section 5. Section 7 gives a number of examples and Section 8 states a test for discerning whether or not the string segment constructed by the formulas of Theorem 4.1 and Theorem 6.1 fall short of the full string of which $\Delta^{\cdot}$ is the principal spectral function.

2. Factorization Identities

The purpose of this section is to sketch a proof of the Gohberg–Krein factorization formula for a special class of Fredholm integral operators and to deduce a number of related identities from it for future use. The operators are assumed to act in the Hilbert space

$$\mathbf{H}_\delta = \mathbf{L}^2([0, \delta], d\sigma)$$

in which δ is finite and $\sigma = \sigma(x)$ is bounded, nondecreasing, and right continuous on the interval $0 \leqslant x \leqslant \delta$, with $\sigma(0-) = 0 < \sigma(x) < \sigma(\delta)$ for $0 < x < \delta$. The last condition is imposed in order to ensure that 0 and δ are both growth points of σ. (Otherwise, you could contract the interval without affecting the Hilbert space.) Define the projection operators

$$P_{\xi+} \equiv P_\xi : f \in \mathbf{H}_\delta \to \begin{cases} f(x) & \text{if } 0 \leqslant x \leqslant \xi \\ 0 & \text{if } \xi < x \leqslant \delta, \end{cases}$$

$$P_{\xi-} : f \in \mathbf{H}_\delta \to \begin{cases} f(x) & \text{if } 0 \leqslant x < \xi \\ 0 & \text{if } \xi \leqslant x \leqslant \delta, \end{cases}$$

and the subspaces

$$\mathbf{H}_{\xi+} \equiv \mathbf{H}_\xi = P_\xi \mathbf{H}_\delta, \qquad \mathbf{H}_{\xi-} = P_{\xi-}\mathbf{H}_\delta,$$

for every growth point ξ of σ, $0 \leqslant \xi \leqslant \delta$. Notice that the subspace $\mathbf{H}_{\xi-}$ is superfluous unless ξ is a jump point of σ, i.e., unless

$$\sigma[\xi] = \sigma(\xi) - \sigma(\xi-) > 0,$$

because otherwise $\mathbf{H}_{\xi-} = \mathbf{H}_\xi$. $\mathbf{H}_{\xi-}$ may even be superfluous at a jump point ξ of σ: if $\sigma(\xi-) = \sigma(\xi_0)$ for some growth point $\xi_0 < \xi$, then $\mathbf{H}_{\xi-} = \mathbf{H}_{\xi_0}$. For example, if σ is a step function with jumps at the points $0 = \xi_0 < \xi_1 < \cdots < \xi_n = \delta$, then the spaces $\mathbf{H}_{\xi_i}$, $i = 0, \ldots, n$, give a complete listing of the distinct nontrivial subspaces of $\mathbf{H}_\delta$ of the indicated form. In this instance $\mathbf{H}_{0-} = 0$ and $\mathbf{H}_{\xi_i -} = \mathbf{H}_{\xi_{i-1}}$ for $i = 1, \ldots, n$.

Let

$$F: f \in \mathbf{H}_\delta \to \int_{0-}^{\delta} F(x, y) f(y)\, d\sigma(y), \qquad 0 \leqslant x \leqslant \delta,$$

denote the compact, self-adjoint integral operator from $\mathbf{H}_\delta$ into itself which is based upon the real, continuous, symmetric kernel $F(x, y)$, and suppose that for every growth point ξ of σ, $0 \leqslant \xi \leqslant \delta$, the restriction of

$$(I + F)_{\xi\pm} = P_{\xi\pm}(I + F)P_{\xi\pm}$$

to the subspace $P_{\xi\pm}\mathbf{H}_\delta$ is a 1–1 mapping of that subspace onto itself, or, what amounts to the same thing, $I + P_{\xi\pm}FP_{\xi\pm}$ is a 1–1 mapping of $\mathbf{H}_\delta$ onto itself.

The assumptions on F will remain in force for the rest of this section. They permit you to solve equations of the form

$$(I + PFP)u = v$$

for $u \in \mathbf{H}_\delta$ given $v \in \mathbf{H}_\delta$ and $P = P_\xi$, for ξ a growth point of σ. This does not determine u pointwise. However, if v is continuous, then u may be specified to be continuous also.

LEMMA 2.1. *If $g(\xi, y)$ is continuous in each variable separately on the square $0 \leqslant \xi, y \leqslant \delta$, then for every choice of $0 \leqslant \xi \leqslant \delta$, growth point or not, the equations*

$$\phi(\xi, y) + \int_{0-}^{\xi} F(y, s)\phi(\xi, s)\, d\sigma(s) = g(\xi, y), \tag{2.1}$$

and

$$\phi(\xi-, y) + \int_{0-}^{\xi-} F(y, s)\phi(\xi-, s)\, d\sigma(s) = g(\xi, y)$$

have unique solutions $\phi(\xi, \cdot)$ and $\phi(\xi-, \cdot)$, respectively, which are continuous in the interval $0 \leqslant y \leqslant \delta$. Moreover,

$$\phi(\xi, \xi) = \phi(\xi-, \xi)\{1 + K(\xi, \xi)\sigma[\xi]\}, \tag{2.2}$$

where $K(\xi, y)$ is the solution of (2.1) with $g(\xi, y) = -F(\xi, y)$:

$$K(\xi, y) + \int_{0-}^{\xi} F(y, s)K(\xi, s)\, d\sigma(s) = -F(\xi, y). \tag{2.3}$$

Proof. Fix $0 \leqslant \xi \leqslant \delta$, let ξ_1 denote the largest growth point of σ which sits to the left of $\xi : \xi_1 \leqslant \xi$, and write P for P_{ξ_1}, $Q = I - P$, ϕ for $\phi(\xi, \cdot)$, and g for $g(\xi, \cdot)$. Then the equation for ϕ can be expressed in the form

$$(I + FP)\phi = g,$$

and the identity

$$(I + FP) = (I + PFP)(I + QFP)$$

displays the fact that $(I + FP)$ is invertible since $I + PFP$ is invertible by assumption and $I + QFP$ is automatically invertible:

$$(I + QFP)^{-1} = I - QFP.$$

In particular,

$$P\phi = P(I + PFP)^{-1}g = (I + PFP)^{-1}Pg \tag{2.4}$$

and

$$\phi = \{I - F(I + PFP)^{-1}P\}g. \tag{2.5}$$

The last two formulas specify $P\phi$ and ϕ as elements of $\mathbf{H}_\delta$. The integral formula

$$\phi(\xi, y) = -\int_{0-}^{\xi_1} F(y, s)\phi(\xi, s)\, d\sigma(s) + g(\xi, y)$$

serves to specify ϕ at *every point* of the square $0 \leqslant \xi, y \leqslant \delta$, and shows clearly that ϕ is continuous in y. The continuity of $\phi(\xi -, y)$ in y is established in much the same way. To prove (2.2) notice that

$$\begin{aligned}\phi(\xi, y) - \phi(\xi -, y) + \int_{0-}^{\xi -} F(y, s)[\phi(\xi, s) - \phi(\xi -, s)]\, d\sigma(s)\\ = -F(\xi, y)\phi(\xi, \xi)\sigma[\xi],\end{aligned}$$

whereas

$$K(\xi -, y) + \int_{0-}^{\xi -} F(y, s)K(\xi -, s)\, d\sigma(s) = -F(\xi, y). \tag{2.6}$$

Since the latter is uniquely solvable it follows that

$$\phi(\xi, y) - \phi(\xi -, y) = K(\xi -, y)\phi(\xi, \xi)\sigma[\xi]. \tag{2.7}$$

In particular,

$$\begin{aligned}\phi(\xi, \xi) &= \phi(\xi -, \xi) + K(\xi -, \xi)\phi(\xi, \xi)\sigma[\xi],\\ K(\xi, \xi) &= K(\xi -, \xi) + K(\xi -, \xi)K(\xi, \xi)\sigma[\xi],\end{aligned}$$

and (2.2) drops out after a short computation.

The chief consequence of the assumption that $I + PFP$ is invertible for every projection P in the chain of projections $P_{\xi\pm}$ associated with growth points ξ of σ, $0 \leqslant \xi \leqslant \delta$, is the fact that $(I + F)^{-1}$ can be factored into a product of triangular operators. The general fact is due to Gohberg and Krein (1970, pp. 156–189). The next theorem presents a form of this result which is appropriate for present needs.

THEOREM 2.1. *$(I + F)^{-1}$ permits the factorization*

$$(I + F)^{-1} = (I + L^*)\Lambda^{-1}(I + L) \tag{2.8}$$

in which L is the triangular integral operator based upon the kernel

$$L(\xi, y) = \begin{cases} K(\xi, y) & \text{if} \quad y \leqslant \xi \\ 0 & \text{if} \quad \xi < y \end{cases}$$

and

$$\Lambda - I = D$$

is the "diagonal" operator defined by the rule

$$Df(x) = K(x, x)f(x)\sigma[x]. \tag{2.9}$$

Proof. Introduce the auxiliary triangular kernel

$$U(\xi, y) = K(\xi, y) - L(\xi, y) = \begin{cases} K(\xi, y) & \text{if} \quad y > \xi \\ 0 & \text{if} \quad y \leqslant \xi \end{cases}$$

and observe that

$$U(\xi, y) + L(\xi, y) + F(\xi, y) + \int_{0-}^{\delta} L(\xi, s)F(s, y)\, d\sigma(s) = 0$$

on the square $0 \leqslant \xi$, $y \leqslant \delta$. But, in the language of operators, this is the same as saying that

$$U + L + F + LF = 0 \tag{2.10}$$

or, equivalently,

$$(I + L)(I + F) = I - U$$

on $\mathbf{H}_\delta$. Now since U is a Volterra operator,[2] $(I - U)^{-1}$ exists and

$$(I + F)^{-1} = (I - U)^{-1}(I + L). \tag{2.11}$$

[2] This depends crucially upon the fact that $U(\xi, \xi) = 0$. In the language of Gohberg and Krein, this forces $(P_\xi - P_{\xi-})U(P_\xi - P_{\xi-}) = 0$ at every jump point ξ of σ, and that suffices to prove that U is a Volterra operator; see Gohberg and Krein (1970, pp. 25–28).

Moreover,

$$(I - U)^{-1}(I + L) = (I + L^*)(I - U^*)^{-1},$$

since F is self-adjoint, and so

$$(I - U)(I + L^*) = (I + L)(I - U^*).$$

Now define

$$\Lambda = (I - U)(I + L^*) = \Lambda^*$$

and

$$D = \Lambda - I = L^* - U - UL^* \tag{2.12}$$

and observe that

$$D(\xi, y) = L(y, \xi) - U(\xi, y) - \int_{0-}^{\delta} U(\xi, s)L(y, s)\, d\sigma(s)$$

vanishes for $y < \xi$ and so for $y > \xi$ too, whereas

$$D(\xi, \xi) = L(\xi, \xi) = K(\xi, \xi).$$

In other words, D is the "diagonal" operator which acts on $f \in \mathbf{H}_\delta$ according to the rule (2.9). Notice, in particular, that D is nonzero only if σ has jumps.

Finally, since Λ is self-adjoint and invertible, you see that

$$(I + L^*)^{-1}(I - U)^{-1} = \Lambda^{-1}.$$

Hence

$$(I - U)^{-1} = (I + L^*)\Lambda^{-1},$$

and (2.8) is now immediate from (2.11). The proof is complete.

Amplification. Notice that the operators L, U, and D enjoy the following relationships with regard to every projection $P = P_{\xi\pm}$ in the chain:

$$PL = PLP, \qquad UP = PUP, \qquad DP = PD.$$

COROLLARY 2.1. *For each growth point ξ of σ, $0 \leqslant \xi \leqslant \delta$, the operator $(I + F_\xi)^{-1}P_\xi$ permits the factorization*

$$(I + F_\xi)^{-1}P_\xi = (I + L^*)P_\xi\Lambda^{-1}(I + L), \tag{2.13}$$

in which $F_\xi = P_\xi F P_\xi$.

Proof. Let $L_\xi = P_\xi L P_\xi$, $U_\xi = P_\xi U P_\xi$, $D_\xi = P_\xi D P_\xi$, and multiply (2.10) and (2.12) from the left and then from the right by P_ξ, taking advantage of the identities

$$P_\xi L = P_\xi L P_\xi = L_\xi,$$
$$L^* P_\xi = P_\xi L^* P_\xi = (L_\xi)^*,$$

to obtain

$$(I + L_\xi)(I + F_\xi) = I - U_\xi$$
$$I + D_\xi = (I - U_\xi)(I + L_\xi{}^*).$$

Now it follows just as in the proof of the theorem that

$$(I + F_\xi)^{-1} = (I + L_\xi{}^*)(I + D_\xi)^{-1}(I + L_\xi),$$

and in order to complete the proof of the corollary you have only to multiply the last equality by P_ξ from the right and to check that

$$(I + L_\xi)P_\xi = P_\xi(I + L)$$
$$(I + D_\xi)^{-1}P_\xi = P_\xi(I + D)^{-1}$$
$$(I + L_\xi{}^*)P_\xi = (I + L^*)P_\xi.$$

COROLLARY 2.2. *Let $f \in \mathbf{C}[0, \delta]$ and for every ξ in the interval $0 \leqslant \xi \leqslant \delta$, growth point or not, let $\phi(\xi, \cdot)$ denote the unique continuous solution specified by the equation*

$$\phi(\xi, y) + \int_{0-}^{\xi} F(y, s)\phi(\xi, s)\, d\sigma(s) = f(y). \tag{2.14}$$

Then the identities

$$\phi(\xi, \xi) = f(\xi) + \int_{0-}^{\xi} K(\xi, s) f(s)\, d\sigma(s) \tag{2.15}$$

and

$$\begin{aligned}\phi(\xi, y) &= f(y) + \int_{0-}^{\xi} K(s-, y)\phi(s, s)\, d\sigma(s)\\ &= \phi(y, y) + \int_{y}^{\xi} K(s-, y)\phi(s, s)\, d\sigma(s),\end{aligned} \tag{2.16}$$

are valid in the square $0 \leqslant \xi, y \leqslant \delta$. Moreover,

$$\begin{aligned}\int_{0-}^{\xi} f(y)\phi(\xi, y)\, d\sigma(y) &= \int_{0-}^{\xi} \psi(y)(\Lambda^{-1}\psi)(y)\, d\sigma(y)\\ &= \int_{0-}^{\xi} \phi(y, y)\phi(y-, y)\, d\sigma(y),\end{aligned} \tag{2.17}$$

in which

$$\psi(y) = \phi(y, y) \tag{2.18}$$

and

$$(\Lambda^{-1}\psi)(y) = \frac{\psi(y)}{1 + K(y, y)\sigma[y]}. \tag{2.19}$$

Proof. (2.15) is proved by multiplying both sides of (2.14) by $K(\xi, y)\,d\sigma(y)$, integrating from $0-$ to ξ, and invoking (2.3). This yields the identity

$$\int_{0-}^{\xi} F(\xi, s)\phi(\xi, s)\,d\sigma(s) = -\int_{0-}^{\xi} K(\xi, s)f(s)\,d\sigma(s)$$

and (2.15) drops out upon putting this into (2.14) with $y = \xi$.

Now, in terms of the definition (2.18), (2.15) states that

$$\psi = (I + L)f.$$

At the same time, if ξ_1 is the largest growth point of σ which sits to the left of $\xi : \xi_1 \leqslant \xi$, then (2.4) and (2.5) adapted to present circumstances state that

$$P_{\xi_1}\phi(\xi, \cdot) = (I + F_{\xi_1})^{-1}P_{\xi_1}f,$$
$$\phi(\xi, \cdot) = [I - F(I + F_{\xi_1})^{-1}P_{\xi_1}]f.$$

Now by Corollary 2.1,

$$\begin{aligned}
\int_{0-}^{\xi} f(s)\phi(\xi, s)\,d\sigma(s) &= \langle f, P_{\xi_1}\phi(\xi, \cdot)\rangle_\sigma \\
&= \langle f, (I + F_{\xi_1})^{-1}P_{\xi_1}f\rangle_\sigma \\
&= \langle P_{\xi_1}(I + L)f, \Lambda^{-1}(I + L)f\rangle_\sigma \\
&= \langle P_{\xi_1}\psi, \Lambda^{-1}\psi\rangle_\sigma \\
&= \int_{0-}^{\xi_1} \psi(y)(\Lambda^{-1}\psi)(y)\,d\sigma(y) \\
&= \int_{0-}^{\xi} \frac{\phi^2(y, y)}{1 + K(y, y)\sigma[y]}\,d\sigma(y) \\
&= \int_{0-}^{\xi} \phi(y, y)\phi(y-, y)\,d\sigma(y);
\end{aligned}$$

the last two lines use (2.19) and (2.2). This completes the proof of (2.17). The proof of (2.16) is based upon the observation that

$$\begin{aligned}
I - F(I + F_{\xi_1})^{-1}P_{\xi_1} &= I - F(I + L^*)P_{\xi_1}\Lambda^{-1}(I + L) \\
&= I + K^*P_{\xi_1}\Lambda^{-1}(I + L).
\end{aligned}$$

From this it follows that

$$\begin{aligned}
\phi(\xi, \cdot) &= f - F(I + F_{\xi_1})^{-1}P_{\xi_1}f \\
&= f + K^*P_{\xi_1}\Lambda^{-1}\psi.
\end{aligned}$$

Hence

$$\phi(\xi, y) = f(y) + \int_{0-}^{\xi_1} K(s, y) \frac{\phi(s, s)}{1 + K(s, s)\sigma[s]} d\sigma(s)$$

$$= f(y) + \int_{0-}^{\xi} K(s-, y)\phi(s, s)\, d\sigma(s).$$

The last line uses (2.7) with $K(\xi, y)$ in place of $\phi(\xi, y)$.

3. Auxiliary Identities

A number of identities are now derived, chiefly for use in the proof of the main theorem (Theorem 4.1), under the assumption that the perturbation $d\tilde{\Delta} = d\Delta^{\cdot} - d\Delta$ is suitably small. The proof of these identities is largely an application of Theorem 2.1 and its corollaries to the Hilbert spaces

$$\mathbf{M}_{\xi+} = \mathbf{M}_{\xi} = \mathbf{H}_{\xi} \qquad \text{with} \quad \sigma(x) = m(x),$$
$$\mathbf{M}_{\xi-} = \mathbf{H}_{\xi-} \qquad \text{with} \quad \sigma(x) = m(x),$$

m being the mass distribution of the starting string, and

$$\mathbf{Z}_{\xi} = \mathbf{H}_{\xi} \qquad \text{with} \quad \sigma(x) = x.$$

In both instances ξ is taken to be a growth point of σ; however, this is not a real restriction in case $\sigma(x) = x$. For growth points ξ of m the subspaces $\mathbf{X}_{\xi}$ of functions $f \in \mathbf{Z}_{\xi}$ which are constant on mass-free intervals[3] play a special role. In particular, the B transform $f \to \int_0^{\xi} f(x)B(x,\gamma)\,dx$ is invertible for $f \in \mathbf{X}_{\xi}$ but not for general $f \in \mathbf{Z}_{\xi}$; see Dym and McKean (1976, §5.7) for additional information on this point.

Now assume that the perturbation $d\tilde{\Delta} = d\Delta^{\cdot} - d\Delta$ is small enough to ensure the existence of a growth point $\delta \leqslant l$ of m, with $m(\delta) + \delta < \infty$, such that:

I. $(1/\pi) \int_{-R}^{R} A(x, \gamma)A(y, \gamma)\, d\tilde{\Delta}(\gamma)$ converges boundedly, as $R \uparrow \infty$, to a limit $F_1(x, y)$ which is continuous in each variable separately on the square $0 \leqslant x, y \leqslant \delta$.

II. $(1/\pi) \int_{-R}^{R} B(x, \gamma)B(y, \gamma)\, d\tilde{\Delta}(\gamma)$ converges boundedly, as $R \uparrow \infty$, to a limit $F_2(x, y)$ which is continuous in each variable separately on the square $0 \leqslant x, y \leqslant \delta$, except at the jump points of m where it is right continuous and has left-hand limits.

III. The operator

$$(I + F_1): f \to f(x) + \int_{0-}^{\delta} F_1(x, y)f(y)\, dm(y), \qquad 0 \leqslant x \leqslant \delta,$$

is a 1–1 mapping of the Hilbert space $\mathbf{M}_{\delta}$ onto itself.

[3] The interval $a \leqslant x < b$ ($a \leqslant x \leqslant b$) is said to be mass-free if $m(b-) = m(a)$ ($m(b) = m(a)$).

IV. The restriction of the operator

$$(I + F_2): f \to f(x) + \int_0^\delta F_2(x, y) f(y)\, dy, \qquad 0 \leqslant x \leqslant \delta,$$

to $\mathbf{X}_\delta$ is a 1–1 mapping of that Hilbert space onto itself.[4]

THEOREM 3.1. *Let the perturbation $d\tilde{\Delta}$ be such that the assumptions* I–IV *hold for some growth point $\delta \leqslant l$ of m with $\delta + m(\delta) < \infty$. Then for each fixed growth point ξ of m, $0 \leqslant \xi \leqslant \delta$, the restriction of $(I + F_1)_{\xi\pm}$ to $\mathbf{M}_{\xi\pm}$ is a* 1–1 *map of $\mathbf{M}_{\xi\pm}$ onto itself, and for every point $0 \leqslant \xi \leqslant \delta$ the restriction of $(I + F_2)_\xi$ to $\mathbf{Z}_\xi$ is a* 1–1 *mapping of $\mathbf{Z}_\xi$ onto itself.*

Proof. The proof proceeds in steps.

STEP 1. The operators $I + F_1$ acting on $\mathbf{M}_\delta$ and $I + F_2$ acting on $\mathbf{Z}_\delta$ are both strictly positive.

Proof of Step 1. Every f in $\mathbf{M}_\delta$ may be recovered from its transform

$$\hat{f}(\gamma) = \int_{0-}^{\delta} f(x) A(x, \gamma)\, dm(x)$$

by the inversion formula

$$f(x) = \lim_{R \uparrow \infty} \frac{1}{\pi} \int_{-R}^{R} \hat{f}(\gamma) A(x, \gamma)\, d\Delta(\gamma),$$

where the limit is taken in $\mathbf{M}_\delta$; for a proof see, for example, Dym and McKean (1976, §5.7). At the same time it is readily checked that

$$\begin{aligned}(F_1 f)(x) &= \int_{0-}^{\delta} F_1(x, y) f(y)\, dm(y) \\ &= \lim_{R \uparrow \infty} \frac{1}{\pi} \int_{-R}^{R} \hat{f}(\gamma) A(x, \gamma)\, d\tilde{\Delta}(\gamma)\end{aligned}$$

with convergence in $\mathbf{M}_\delta$. Therefore,

$$\begin{aligned}\langle (I + F_1) f, f \rangle_m &= \lim_{R \uparrow \infty} \frac{1}{\pi} \int_{0-}^{\delta} \int_{-R}^{R} \hat{f}(\gamma) A(x, \gamma)\, d\Delta^{\cdot}(\gamma) f^*(x)\, dm(x) \\ &= \lim_{R \uparrow \infty} \frac{1}{\pi} \int_{-R}^{R} |\hat{f}(\gamma)|^2\, d\Delta^{\cdot}(\gamma). \qquad (3.1)\end{aligned}$$

The identity

$$\langle (I + F_2) f, f \rangle_x = \lim_{R \uparrow \infty} \frac{1}{\pi} \int_{-R}^{R} \left| \int_0^\delta f(x) B(x, \gamma)\, dx \right|^2 d\Delta^{\cdot}(\gamma) \qquad (3.2)$$

[4] F_2 maps $\mathbf{Z}_\delta$ into $\mathbf{X}_\delta$ because $F_2(x, y)$ is constant on mass-free intervals.

is established in much the same way for $f \in \mathbf{X}_\delta$. Formula (3.1) implies that $I + F_1$ is nonnegative on $\mathbf{M}_\delta$, whereas (3.2) implies that $I + F_2$ is nonnegative on $\mathbf{X}_\delta$. $I + F_2$ is in fact nonnegative on the larger space $\mathbf{Z}_\delta$ also. To check this, split $f \in \mathbf{Z}_\delta$ into the sum of a piece f_1 belonging to $\mathbf{X}_\delta$ and a piece f_2 perpendicular to $\mathbf{X}_\delta$ and observe that

$$\int_0^\delta F_2(x, y) f_2(y)\, dy = 0,$$

since $F_2(x, \cdot)^* = F_2(x, \cdot) \in \mathbf{X}_\delta$, whence

$$\langle (I + F_2)f, f\rangle_x = \langle (I + F_2)f_1, f_1\rangle_x + \langle f_2, f_2\rangle_x. \tag{3.3}$$

But now it is a general fact that a nonnegative operator Q has a nonnegative square root $R : Q = R^2$, and the identity

$$\langle Qf, f\rangle = \langle Rf, Rf\rangle$$

makes it plain that such a nonnegative operator is strictly positive if and only if it is 1–1. This completes the proof of step 1 in view of assumptions III and IV and the easily verified fact that $I + F_2$ is 1–1 on $\mathbf{Z}_\delta$ if and only if it is 1–1 on $\mathbf{X}_\delta$.

STEP 2. $(I + F_1)_{\xi\pm}$ is a 1–1 map of $\mathbf{M}_{\xi\pm}$ onto itself for every growth point ξ of m, $0 \leqslant \xi \leqslant \delta$, and $(I + F_2)_\xi$ is a 1–1 map of $\mathbf{Z}_\xi$ onto itself for every point ξ, $0 \leqslant \xi \leqslant \delta$.

Proof of Step 2. Suppose, for example, that $(I + F_1)_\xi f = 0$ for some choice of $f \in \mathbf{M}_\xi = P_\xi \mathbf{M}_\delta$. Then

$$\langle (I + F_1)_\xi f, f\rangle = \langle (I + F_1)P_\xi f, P_\xi f\rangle = 0$$

and so, by step 1, $f = P_\xi f = 0$. This proves that $(I + F_1)_\xi$ is 1–1 on $\mathbf{M}_\xi$. The Fredholm alternative guarantees that it is also onto, F_1 being compact. The proof of the second half of step 2 is carried out in exactly the same way.

COROLLARY 3.1. *For every point* ξ, $0 \leqslant \xi \leqslant \delta$, *the integral equations*

$$\begin{aligned} K_1(\xi, y) + \int_{0-}^{\xi} F_1(y, s) K_1(\xi, s)\, dm(s) &= -F_1(\xi, y) \\ K_1(\xi-, y) + \int_{0-}^{\xi-} F_1(y, s) K_1(\xi-, s)\, dm(s) &= -F_1(\xi, y) \end{aligned} \tag{3.4}$$

have unique solutions $K_1(\xi, \cdot)$ *and* $K_1(\xi-, \cdot)$, *respectively, which are continuous in the interval* $0 \leqslant y \leqslant \delta$, *and the integral equation*

$$K_2(\xi, y) + \int_0^\xi F_2(y, s) K_2(\xi, s)\, ds = -F_2(\xi, y) \tag{3.5}$$

has a unique solution $K_2(\xi, \cdot)$ *which has left- and right-hand limits at every point* y *in the interval* $0 \leqslant y \leqslant \delta$. *The limits agree except at the jump points of* m. *At the jump points* $K_2(\xi, \cdot)$ *is right continuous. Moreover, for* $i = 1, 2,$

$$\lim_{h \downarrow 0} K_i(\xi + h, y) = K_i(\xi, y)$$

for $0 \leqslant \xi < \delta$, *and*

$$\lim_{h \downarrow 0} K_i(\xi - h, y) = K_i(\xi -, y)$$

for $0 < \xi \leqslant \delta$, *in which* $K_1(\xi -, \cdot)$ *is the solution of the second equation of* (3.4) *and* $K_2(\xi -, \cdot)$ *is the solution of* (3.5) *with* ξ *replaced by* $\xi -$.[5] *The indicated convergence takes place both pointwise, i.e., for each fixed* y, $0 \leqslant y \leqslant \delta$, *and in norm.*

Proof. The first part of the corollary is easily established with the proof of Lemma 2.1 as a guide. It remains to check the behavior of $K_i(\xi, y)$ as a function of ξ. Consider, for example, $K_2(\xi, \cdot) - K_2(\xi_0, \cdot)$ with $\xi > \xi_0$. It is a solution of the equation

$$\begin{aligned} K_2(\xi, y) - K_2(\xi_0, y) + \int_0^\xi F_2(y, s)[K_2(\xi, s) - K_2(\xi_0, s)]\, ds \\ = -\int_{\xi_0}^\xi F_2(y, s) K_2(\xi_0, s)\, ds - F_2(\xi, y) + F_2(\xi_0, y), \end{aligned}$$

i.e.,

$$K_2(\xi, \cdot) - K_2(\xi_0, \cdot) = -(I + F_2 P_\xi)^{-1}\{F_2 \Delta P K_2(\xi_0, \cdot) + F_2(\xi, \cdot) - F_2(\xi_0, \cdot)\}$$

where

$$\Delta P = P_\xi - P_{\xi_0}.$$

But now as $\xi \downarrow \xi_0$ you see that

$$K_2(\xi, \cdot) - K_2(\xi_0, \cdot) \to 0 \quad \text{in} \quad \mathbf{Z}_\delta$$

since

$$F_2(\xi, \cdot) - F_2(\xi_0, \cdot) \to 0 \quad \text{in} \quad \mathbf{Z}_\delta,$$

$$\|F_2 \Delta P\|^2 \leqslant \int_0^\delta \int_{\xi_0}^\xi |F_2(y, s)|^2\, ds\, dy \downarrow 0,$$

and the operators $(I + F_2 P_\xi)^{-1}$ stay bounded in norm. To clarify the last point set $P = P_{\xi_0}$, for short, and express

$$\begin{aligned} [I + F_2 P_\xi]^{-1} &= [I + F_2(P + \Delta P)]^{-1} \\ &= [(I + F_2 P)(I + (I + F_2 P)^{-1} F_2 \Delta P)]^{-1} \\ &= (I + (I + F_2 P)^{-1} F_2 \Delta P)^{-1}(I + F_2 P)^{-1}, \end{aligned}$$

[5] The point is that $F_2(\xi -, y)$ differs from $F_2(\xi, y)$ if ξ is a jump point of m.

bearing in mind that $\|F_2 \Delta P\| \downarrow 0$ as $\xi \downarrow \xi_0$. The pointwise convergence is now easily deduced from the equation for $K_2(\xi, \cdot) - K_2(\xi_0, \cdot)$. The remaining statements are proved in much the same way.

Amplification. If ξ sits inside the mass free interval bounded by the growth points ξ_1, ξ_2, i.e., if $\xi_1 < \xi < \xi_2$ and $m(\xi_2 -) = m(\xi_1)$, then $K_1(\xi, \cdot)$ is the solution of

$$K_1(\xi, y) + \int_{0-}^{\xi_1} F_1(y, s) K_1(\xi, s)\, dm(s) = -F_1(\xi, y),$$

whereas $K_2(\xi, \cdot)$ is the solution of

$$K_2(\xi, y) + \int_0^{\xi_1} F_2(y, s) K_2(\xi, s)\, ds = -F_2(\xi_1, y)[1 + (\xi - \xi_1) K_2(\xi, \xi_1)].$$

The latter takes advantage of the fact that $K_2(\xi, \cdot)$ is both right continuous and constant on mass-free intervals together with $F_2(\xi, \cdot) = F_2(\xi_1, \cdot)$. Now as equation (3.5) is uniquely solvable, it follows that

$$K_2(\xi, y) = [1 + (\xi - \xi_1) K_2(\xi, \xi_1)] K_2(\xi_1, y)$$

for all points $0 \leqslant y \leqslant \delta$. In particular,

$$K_2(\xi, \xi_1) = [1 + (\xi - \xi_1) K_2(\xi, \xi_1)] K_2(\xi_1, \xi_1)$$

or, what amounts to the same thing,

$$[1 + (\xi - \xi_1) K_2(\xi, \xi_1)][1 - (\xi - \xi_1) K_2(\xi_1, \xi_1)] = 1.$$

Hence,

$$1 - (\xi - \xi_1) K_2(\xi_1, \xi_1) \neq 0$$

for $\xi_1 \leqslant \xi < \xi_2$, and

$$K_2(\xi, y) = K_2(\xi_1, y)/[1 - (\xi - \xi_1) K_2(\xi_1, \xi_1)].$$

Theorem 2.1 and its corollaries are now seen to be directly applicable to $I + F_1$ acting on $\mathbf{M}_\delta$ and almost directly applicable to $I + F_2$ acting on $\mathbf{Z}_\delta$: you have only to write m, F_1, K_1, L_1, and Λ_1 in place of σ, F, K, L, and Λ, respectively, in the first case, and x, F_2, K_2, and L_2 in place of σ, F, K, and L in the second case. The fact that F_2 is not continuous at the jumps of m does not cause any harm since $\sigma(x) = x$ is jump free. The diagonal operator D does not intervene in the second case for the same reason, i.e., $\Lambda_2 = I$.

It will come as no surprise that there is a complete analog of Corollary 2.2 for integral equations based on the kernel F_2. A partial restatement follows.

COROLLARY 3.2. *Let f be a bounded right continuous function on the interval $[0, \delta]$, and suppose that f is constant on mass-free intervals. Let $\phi(\xi, \cdot)$ denote the unique solution of the equation*

$$\phi(\xi, y) + \int_0^\xi F_2(y, s)\phi(\xi, s)\,ds = f(y), \qquad 0 \leqslant y \leqslant \delta,$$

which enjoys the same degree of smoothness as f. Then

$$\phi(\xi, y) = \phi(y, y) + \int_y^\xi K_2(s, y)\phi(s, s)\,ds \tag{3.6}$$

for every point in the square $0 \leqslant \xi, y \leqslant \delta$, and

$$\int_0^\xi f(y)\phi(\xi, y)\,dy = \int_0^\xi [\phi(y, y)]^2\,dy \tag{3.7}$$

for every point ξ in the interval $0 \leqslant \xi \leqslant \delta$.

Now define

$$\phi_1(x, y) = 1 + \int_y^x K_1(x, s)\,dm(s) \tag{3.8}$$

and

$$\phi_2(x, y) = 1 + \int_y^x K_2(x, s)\,ds \tag{3.9}$$

for all points $0 \leqslant x, y \leqslant \delta$.

THEOREM 3.2. *Let $\delta \leqslant l$ be a growth point of m such that $\delta + m(\delta) < \infty$ and such that I–IV hold. Then*

$$\phi_1(x, 0-) = 1 + \int_{0-}^x K_1(x, s)\,dm(s)$$

is never zero on the interval $0 \leqslant x \leqslant \delta$ and the following identities are valid on the triangle[6] *$0 \leqslant y \leqslant x \leqslant \delta$:*

(a) $\int_{0-}^x [F_1(y, s) - F_1(0, s)]\,dm(s) = -\int_0^y F_2(x, s)\,ds,$

(b) $\phi_1(x, y) + \int_0^x F_2(y, s)\phi_1(x, s)\,ds = \phi_1(x, 0-)\phi_1(0-, y),$

(c) $\phi_2(x, y) + \int_{0-}^x F_1(y, s)\phi_2(x, s)\,dm(s) = \phi_2(0-, y)/\phi_1(x, 0-) = 1/\phi_1(x, 0-),$

(d) $\phi_1(x, y)/\phi_1(x, 0-) = [1/\phi_1(y, 0-)] + \int_y^x K_2(s, y)[1/\phi_1(s, 0-)]\,ds,$

(e) $\phi_1(x, 0-)\phi_2(x, y) = \phi_1(y, 0-) + \int_y^x K_1(s-, y)\phi_1(s, 0-)\,dm(s),$

(f) $\int_y^x [\phi_2(s, y)/\phi_1(s, 0-)]\,ds = \int_y^x [\phi_1(x, s)/\phi_1(x, 0-)]\,ds,$

[6] (a), (b), and (c) are in fact valid on the full square $0 \leqslant x, y \leqslant \delta$.

(g) $\int_0^x [\phi_1(x, s)\phi_1(0-, s)/\phi_1(x, 0-)]\, ds = \int_0^x ds/[\phi_1(s, 0-)]^2$,

(h) $\int_{0-}^x \phi_1(x, 0-)\phi_2(x, s)\, dm(s) = \int_{0-}^x \phi_1(s, 0-)\phi_1(s-, 0-)\, dm(s)$.

Proof of (a).

$$\begin{aligned}\int_{0-}^x F_1(y, s)\, dm(s) &= \int_{0-}^x \left[\lim_{R\uparrow\infty} \frac{1}{\pi}\int_{-R}^R A(y, \gamma)A(s, \gamma)\, d\tilde{\Delta}(\gamma)\right] dm(s)\\ &= \lim_{R\uparrow\infty} \int_{0-}^x \left[\frac{1}{\pi}\int_{-R}^R A(y, \gamma)A(s, \gamma)\, d\tilde{\Delta}(\gamma)\right] dm(s)\\ &= \lim_{R\uparrow\infty} \frac{1}{\pi}\int_{-R}^R A(y, \gamma)\int_{0-}^x A(s, \gamma)\, dm(s)\, d\tilde{\Delta}(\gamma)\\ &= \lim_{R\uparrow\infty} \frac{1}{\pi}\int_{-R}^R A(y, \gamma)\frac{B(x, \gamma)}{\gamma}\, d\tilde{\Delta}(\gamma),\end{aligned}$$

the passage from line 1 to line 2 being justified by dominated convergence.[7] Therefore,

$$\begin{aligned}\int_{0-}^x [F_1(y, s) - F_1(0, s)]\, dm(s) &= \lim_{R\uparrow\infty} \frac{1}{\pi}\int_{-R}^R B(x, \gamma)\left[\frac{A(y, \gamma) - 1}{\gamma}\right] d\tilde{\Delta}(\gamma)\\ &= -\lim_{R\uparrow\infty} \frac{1}{\pi}\int_{-R}^R B(x, \gamma)\int_0^y B(s, \gamma)\, ds\, d\tilde{\Delta}(\gamma)\\ &= -\lim_{R\uparrow\infty} \int_0^y \left[\frac{1}{\pi}\int_{-R}^R B(x, \gamma)B(s, \gamma)\, d\tilde{\Delta}(\gamma)\right] ds\\ &= -\int_0^y F_2(x, s)\, ds.\end{aligned}$$

The interchange in the last line is again justified by dominated convergence.

Proof of (b). The first step is to integrate the identity (see (3.4))

$$K_1(x, t) + \int_{0-}^x K_1(x, s)F_1(s, t)\, dm(s) = -F_1(x, t)$$

with respect to $dm(t)$ across the interval $0 \leqslant t \leqslant y$ to get

$$\begin{aligned}&\int_{0-}^y K_1(x, t)\, dm(t) + \int_{0-}^x K_1(x, s)\left[\int_{0-}^y F_1(s, t)\, dm(t)\right] dm(s)\\ &\quad = -\int_{0-}^y F_1(x, t)\, dm(t).\end{aligned}$$

[7] In checking the passage from line 2 to line 3 you should bear in mind that $\tilde{\Delta} = \Delta^{\cdot} - \Delta$ is the difference of two spectral functions and hence is of bounded variation on compact subintervals of the line.

Now invoking (a) you find that

$$\int_{0-}^{y} K_1(x, t)\, dm(t) + \int_{0-}^{x} K_1(x, s)\left[\int_{0-}^{y} F_1(t, 0)\, dm(t) - \int_0^s F_2(y, t)\, dt\right] dm(s)$$

$$= -\int_{0-}^{y} F_1(0, t)\, dm(t) + \int_0^x F_2(y, t)\, dt.$$

Therefore,

$$\int_{0-}^{y} K_1(x, t)\, dm(t) + \int_{0-}^{x} K_1(x, s)\, dm(s) \int_{0-}^{y} F_1(t, 0)\, dm(t) + \int_{0-}^{y} F_1(0, t)\, dm(t)$$

$$= \int_0^x F_2(y, t)\left[1 + \int_t^x K_1(x, s)\, dm(s)\right] dt,$$

and so, upon adding

$$\phi_1(x, y) = 1 + \int_y^x K_1(x, t)\, dm(t)$$

to both sides, you see that

$$\phi_1(x, 0-)\left[1 + \int_{0-}^{y} F_1(0, t)\, dm(t)\right] = \phi_1(x, y) + \int_0^x F_2(y, t)\phi_1(x, t)\, dt,$$

which proves (b) since

$$1 + \int_{0-}^{y} F_1(0, t)\, dm(t) = 1 - \int_{0-}^{y} K_1(0-, t)\, dm(t) = \phi_1(0-, y).$$

Proof of (c) *and* $\phi_1(x, 0-) \neq 0$. Integrate the identity

$$-F_2(x, t) = \int_0^x K_2(x, s)F_2(s, t)\, ds + K_2(x, t)$$

with respect to dt across the interval $0 \leqslant t \leqslant y$ to get

$$-\int_0^y F_2(x, t)\, dt = \int_0^x K_2(x, s)\left[\int_0^y F_2(s, t)\, dt\right] ds + \int_0^y K_2(x, t)\, dt.$$

But, by identity (a), this is the same as

$$\int_{0-}^{x} [F_1(y, t) - F_1(0, t)]\, dm(t)$$

$$= \int_0^x K_2(x, s)\left\{\int_{0-}^{s} [F_1(0, t) - F_1(y, t)]\, dm(t)\right\} ds + \int_0^y K_2(x, t)\, dt$$

$$= \int_0^x [F_1(0, t) - F_1(y, t)]\left[\int_t^x K_2(x, s)\, ds\right] dm(t) + \int_0^y K_2(x, t)\, dt.$$

Therefore,

$$\int_{0-}^{x} F_1(y,t)\left[1+\int_t^x K_2(x,s)\,ds\right]dm(t)$$

$$=\int_{0-}^{x} F_1(0,t)\left[1+\int_t^x K_2(x,s)\,ds\right]dm(t)+\int_0^y K_2(x,t)\,dt,$$

and so, upon adding

$$\phi_2(x,y)=1+\int_y^x K_2(x,s)\,ds$$

to both sides, you see that

$$\phi_2(x,y)+\int_{0-}^{x} F_1(y,t)\phi_2(x,t)\,dm(t)$$

$$=\phi_2(x,0)+\int_{0-}^{x} F_1(0,t)\phi_2(x,t)\,dm(t)$$

for all $0 \leqslant y \leqslant x$. The constant which appears on the right is evaluated by first invoking formula (2.15) of Corollary 2.2 to get

$$1=\phi_2(x,x)$$

$$=\left[1+\int_{0-}^{x} K_1(x,s)\,dm(s)\right]\left[\phi_2(x,0)+\int_{0-}^{x} F_1(0,t)\phi_2(x,t)\,dm(t)\right]$$

$$=\phi_1(x,0-)\left[\phi_2(x,0)+\int_{0-}^{x} F_1(0,t)\phi_2(x,t)\,dm(t)\right].$$

This proves that $\phi_1(x,0-)$ is nonzero and that the constant in question is the same as

$$1/\phi_1(x,0-)=\phi_2(0-,y)/\phi_1(x,0-),$$

which completes the proof of (c).

Proof of (d). In view of (b) you may invoke Corollary 3.2 with $f(y)=\phi_1(0-,y)$ and

$$\phi(x,y)=\phi_1(x,y)/\phi_1(x,0-).$$

(d) now drops out from (3.6); just bear in mind that $\phi_1(y,y)=1$.

Proof of (e). Apply formula (2.9) to (c) with

$$F=F_1,\ \phi(x,y)=\phi_2(x,y)\phi_1(x,0-),\qquad \text{and}\qquad f(y)=\phi_2(0-,y)=1.$$

Proof of (f).

$$\int_y^x \frac{\phi_2(s, y)}{\phi_1(s, 0-)} ds = \int_y^x \frac{[1 + \int_y^s K_2(s, t)\, dt]}{\phi_1(s, 0-)} ds$$

$$= \int_y^x \left[\frac{1}{\phi_1(s, 0-)} + \int_s^x K_2(t, s) \frac{1}{\phi_1(t, 0-)} dt \right] ds,$$

as follows upon interchanging both the order of integration and the notation. But by (d) the last line is seen to be the same as

$$\int_y^x \frac{\phi_1(x, s)}{\phi_1(x, 0-)} ds$$

and the proof is complete.

Proof of (g) *and* (h). (g) is immediate from (b) and (3.7); (h) is immediate from (c) and (2.17).

4. The Main Theorem

THEOREM 4.1. *Suppose the perturbation of $d\tilde{\Delta}$ is small enough to ensure the existence of a growth point $\delta \leqslant l$ of m with $\delta + m(\delta) < \infty$ such that assumptions* I–IV *of Section* 3 *hold, and let*

$$a(x, \gamma) = A(x, \gamma) + \int_{0-}^x K_1(x, s) A(s, \gamma)\, dm(s) \tag{4.1}$$

for every point x in the interval $0 \leqslant x \leqslant \delta$. Then:

(a) *$a(x, \gamma)$ is a right continuous function of x, with left-hand limits $a(x-, \gamma)$ at every point $0 \leqslant x \leqslant \delta$,*

$$a(x, \gamma) = a(x-, \gamma)\{1 + K_1(x, x) m[x]\}, \tag{4.2}$$

and

$$a(x, 0) = 1 + \int_{0-}^x K_1(x, s)\, dm(s)$$

is never zero on the interval $0 \leqslant x \leqslant \delta$;

(b)
$$A^{\cdot}(x, \gamma) = a(x, \gamma)/a(x, 0) \tag{4.3}$$

is the unique solution of the string equation

$$A^{\cdot}(x, \gamma) = 1 - \gamma^2 \int_0^x dx^{\cdot}(\xi) \int_{0-}^{\xi} A^{\cdot}(\eta, \gamma)\, dm^{\cdot}(\eta)$$

for the string segment parameterized by

$$x^{\cdot}(x) = \int_0^x [a(s, 0)]^{-2}\, ds,$$
$$m^{\cdot}(x) = \int_{0-}^x a(s, 0)a(s-, 0)\, dm(s), \tag{4.4}$$

$0 \leqslant x \leqslant \delta$;

(c) Δ *is a spectral function for the string segment described in* (b):

$$(1/\pi) \int_{-\infty}^{\infty} \left| \int_{0-}^{\delta} f(x)A^{\cdot}(x, \gamma)\, dm^{\cdot}(x) \right|^2 d\Delta^{\cdot}(\gamma) = \int_{0-}^{\delta} |f(x)|^2\, dm^{\cdot}(x) \tag{4.5}$$

for $f \in \mathbf{L}^2([0, \delta], dm^{\cdot})$.

Proof of (a). The fact that $a(x, 0)$ is never zero on the interval $0 \leqslant x \leqslant \delta$ is proved in Theorem 3.2; you have only to notice that

$$a(x, 0) = \phi_1(x, 0-).$$

The right continuity of $a(x, \gamma)$ and the existence of left-hand limits $a(x-, \gamma)$ is an immediate consequence of the fact that $m(x)$ and $K_1(x, s)$ enjoy similar properties, whereas (4.2) follows from (2.7) upon making the appropriate identifications:

$$\begin{aligned} a(x, \gamma) - a(x-, \gamma) &= K_1(x, x)A(x, \gamma)m[x] + \int_{0-}^{x-} [K_1(x, s) - K_1(x-, s)] \\ &\quad \times A(s, \gamma)\, dm(s) \\ &= K_1(x, x)m[x]\left\{A(x, \gamma) + \int_{0-}^{x-} K_1(x-, s)A(s, \gamma)\, dm(s)\right\} \\ &= K_1(x, x)m[x]a(x-, \gamma). \end{aligned}$$

Proof of (b). It suffices to show that $a(x, \gamma)/a(x, 0)$ is a solution of the indicated string equation. Consider therefore

$$\begin{aligned} 1 - \gamma^2 &\int_0^x \frac{d\xi}{a^2(\xi, 0)} \int_{0-}^{\xi} \frac{a(\eta, \gamma)}{a(\eta, 0)}\, a(\eta, 0)a(\eta-, 0)\, dm(\eta) \\ &= 1 - \gamma^2 \int_0^x \frac{d\xi}{a^2(\xi, 0)} \int_{0-}^{\xi} [A(\eta, \gamma) \\ &\quad + \int_{0-}^{\eta} K_1(\eta, s)A(s, \gamma)\, dm(s)]a(\eta-, 0)\, dm(\eta). \end{aligned}$$

But with the help of identity (e) of Theorem 3.2 the inner integral

$$\int_{0-}^{\xi} [\cdots] a(\eta-, 0)\, dm(\eta)$$

is seen to be the same as

$$\begin{aligned}
&\int_{0-}^{\xi} A(\eta, \gamma)\left[a(\eta-, 0) + \int_{\eta-}^{\xi} K_1(s, \eta) a(s-, 0)\, dm(s)\right] dm(\eta) \\
&= \int_{0-}^{\xi} A(\eta, \gamma)\left[a(\eta-, 0) + K_1(\eta, \eta) a(\eta-, 0) m[\eta]\right. \\
&\quad \left. + \int_{\eta}^{\xi} K_1(s, \eta) a(s-, 0)\, dm(s)\right] dm(\eta) \\
&= \int_{0-}^{\xi} A(\eta, \gamma)\left[a(\eta, 0) + \int_{\eta}^{\xi} K_1(s-, \eta) a(s, 0)\, dm(s)\right] dm(\eta) \\
&= \int_{0-}^{\xi} A(\eta, \gamma)\phi_2(\xi, \eta) a(\xi, 0)\, dm(\eta),
\end{aligned}$$

and so the full expression is equal to

$$\begin{aligned}
&1 - \gamma^2 \int_0^x \frac{d\xi}{a(\xi, 0)} \int_{0-}^{\xi} A(\eta, \gamma)\phi_2(\xi, \eta)\, dm(\eta) \\
&\quad = 1 - \gamma^2 \int_{0-}^{x} \left[\int_{\eta}^{x} \frac{\phi_2(\xi, \eta)}{a(\xi, 0)}\, d\xi\right] A(\eta, \gamma)\, dm(\eta).
\end{aligned}$$

Finally, invoking identity (f) of Theorem 3.2 and the definition of $\phi_1(x, \xi)$, you see that this is the same as

$$\begin{aligned}
&1 - \gamma^2 \int_{0-}^{x} \left[\int_{\eta}^{x} \frac{\phi_1(x, \xi)}{a(x, 0)}\, d\xi\right] A(\eta, \gamma)\, dm(\eta) \\
&\quad = 1 + \frac{A(x, \gamma) - 1}{a(x, 0)} + \frac{\int_{0-}^{x} K_1(x, s)[A(s, \gamma) - 1]\, dm(s)}{a(x, 0)} \\
&\quad = \frac{A(x, \gamma) + \int_{0-}^{x} K_1(x, s) A(s, \gamma)\, dm(s)}{a(x, 0)} \\
&\quad = a(x, \gamma)/a(x, 0),
\end{aligned}$$

and the proof is complete.

Proof of (c).

$$\int_{0-}^{\delta} f(x)A^{\cdot}(x, \gamma)\, dm^{\cdot}(x)$$

$$= \int_{0-}^{\delta} f(x)a(x-, 0)a(x, \gamma)\, dm(x)$$

$$= \int_{0-}^{\delta} f(x)a(x-, 0)\left[A(x, \gamma) + \int_{0-}^{x} K_1(x, s)A(s, \gamma)\, dm(s)\right] dm(x).$$

But now setting

$$g(x) = f(x)a(x-, 0) \qquad \text{and} \qquad h = (I + L_1{}^*)g$$

you see that the last integral can be expressed as the inner product

$$\langle g, (I + L_1)A\rangle_m = \langle (I + L_1{}^*)g, A\rangle_m = \langle h, A\rangle_m.$$

Consequently,

$$(1/\pi)\int \left|\int_{0-}^{\delta} f(x)A^{\cdot}(x, \gamma)\, dm^{\cdot}(x)\right|^2 d\Delta^{\cdot}(\gamma)$$

$$= (1/\pi)\int \left|\int_{0-}^{\delta} h(x)A(x, \gamma)\, dm(x)\right|^2 d\Delta^{\cdot}(\gamma)$$

$$= \langle (I + F_1)_\delta h, h\rangle_m,$$

in view of the evaluation derived in step 1 of the proof of Theorem 3.1. But now it follows easily from the factorization identity (2.8) that

$$(I + L_1)(I + F_1)(I + L_1{}^*) = \Lambda_1,$$

and so the above can be written as

$$\langle \Lambda_1 g, g\rangle_m = \int_{0-}^{\delta} \{1 + K_1(x, x)m[x]\}|f(x)a(x-, 0)|^2\, dm(x)$$

$$= \int_{0-}^{\delta} |f(x)|^2 a(x, 0)a(x-, 0)\, dm(x),$$

because of (4.2). The proof is complete.

COROLLARY 4.1. *Let*

$$b(x, \gamma) = B(x, \gamma) + \int_0^x K_2(x, s)B(s, \gamma)\, ds. \tag{4.6}$$

Then

$$B^{\cdot}(x, \gamma) = a(x, 0)b(x, \gamma) \tag{4.7}$$

for every point x in the interval $0 \leqslant x \leqslant \delta$.

Proof. It suffices to show that

$$a(x, 0)b(x, \gamma) = \gamma \int_{0-}^{x} A^{\cdot}(s, \gamma)\, dm^{\cdot}(s).$$

Now by (4.6) and the definition of B you see that

$$\begin{aligned} &a(x, 0)b(x, \gamma) \\ &\quad = a(x, 0)[B(x, \gamma) + \int_{0}^{x} K_2(x, s)B(s, \gamma)\, ds] \\ &\quad = \gamma a(x, 0)[\int_{0-}^{x} A(y, \gamma)\, dm(y) + \int_{0}^{x} K_2(x, s) \int_{0-}^{s} A(y, \gamma)\, dm(y)\, ds] \\ &\quad = \gamma a(x, 0) \int_{0-}^{x} A(y, \gamma)[1 + \int_{y}^{x} K_2(x, s)\, ds]\, dm(y). \end{aligned}$$

The next step is to invoke identity (e) of Theorem 3.2 which shows that

$$a(x, 0)\left[1 + \int_{y}^{x} K_2(x, s)\, ds\right] = a(y, 0) + \int_{y}^{x} K_1(s-, y)a(s, 0)\, dm(s).$$

This permits you to reexpress

$$\begin{aligned} &a(x, 0)b(x, \gamma) \\ &\quad = \gamma \int_{0-}^{x} A(y, \gamma)\left[a(y, 0) + \int_{y}^{x} K_1(s-, y)a(s, 0)\, dm(s)\right] dm(y) \\ &\quad = \gamma \int_{0-}^{x} a(y, 0)\left[A(y, \gamma) + \int_{0-}^{y-} K_1(y-, s)A(s, \gamma)\, dm(s)\right] dm(y) \\ &\quad = \gamma \int_{0-}^{x} a(y, 0)a(y-, \gamma)\, dm(y) \\ &\quad = \gamma \int_{0-}^{x} [a(y-, \gamma)/a(y-, 0)]a(y, 0)a(y-, 0)\, dm(y). \end{aligned}$$

But that completes the proof since

$$A^{\cdot}(y-, \gamma) = a(y-, \gamma)/a(y-, 0) = A^{\cdot}(y, \gamma)$$

and

$$a(y, 0)a(y-, 0)\, dm(y) = dm^{\cdot}(y)$$

by Theorem 4.1.

5. The Gelfand–Levitan Construction

The purpose of this section is to indicate the connection between Theorem 4.1 and work of Gelfand and Levitan (1951). The basic example treated by them corresponds to the string with $m(x) = x$ and $l = \infty$. For this string

$A(x, \gamma) = \cos \gamma x$, $B(x, \gamma) = \sin \gamma x$, and the principal spectral function $\Delta(\gamma) = \gamma$. This example will be considered under the assumption that the perturbation $\tilde{\Delta}$ is such that conditions I and III of Section 3 hold for every finite choice of $\delta > 0$. That conforms to the type of conditions imposed by Levitan and Gasymov (1964) in their study of the Gelfand–Levitan procedure.[8] The first point to check is that under I and III the conditions II and IV will be fulfilled automatically.

The fact that I implies II is an easy consequence of the double angle formulas

$$2 \cos \gamma x \cos \gamma y = \cos \gamma(x - y) + \cos \gamma(x + y),$$
$$2 \sin \gamma x \sin \gamma y = \cos \gamma(x - y) - \cos \gamma(x + y):$$

I is equivalent to assumption I′,

I′. $\int_{-R}^{R} \cos \gamma x \, d\tilde{\Delta}(\gamma)$ tends boundedly to a continuous limit $\Phi(x)$ on the interval $0 \leqslant x \leqslant 2\delta$, as $R \uparrow \infty$, and by another application of the double angle formulas you see that I′ implies II.

It remains to check that if III is valid for every δ, then so is IV. Suppose to the contrary that for some choice of δ there is a nontrivial $f \in \mathbf{L}^2([0, \delta], dx)$ with

$$\int \left| \int_0^\delta f(x) \sin \gamma x \, dx \right|^2 d\Delta^{\cdot}(\gamma) = 0.$$

Then, for any $\epsilon > 0$,

$$\sin \epsilon\gamma \int_0^\delta f(x) \sin \gamma x \, dx$$

is both equivalent to zero in $\mathbf{L}^2(R^1, d\Delta^{\cdot})$ and expressible in the form

$$\sin \epsilon\gamma \int_0^\delta f(x) \sin \gamma x \, dx = \int_0^{\delta+\epsilon} g(x) \cos \gamma x \, dx$$

for some nontrivial $g \in \mathbf{L}^2([0, \delta + \epsilon], dx)$, as follows from either the double angle formulas or the Paley–Wiener theorem. But that conclusion,

$$\int \left| \int_0^{\delta+\epsilon} g(x) \cos \gamma x \, dx \right|^2 d\Delta^{\cdot}(\gamma) = 0,$$

with $g \not\equiv 0$, stands in contradiction to III.

Theorem 4.1 is now seen to be applicable. It implies that

$$p(x) = [a(x, 0)]^2 = \left[1 + \int_0^x K_1(x, s) \, ds \right]^2$$

[8] The present conditions (for the existence of a string equation) are actually less restrictive in the smoothness requirements put on $F_1(x, y)$ than those imposed by Levitan and Gasymov; however, only differential operators with nonnegative spectrum are considered.

is continuous and strictly positive on the interval $0 \leqslant x < \infty$ and that, at least formally,

$$a = a(x, \gamma) = [p(x)]^{1/2} A^{\cdot}(x, \gamma)$$

is a solution of the differential equation

$$\frac{1}{p}\frac{d}{dx} p \frac{d}{dx} p^{-1/2} a = -\gamma^2 p^{-1/2} a,$$

$0 < x < \infty$. But this in turn implies that a is a solution of the Schrödinger equation

$$a'' + (\gamma^2 - V)a = 0$$

with potential

$$V = \frac{p''}{2p} - \left(\frac{p'}{2p}\right)^2 = a''(x, 0)/a(x, 0).$$

Moreover, it follows from the representation (4.1),

$$a(x, \gamma) = \cos \gamma x + \int_0^x K_1(x, s) \cos \gamma s \, ds,$$

that the potential

$$V(x) = 2(d/dx)K_1(x, x)$$

and that a meets the initial conditions

$$a(0, \gamma) = 1, \qquad a'(0, \gamma) = K_1(0, 0) = -F_1(0, 0) = -(1/\pi) \int d\tilde{\Delta}(\gamma).$$

The formal differentiations carried out above can be fully justified if p is sufficiently smooth. That is achieved by imposing additional smoothness constraints on Φ in I'; see Levitan and Gasymov (1964; pp. 6–22) for more details on this.

It should be noted that $\Delta^{\cdot}$ is a spectral function for the Schrödinger equation (as well as for the string equation):

$$\begin{aligned}
&\frac{1}{\pi} \int \left| \int_0^\infty f(x) a(x, \gamma) \, dx \right|^2 d\Delta^{\cdot}(\gamma) \\
&\quad = \frac{1}{\pi} \int \left| \int_0^\infty \frac{f(x)}{a(x, 0)} A^{\cdot}(x, \gamma) \, dm^{\cdot}(x) \right|^2 d\Delta^{\cdot}(\gamma) \\
&\quad = \int_0^\infty \left| \frac{f(x)}{a(x, 0)} \right|^2 dm^{\cdot}(x) = \int_0^\infty |f(x)|^2 \, dx.
\end{aligned}$$

Amplification 1. The inequality

$$\xi = \int_0^\xi dx \leqslant \left\{\int_0^\xi [a(x,0)]^2\,dx \int_0^\xi [a(x,0)]^{-2}\,dx\right\}^{1/2}$$
$$= [m^{\cdot}(\xi)x^{\cdot}(\xi)]^{1/2} \leqslant [m^{\cdot}(\xi) + x^{\cdot}(\xi)]/2$$

implies that the spot string is long:

$$x^{\cdot}(\xi) + m^{\cdot}(\xi) \uparrow \infty \qquad \text{as} \qquad \xi \uparrow l = \infty.$$

Hence, the parametrized string $x^{\cdot}(\xi)$, $m^{\cdot}(\xi)$, $0 \leqslant \xi < \infty$, is the full string of which Δ is the principal spectral function and not just the initial segment; for more information on this point see Section 8.

Amplification 2. The inversion formula

$$\cos \gamma x = a(x, \gamma) - \int_0^x K_1(s, x)a(s, \gamma)\,ds$$

is an immediate byproduct of the identity

$$(I - U^*) = (I + L)^{-1}\Lambda \tag{5.1}$$

established within the proof of Theorem 2.1; $\Lambda = I$ in the present circumstances since $m(x) = x$ is jump free.

6. Ramifications

The purpose of this section is to present a variant of the main theorem which lends itself more readily to applications than Theorem 4.1. A number of illustrative examples will be presented in the next section. The notation

$$J_\beta{}^x(\gamma) = (1/\pi) \int_{0-}^x A(s, \beta^*)A(s, \gamma)\,dm(s)$$

will be employed. $J_\beta{}^x$ is a reproducing kernel for the closed subspace

$$\hat{\mathbf{M}}_x = \left\{\int_{0-}^x f(s)A(s, \gamma)\,dm(s) : f \in \mathbf{M}_x\right\}$$

of $\mathbf{L}^2(R^1, d\Delta)$, although that aspect of things will not be entered into here.

The main result rests upon the observation that if x is a growth point of m and $K_1(x, \cdot)$ is the unique solution specified by the equation

$$K_1(x, y) + F_1(x, y) + \int_{0-}^x K_1(x, s)F_1(s, y)\,dm(s) = 0, \tag{6.1}$$

then

$$K_1(x, y) = -\lim_{R \uparrow \infty} \frac{1}{\pi} \int_{-R}^{R} \left[A(x, \lambda) + \int_{0-}^{x} K_1(x, s) A(s, \lambda)\, dm(s) \right] \times A(y, \lambda)\, d\tilde{\Delta}(\lambda)$$

$$= -\lim_{R \uparrow \infty} \frac{1}{\pi} \int_{-R}^{R} a(x, \lambda) A(y, \lambda)\, d\tilde{\Delta}(\lambda). \tag{6.2}$$

where the right-hand side remains bounded uniformly in y $(0 \leqslant y \leqslant \delta)$, as $R \uparrow \infty$. It follows at once that

$$a(x, \gamma) = A(x, \gamma) + \int_{0-}^{x} K_1(x, s) A(s, \gamma)\, dm(s) \tag{6.3}$$

is a solution of

$$a(x, \gamma) + \lim_{R \uparrow \infty} \int_{-R}^{R} a(x, \lambda) J_\lambda{}^x(\gamma)\, d\tilde{\Delta}(\lambda) = A(x, \gamma). \tag{6.4}$$

In fact $a(x, \cdot)$ is the only even analytic solution of (6.4) of the form

$$a = A + g \tag{6.5}$$

with $g \in \hat{\mathbf{M}}_x$: if there were two such solutions, then the difference $h(x, \gamma)$ would be a solution of the homogeneous equation

$$h(x, \gamma) + \lim_{R \uparrow \infty} \int_{-R}^{R} h(x, \lambda) J_\lambda{}^x(\gamma)\, d\tilde{\Delta}(\lambda) = 0$$

and

$$H(x, y) = -\lim_{R \uparrow \infty} \frac{1}{\pi} \int_{-R}^{R} h(x, \lambda) A(y, \lambda)\, d\tilde{\Delta}(\lambda)^9$$

would be a nontrivial solution of the homogeneous equation

$$H(x, y) + \int_{0-}^{x} H(x, s) F_1(s, y)\, dm(s) = 0,$$

contrary to assumption.

The argument sketched above indicates that[10] *if* $K_1(x, \cdot)$ *is the only solution of* (6.1) *belonging to* $\mathbf{M}_x$, *then* $a(x, \cdot)$, *as given by* (6.3), *is the only even analytic solution of* (6.4) *of the specified form* [(6.5)]. *The converse is also true*: *If* $a(x, \cdot)$ *is the only even analytic solution of* (6.4) *of the form* (6.5), *then* $K_1(x, \cdot)$, *as given by* (6.2), *is the only solution of* (6.1) *belonging to* $\mathbf{M}_x$. The converse

[9] The limit exists in view of assumption I, since $h(x, \cdot) \in \hat{\mathbf{M}}_x$.

[10] x is presumed to be a growth point of m, and the statement $K_1(x, \cdot) \in \mathbf{M}_x$ is meant only in the sense that $K_1(x, y)$ agrees with a function of class $\mathbf{M}_x$ for $0 \leqslant y \leqslant x$.

statement is proved in step 1 of the next theorem for the special case $x = \delta$. However, the same proof works for any growth point x of m.

THEOREM 6.1. *Let $\delta \leqslant l$ be a point of growth of m such that $\delta + m(\delta) < \infty$ and assumptions* I *and* II *of Section* 3 *are in force. Suppose further that $a(\delta, \cdot)$ is the unique even analytic solution of equation* (6.4) *with $x = \delta$ of the form*

$$a(\delta, \gamma) = A(\delta, \gamma) + \int_{0-}^{\delta} G(\delta, s)A(s, \gamma)\, dm(s) \tag{6.6}$$

with $G(\delta, \cdot) \in \mathbf{M}_\delta$ and that

$$a(\delta, 0) \neq 0. \tag{6.7}$$

Then (6.4) *has a unique even analytic solution $a(x, \cdot)$ of the form* (6.5) *with $a(x, 0) \neq 0$ for every choice of x in the interval $0 \leqslant x \leqslant \delta$, and the initial segment of the string with spectral function $\Delta^{\cdot}$ is given by equation* (4.4).

Proof. The proof is in steps.

STEP 1. Show that $(I + F_1)_\delta$ is a 1–1 map of $\mathbf{M}_\delta$ onto itself.

Proof of Step 1. The representation (6.6) ensures that

$$-(1/\pi) \int_{-R}^{R} a(\delta, \lambda)A(y, \lambda)d\tilde{\Delta}(\lambda)$$

tends boundedly to a limit on the interval $0 \leqslant y \leqslant \delta$ as $R \uparrow \infty$. Moreover, the fact that $a(\delta, \cdot)$ is a solution of (6.4) (with $x = \delta$) permits you to identify the limit as $G(\delta, y)$:

$$G(\delta, y) = -\lim_{R \uparrow \infty} \frac{1}{\pi} \int_{-R}^{R} a(\delta, \lambda)A(y, \lambda)\, d\tilde{\Delta}(\lambda).$$

Now multiply both sides of (6.6) by $A(y, \gamma)\, d\tilde{\Delta}(\gamma)$ and integrate to get

$$G(\delta, y) + F_1(\delta, y) + \int_{0-}^{\delta} G(\delta, s)F_1(s, y)\, dm(s) = 0.$$

This exhibits $G(\delta, \cdot)$ as a solution of (6.1). Now if there were more than one such solution, then you could find a nontrivial solution $H(\delta, \cdot)$ of the homogeneous equation

$$H(\delta, y) + \int_{0-}^{\delta} H(\delta, s)F_1(s, y)\, dm(s) = 0.$$

and a corresponding nontrivial solution

$$h(\delta, \gamma) = \int_{0-}^{\delta} H(\delta, s)A(s, \gamma)\, dm(s)$$

of the homogeneous equation

$$h(\delta, \gamma) + \lim_{R \uparrow \infty} \int_{-R}^{R} h(\delta, \lambda) J_\lambda{}^\delta(\gamma)\, d\tilde{\Delta}(\lambda) = 0,$$

contrary to assumption.

STEP 2. Deduce that $(I + F_2)_\delta$ is a 1–1 mapping of $\mathbf{X}_\delta$ onto itself from step 1 and (6.7).

Proof of Step 2. Suppose, to the contrary, that there exists a nontrivial $f \in \mathbf{X}_\delta$ such that

$$f(x) = -\int_0^\delta F_2(x, y) f(y)\, dy, \qquad 0 \leqslant x \leqslant \delta,$$

and let

$$\psi(y) = \int_y^\delta f(s)\, ds.$$

Then

$$\begin{aligned}
&\int_{0-}^{\delta} [F_1(x, y) - F_1(0, y)]\psi(y)\, dm(y) \\
&\quad = \int_0^\delta f(s) \int_{0-}^{s} [F_1(x, y) - F_1(0, y)]\, dm(y)\, ds \\
&\quad = -\int_0^\delta f(s) \int_0^x F_2(s, t)\, dt\, ds \\
&\quad = -\int_0^x \left[\int_0^\delta F_2(t, s) f(s)\, ds \right] dt \\
&\quad = \int_0^x f(t)\, dt,
\end{aligned}$$

in view of identity (a) of Theorem 3.2 and the assumptions on f. But this implies that

$$\psi(x) + \int_{0-}^{\delta} F_1(x, y)\psi(y)\, dm(y) = \psi(0) + \int_{0-}^{\delta} F_1(0, y)\psi(y)\, dm(y)$$

independently of x. In other words,

$$(I + F_1)_\delta \psi = c,$$

a constant. Therefore, by formula (2.15) of Corollary 2.2,

$$0 = \psi(\delta) = c\left[1 + \int_{0-}^{\delta} K_1(\delta, s)\, dm(s) \right] = ca(\delta, 0).$$

But in light of (6.7), this means that $c = 0$, and so exhibits ψ as a nontrivial solution of the homogeneous equation

$$\psi(x) + \int_{0-}^{\delta} F_1(x, s)\psi(s)\, dm(s) = 0,$$

contrary to step 1.

STEP 3. Simply notice that assumptions I–IV of Section 3 are now in force, and hence conclude that the solution $a(x, \gamma)$ assured by Theorem 4.1 is the one and only solution of (6.4) of the indicated form [(6.5)] and that it enjoys all the properties attributed to it in the statement of the theorem.

Amplification. The computational advantage of Theorem 6.1 is that it gives you $a(x, \gamma)$ directly without having to go through the computation of K_1. *If* $d\tilde{\Delta}$ *is a nonnegative measure, then conditions* III *and* IV *of Section* 3 *are automatically fulfilled*: $(I + F_1)_\delta$ *and* $(I + F_2)_\delta$ *are both* 1–1 *operators.* In these circumstances $a(\delta, 0) \neq 0$ and (6.4) is uniquely solvable.

7. EXAMPLES

The present section contains a number of examples to illustrate the application of Theorems 4.1 and 6.1. The calculcations are not always complete, but are carried out as far as is deemed instructive.

EXAMPLE 1. $\tilde{\Delta}$ has a jump of height

$$q \geqslant -\Delta[\mu] \qquad \text{at the points} \qquad \pm\mu(\mu \neq 0).$$[11]

Solution. For this choice of $\tilde{\Delta}$, assumptions I and II of Section 3 are clearly fulfilled for every $\delta \leqslant l$ with $\delta + m(\delta) < \infty$ and equation (6.4) is easily solved:

$$a(x, \gamma) + a(x, \mu)(2q/\pi) \int_{0-}^{x} A(s, \mu)A(s, \gamma)\, dm(s) = A(x, \gamma)$$

for every complex γ, and so, in particular,

$$a(x, \mu) = A(x, \mu)/\left[1 + (2q/\pi) \int_{0-}^{x} [A(s, \mu)]^2\, dm(s)\right].$$

Therefore,

$$a(x, \gamma) = A(x, \gamma) - \frac{A(x, \mu)(2q/\pi) \int_{0-}^{x} A(s, \mu)A(s, \gamma)\, dm(s)}{1 + (2q/\pi) \int_{0-}^{x} [A(s, \mu)]^2\, dm(s)},$$

[11] $\Delta[\mu] \equiv \Delta(\mu+) - \Delta(\mu-)$ stands for the jump at μ, if any; $\Delta[\mu] = \Delta[-\mu]$ since Δ is odd.

which exhibits a in the requisite form [(6.5)]. Moreover,

$$a(x, 0) = \frac{1 + (2q/\pi)\{\int_{0-}^{x} [A(s, \mu)]^2 \, dm(s) - [A(x, \mu)B(x, \mu)/\mu]\}}{1 + (2q/\pi) \int_{0-}^{x} [A(s, \mu)]^2 \, dm(s)}$$

$$= \frac{1 + (2q/\pi) \int_{0}^{x} [B(s, \mu)]^2 \, ds}{1 + (2q/\pi) \int_{0-}^{x} [A(s, \mu)]^2 \, dm(s)}. \tag{7.1}$$

The last line is based upon the identity

$$\int_{0-}^{x} [A(s, \mu)]^2 \, dm(s) - \int_{0}^{x} [B(s, \mu)]^2 \, ds = A(x, \mu)B(x, \mu)/\mu, \tag{7.2}$$

which is verified by integration by parts; you have only to bear in mind that $dA = -\mu B \, dx$ and $dB = \mu A \, dm$ for $A = A(x, \mu)$ and $B = B(x, \mu)$. Now if $q > 0$, then $a(\delta, 0)$ is nonzero for every growth point $\delta \leqslant l$ with $\delta + m(\delta) < \infty$, and so, by Theorem 6.1, formulas (4.4) are applicable for all points $x \leqslant \delta$. On the other hand, if $q < 0$, then the constraint

$$q \geqslant -\Delta[\mu] = -\pi \left\{ 2 \int_{0-}^{l} [A(s, \mu)]^2 \, dm(s) \right\}^{-1}$$

is in force and you see that the denominator in (7.1) does not vanish for $x < l$. It remains to check the numerator. Assume first that $l + m(l-) < \infty$. Then the principal spectral function Δ of the starting string lmk has jumps precisely at the roots of the equation

$$kA^+(l, \gamma) + A(l, \gamma) = 0,$$

as was spelled out in Section 1. Therefore, since $\Delta[\mu] \geqslant -q > 0$ in the present instance, you see that μ must coincide with one of these roots. Hence

$$\int_{0-}^{l} [A(s, \mu)]^2 \, dm(s) - \int_{0}^{l} [B(s, \mu)]^2 \, ds$$
$$= -A(l, \mu)A^+(l, \mu)/\mu^2 = k[A^+(l, \mu)/\mu]^2 \geqslant 0.$$

This insures that $a(\delta, 0)) \neq 0$ for every growth point $\delta < l$ of m and formulas (4.4) are again applicable for every point $x \leqslant \delta$. If $q < 0$ and $l + m(l-) = \infty$, then the same conclusions hold, only the argument for the nonvanishing of $a(x, 0)$ is a bit more delicate. The crux of the matter is to show that

$$\lim_{x \uparrow l} A(x, \mu)A^+(x, \mu) = 0.$$

A prototype of this fact is proved in Dym and McKean (1976:§5.2) and will not be repeated here. Granting this point, you see that

$$\int_{0}^{x} [B(s, \mu)]^2 \, ds < \int_{0}^{l} [B(s, \mu)]^2 \, ds = \int_{0-}^{l} [A(s, \mu)]^2 \, ds$$

for $x < l$ (and $\mu \neq 0$) and the fact that $a(x, 0) \neq 0$ for $x < l$ drops out.

EXAMPLE 2. Let $m(x) = x$, $l = \pi$, and $k = \infty$, so that $A(x, \gamma) = \cos \gamma x$, $B(x, \gamma) = \sin \gamma x$, and Δ has jumps of height 1 at every integer, and choose $\tilde{\Delta}$ to have jumps of height -1 at the points $\pm\mu_1, \pm\mu_2, \ldots, \pm\mu_n$ where the μ_i are positive integers.

Solution. For this example assumptions I and II of Section 3 are in force for every $l \leqslant \pi$. However, equation (6.4),

$$a(x, \gamma) - (2/\pi) \sum_{i=1}^{n} a(x, \mu_i) \int_0^x \cos \mu_i s \cos \gamma s \, ds = \cos \gamma x,$$

is uniquely solvable for $a(x, \gamma)$ if and only if the $n \times n$ matrix

$$I - Q$$

with entries

$$Q_{ij} = (2/\pi) \int_0^x \cos \mu_i s \cos \mu_j s \, ds$$

is nonsingular. That is the case for every choice of x in the interval $0 \leqslant x < \pi$. Moreover, $a(x, 0) \neq 0$ for $0 \leqslant x < \pi$. An indirect proof of these facts can be based upon Theorem 4.1: It suffices to show that the operators $(I + F_1)_x$ and $(I + F_2)_x$ are both strictly positive operators. Suppose, to the contrary, that $(I + F_1)_x$ is not strictly positive. Then the identity

$$\langle (I + F_1)_x f, f \rangle = \frac{1}{\pi} \int \left| \int_0^x f(s) \cos \gamma s \, ds \right|^2 d\Delta^{\cdot}(\gamma)$$

$$= \frac{1}{\pi} \sum_{\substack{n=-\infty \\ n \neq \pm\mu_i}}^{\infty} \left| \int_0^x f(s) \cos ns \, ds \right|^2$$

implies that there exists a function $f \not\equiv 0$ of class $\mathsf{L}^2([0, \pi], ds)$ such that

(1) $f(y) = 0 \qquad \text{for} \quad x \leqslant y \leqslant \pi$

and

(2) $\int_0^\pi f(y) \cos vy \, dy = 0 \qquad \text{for} \quad v \neq \pm\mu_i$

hold. However, (2) forces f to be of the form

$$f(y) = \sum_{i=1}^{n} c_i \cos \mu_i y$$

and that is not consistent with (1) unless $f \equiv 0$. The fact that $(I + F_2)_x$ is 1–1 is proved in exactly the same way via the identity

$$\langle (I + F_2)_x f, f \rangle = \frac{1}{\pi} \sum_{\substack{n=-\infty \\ n \neq \pm\mu_i}}^{\infty} \left| \int_0^x f(s) \sin ns \, ds \right|^2.$$

The analysis to this point shows that equation (6.4) is uniquely solvable for $a(x, \gamma)$ for $0 \leqslant x < \pi$. The rest is pure computation. For example, if $\tilde{\Delta}$ has a jump of height -1 at the pair of points $\pm\mu$ ($\mu \neq 0$) only, then

$$a(x, \gamma) = \cos \gamma x + \frac{(\cos \mu x)(2/\pi) \int_0^x \cos \mu s \cos \gamma s \, ds}{1 - (2/\pi) \int_0^x \cos^2 \mu s \, ds}$$

and

$$a(x, 0) = \frac{1 - (2/\pi) \int_0^x \sin^2 \mu s \, ds}{1 - (2/\pi) \int_0^x \cos^2 \mu s \, ds}.$$

It follows readily that

$$x^{\cdot}(x) \uparrow \infty$$

and

$$m^{\cdot}(x) \uparrow \int_0^\pi [a(s, 0)]^2 \, ds < \infty$$

as $x \uparrow \pi$, since

$$a(x, 0) = O((\pi - x)^2) \qquad \text{as} \quad x \uparrow \pi.$$

On the other hand, if $\tilde{\Delta}$ has a single jump of height -1 at $\mu = 0$, then the limiting behavior of $x^{\cdot}(x)$ and $m^{\cdot}(x)$ as $x \uparrow \pi$ is completely reversed:

$$a(x, \gamma) = \cos \gamma x + \frac{\sin \gamma x}{\gamma(\pi - x)}$$

and

$$a(x, 0) = 1 + \frac{x}{\pi - x} = \frac{\pi}{\pi - x},$$

and consequently

$$m^{\cdot}(x) \uparrow \infty$$

and

$$x^{\cdot}(x) \uparrow \int_0^\pi \left(\frac{\pi - x}{\pi}\right)^2 dx$$

as $x \uparrow \pi$.

EXAMPLE 3. Let $m(x) = x$ and $l = \infty$, so that $A(x, \gamma) = \cos \gamma x$, $B(x, \gamma) = \sin \gamma x$, and $d\Delta(\gamma) = d\gamma$, and choose $d\tilde{\Delta}(\gamma) = -d\gamma/(\gamma^2 + 1)$.

Solution. For this example assumptions I–IV of Section 3 hold for every choice of $\delta < \infty$. In fact, as follows from the discussion of the Gelfand–Levitan construction in Section 5, it suffices to check that I and III hold for every choice of $\delta < \infty$, because then II and IV will be fulfilled automatically. The only real issue is the proof of III. But that, too, is easy. You have only to notice that if

$$\langle (I + F_1)_\delta f, f \rangle_m = (1/\pi) \int \left| \int_0^\delta f(s) \cos \gamma s \, ds \right|^2 d\Delta^{\cdot}(\gamma) = 0,$$

then $\int_0^\delta f(s) \cos \gamma s \, ds$ is an integral function of exponential type δ which is equivalent to zero in $\mathbf{L}^2(R^1, d\Delta^{\cdot})$. However, in the case at hand $(d\Delta^{\cdot} = \gamma^2 \, d\gamma / \gamma^2 + 1)$, this clearly means that

$$\int_0^\delta f(s) \cos \gamma s \, ds = 0$$

for all points $\gamma \in R^1$, and so too $f \equiv 0$ by the Fourier inversion formula.

Amplification. The point underlying the verification of III furnished just above is that if

$$\int \left| \int_0^\delta f(s) A(s, \gamma) \, dm(s) \right|^2 d\Delta^{\cdot}(\gamma) = 0,$$

then the analytic function $\int_0^\delta f(s) A(s, \gamma) \, dm(s)$ vanishes on the support of $d\Delta^{\cdot}$. Therefore, if the latter is a sufficiently rich set (e.g., contains an accumulation point), then $\int_0^\delta f(s) A(s, \gamma) \, dm(s)$ must vanish identically. The same idea can be used to justify the validity of III in Example 2: $\int_0^\delta f(s) \cos \gamma s \, ds$ is an integral function of exponential type $\leqslant \delta$, and for $\delta < \pi$ such a function cannot vanish on all but finitely many integers unless it vanishes identically, as follows easily from Carlson's theorem. (For the statement and proof of the latter see, for example, Dym and McKean (1976, §1.7).) The same kind of argument lends itself to more general circumstances since $\int_0^\delta f(s) A(s, \gamma) \, dm(s)$ is an integral function of exponential type $\leqslant \int_0^\delta [m'(s)]^{1/2} \, ds$, where m' stands for the Radon–Nikodym derivative of m with respect to Lebesgue measure; see Dym and McKean (1976, §6.3) for more information on the type bound.

It now follows, from the discussion in Section 6, that equation (6.4),

$$a(x, \gamma) - \lim_{R \uparrow \infty} \frac{1}{2\pi i} \int_{-R}^{R} a(x, \lambda) \frac{e^{ix(\lambda - \gamma)} - e^{-ix(\lambda - \gamma)}}{\lambda - \gamma} \frac{d\lambda}{\lambda^2 + 1} = \cos \gamma x,$$

has a unique even analytic solution of the form

$$a(x, \gamma) = \cos \gamma x + \int_0^x g(s) \cos \gamma s \, ds \tag{7.3}$$

with $g \in \mathsf{L}^2([0, x], ds)$. The latter yields the pointwise estimate

$$|a(x, u + iv)| \leqslant \text{constant} \times e^{|v|x}, \tag{7.4}$$

which permits you to evaluate the integral in (6.4) by residues. If, in particular, γ belongs to the open upper half-plane, then

$$\lim_{R \uparrow \infty} \frac{1}{2\pi i} \int_{-R}^{R} a(x, \lambda) \frac{e^{ix(\lambda - \gamma)}}{\lambda - \gamma} \frac{d\lambda}{\gamma^2 + 1} = \frac{a(x, \gamma)}{\lambda^2 + 1} + \frac{a(x, i)e^{ix(i-\gamma)}}{2i(i - \gamma)}$$

and

$$\lim_{R \uparrow \infty} \frac{1}{2\pi i} \int_{-R}^{R} a(x, \lambda) \frac{e^{-ix(\lambda - \gamma)}}{\lambda - \gamma} \frac{d\lambda}{\lambda^2 + 1} = -\frac{a(x, -i)e^{ix(i+\gamma)}}{2i(i + \gamma)}.$$

Thus

$$\begin{aligned} \gamma^2 a(x, \gamma) = e^{-i\gamma x} &\left[\frac{\gamma^2 + 1}{2} - \frac{(\gamma + i)a(x, i)e^{-x}}{2i}\right] \\ &+ e^{i\gamma x}\left[\frac{\gamma^2 + 1}{2} + \frac{(\gamma - i)a(x, -i)e^{-x}}{2i}\right] \end{aligned}$$

for all complex γ, and the unknown constant

$$a(x, i) = a(x, -i)$$

is determined from the fact that the right-hand side must vanish at the point $\gamma = 0$ in order for a to be analytic. An elementary computation yields the evaluation

$$a(x, i) = e^x,$$

and the rest is routine:

$$\begin{aligned} a(x, \gamma) &= \cos \gamma x + (\sin \gamma x/\gamma), \\ a(x, 0) &= 1 + x, \\ x^{\cdot}(x) &= \int_0^x (1 + s)^{-2}\, ds \uparrow 1 \qquad \text{as} \quad x \uparrow \infty, \end{aligned}$$

and

$$m^{\cdot}(x) = \int_0^x (1 + s)^2\, ds \uparrow \infty \qquad \text{as} \quad x \uparrow \infty.$$

EXAMPLE 4. Let $m(x) = x$ and $l = \infty$, so that $A(x, \gamma) = \cos \gamma x$, $B(x, \gamma) = \sin \gamma x$, and $d\Delta(\gamma)^{\cdot} = d\gamma$, and choose $d\tilde{\Delta}(\gamma) = -(\sin \gamma/\gamma)^2\, d\gamma$.

Solution. I–IV are valid for every choice of $\delta < \infty$, just as in Example 3; exactly the same arguments work. Therefore, equation (6.4) has a unique even analytic solution $a(x, \gamma)$ of the form (7.3) which is subject to the bound (7.4). It remains to solve (6.4):

$$a(x, \gamma) - \lim_{R \uparrow \infty} \int_{-R}^{R} a(x, \lambda) \frac{\sin[x(\lambda - \gamma)]}{\pi(\lambda - \gamma)} \left(\frac{\sin \lambda}{\lambda}\right)^2 d\lambda = \cos \gamma x.$$

The computation of the integral is carried out much as in Example 3 but is more complicated because of the exponential growth of the term $\sin^2 \lambda$ off the real axis. Only the simplest case, $0 \leqslant x \leqslant 1$, will be considered. For such x,

$$\begin{aligned}
&\lim_{R \uparrow \infty} \int_{-R}^{R} \frac{a(x, \lambda) - a(x, 0)}{\lambda^2} \frac{\sin[x(\lambda - \gamma)]}{\pi(\lambda - \gamma)} \frac{e^{2i\lambda} + e^{-2i\lambda} - 2}{-4} d\lambda \\
&\quad = \lim_{R \uparrow \infty} \frac{1}{2\pi} \int_{-R}^{R} \frac{a(x, \lambda) - a(x, 0)}{\lambda^2} \frac{\sin[x(\lambda - \gamma)]}{(\lambda - \gamma)} d\lambda \\
&\quad = \lim_{R \uparrow \infty} \frac{1}{2\pi} \int_{-R}^{R} \frac{a(x, \lambda) - a(x, 0)}{\lambda^2} \frac{e^{i(\lambda - \gamma)x} - e^{-i(\lambda - \gamma)x}}{2i(\lambda - \gamma)} d\lambda \\
&\quad = \frac{a(x, \gamma) - a(x, 0)}{2\gamma^2}.
\end{aligned}$$

That, together with the evaluation

$$\begin{aligned}
&\lim_{R \uparrow \infty} \int_{-R}^{R} \frac{\sin[x(\lambda - \gamma)]}{\pi(\lambda - \gamma)} \left(\frac{\sin \lambda}{\lambda}\right)^2 d\lambda \\
&\quad = \lim_{R \uparrow \infty} \int_{-R}^{R} \left[\frac{1}{\pi} \int_0^x \cos \lambda s \cos \gamma s \, ds\right] \left(\frac{\sin \lambda}{\lambda}\right)^2 d\lambda \\
&\quad = \frac{1}{\pi} \int_0^x \cos \gamma s \left[\int_{-\infty}^{\infty} \cos \lambda s \left(\frac{\sin \lambda}{\lambda}\right)^2 d\lambda\right] ds \\
&\quad = \begin{cases} \displaystyle\int_0^x \cos \gamma s \left[\frac{2 - s}{2}\right] ds & \text{for} \quad x < 2 \\ \displaystyle\int_0^2 \cos \gamma s \left[\frac{2 - s}{2}\right] ds & \text{for} \quad x \geqslant 2 \end{cases} \\
&\quad = \begin{cases} \displaystyle(2 - x) \frac{\sin \gamma x}{2\gamma} + \frac{1 - \cos \gamma x}{2\gamma^2} & \text{for} \quad x < 2 \\ \displaystyle\frac{1 - \cos 2\gamma}{2\gamma^2} = \left(\frac{\sin \gamma}{\gamma}\right)^2 & \text{for} \quad x \geqslant 2, \end{cases}
\end{aligned}$$

enables you to put the equation for a into the form

$$a(x, \gamma) = \frac{2\gamma^2}{2\gamma^2 - 1}\left\{a(x, 0)\left[(2 - x)\frac{\sin \gamma x}{2\gamma} - \frac{\cos \gamma x}{2\gamma^2}\right] + \cos \gamma x\right\}.$$

Now in order for a to be analytic, the numerator of the right-hand side must vanish for $\gamma^2 = \frac{1}{2}$. This condition enables you to evaluate

$$a(x, 0) = \left[1 + \frac{x - 2}{\sqrt{2}} \tan\left(\frac{x}{\sqrt{2}}\right)\right]^{-1}$$

for $0 \leqslant x \leqslant 1$.

8. Extensions of the Construction

The main results of this paper, Theorems 4.1 and 6.1, give an algorithm for computing the initial segment $x^{\cdot}(x)$, $m^{\cdot}(x)$, $0 \leqslant x \leqslant \delta$, of the string with principal spectral function $\Delta^{\cdot}$. It remains to check whether the constructed initial segment falls short of the full string or not. If $x^{\cdot}(\delta) + m^{\cdot}(\delta) = \infty$, then the constructed segment must coincide with the full string, in view of the fact there is a unique spectral function associated with each long string. If $x^{\cdot}(\delta) + m^{\cdot}(\delta) < \infty$, then the constructed segment may either fall short of the full string or coincide with the full string.

The purpose of the present section is to state a test for deciding which of the two alternatives is in effect in case $x^{\cdot}(\delta) + m^{\cdot}(\delta) < \infty$, and to draw some conclusions from it. The test is based on a formula of Krein (1952b); see also Dym and McKean (1976, §5.10). It is stated most conveniently for present purposes in terms of the function

$$D^{\cdot}(x, ic) = \frac{1}{\pi}\int \frac{A^{\cdot}(x, \gamma)}{\gamma^2 + c^2}\, d\Delta^{\cdot}(\gamma), \qquad c > 0, \tag{8.1}$$

and its right-hand derivative with respect to $x^{\cdot}$:

$$\begin{aligned} D^{\cdot +}(x, ic) &= \lim_{h \downarrow 0} \frac{D^{\cdot}(x + h, ic) - D^{\cdot}(x, ic)}{x^{\cdot}(x + h) - x^{\cdot}(h)} \\ &= \frac{c^2}{\pi}\int \frac{B^{\cdot}(x, \gamma)}{\gamma(\gamma^2 + c^2)}\, d\Delta^{\cdot}(\gamma) - 1. \end{aligned} \tag{8.2}$$

Amplification. $D^{\cdot}$ is the positive decreasing solution of the basic generalized second-order differential equation

$$\frac{d^2 u}{dm^{\cdot}\, dx^{\cdot}} = c^2 u, \qquad c > 0,$$

which is square summable $dm^{\cdot}$, meets the boundary conditions imposed at the right, if any, and is normalized so as to have its left derivative equal to -1 at the origin:

$$D^{\cdot}(x, ic) = D^{\cdot}(0, ic) - x^{\cdot}(x) + c^2 \int_0^x dx^{\cdot}(\xi) \int_{0-}^{\xi} D^{\cdot}(\eta, ic)\, dm^{\cdot}(\eta).$$

Formula (8.1) follows from the spectral representation for the Green function,

$$A^{\cdot}(y, ic)D^{\cdot}(x, ic) = \frac{1}{\pi} \int \frac{A^{\cdot}(y, \gamma)A^{\cdot}(x, ic)}{\gamma^2 + c^2}\, d\Delta^{\cdot}(\gamma),$$

which is valid for growth points $y \leqslant x$, upon setting $y = 0$. Formula (8.2) is deduced from (8.1) by noticing that

$$D^{\cdot +}(x, ic) = c^2 \int_{0-}^{x} D^{\cdot}(\eta, ic)\, dm^{\cdot}(\eta) - 1.$$

Krein's formula, adapted to the present circumstances, states that if assumptions I–IV of Section 3 hold for some growth point $\delta \leqslant l$ of m with $\delta + m(\delta) < \infty$, then

$$\frac{D^{\cdot}(\delta, ic)}{-D^{\cdot +}(\delta, ic)} = k_1 + \frac{1}{\pi} \int \frac{d\Delta_1(\gamma)}{\gamma^2 + c^2}, \tag{8.3}$$

where $k_1 \geqslant 0$ is a constant independent of c ($k_1 = \infty$ is permitted, only then $d\Delta_1 = 0$) and Δ_1 is the principal spectral function of the string segment which sits to the right of the point $x^{\cdot}(\delta) + k_1$. This leads in particular, to the following

TEST. *The constructed string segment $x^{\cdot}(x)$, $m^{\cdot}(x)$, $0 \leqslant x \leqslant \delta$, coincides with the full string having principal spectral function $\Delta^{\cdot}$ if and only if $-D^{\cdot}(\delta, ic)/D^{\cdot +}(\delta, ic)$ is constant independent of $c > 0$.*

In order to apply the test, it is desirable to reexpress formulas (8.1) and (8.2) for $D^{\cdot}$ and $D^{\cdot +}$, respectively, in more transparent form. This is the subject of the next lemma.

LEMMA 8.1. *Let $\delta \leqslant l$ be a growth point of m such that $\delta + m(\delta) < \infty$ and assumptions* I–IV *of Section* 3 *hold. Then*

$$D^{\cdot}(x, ic)$$
$$= \left[\frac{1}{a(x,0)}\right]\left[D(x, ic) - \frac{1}{\pi}\int a(x,\gamma) \frac{D^{+}(x, ic)A(x,\gamma) - D(x, ic)A^{+}(x,\gamma)}{\gamma^2 + c^2}\, d\tilde{\Delta}(\gamma)\right] \tag{8.4}$$

and

$$D^{\cdot +}(x, ic) = a(x, 0)\Big\{D^{+}(x, ic) + \frac{1}{\pi}\int \gamma b(x, \gamma)\frac{D^{+}(x, ic)A(x, \gamma) - D(x, ic)A^{+}(x, \gamma)}{\gamma^2 + c^2}\, d\tilde{\Delta}(\gamma) + D(x, ic)K_2(x, x)\Big\}, \tag{8.5}$$

for every point x in the interval $0 \leqslant x \leqslant \delta$.

Proof of (8.4). Substitute (4.1) and (4.3) into (8.1) to get

$$\begin{aligned} &a(x, 0)D^{\cdot}(x, ic) \\ &= \frac{1}{\pi}\int \frac{A(x, \gamma) + \int_{0-}^{x} K_1(x, s)A(s, \gamma)\, dm(s)}{\gamma^2 + c^2}\, d\Delta(\gamma) + \frac{1}{\pi}\int \frac{a(x, \gamma)}{\gamma^2 + c^2}\, d\tilde{\Delta}(\gamma) \\ &= D(x, ic) + \int_{0-}^{x} K_1(x, s)D(s, ic)\, dm(s) + \frac{1}{\pi}\int \frac{a(x, \gamma)}{\gamma^2 + c^2}\, d\tilde{\Delta}(\gamma) \\ &= D(x, ic) - \frac{1}{\pi}\int a(x, \gamma)\left[\int_{0-}^{x} A(s, \gamma)D(s, ic)\, dm(s)\right] d\tilde{\Delta}(\gamma) \\ &\quad + \frac{1}{\pi}\int \frac{a(x, \gamma)}{\gamma^2 + c^2}\, d\tilde{\Delta}(\gamma). \end{aligned}$$

The calculation is easily completed with the help of the evaluation

$$\int_{0-}^{x} A(s, \gamma)D(s, ic)\, dm(s) = \frac{D^{+}(x, ic)A(x, \gamma) - D(x, ic)A^{+}(x, \gamma) + 1}{\gamma^2 + c^2}.$$

Proof of (8.5). Substitute (4.6) and (4.7) into (8.2) to get

$$\begin{aligned} D^{\cdot +}(x, ic) &= \frac{c^2}{\pi}\int a(x, 0)\frac{B(x, \gamma) + \int_0^x K_2(x, s)B(s, \gamma)\, ds}{\gamma(\gamma^2 + c^2)}\, d\Delta(\gamma) \\ &\quad + \frac{c^2}{\pi}\int a(x, 0)\frac{b(x, \gamma)}{\gamma}\frac{d\tilde{\Delta}(\gamma)}{\gamma^2 + c^2} - 1 \\ &= a(x, 0)\Big\{D^{+}(x, ic) + 1 + \int_0^x K_2(x, s)[D^{+}(s, ic) + 1]\, ds \\ &\quad + \frac{c^2}{\pi}\int \frac{b(x, \gamma)}{\gamma}\frac{d\tilde{\Delta}(\gamma)}{\gamma^2 + c^2}\Big\} - 1. \end{aligned}$$

Now since

$$\int_0^x K_2(x, s)[D^+(s, ic) + 1]\, ds$$

$$= -\frac{1}{\pi} \int b(x, \gamma) \int_0^x B(s, \gamma)[D^+(s, ic) + 1]\, ds\, d\tilde{\Delta}(\gamma)$$

$$= -\frac{1}{\pi} \int b(x, \gamma)\left[\int_0^x B(s, \gamma) D^+(s, ic)\, ds + \frac{1 - A(x, \gamma)}{\gamma}\right] d\tilde{\Delta}(\gamma)$$

and

$$\int_0^x B(s, \gamma) D^+(s, ic)\, ds$$

$$= B(x, \gamma) D(x, ic) - \gamma \int_{0-}^x A(s, \gamma) D(s, ic)\, dm(s)$$

$$= B(x, \gamma) D(x, ic) - \gamma \left\{\frac{D^+(x, ic) A(x, \gamma) - D(x, ic) A^+(x, \gamma) + 1}{\gamma^2 + c^2}\right\}$$

it follows that

$$D^{\cdot +}(x, ic) = a(x, 0)\Biggl\{D^+(x, ic) + 1$$

$$+ \frac{1}{\pi} \int \gamma b(x, \gamma) \frac{D^+(x, ic) A(x, \gamma) - D(x, ic) A^+(x, \gamma)}{\gamma^2 + c^2}\, d\tilde{\Delta}(\gamma)$$

$$- \frac{1}{\pi} \int b(x, \gamma) B(x, \gamma) D(x, ic)\, d\tilde{\Delta}(\gamma)$$

$$+ \frac{1}{\pi} \int \frac{b(x, \gamma)}{\gamma} A(x, \gamma)\, d\tilde{\Delta}(\gamma)\Biggr\} - 1.$$

But this is the same as formula (8.5) since

$$a(x, 0)\left\{1 + \frac{1}{\pi} \int \frac{b(x, \gamma)}{\gamma} A(x, \gamma)\, d\tilde{\Delta}(\gamma)\right\}$$

$$= a(x, 0) + \frac{1}{\pi} \int A(x, \gamma) \int_{0-}^x A^{\cdot}(s, \gamma)\, dm^{\cdot}(s)\, d\tilde{\Delta}(\gamma)$$

$$= a(x, 0) - \frac{1}{\pi} \int_{0-}^x K_1(s, x) a(s-, 0)\, dm(s)$$

$$= A(x, 0) = 1,$$

as follows by (6.2) and the inversion formula (5.1).

To illustrate the application of the test consider the case of a short string lmk with principal spectral function Δ subject to a perturbation $\tilde{\Delta}$ which is such that assumptions I–IV of Section 3 hold for $\delta = l$. Then, because D meets the boundary condition

$$kD^+(l, ic) + D(l, ic) = 0$$

for all $c > 0$, it follows from (8.4) and (8.5) that

$$\frac{D^{\cdot}(l, ic)}{-D^{\cdot+}(l, ic)} = \frac{k + \dfrac{1}{\pi}\displaystyle\int a(l, \gamma)\frac{A(l, \gamma) + kA^+(l, \gamma)}{\gamma^2 + c^2}\, d\tilde{\Delta}(\gamma)}{[a(l, 0)]^2\left[1 - kK_2(l, l) + \dfrac{1}{\pi}\displaystyle\int \gamma b(l, \gamma)\frac{A(l, \gamma) + kA^+(l, \gamma)}{\gamma^2 + c^2}\, d\tilde{\Delta}(\gamma)\right]} \tag{8.6}$$

with the obvious conventions for $k = \infty$. The formula can be expressed more succinctly by defining $m(x) = m(l)$ for $l \leqslant x < l + k$ and extending the domain of definitions of A, A^+, a, and b to include this interval. For example, if $0 < k < \infty$, and $1 - kK_2(l, l) \neq 0$, then

$$\begin{aligned} A(l + k, \gamma) &= A(l, \gamma) + kA^+(l, \gamma), \\ a(l + k, 0) &= a(l, 0)[1 - kK_2(l, l)], \\ a(l + k, \gamma) &= a(l + k, 0)[a(l, \gamma)/a(l, 0)] - \gamma k b(l, \gamma), \end{aligned}$$

and

$$\frac{D^{\cdot}(l, ic)}{-D^{\cdot+}(l, ic)} = \frac{k}{a(l, 0)a(l + k, 0)} + \frac{\dfrac{1}{\pi}\displaystyle\int \frac{A^{\cdot}(l + k, \gamma)A(l + k, \gamma)}{\gamma^2 + c^2}\, d\tilde{\Delta}(\gamma)}{a(l + k, 0) + \dfrac{1}{\pi}\displaystyle\int \frac{\gamma B^{\cdot}(l, \gamma)A(l + k, \gamma)}{\gamma^2 + c^2}\, d\tilde{\Delta}(\gamma)}.$$

The new formulation of (8.6) is mentioned just to indicate the possibilities and shall neither be justified nor used in the remaining few lines of this paper. (8.6) itself permits the following interesting conclusion: *If $\tilde{\Delta}$ is a step function with jumps at (one or more of) the jump points of Δ only, then*

$$\frac{D^{\cdot}(l, ic)}{-D^{\cdot+}(l, ic)} = \frac{k}{[a(l, 0)]^2\{1 - kK_2(l, l)\}} \equiv k^{\cdot}.$$

In this instance the segment $x^{\cdot}(x)$, $m^{\cdot}(x)$, $0 \leqslant x \leqslant l$, is the full string with principal spectral function $\Delta^{\cdot}$ and the number $k^{\cdot}$ is the constant which enters into the boundary condition which is imposed at the right-hand end of the spot string. On the other hand, if $\tilde{\Delta}$ has support at points other than the jump points *of* Δ, *then the segment* $x^{\cdot}(x)$, $m^{\cdot}(x)$, $0 \leqslant x \leqslant l$, *falls short of the full string of which* $\Delta^{\cdot}$ *is the principal spectral function.*

The special case in which $\tilde{\Delta}$ has a finite number of jumps of height $q_i > 0$ placed at the points $\pm\mu_i$, $i = 1, \ldots, n$, none of which are jump points of Δ, is particularly easy to describe. In that case $-D^{\cdot}/D^{\cdot +}$ can be expanded as a continued fraction,

$$-\frac{D^{\cdot}(l, ic)}{D^{\cdot +}(l, ic)} = \beta_0 + \frac{1\,|}{|\alpha_1 c^2} + \frac{1|}{|\beta_1} + \cdots + \frac{1\,|}{|\alpha_n c^2} + \frac{1|}{|\beta_n},$$

and the augmented segment which sits to the right of $x^{\cdot}(l)$ consists of n point masses of magnitude $\alpha_1, \ldots, \alpha_n$ placed at the points $x^{\cdot}(l) + \beta_0$, $x^{\cdot}(l) + \beta_0 + \beta_1, \ldots, x^{\cdot}(l) + \beta_0 + \cdots + \beta_{n-1} = l^{\cdot}$, respectively. The last coefficient β_n intervenes in the boundary condition:

$$\beta_n D^{\cdot +}(l, ic) + D^{\cdot}(l, ic) = 0,$$

i.e., $\beta_n = k^{\cdot}$. The details of this construction will not be filled in here. A proof can be modeled upon the proof of Krein's formula given in Dym and McKean (1976, §5–19); §5.9 of the same reference may be consulted for additional information on the role of continued fractions.

References

H. Dym and H. P. McKean, "Gaussian Processes, Function Theory and the Inverse Spectral Problem," Academic Press, New York, 1976.

I. M. Gel'fand and B. M. Levitan, On the determination of a differential equation from its spectral function, *Izv. Akad. Nauk SSSR* **15** (1951), 399–360; English transl. *Amer. Math. Soc. Transl.* **1**(2) (1955), 253–304.

I. C. Gohberg and M. G. Krein, "Theory and Applications of Volterra Operators in Hilbert Space," Transl. of Monographs No. 24, Amer. Math. Soc., Providence, Rhode Island, 1970.

I. S. Kac and M. G. Krein, On the spectral functions of the string, *in* [Russian translation of] F. V. Atkinson, "Discrete and Continuous boundary Value Problems," Supplement II, pp. 648–737, Mir, Moscow, 1968; English transl. *Amer. Math. Soc. Transl.* **103**(2) (1974), 19–102.

M. G. Krein, On inverse problems for a non-homogeneous cord, *Dokl. Akad. Nauk SSSR* **82** (1952a), 669–672.

M. G. Krein, On a generalization of an investigation of Stieltjes, *Dokl. Akad. Nauk SSSR* **87** (1952b), 881–894.

M. G. Krein, On some cases of the effective determination of the density of a non-homogeneous string from its spectral function, *Dokl. Akad. Nauk SSSR* **93** (1953a), 617–620.

M. G. Krein, On a fundamental approximation problem in the theory of extrapolation and filtration of stationary random processes. *Dokl. Akad. Nauk SSSR* **94** (1953b), 13–16; English transl. *Amer. Math. Soc. Selected Transl. Math. Statist. Probability* **4** (1964), 127–131.

M. G. Krein, On a method for the effective solution of the inverse boundary-value problem, *Dokl. Akad. Nauk SSSR* **94** (1954a), 987–990.

M. G. KREIN, On integral equations, which generate differential equations of second order, *Dokl. Akad. Nauk. SSSR* **97** (1954b), 21–24.

B. M. LEVITAN AND M. G. GASYMOV, Determination of a differential equation by two of its spectra, *Uspekhi Mat. Nauk.* **19**(2) (1964), 3–64; English transl. *Russian Math. Surveys* **19**(2) (1964), 1–63.

AMS (MOS) subject classifications: 34B25, 45B05, 47A10, 47A55, 47E05.

TOPICS IN FUNCTIONAL ANALYSIS
ADVANCES IN MATHEMATICS SUPPLEMENTARY STUDIES, VOL. 3

Spectral Analysis of Families of Operator Polynomials and a Generalized Vandermonde Matrix, 1. The Finite-Dimensional Case

I. GOHBERG†

Department of Mathematical Sciences
Tel Aviv University
Tel Aviv, Israel

AND

Department of Pure Mathematics
The Weizmann Institute of Science
Rehovot, Israel

M. A. KAASHOEK

Wiskundig Seminarium der Vrije Universiteit
Amsterdam, The Netherlands

AND

L. RODMAN

Department of Mathematical Sciences
Tel Aviv University
Tel Aviv, Israel

DEDICATED TO M. G. KREIN ON THE OCCASION OF HIS SEVENTIETH BIRTHDAY WITH ADMIRATION AND GRATITUDE

I. INTRODUCTION AND PRELIMINARIES

1. *Introduction*

For a family of $n \times n$ matrix polynomials, problems concerning the existence of common (left) multiples and common (right) divisors of different types and their corresponding characteristics (degrees, number of common

† This paper was written while the first author was a visiting professor of the Vrije Universiteit at Amsterdam.

ISBN 0-12-287150-2

roots, multiplicities, etc.) arise naturally in different contexts. In the case of scalar polynomials there exist basically two ways to deal with these problems: (1) an effective method of direct computation based, for example, on the Euclid algorithm, (2) a more indirect procedure, where all computations are made through the roots of the polynomials. For matrix polynomials a somewhat analogous version of the first method is available (e.g., see [8, §23]). One of our aims is to develop the analog of the second method.

To do this it is necessary to consider not only the scalar roots (that is, the eigenvalues) of the matrix polynomials and their multiplicities, but also the chains of generalized eigenvectors arranged in a canonical way. In other words, it is necessary to take into account the full Jordan structure of the matrix polynomials involved. It turns out that in this way solutions of the problems referred to above can be obtained through explicit operations of geometrical character on the Jordan structures of the different matrix polynomials, which make the internal relations very transparent. In particular, with our method it is possible to solve the "inverse" problem of constructing a monic matrix polynomial with a given arbitrary system of complex numbers and vectors as eigenvalues, eigenvectors, and generalized eigenvectors.

This paper can be viewed as a continuation of the papers [3, 4], in which, on the basis of a theory of canonical forms, the inverse problem mentioned above has been solved for so-called canonical systems. Further, we use the theory of polynomial divisors, as developed in [3, 4], to see to what extent the properties of a common multiple are determined by those of the original polynomial. The fact that polynomial divisors may be characterized in terms of invariant subspaces of certain linear operators [3, Theorem 8] allows us to deal with these problems in terms of spectral analysis.

One of our main tools is a generalized Vandermonde matrix for a finite family of monic operator polynomials. It turns out that after appropriate modifications the Vandermonde matrix epitomizes all the information about the family. In this paper (Part I of a two part work) we show this for those problems which are in the framework of monic matrix polynomials. (Part II will deal with the infinite-dimensional case. When properly formulated, most results of this paper also hold for the nonmonic case. This will be the topic of a future publication.)

For monic operator polynomials of degree one the notion of a generalized Vandermonde matrix has been introduced by Markus and Mereutsa in [9], where it was used in a very effective way to study operator polynomials in terms of their roots. The first results in this direction for two roots of a quadratic operator polynomial were obtained by Kreĭn and Langer in their fundamental paper [5].

The present paper is divided into three sections and thirteen subsections. Section I consists of Introduction and Preliminaries. In Section II the generalized Vandermonde matrix $V(L_1, \ldots, L_r)$ is introduced and studied for a finite family of right divisors $L_1, \ldots, L_r$ of a given monic matrix polynomial L. Our results show that to a large extent L is determined by $L_1, \ldots, L_r$ whenever $V(L_1, \ldots, L_r)$ is invertible. Invertibility tests for $V(L_1, \ldots, L_r)$ are given in terms of the supporting subspaces of $L_1, \ldots, L_r$ and in terms of spectral conditions. In Section III the main problem is to determine the kernel of a modified Vandermonde matrix. This leads to the study of certain infinite systems of equations. The results obtained here are used in the last section to construct for an arbitrary finite family of monic matrix polynomials $L_1, \ldots, L_r$ monic common left multiples of minimal degree. Further, if the family $L_1, \ldots, L_r$ satisfies a certain minimality condition, we show that a number of monic matrix polynomials can be constructed such that the extended family has an invertible Vandermonde matrix.

2. *Preliminaries*

In this subsection we collect together some definitions, notations, and results concerning operator polynomials, which will be relevant throughout this paper.

First some notation. We shall write $\operatorname{col}(T_j)_{j=1}^{\ell}$ to denote the column operator matrix

$$\begin{bmatrix} T_1 \\ T_2 \\ \vdots \\ T_\ell \end{bmatrix}.$$

Further, we shall write $\operatorname{row}(S_j)_{j=1}^{\ell}$ for the one row operator matrix $(S_1 S_2 \cdots S_\ell)$. Finally, $\operatorname{diag}(U_j)_{j=1}^{\ell}$ will denote the square block diagonal operator matrix whose main diagonal is given by $U_1, U_2, \ldots, U_\ell$. Sometimes the latter operator will also be denoted by $U_1 \oplus U_2 \oplus \cdots \oplus U_\ell$.

Instead of $n \times n$ matrices we shall deal with linear operators on $\mathbb{C}_n$. So the polynomials we shall be considering will be of the form

$$L(\lambda) = A_0 + A_1\lambda + \cdots + A_\ell\lambda^\ell, \tag{1}$$

where the coefficients $A_0, \ldots, A_\ell$ are linear operators on $\mathbb{C}_n$. If in (1) the operator A_ℓ is nonzero, then ℓ is said to be the *degree* of L. If $A_\ell = I$, the identity operator on $\mathbb{C}_n$, then L is called *monic*.

A pair of linear operators $X:\mathbb{C}_{n\ell} \to \mathbb{C}_n$ and $T:\mathbb{C}_{n\ell} \to \mathbb{C}_{n\ell}$ will be called a *standard pair* of *degree* ℓ if the map

$$A = \mathrm{col}(XT^{j-1})_{j=1}^{\ell}:\mathbb{C}_{n\ell} \to \mathbb{C}_{n\ell}$$

is invertible. If L is a monic operator polynomial on $\mathbb{C}_n$ of degree ℓ, then one can find (see [3, §3]) a standard pair (X, T) of degree ℓ such that

$$L(\lambda) = \lambda^{\ell} I - XT^{\ell}(V_1 + V_2\lambda + \cdots + V_{\ell}\lambda^{\ell-1}),$$

where

$$\mathrm{row}(V_j)_{j=1}^{\ell} = [\mathrm{col}(XT^{j-1})_{j=1}^{\ell}]^{-1}.$$

In that case we say that L is *represented* by (X, T). For given L, to get a representing pair one may take T to be the Jordan matrix associated with L and for X one may take the $n \times n\ell$ matrix of eigenvectors and generalized eigenvectors of L organized in an appropriate way. But, as the next proposition shows, many other choices of X and T are possible too.

PROPOSITION 2.1. *Two standard pairs* (X, T) *and* (X_1, T_1) *represent the same monic operator polynomial* L *on* $\mathbb{C}_n$ *if and only if they have the same degree,* ℓ *say, and there exists a bijective linear operator* S *on* $\mathbb{C}_{n\ell}$ *such that*

$$X_1 = XS^{-1}, \qquad T_1 = STS^{-1}. \tag{2}$$

Proof. Suppose that (X, T) and (X_1, T_1) have the same degree ℓ, and let S be as in formula (2). Then

$$\begin{bmatrix} X_1 \\ X_1T_1 \\ \vdots \\ X_1T_1^{\ell-1} \end{bmatrix} = \begin{bmatrix} XS^{-1} \\ XS^{-1}(STS^{-1}) \\ \vdots \\ XS^{-1}(STS^{-1})^{\ell-1} \end{bmatrix} = \begin{bmatrix} X_1 \\ X_1T_1 \\ \vdots \\ X_1T_1^{\ell-1} \end{bmatrix} S^{-1}. \tag{3}$$

So if $[\mathrm{col}(XT^{j-1})_{j=1}^{\ell}]^{-1} = \mathrm{row}(V_j)_{j=1}^{\ell}$, then the inverse of the left-hand side of (3) is equal to $\mathrm{row}(SV_j)_{j=1}^{\ell}$. As

$$X_1T_1^{\ell}(SV_j) = XS^{-1}(STS^{-1})^{\ell}SV_j = XT^{\ell}V_j$$

for $1 \leqslant j \leqslant \ell$, it follows that the pairs (X, T) and (X_1, T_1) represent the same polynomial.

Conversely, suppose that (X, T) and (X_1, T_1) represent the same polynomial L. Let ℓ be the degree of L. Then the degrees of (X, T) and (X_1, T_1) are both equal to ℓ and the operators

$$A = \mathrm{col}(XT^{j-1})_{j=1}^{\ell}, \qquad A_1 = \mathrm{col}(X_1T_1^{j-1})_{j=1}^{\ell}$$

are invertible. As (X, T), as well as (X_1, T_1), represents L, we have $AT = CA$ and $A_1 T_1 = CA_1$, where C is the first companion operator associated with L (see [3], §2.2). Hence

$$T_1 = A_1^{-1} C A_1 = A_1^{-1} A T A^{-1} A_1 = (A_1^{-1} A) T (A_1^{-1} A)^{-1}.$$

Put $S = A_1^{-1} A$. Then S is a bijective linear operator on $\mathbb{C}_{n\ell}$, $T_1 = STS^{-1}$, and from $A_1 = AS^{-1}$ it follows that $X_1 = XS^{-1}$. This completes the proof.

Let $L(\lambda) = A_0 + A_1\lambda + \cdots + A_{\ell-1}\lambda^{\ell-1} + I\lambda^\ell$ be represented by the standard pair (X, T), and let

$$\operatorname{row}(V_j)_{j=1}^{\ell} = [\operatorname{col}(XT^{j-1})_{j=1}^{\ell}]^{-1}.$$

From this formula it is clear that

$$XT^\alpha V_\beta = \delta_{\alpha,\beta-1} I \qquad (0 \leqslant \alpha \leqslant \ell - 1, 1 \leqslant \beta \leqslant \ell). \tag{4}$$

(Here, as well as in the sequel, δ_{ij} denotes the Kronecker delta.) Further, as L is represented by (X, T), we have

$$XT^\ell V_\beta = -A_{\beta-1} \qquad (1 \leqslant \beta \leqslant \ell). \tag{5}$$

For later purposes (see the definition of the Vandermonde operator in Subsection 3) it is important to know that for $\alpha > \ell$ the operator $XT^\alpha V_\beta$ can be expressed in terms of the coefficients of L too.

THEOREM 2.2. *Let $L(\lambda) = A_0 + A_1\lambda + \cdots + A_{\ell-1}\lambda^{\ell-1} + I\lambda^\ell$ be represented by the standard pair (X, T), and let*

$$\operatorname{row}(V_j)_{j=1}^{\ell} = [\operatorname{col}(XT^{j-1})_{j=1}^{\ell}]^{-1}.$$

Then for $\rho \geqslant 1$ and $1 \leqslant \beta \leqslant \ell$, we have

$$XT^{\ell+\rho} V_\beta = \sum_{k=1}^{\rho} \left[\sum_{q=1}^{k} \sum_{\substack{i_1+\cdots+i_q=k \\ i_j>0}} \prod_{j=1}^{q} (-A_{\ell-i_j}) \right] \cdot (-A_{\beta+k-\rho-1}) + (-A_{\beta-\rho-1}),$$

where, by definition, $A_i = 0$ for $i < 0$.

Proof. Put $V_0 = 0$. From formula (5) and [3], §4 Lemma we know that

$$XT^{\alpha+1} V_\beta = (XT^\alpha V_\ell)(-A_{\beta-1}) + XT^\alpha V_{\beta-1} \tag{6}$$

for $1 \leqslant \beta \leqslant \ell$ and $\alpha \geqslant 0$. If in (6) we take $\alpha = \ell$ and use formula (5) once again, we obtain

$$XT^{\ell+1} V_\beta = (-A_{\ell-1})(-A_{\beta-1}) + (-A_{\beta-2}).$$

It follows that the lemma is proved for $\beta = 1$ and $1 \leqslant \beta \leqslant \ell$.

Next we observe that

$$B = \sum_{k=1}^{\rho} \left[\sum_{q=1}^{k} \sum_{\substack{i_1 + \cdots + i_q = k \\ i_j > 0}} \prod_{j=1}^{q} (-A_{\ell - i_j}) \right] \cdot (-A_{\ell + k - \rho - 1})$$

$$= \sum_{k=1}^{\rho} \sum_{q=1}^{k} \sum_{\substack{i_1 + \cdots + i_q + i_{q+1} = \rho + 1 \\ i_{q+1} = (\rho+1) - k \\ i_j > 0}} \prod_{j=1}^{q+1} (-A_{\ell - i_j}).$$

By interchanging the order of the first two summations in the last part of this identity and replacing $q + 1$ by q, we see that

$$B = \left[\sum_{q=1}^{\rho+1} \sum_{\substack{i_1 + \cdots + i_q = \rho + 1 \\ i_j > 0}} \prod_{j=1}^{q} (-A_{\ell - i_j}) \right] - (-A_{\ell - \rho - 1}). \tag{7}$$

The proof is completed by induction on ρ. Suppose the lemma is true for some $\rho \geqslant 1$ and $1 \leqslant \beta \leqslant \ell$. Let B be as above. Using formula (6), we have $XT^{\ell+\rho+1}V_\beta$ is equal to

$$B(-A_{\beta-1}) + \sum_{k=1}^{\rho} \left[\sum_{q=1}^{k} \sum_{\substack{i_1 + \cdots + i_q = k \\ i_j > 0}} \prod_{j=1}^{q} (-A_{\ell - i_j}) \right] \cdot (-A_{\beta + k - \rho - 2}) + (A_{\beta - \rho - 2}).$$

Inserting the expression for B given by (7), we obtain the desired formula for $XT^{\ell+\rho+1}V_\beta$.

Note that in the previous theorem, as well as in its proof, the product $(-A_{\ell-i_1})(-A_{\ell-i_2}) \cdots (-A_{\ell-i_q})$ is denoted by $\prod_{j=1}^{q} (-A_{\ell-i_j})$.

II. The Generalized Vandermonde Matrix

3. *The Vandermonde Operator for a Family of Polynomials*

Let $L_1, \ldots, L_r$ be monic operator polynomials on $\mathbb{C}_n$ with degrees $\ell_1, \ldots, \ell_r$, respectively. For each j let (X_j, T_j) be a standard pair representing L_j, i.e.,

$$L_j(\lambda) = \lambda^{\ell_j} I - X_j T_j^{\ell_j} (U_{j1} + \cdots + U_{j\ell_j} \lambda^{\ell_j - 1}),$$

where

$$U_j = \operatorname{row}(U_{ji})_{i=1}^{\ell_j} = [\operatorname{col}(X_j T_j^{i-1})_{i=1}^{\ell_j}]^{-1}. \tag{1}$$

Put $m = \ell_1 + \ell_2 + \cdots + \ell_2$. By definition the *Vandermonde operator* of the family $\{L_1, \ldots, L_r\}$ is the operator $V = V(L_1, \ldots, L_r)$ given by the following operator matrix:

$$V = \begin{bmatrix} X_1U_1 & X_2U_2 & \cdots & X_rU_r \\ X_1T_1U_1 & X_2T_2U_2 & \cdots & X_rT_rU_r \\ \vdots & \vdots & & \vdots \\ X_1T_1^{m-1}U_1 & X_2T_2^{m-1}U_2 & \cdots & X_rT_r^{m-1}U_r \end{bmatrix}. \tag{2}$$

Here the operators $U_1, \ldots, U_r$ are as in formula (1). Note that V is an operator on $\mathbb{C}_{nm}$. The definition of the Vandermonde operator does not depend on the special choice of the standard pairs $(X_1, T_1), \ldots, (X_r, T_r)$. Indeed, if $(\tilde{X}_j, \tilde{T}_j)$ is a second pair representing L_j, then there exists an invertible operator S_j on $\mathbb{C}_{nk_j}$ (cf. Proposition 2.1) such that $\tilde{X}_j = X_jS_j^{-1}$, $\tilde{T}_j = S_jT_jS_j^{-1}$, and hence in that case

$$\tilde{U}_j = \operatorname{col}(\tilde{X}_j\tilde{T}_j^{i-1})_{i=1}^{k_j} = SU_j.$$

It follows that for each $\alpha \geqslant 0$ we have $XT_j^{\alpha}U_j = \tilde{X}\tilde{T}_j^{\alpha}\tilde{U}_j$.

There is another way to see that the definition of the Vandermonde operator $V(L_1, \ldots, L_r)$ does not depend on the choice of the representing pairs. Note that jth column V_j in (2) can be written as

$$V_j = \begin{bmatrix} X_jU_{j1} & X_jU_{j2} & \cdots & X_jU_{jk_j} \\ X_jT_jU_{j1} & X_jT_jU_{j2} & \cdots & X_jT_jU_{jk_j} \\ \vdots & \vdots & & \vdots \\ X_jT_j^{m-1}U_{j1} & X_jT_j^{m-1}U_{j2} & \cdots & X_jT_j^{m-1}U_{jk_j} \end{bmatrix}.$$

Now by Theorem 2.2 and formulas (4) and (5) in Subsection 2, each operator $X_jT_j^{\alpha}U_{j\beta}$ can be expressed in terms of the coefficients of L_j, and this expression does not depend on the special choice of the representing pairs (X_j, T_j). It follows that $V(L_1, \ldots, L_r)$ can be written as an $m \times m$ matrix whose entries are operators on $\mathbb{C}_n$ which are uniquely determined by the coefficients of $L_1, \ldots, L_r$. To illustrate this fact we compute the Vandermonde operator for three scalar polynomials of degree 2.

EXAMPLE 3.1. Consider the scalar polynomials $\lambda^2 + a_1\lambda + a_0$, $\lambda^2 + b_1\lambda + b_0$, and $\lambda^2 + c_1\lambda + c_0$. In this case the Vandermonde operator V acts on $\mathbb{C}_6$ and the matrix of V with respect to the canonical basis of $\mathbb{C}_6$ is equal to

$$\begin{bmatrix} 1 & 0 & 1 & 0 & 1 & 0 \\ 0 & 1 & 0 & 1 & 0 & 1 \\ -a_0 & -a_1 & -b_0 & -b_1 & -c_0 & -c_1 \\ a_1a_0 & -a_0 + a_1^2 & b_1b_0 & -b_0 + b_1^2 & c_1c_0 & -c_0 + c_1^2 \\ a_0^2 - a_1^2a_0 & 2a_1a_0 - a_1^3 & b_0^2 - b_1^2b_0 & 2b_1b_0 - b_1^3 & c_0^2 - c_1^2c_0 & 2c_1c_0 - c_1^3 \\ -2a_1a_0^2 - a_1^3a_0 & a_0^2 - 3a_1^2a_0 + a_1^4 & -2b_1b_0^2 - b_1^3b_0 & b_0^2 - 3b_1^2b_0 + b_1^4 & -2c_1c_0^2 - c_1^3c_0 & c_0^2 - 3c_1^2c_0 + c_1^4 \end{bmatrix}$$

To give some further motivation for the definition of the Vandermonde operator we consider two special cases. First we assume that all operator polynomials $L_1, \ldots, L_r$ have degree 1, i.e., $L_j(\lambda) = \lambda I - T_j$ for $j = 1, \ldots, r$. In this case

$$V(L_1, \ldots, L_r) = (T_j^{i-1})_{i,j=1}^r. \tag{3}$$

To see this one can use the method described in the paragraph preceding the above example. Reversing the order of the rows in the matrix of formula (3), we obtain the Vandermonde operator of the operators $T_1, \ldots, T_r$, which has been defined and studied by Markus and Mereutsa in [9]. See also [1].

Next we assume that $n = 1$, i.e., all polynomials $L_1, \ldots, L_r$ are scalar polynomials. In that case we may suppose that for each j the polynomial L_j is given by $L_j(\lambda) = \prod_{i=1}^{k_j} (\lambda - \lambda_{ji})$. Then

$$\det[V(L_1, \ldots, L_r)] = \prod_{1 \leqslant j_1 < j_2 \leqslant r} (\lambda_{j_2 i_2} - \lambda_{j_1 i_1}).$$

Here the product is taken over all possible pairs (j_1, i_1) and (j_2, i_2) such that $j_1 < j_2$. This result will appear as a special case of Theorem 4.3 (see Corollary 4.4).

One of the topics of this paper is the analysis of common left multiples of a given family $L_1, \ldots, L_r$ of monic operator polynomials. Recall that a monic operator polynomial L is said to be a *common left multiple* of $L_1, \ldots, L_r$ whenever $L_1, \ldots, L_r$ are right divisors of L. If the Vandermonde operator $V(L_1, \ldots, L_r)$ is invertible, then the next theorem guarantees the existence (and its proof describes the structure) of a common left multiple of minimal degree of $L_1, \ldots, L_r$.

THEOREM 3.2. *Let $L_1, \ldots, L_r$ be monic operator polynomials on $\mathbb{C}_n$ with degrees $k_1, \ldots, k_r$, respectively. Suppose that the Vandermonde operator $V(L_1, \ldots, L_r)$ is invertible. Then there exists a monic operator polynomial L of degree $m = k_1 + \cdots + k_r$ such that $L_1, \ldots, L_r$ are right divisors of L.*

Proof. Let $(X_1, T_1), \ldots, (X_r, T_r)$ be standard pairs representing $L_1, \ldots, L_r$, respectively. Define operators $X: \mathbb{C}_{nm} \to \mathbb{C}_n$ and $T: \mathbb{C}_{nm} \to \mathbb{C}_{nm}$ by setting $X = \operatorname{row}(X_j)_{j=1}^r$ and $T = \operatorname{diag}(T_j)_{j=1}^r$. Note that $XT^i = \operatorname{row}(X_j T_j^i)_{j=1}^r$ for $i \geqslant 0$. Thus

$$\operatorname{col}(XT^{i-1})_{i=1}^m = V(L_1, \ldots, L_r) \operatorname{diag}(U_j^{-1})_{j=1}^r,$$

where $U_1, U_2, \ldots, U_r$ are defined by formula (1). It follows that (X, T) is a standard pair. Let L be the monic operator polynomial represented by (X, T). Note that the degree of L is equal to $m = k_1 + k_2 + \cdots + k_r$. We shall prove that L has the right properties.

For each j, let τ_j be the canonical embedding of $\mathbb{C}_{nk_j}$ into the space $\mathbb{C}_{nm} = \mathbb{C}_{nk_1} \oplus \cdots \oplus \mathbb{C}_{nk_j} \oplus \cdots \oplus \mathbb{C}_{nk_r}$. Put $\mathscr{M}_j = \operatorname{Im} \tau_j$. Then $\mathscr{M}_j$ is an nk_j-dimensional T-invariant subspace of $\mathbb{C}_{nm}$ and the map

$$[\operatorname{col}(XT^{i-1})_{i=1}^{k_j}] \circ \tau_j = \operatorname{col}(X_j T_j^{i-1})_{i=1}^{k_j}$$

is invertible. Thus $\mathscr{M}_j$ is a supporting subspace for L. Note that

$$\{[\operatorname{col}(XT^{i-1})_{i=1}^{k_j}]|_{\mathscr{M}_j}\}^{-1} = \tau_j[\operatorname{col}(X_j T_j^{i-1})_{i=1}^{k_j}]^{-1} = \tau_j \operatorname{row}(U_{ji})_{i=1}^{k_j}.$$

Thus, by Theorem 8 in [3], the operator polynomial

$$S_j(\lambda) = \lambda^{k_j} I - XT^{k_j}(\tau_j U_{j1} + \cdots + \tau_j U_{jk_j} \lambda^{k_j - 1})$$

is a right divisor of L. As $T^\alpha \tau_j = T_j^\alpha$ for $\alpha \geqslant 0$, we see that $L_j = S_j$, and the theorem is proved.

Remark 3.3. In Subsection 4 we shall prove that in the previous theorem the operator polynomial L is uniquely determined by $L_1, \ldots, L_r$ (cf. Theorem 4.1).

Without the invertibility condition on $V(L_1, \ldots, L_r)$, Theorem 3.2 does not hold (cf. Remark 1.1 in [9] and Subsection 13). On the other hand, in Subsection 13 it will be proved that any finite family of monic operator polynomials $L_1, \ldots, L_r$ on $\mathbb{C}_n$ has a monic common left multiple L. However, the degree of L may be strictly larger than $\sum_j \operatorname{degree}(L_j)$.

Theorem 3.2 may be viewed as a special case of a more general statement about division with remainders. Let $L_1, \ldots, L_r$ be monic operator polynomials on $\mathbb{C}_n$ with degrees $k_1, \ldots, k_2$, respectively. Further, for each j let R_j be an operator polynomial on $\mathbb{C}_n$ of degree at most $k_j - 1$. Now one may ask the question whether there exists a monic operator polynomial L on $\mathbb{C}_n$ such that L after division on the right by L_j yields R_j as a remainder. If all $R_1, \ldots, R_r$ are zero, then this question reduces to the problem concerning the existence of a monic common left multiple for $L_1, \ldots, L_r$. In case $L_1, \ldots, L_r$ all have degree 1 the question is related to the interpolation problem discussed by Dennis *et al.* [1, §5]. The above question has an affirmative answer if the Vandermonde operator $V(L_1, \ldots, L_r)$ is invertible. In fact, in that case one can use the arguments employed in the proof of Theorem 3.2 to show that there exists a unique L of degree $k_1 + \cdots + k_r$ with the desired properties. The proof of this and related results will appear elsewhere.

To define the Vandermonde operator $V(L_1, \ldots, L_r)$ we have used the right standard forms of the operator polynomials $L_1, \ldots, L_r$. Using left standard forms, the left Vandermonde operator V_{left} of the family $L_1, \ldots, L_r$

can be introduced. Indeed, suppose that for each j

$$L_j(\lambda) = \lambda^{k_j} I - (Z_{j1} + \cdots + Z_{jk_j}\lambda^{k_j - 1})T_j^{k_j} Y_j \tag{4}$$

is written in left standard form (cf. [3, §3]), i.e.,

$$Z_j = \mathrm{col}(Z_{ji})_{i=1}^{k_j} = [\mathrm{row}(T^{i-1} Y_j)_{i=1}^{k_j}]^{-1}.$$

Then the left Vandermonde operator of the family $L_1, \ldots, L_r$ is given by the following operator matrix:

$$V_{\mathrm{left}} = (Z_j T_j^{i-1} Y_j)_{i=1,\,j=1}^{m\quad r}.$$

Here, as before, $m = k_1 + \cdots + k_r$. If we take the transpose of both sides of (4), then we obtain a representation of L_j^T in right standard form. From this observation it follows easily that

$$V_{\mathrm{left}} = V(L_1^T, \ldots, L_r^T)^T,$$

and hence V_{left} is connected with left divisors in exactly the same way as $V(L_1, \ldots, L_r)$ is connected with right divisors.

4. *The Vandermonde Operator for Right Divisors* (*Case* $\ell = m$)

Throughout this subsection $L, L_1, \ldots, L_r$ are monic operator polynomials on $\mathbb{C}_n$ with degrees $\ell, k_1, \ldots, k_r$, respectively. Further, we suppose that for each j the polynomial L_j is a right divisor of L. Write L in right standard form, i.e.,

$$L(\lambda) = \lambda^{\ell} I - X T^{\ell}(W_1 + W_2\lambda + \cdots + W_{\ell}\lambda^{\ell-1}),$$

where

$$\mathrm{row}(W_j)_{j=1}^{\ell} = [\mathrm{col}(X T^{i-1})_{i=1}^{\ell}]^{-1}. \tag{1}$$

Let $(X_1, T_1), \ldots, (X_r, T_r)$ be standard pairs representing $L_1, \ldots, L_r$, respectively. For each j define

$$K_j = [\mathrm{row}(W_j)_{j=1}^{\ell}] \cdot [\mathrm{col}(X_j T_j^{i-1})_{i=1}^{\ell}] : \mathbb{C}_{nk_j} \to \mathbb{C}_{n\ell},$$

and put $\mathcal{M}_j = \mathrm{Im}\, K_j$. The subspace $\mathcal{M}_j$ is the so-called supporting subspace for the divisor L_j. We recall some facts about K_j and $\mathcal{M}_j$ (see [3, Section 5]):

(i) The map K_j is left invertible.
(ii) $\mathcal{M}_j$ is invariant under T and $TK_j = K_j T_j$.
(iii) $X T^{\alpha} K_j = X_j T_j^{\alpha}$ for $\alpha = 0, 1, 2, \ldots$.
(iv) The map

$$[\mathrm{col}(X T^{i-1})_{i=1}^{k_j}]|_{\mathcal{M}_j} : \mathcal{M}_j \to \mathbb{C}_{nk_j}$$

is invertible and its inverse is given by K_jU_j, where U_j is as in formula (1) of Subsection 3.

Throughout this subsection we shall use the above notation without further explanation. Observe that by property (iii) mentioned above the Vandermonde operator admits the following factorization:

$$V(L_1, \ldots, L_r) = [\operatorname{col}(XT^{i-1})_{i=1}^{m}] \cdot [\operatorname{row}(K_j)_{j=1}^{r}] \cdot \operatorname{diag}(U_j)_{j=1}^{r}. \qquad (2)$$

Here $m = k_1 + k_2 + \cdots + k_r$. We shall first deal with the case that $m = \ell$.

THEOREM 4.1. *Let $L_1, \ldots, L_r$ be right divisors of* L, and suppose that $\ell = k_1 + \cdots + k_r$. *Then*

$$CV(L_1, \ldots, L_r) = V(L_1, \ldots, L_r) \cdot \operatorname{diag}(C_j)_{j=1}^{r},$$

where $C, C_1, \ldots, C_r$ are the first companion operators of $L, L_1, \ldots, L_r$, respectively. In particular, if $V(L_1, \ldots, L_r)$ is invertible, then the first companion operator of L is similar to $C_1 \oplus \cdots \oplus C_r$ and L is uniquely determined by $L_1, \ldots, L_r$.

Proof. Let $W = \operatorname{row}(W_j)_{j=1}^{\ell}$ be given by formula (1). We know that $W^{-1}TW = C$ (see [3, §2.2]). Similarly,

$$U_j^{-1}T_jU_j = C_j \qquad (1 \leqslant j \leqslant r). \qquad (3)$$

Using formula (2) it follows that $CV(L_1, \ldots, L_r)$ is equal to

$$\begin{aligned} &W^{-1}TW[\operatorname{col}(XT^{i-1})_{i=1}^{\ell}] \cdot [\operatorname{row}(K_j)_{j=1}^{r}] \cdot \operatorname{diag}(U_j)_{j=1}^{r} \\ &= [\operatorname{col}(XT^{i})_{i=1}^{\ell}] \cdot [\operatorname{row}(K_j)_{j=1}^{r}] \cdot \operatorname{diag}(U_j)_{j=1}^{r} \\ &= [(X_jT_j^{\,i})_{i=1,\,j=1}^{\ell\quad r}] \cdot \operatorname{diag}(U_j)_{j=1}^{r} \\ &= [(X_jT_j^{i-1})_{i=1,\,j=1}^{\ell\quad r}] \cdot [\operatorname{diag}(T_j)_{j=1}^{r}] \cdot \operatorname{diag}(U_j)_{j=1}^{r} \\ &= V(L_1, \ldots, L_r) \cdot [\operatorname{diag}(U_j^{-1})_{j=1}^{r}] \cdot [\operatorname{diag}(T_j)_{j=1}^{r}] \cdot \operatorname{diag}(U_j)_{j=1}^{r} \\ &= V(L_1, \ldots, L_r) \cdot \operatorname{diag}(C_j)_{j=1}^{r}, \end{aligned}$$

where the last equality follows from formula (3). This proves the first part of the theorem. The second part is an immediate consequence of the first part and the fact that $V(L_1, \ldots, L_r)$ is uniquely determined by $L_1, \ldots, L_r$.

As $TK_j = K_jT_j$ for each j, we have

$$T \cdot \operatorname{row}(K_j)_{j=1}^{r} = [\operatorname{row}(K_j)_{j=1}^{r}] \cdot \operatorname{diag}(T_j)_{j=1}^{r}. \qquad (4)$$

If the map $\Omega = \operatorname{row}(K_j)_{j=1}^{r} : \mathbb{C}_{nm} \to \mathbb{C}_{n\ell}$ is invertible, then formula (4) can be used to show that T and $T_1 \oplus \cdots \oplus T_r$ are similar. The proof of the next theorem shows that Ω is invertible is equivalent to the requirement that $\ell = k_1 + \cdots + k_r$ and $V(L_1, \ldots, L_r)$ is invertible.

THEOREM 4.2. *Let $L_1, \ldots, L_r$ be right divisors of L, and suppose that $\ell = k_1 + \cdots + k_r$. Then*

$$\operatorname{rank} V(L_1, \ldots, L_r) = \dim\{\mathcal{M}_1 + \cdots + \mathcal{M}_r\}.$$

In particular, $V(L_1, \ldots, L_r)$ is invertible if and only if $\mathbb{C}_{n\ell} = \mathcal{M}_1 \oplus \cdots \oplus \mathcal{M}_r$.

Proof. As $m = k_1 + \cdots + k_r = \ell$, the first factor in the right-hand side of formula (2) is invertible. By definition, the operators $U_1, \ldots, U_r$ are invertible. Hence, the third factor in the right hand side of (2) is invertible too. This shows that rank $V(L_1, \ldots, L_r)$ is equal to rank Ω, where $\Omega = \operatorname{row}(K_j)_{j=1}^r$. But $\operatorname{Im} \Omega = \mathcal{M}_1 + \cdots + \mathcal{M}_r$, and hence the first part of the theorem is proved.

To prove the second part, note that (by what has just been proved) $V(L_1, \ldots, L_r)$ is invertible if and only if $\mathbb{C}_{n\ell} = \mathcal{M}_1 + \cdots + \mathcal{M}_r$. As $\dim \mathcal{M}_j = nk_j$ for $1 \leqslant j \leqslant r$ and $\ell = k_1 + \cdots + k_r$, it follows that $\mathbb{C}_{n\ell}$ must be the direct sum of the supporting subspaces $\mathcal{M}_1, \ldots, \mathcal{M}_r$.

Let $L_1, \ldots, L_r$ be right divisors of L as before, and suppose that $\ell = m = k_1 + \cdots + k_r$. Let $W = \operatorname{row}(W_j)_{j=1}^{\ell}$ be as in formula (1). Using the expression for $V(L_1, \ldots, L_r)$ given in formula (2), we see that

$$WV(L_1, \ldots, L_r) = [\operatorname{row}(K_j)_{j=1}^r] \cdot \operatorname{diag}(U_j)_{j=1}^r.$$

Hence for any $p \geqslant 1$ we have

$$[\operatorname{col}(XT^{i-1}W)_{i=1}^{p}]V(L_1, \ldots, L_r) = (X_jT_j^{i-1}U_j)_{i=1,\, j=1}^{p,\ \ r}. \tag{5}$$

Now suppose that L is the kth element in a family $R_1, \ldots, R_s$ of monic operator polynomials on $\mathbb{C}_n$. Our aim is to study the relation between $V(R_1, \ldots, R_s)$ and

$$\tilde{V} = V(R_1, \ldots, R_{k-1}, L_1, \ldots, L_r, R_{k+1}, \ldots, R_s).$$

Let $\ell_1, \ldots, \ell_s$ be the degrees of $R_1, \ldots, R_s$. Put $p = \ell_1 + \cdots + \ell_s$. Further, for each j let (Y_j, S_j) be a standard pair representing R_j, and define $Q_j = [\operatorname{col}(Y_jS_j^{i-1})_{i=1}^{\ell_j}]^{-1}$. Then, by definition,

$$V(R_1, \ldots, R_s) = (Y_jS_j^{i-1}Q_j)_{i=1,\, j=1}^{p,\ \ s}. \tag{6}$$

As $R_k = L$, the kth column in the operator matrix (6) representing $V(R_1, \cdots, R_s)$ is equal to $\operatorname{col}(XT^{i-1}W)_{i=1}^{p}$, that is, this kth column is equal to the first factor in the left-hand side of (5). Note that

$$\ell_1 + \cdots + \ell_{k-1} + k_1 + \cdots + k_r + \ell_{k+1} + \cdots + \ell_s = \sum_{j=1}^{s} \ell_j = p.$$

So, if in the operator matrix representing $V(R_1, \ldots, R_s)$ (see formula (6)) the kth column is replaced by the right-hand side of (5), we obtain the operator matrix representing $\tilde{V}$. It follows that

$$\tilde{V} = V(R_1, \ldots, R_s) \cdot \{I \oplus \cdots \oplus I \oplus V(L_1, \ldots, L_r) \oplus I \oplus \cdots \oplus I\}.$$

THEOREM 4.3. *Let $R_1, \ldots, R_s$ be monic operator polynomials on $\mathbb{C}_n$, and for each j let $L_{j1}, \ldots, L_{jr_j}$ be a family of monic right divisors of R_j. Suppose that*

$$\text{degree}(R_j) = \sum_{i=1}^{r_j} \text{degree}(L_{ji}) \qquad (1 \leqslant j \leqslant s).$$

Then $V(L_{11}, \ldots, L_{1r_1}, L_{21}, \ldots, L_{2r_2}, \ldots, L_{s1}, \ldots, L_{sr_s})$ is equal to

$$V(R_1, \ldots, R_s) \cdot \text{diag}(V(L_{j1}, \ldots, L_{jr_j}))_{j=1}^{s}.$$

Proof. Apply to each polynomial R_j the argument described in the paragraph preceding this theorem.

COROLLARY 4.4. *Let $R_1, \ldots, R_s$ be monic scalar polynomials, say $R_j(\lambda) = \prod_{i=1}^{r_j} (\lambda - \lambda_{ji})$, $1 \leqslant j \leqslant s$. Then*

$$\det[V(R_1, \ldots, R_s)] = \prod_{1 \leqslant j_1 < j_2 \leqslant s} (\lambda_{j_2 i_2} - \lambda_{j_1 i_1}),$$

where the product is taken over all possible pairs (j_1, i_1) and (j_2, i_2) such that $j_1 < j_2$.

Proof. For each $1 \leqslant j \leqslant s$, let $L_{ji}(\lambda) = (\lambda - \lambda_{ji})$. Then $L_{j1}, \ldots, L_{jr_j}$ are monic right divisors of R_j and $\text{degree}(R_j) = \sum_{i=1}^{r_j} \text{degree}(L_{ji})$. So one can apply Theorem 4.3. As $L_{j1}, \ldots, L_{jr_j}$ are polynomials of degree 1, we have

$$V(L_{j1}, \ldots, L_{jr_j}) = (\lambda_{jh}^{i-1})_{i,h=1}^{r_j}.$$

It follows that

$$\det[V(L_{j1}, \ldots, L_{jr_j})] = \prod_{1 \leqslant i_1 < i_2 \leqslant r_j} (\lambda_{ji_2} - \lambda_{ji_1}).$$

In a similar way one computes $\det[V(L_{11}, \ldots, L_{ji}, \ldots, L_{sr_s})]$. Finally, one uses the formula for $V(L_{11}, \ldots, L_{ji}, \ldots, L_{sr_s})$ given in the previous theorem to get the desired result.

5. *The Vandermonde Operator for Right Divisors (Case $\ell \neq m$)*

Throughout this subsection L is a monic operator polynomial on $\mathbb{C}_n$ and $L_1, \ldots, L_r$ are monic right divisors of L. The degrees of $L, L_1, \ldots, L_r$

will be denoted by $\ell, k_1, \ldots, k_r$, respectively. In this subsection the case that $m = k_1 + \cdots + k_r \neq \ell$ will be considered. We shall make use of the notation introduced in the first paragraph of Subsection 4. So (X, T), $(X_1, T_1), \ldots, (X_r, T_r)$ will be standard pairs representing $L, L_1, \ldots, L_r$, respectively. Further, $\mathcal{M}_1, \ldots, \mathcal{M}_r$ will denote the supporting subspaces corresponding to $L_1, \ldots, L_r$.

The family of subspaces $\mathcal{M}_1, \ldots, \mathcal{M}_r$ is said to be *linearly independent* if the condition $x_1 + \cdots + x_r = 0$, $x_j \in \mathcal{M}_j$ for $1 \leqslant j \leqslant r$, implies that $x_1 = \cdots = x_r = 0$. Note that this is equivalent to the requirement that

$$\dim\{\mathcal{M}_1 + \cdots + \mathcal{M}_r\} = n(k_1 + \cdots + k_r).$$

THEOREM 5.1. *Let $L_1, \ldots, L_r$ be right divisors of L. Then $V(L_1, \ldots, L_r)$ is invertible if and only if the supporting subspaces $\mathcal{M}_1, \ldots, \mathcal{M}_r$ are linearly independent and the subspace $\mathcal{L} = \mathcal{M}_1 + \cdots + \mathcal{M}_r$ is a supporting subspace for L.*

Proof. As the third factor in the right-hand side of formula (2) in Subsection 4 is invertible, the Vandermonde operator $V(L_1, \ldots, L_r)$ is invertible if and only if

$$\operatorname{Ker}\{[\operatorname{col}(XT^{i-1})_{i=1}^{m}] \cdot [\operatorname{row}(K_j)_{j=1}^{r}]\} = \{0\}. \tag{1}$$

Recall that $\Omega = \operatorname{row}(K_j)_{j=1}^{r} : \mathbb{C}_{nm} \to \mathbb{C}_{n\ell}$ and $\operatorname{Im}\Omega = \mathcal{L}$. Thus the map Ω is injective if and only if the spaces $\mathcal{M}_1, \ldots, \mathcal{M}_r$ are linearly independent.

Suppose that $V(L_1, \ldots, L_r)$ is invertible. Then (1) holds true. Thus the map Ω and the map

$$\operatorname{col}(XT^{i-1})_{i=1}^{m}\big|_{\mathcal{L}} : \mathcal{L} \to \mathbb{C}_{nm} \tag{2}$$

are injective. It follows that $\mathcal{M}_1, \ldots, \mathcal{M}_r$ are linearly independent and $\dim \mathcal{L} = n(k_1 + \cdots + k_r) = nm$. But then the map (2) must be invertible, and we can apply [3], Theorem 8 to show that $\mathcal{L}$ is a supporting subspace for L.

Conversely, suppose that $\mathcal{M}_1, \ldots, \mathcal{M}_r$ are linearly independent, and let $\mathcal{L} = \mathcal{M}_1 + \cdots + \mathcal{M}_r$ be a supporting subspace for L. Then Ω is injective, and, as $\dim \mathcal{L} = nm$, the map (2) is invertible. It follows that (1) holds, but then $V(L_1, \ldots, L_r)$ is invertible.

Suppose $V(L_1, \ldots, L_r)$ is invertible. In that case there exists a unique monic operator polynomial L_o on $\mathbb{C}_n$ (see Theorem 3.2 and Remark 3.3) such that $L_1, \ldots, L_r$ are right divisors of L_o and degree $L_o = k_1 + \cdots + k_r$. The next theorem shows that L_o is a right divisor of L and the supporting subspace corresponding to L_o is equal to $\mathcal{M}_1 + \cdots + \mathcal{M}_r$.

THEOREM 5.2. *Let $L_1, \ldots, L_r$ be right divisors of L, and suppose that $V(L_1, \ldots, L_r)$ is invertible. Let L_o be the right divisor of L corresponding to the supporting subspace $\mathscr{L} = \mathscr{M}_1 + \cdots + \mathscr{M}_r$. Then $L_1, \ldots, L_r$ are right divisors of L_o and* degree$(L_o) = k_1 + \cdots + k_r$.

Proof. By Theorem 5.1, the space $\mathscr{L}$ is a supporting subspace for L and $\dim \mathscr{L} = n(k_1 + \cdots + k_r)$. Let L_o be the right divisor of L corresponding to $\mathscr{L}$. Then, by [3, Theorem 8], degree$(L_o) = k_1 + \cdots + k_r$, and by [3, Theorem 10], the polynomials $L_1, \ldots, L_r$ are right divisors of L_0.

6. *Operator Differential and Difference Equations*

Throughout this subsection L is a monic operator polynomial on $\mathbb{C}_n$, and $L_1, \ldots, L_r$ are monic right divisors of L. To illustrate the previous theory, let us consider the following operator differential equations:

$$(\alpha) \quad L(d/dx)\varphi = 0, \qquad (\beta) \quad L_j(d/dx)\varphi_j = 0.$$

The fact that L_j is a right divisor of L implies that each solution φ_j of equation (β) is a solution of equation (α). Hence it follows that all functions of the form

$$\varphi_1 + \varphi_2 + \cdots + \varphi_r, \tag{1}$$

where for each j the function φ_j is a solution of equation (β), are solutions of (α). We call the family $L_1, \ldots, L_r$ *complete* relative to L if each solution of equation (α) is of the form (1). We say that the family $L_1, \ldots, L_r$ is *minimal* if each solution φ of (α), which can be written in the form (1), admits only one such representation. In other words, $L_1, \ldots, L_r$ is minimal if and only if $\varphi_1 + \varphi_2 + \cdots + \varphi_r = 0$ implies $\varphi_1 = \varphi_2 = \cdots = \varphi_r = 0$. (Note that the last characterization of minimality does not use the fact that $L_1, \ldots, L_r$ are right divisors of L, and hence the notion of minimality makes sense for any arbitrary finite family of monic operator polynomials on $\mathbb{C}_n$. We shall use this later in Subsection 13). If the family $L_1, \ldots, L_r$ is complete and minimal relative to L, then each solution of equation (α) can be written in the form (1) in one and only one way. We shall now characterize the notions of completeness and minimality in terms of the supporting subspaces corresponding to $L_1, \ldots, L_r$.

THEOREM 6.1. *Let $L_1, \ldots, L_r$ be right divisors of L, let $\ell =$* degree(L), *and let $\mathscr{M}_1, \ldots, \mathscr{M}_r$ be supporting subspaces corresponding to $L_1, \ldots, L_r$. The family $L_1, \ldots, L_r$ is complete relative to L if and only if*

$$\mathbb{C}_{n\ell} = \mathscr{M}_1 + \mathscr{M}_2 + \cdots + \mathscr{M}_r,$$

and the family $L_1, \ldots, L_r$ is minimal if and only if the spaces $\mathcal{M}_1, \mathcal{M}_2, \ldots, \mathcal{M}_r$ are linearly independent. In particular, the family $L_1, \ldots, L_r$ is complete and minimal relative to L if and only if $\ell = \sum_{j=1}^{r}$ degree(L_j) and the Vandermonde operator $V(L_1, \ldots, L_r)$ is invertible.

Proof. We use the notation introduced in the first paragraph of Subsection 4. Recall that φ is a solution of the equation

$$L(d/dx)\varphi = 0 \tag{2}$$

if and only if $\varphi(t) = Xe^{tT}x$, where $x \in \mathbb{C}_{n\ell}$ and x is uniquely determined by φ. Similarly, the general solution of

$$L_j(d/dx)\varphi_j = 0 \tag{3}$$

is of the form $\varphi_j(t) = X_j e^{tT_j} x_j$, where x_j is an arbitrary vector in $\mathbb{C}_{nk_j}$ (cf. [6, Theorem 1]). But L_j is a right divisor of L. Hence $X_j T_j^{\alpha} = XT^{\alpha}K_j$ for $\alpha = 0, 1, 1, \ldots$. It follows that the general solution of (3) can be written as

$$\varphi_j(t) = Xe^{tT}K_j x_j,$$

where, as before, $x_j \in \mathbb{C}_{nk_j}$. It follows that the family $L_1, \ldots, L_r$ is complete relative to L if and only if $\operatorname{Im}[\operatorname{row}(K_j)_{j=1}^{r}] = \mathbb{C}_{n\ell}$, and $L_1, \ldots, L_r$ is minimal if and only if $\operatorname{Ker}[\operatorname{row}(K_j)_{j=1}^{r}]$ contains the zero element only. But

$$\operatorname{Im}[\operatorname{row}(K_j)_{j=1}^{r}] = \mathcal{M}_1 + \mathcal{M}_2 + \cdots + \mathcal{M}_r,$$

and $\operatorname{Ker}[\operatorname{row}(K_j)_{j=1}^{r}] = \{0\}$ if and only if the spaces $\mathcal{M}_1, \mathcal{M}_2, \ldots, \mathcal{M}_r$ are linearly independent. This proves the first part of the theorem. The second part of the theorem is an immediate consequence of the first part and Theorem 4.2.

For operator difference equations similar results hold true. For a sequence $u = \{u(1), u(2), \ldots\}$ of vectors in $\mathbb{C}_n$, we introduce the difference operator Δ by $(\Delta u)(r) = u(r+1)$, $r = 1, 2, \ldots$. Note that Δu is a sequence of vectors in $\mathbb{C}_n$ too. We use this to write the operator difference equation

$$B_0 u(r) + B_1 u(r+1) + \cdots + B_p u(r+p) = 0 \qquad (r = 1, 2, \ldots)$$

as $(B_0 + B_1\Delta + \cdots + B_p\Delta^p)u = 0$.

THEOREM 6.2. *Let $L_1, \ldots, L_r$ be right divisors of L. Then* degree$(L) = \sum_{j=1}^{r}$ degree(L_j) *and $V(L_1, \ldots, L_r)$ is invertible if and only if each solution u of*

$$L(\Delta)u = 0 \tag{4}$$

can be written in a unique way as $u = u_1 + \cdots + u_r$, where for each j the sequence u_j is a solution of $L_j(\Delta)u_j = 0$.

Proof. We shall use the notation of the first paragraph of Subsection 4. The general solution u of (4) is of the form

$$u = (XT^{r-1}x)_{r=1}^{\infty}$$

where $x \in \mathbb{C}_{n\ell}$ and x is uniquely determined by u (see [7, Theorem 1]). Similarly, using the fact that L_j is a right divisor of L, the general solution of $L_j(\Delta)u_j = 0$ is of the form

$$u_j = (XT^{r-1}K_jx_j)_{r=1}^{\infty},$$

where $x_j \in \mathbb{C}_{nk_j}$ and x_j is uniquely determined by u_j. Now the proof can be completed by using arguments similar to those employed in the proof of the previous theorem.

7. *Spectral Properties and Spectral Divisors*

Throughout this subsection $L, L_1, \ldots, L_r$ are monic operator polynomials on $\mathbb{C}_n$ with degrees $\ell, k_1, \ldots, k_r$, respectively, and $L_1, \ldots, L_r$ are right divisors of L. Our aim is to find criteria for the invertibility of the Vandermonde operator $V(L_1, \ldots, L_r)$ in terms of the spectra of $L_1, \ldots, L_r$.

Given an arbitrary operator polynomial R on $\mathbb{C}_n$, recall that the spectrum of R is equal to the set

$$\sigma(R) = \{\lambda \in \mathbb{C} \mid R(\lambda) \text{ is not invertible}\}.$$

In other words, $\sigma(R)$ is the set of all $\lambda \in \mathbb{C}$ such that $\det R(\lambda) = 0$. If $R(\lambda) = \lambda I - T$, then the spectrum of R coincides with the spectrum $\sigma(T)$ of the operator T.

THEOREM 7.1. *Let $L_1, \ldots, L_r$ be right divisors of L, and suppose that the spectra of $L_1, \ldots, L_r$ are mutually disjoint. Then the family $L_1, \ldots, L_r$ is minimal. If, in addition, $\ell = k_1 + \cdots + k_r$, then $V(L_1, \ldots, L_r)$ is invertible.*

Proof. Let $(X, T), (X_1, T_1), \ldots, (X_r, T_r)$ be representing pairs for $L, L_1, \ldots, L_r$, respectively, and let $\mathcal{M}_1, \ldots, \mathcal{M}_r$ be the corresponding supporting subspaces. By Theorem 6.1 we have to show that the spaces $\mathcal{M}_1, \ldots, \mathcal{M}_r$ are linearly independent.

Take $1 \leqslant j \leqslant r$, and suppose that $\mathcal{M}_1, \ldots, \mathcal{M}_j$ are linearly independent. Put $\mathcal{L} = \mathcal{M}_1 \oplus \cdots \oplus \mathcal{M}_j$. Then $\mathcal{L}$ is invariant under T and

$$\sigma(T|_{\mathcal{L}}) = \sigma(T|_{\mathcal{M}_1}) \cup \cdots \cup \sigma(T|_{\mathcal{M}_j}).$$

For each j the operators T_j and $T|_{\mathcal{M}_j}$ are similar (cf. [3, Section 5]). Hence

$$\sigma(T_j) = \sigma(T|_{\mathcal{M}_j}).$$

Further, $\sigma(T_j)$ is equal to $\sigma(L_j)$. It follows that

$$\sigma(T|_{\mathcal{L}}) = \bigcup_{i=1}^{j} \sigma(L_i).$$

Now suppose that $\mathcal{M}_1, \ldots, \mathcal{M}_j, \mathcal{M}_{j+1}$ are not linearly independent. Then

$$\mathcal{L}_0 = \mathcal{L} \cap \mathcal{M}_{j+1} = (\mathcal{M}_1 \oplus \cdots \oplus \mathcal{M}_j) \cap \mathcal{M}_{j+1}$$

contains nonzero elements. Observe that $\mathcal{L}_0$ is a T-invariant space and

$$\sigma(T|_{\mathcal{L}_0}) \subset \{\sigma(T|_{\mathcal{L}}) \cap \sigma(L_{j+1})\}.$$

It follows that $\{\bigcup_{i=1}^{j} \sigma(L_i)\} \cap \sigma(L_{j+1})$ is not empty. But this contradicts our hypothesis on the spectra of $L_1, \ldots, L_r$. So $\mathcal{M}_1, \ldots, \mathcal{M}_j, \mathcal{M}_{j+1}$ are linearly independent and, by induction, the first part of the theorem is proved.

From what we proved above and the condition $\ell = k_1 + \cdots + k_r$ it follows that $\mathcal{M}_1 \oplus \cdots \oplus \mathcal{M}_r = \mathbb{C}_{n\ell}$. But then we can apply Theorem 4.2 to get the second part of the theorem.

The invertibility of the Vandermonde operator $V(L_1, \ldots, L_r)$ does not imply that the spectra of $L_1, \ldots, L_r$ are disjoint. For instance, let

$$L_1(\lambda) = \begin{bmatrix} \lambda & 1 \\ 0 & \lambda \end{bmatrix}, \qquad L_2(\lambda) = \begin{bmatrix} \lambda & 0 \\ 1 & \lambda \end{bmatrix}.$$

Then L_1 and L_2 are both right divisors of $L(\lambda) = \lambda^2 I$ and the Vandermonde operator $V(L_1, L_2)$ is invertible, but $\sigma(L_1) = \sigma(L_2) = \{0\}$. We shall see that under certain extra conditions on $L_1, \ldots, L_r$ the converse of the last part of the previous theorem holds true (see Theorem 7.4 below).

A monic right divisor R of L is said to be a *spectral* right divisor of L if the spectrum of the left divisor, or quotient, corresponding to R is disjoint with the spectrum of R, in other words, $L(\lambda) = Q(\lambda)R(\lambda)$ and

$$\sigma(Q) \cap \sigma(R) = \emptyset.$$

Note that in this case $\sigma(L)$ is the disjoint union of $\sigma(Q)$ and $\sigma(R)$.

LEMMA 7.2. *Let $L_1, \ldots, L_r$ be right divisors of L, and for each j let Q_j be the quotient corresponding to L_j. Suppose that $\ell = k_1 + \cdots + k_r$, and let the Vandermonde operator $V(L_1, \ldots, L_r)$ be invertible. Then*

$$\sigma(Q_j) = \bigcup_{i \neq j} \sigma(L_i) \qquad (1 \leqslant j \leqslant r).$$

Proof. Let $(X, T), (X_1, T_1), \ldots, (X_r, T_r)$ be representing pairs for $L, L_1, \ldots, L_r$, respectively, and let $\mathcal{M}_1, \ldots, \mathcal{M}_r$ be the corresponding supporting subspaces. By Theorem 4.2,

$$\mathbb{C}_{n\ell} = \mathcal{M}_1 \oplus \cdots \oplus \mathcal{M}_r.$$

Take a fixed integer j, $1 \leqslant j \leqslant r$. Put

$$\mathcal{L} = \mathcal{M}_1 \oplus \cdots \oplus \mathcal{M}_{j-1} \oplus \mathcal{M}_{j+1} \oplus \cdots \oplus \mathcal{M}_r.$$

Then $\mathcal{L}$ is a T-invariant complementary subspace for $\mathcal{M}_j$ and

$$\sigma(T|_{\mathcal{L}}) = \bigcup_{i \neq j} \sigma(T|_{\mathcal{M}_i}).$$

As in the proof of Theorem 7.1, we have

$$\sigma(T|_{\mathcal{M}_i}) = \sigma(T_i) = \sigma(L_i) \qquad (1 \leqslant i \leqslant r).$$

Thus $\sigma(T|_{\mathcal{L}}) = \bigcup_{i \neq j} \sigma(L_i)$.

Let (Z, S) be a representing pair for the quotient Q_j. Note that $\sigma(Q_j) = \sigma(S)$. As $\mathcal{L}$ is a complementary subspace for $\mathcal{M}_j$, it follows from [3, Section 6] that S and $T|_{\mathcal{L}}$ are similar. Hence

$$\sigma(Q_j) = \sigma(S) = \sigma(T|_{\mathcal{L}}) = \bigcup_{i \neq j} \sigma(L_i),$$

and the proof is complete.

COROLLARY 7.3. *Let $L_1, \ldots, L_r$ be right divisors of L, and suppose that $\ell = k_1 + \cdots + k_r$. Further suppose that the spectra of $L_1, \ldots, L_r$ are mutually disjoint. Then $L_1, \ldots, L_r$ are spectral right divisors of L.*

Proof. By Theorem 7.1 the Vandermonde operator $V(L_1, \ldots, L_r)$ is invertible. But then we can apply the previous lemma to get the desired result.

THEOREM 7.4. *Let $L_1, \ldots, L_r$ be spectral right divisors of L, and suppose that $\ell = k_1 + \cdots + k_r$. Then the Vandermonde operator $V(L_1, \ldots, L_r)$ is invertible if and only if the spectra of $L_1, \ldots, L_r$ are disjoint.*

Proof. Let $V(L_1, \ldots, L_r)$ be invertible. For each j let Q_j be the quotient corresponding to L_j. Then, by Lemma 7.2,

$$\sigma(Q_j) = \bigcup_{i \neq j} \sigma(L_i).$$

As L_j is a spectral right divisor of L, the set $\sigma(Q_j) \cap \sigma(L_j)$ is empty. Thus

$$\sigma(L_j) \cap \left\{ \bigcup_{i \neq j} \sigma(L_i) \right\} = \varnothing \qquad (1 \leqslant j \leqslant r),$$

that is, $\sigma(L_1), \ldots, \sigma(L_r)$ are mutually disjoint. The reverse implication is covered by Theorem 7.1.

For linear divisors the results of this subsection have been proved by Markus and Mereutsa in [9].[1]

8. *Equivalence and Decomposition of Operator Polynomials*

Let L_1 and L_2 be monic operator polynomials on $\mathbb{C}_n$. Recall that L_1 and L_2 are said to be (polynomially) *equivalent* whenever

$$L_1(\lambda) = E(\lambda)L_2(\lambda)F(\lambda) \qquad (\lambda \in \mathbb{C}),$$

where $E(\lambda)$ and $F(\lambda)$ are operator polynomials on $\mathbb{C}_n$ such that $\det E(\lambda)$ and $\det F(\lambda)$ are nonzero constants. It is well known that this condition is equivalent to the requirement that L_1 and L_2 have the same Smith canonical form (cf. [2, Section 21]). In other words, L_1 and L_2 are equivalent if and only if the corresponding Jordan matrices J_{L_1} and J_{L_2} are equal.

Now let (X_1, T_1) and (X_2, T_2) be standard pairs representing L_1 and L_2, respectively. Then T_1 is similar to J_{L_1} and T_2 is similar J_{L_2}. It follows that two standard pairs (X_1, T_1) and (X_2, T_2) represent equivalent monic operator polynomials if and only if T_1 and T_2 are similar.

For polynomials of degree one the result of the previous paragraph yields the following well-known theorem: Two linear polynomials $\lambda I - T_1$ and $\lambda I - T_2$ are equivalent if and only if there exists an invertible operator S such that

$$\lambda I - T_1 = S(\lambda I - T_2)S^{-1} \qquad (\lambda \in \mathbb{C})$$

(cf. [2, Section 21]). The obvious generalization of this statement to polynomials of higher degree does not hold true. This follows from the next example.

EXAMPLE 8.1. For each $\alpha \in \mathbb{C}$ let

$$L_\alpha(\lambda) = \lambda^2 \begin{bmatrix} 1 & 0 \\ 0 & 1 \end{bmatrix} + \lambda \begin{bmatrix} 0 & \alpha \\ 0 & 0 \end{bmatrix} + \begin{bmatrix} 0 & 1 \\ 0 & 0 \end{bmatrix}.$$

[1] See the Note at the end of this paper.

Then one easily sees that

$$J_{L_\alpha} = \begin{bmatrix} 0 & 1 & 0 & 0 \\ 0 & 0 & 1 & 0 \\ 0 & 0 & 0 & 1 \\ 0 & 0 & 0 & 0 \end{bmatrix},$$

and hence it follows that all L_α are mutually equivalent. But $L_\alpha(\lambda) = SL_\beta(\lambda)S^{-1}$ for all λ if and only if $\alpha = \beta$.

THEOREM 8.2. *Let $L, L_1, \ldots, L_r$ be monic operator polynomials on $\mathbb{C}_n$ of degree $\ell, k_1, \ldots, k_r$, and suppose that $\ell = k_1 + \cdots + k_r$. Further, suppose that $L_1, \ldots, L_r$ are right divisors of L, and for each $1 \leqslant j \leqslant r$ let the Vandermonde operator $V(L_1, \ldots, L_j)$ be invertible. Then there exist monic operator polynomials $Q_2, \ldots, Q_r$ on $\mathbb{C}_n$ such that*

$$L(\lambda) = Q_r(\lambda)Q_{r-1}(\lambda) \cdots Q_2(\lambda)L_1(\lambda)$$

and Q_j is equivalent to L_j for $2 \leqslant j \leqslant r$.

Proof. First we assume that $r = 2$. In that case we take Q_2 to be the left divisor, or quotient, corresponding to the right divisor L_1 of L. In other words, Q_2 is the unique monic operator polynomial on $\mathbb{C}_n$ such that

$$L(\lambda) = Q_2(\lambda)L_1(\lambda).$$

Let (X, T), (X_1, T_1), (X_2, T_2), and (Z, S) be representing pairs for L, L_1, L_2, and Q_2, respectively. Further, let $\mathcal{M}_1$ and $\mathcal{M}_2$ be supporting subspaces for L corresponding to L_1 and L_2 (as in the first paragraph of subsection 4). As $V(L_1, L_2)$ is invertible and $\ell = k_1 + k_2$, we know (cf. Theorem 4.2) that

$$\mathcal{M}_1 \oplus \mathcal{M}_2 = \mathbb{C}_{n\ell}.$$

In particular, $\mathcal{M}_2$ is a complementary subspace for $\mathcal{M}_1$. Further, from the description of the quotient given in [3, Section 6], it follows that the restriction $T_{\mathcal{M}_2}$ of T to $\mathcal{M}_2$ is similar to S. On the other hand, by [3, Section 5, Remark 2], the operator $T_{\mathcal{M}_2}$ is similar to T_2. So T_2 and S are similar, and hence the polynomials L_2 and Q_2 are equivalent. This proves the theorem for the case $r = 2$.

Next we consider the case $r > 2$. As $V(L_1, \ldots, L_{r-1})$ is invertible, we can apply Theorem 5.2 to show that there exists a monic right divisor L_o of L such that $L_1, \ldots, L_{r-1}$ are right divisors of L_o and degree$(L_o) = k_1 + \cdots + k_{r-1}$. Now L_o, L_r are right divisors of L, degree(L_o) + degree$(L) = k_1 + \cdots + k_{r-1} + k_r = \ell$ and, by Theorem 4.3, $V(L_o, L_r)$ is invertible. So we can apply the result of the previous paragraph to the pair L_o, L_r. It follows that there exists a monic operator polynomial Q_r on $\mathbb{C}_n$ such that

$$L(\lambda) = Q_r(\lambda)L_o(\lambda)$$

and Q_r and L_r are equivalent. Hence it remains to prove the theorem for L_o instead of L. But L_o has $r-1$ right divisors of the right kind, and hence the proof can be completed by induction.

If $V(L_1, \ldots, L_r)$ is invertible, then in general it does not follow that $V(L_1, \ldots, L_j)$ is invertible for each $1 \leqslant j < r$. This can be seen from the following example. Take $n = 2$, and let

$$A_1 = \begin{bmatrix} 0 & 1 \\ 0 & 0 \end{bmatrix}, \qquad A_2 = \begin{bmatrix} 1 & 0 \\ 0 & 0 \end{bmatrix}, \qquad A_3 = \begin{bmatrix} 1 & 0 \\ 1 & 1 \end{bmatrix}.$$

Put $L_j(\lambda) = \lambda I - A_j$. Then $V(L_1, L_2, L_3)$ is invertible, but $V(L_1, L_2)$ is not invertible.

For linear divisors Theorem 8.2 has been proved by Mereutsa [10].

III. Systems of Equations

9. *Systems of Equations Associated with Operator Polynomials*

To determine the kernel of the Vandermonde operator of two operator polynomials L_1 and L_2, we are led to consider the following system of equations:

$$X_1 T_1^{\alpha} U_1 x_1 + X_2 T_2^{\alpha} U_2 x_2 = 0 \qquad (0 \leqslant \alpha \leqslant m-1). \tag{1}$$

As the operators U_1 and U_2 are invertible, we may omit U_1 and U_2 in (1). Further we shall have to consider (1) for different positive integers m. In particular, we shall be interested in the following infinite system:

$$X_1 T_1^{\alpha} x_1 + X_2 T_2^{\alpha} x_2 = 0 \qquad (\alpha = 0, 1, 2, \ldots). \tag{2}$$

Throughout this subsection n will be a fixed positive integer. Two linear operators X and T are said to form an *admissible* pair of *order p* if $X: \mathbb{C}_p \to \mathbb{C}_n$ and $T: \mathbb{C}_p \to \mathbb{C}_p$. Each standard pair is admissible and the order of a standard pair is n times its degree. Clearly the system of equations (2) makes sense whenever (X_1, T_1) and (X_2, T_2) are admissible pairs.

For an admissible pair (X, T) of order p we put

$$\operatorname{Ker}(X, T) = \{x \in \mathbb{C}_p \,|\, X T^{\alpha} x = 0, \alpha = 0, 1, 2, \ldots\}.$$

Similarly, we let $\operatorname{Ker}\{(X_1, T_1), (X_2, T_2)\}$ denote the solution space of the system (2). In other words

$$\operatorname{Ker}\{(X_1, T_1), (X_2, T_2)\} = \operatorname{Ker}(X, T), \tag{3}$$

where $X = (X_1 X_2)$ and $T = T_1 \oplus T_2$. If p_1 and p_2 are the orders of the pairs (X_1, T_1) and (X_2, T_2), then the space given by formula (3) is a linear subspace of $\mathbb{C}_{p_1+p_2} = \mathbb{C}_{p_1} \oplus \mathbb{C}_{p_2}$.

Let us describe $\operatorname{Ker}\{(X_1, T_1), (X_2, T_2)\}$ for the case that (X_1, T_1) and (X_2, T_2) are standard pairs representing right divisors L_1 and L_2 of a given monic operator polynomial L. Let (X, T) be a standard pair representing L, and let ℓ, k_1, and k_2 be the degrees of L, L_1, and L_2, respectively. Using the notation introduced in the first paragraph of Subsection 4, we have injective linear maps

$$K_1 : \mathbb{C}_{nk_1} \to \mathbb{C}_{n\ell}, \qquad K_2 : \mathbb{C}_{nk_2} \to \mathbb{C}_{n\ell}$$

such that for $j = 1, 2$ the image of K_j is equal to the supporting subspace $\mathcal{M}_j$ of L_j. As $\mathcal{M}_1$ and $\mathcal{M}_2$ are both T-invariant, the same is true for $\mathcal{M}_1 \cap \mathcal{M}_2$. Put $\mathcal{L} = K_1^{-1}[\mathcal{M}_1 \cap \mathcal{M}_2]$, and let S be the restriction of $K_2^{-1}K_1$ to $\mathcal{L}$. Then $\mathcal{L}$ is the largest T_1-invariant subspace of $\mathbb{C}_{nk_1}$ such that

$$X_1|_{\mathcal{L}} = X_2 S, \qquad ST_1|_{\mathcal{L}} = T_2 S.$$

Furthermore, $\operatorname{Ker}\{(X_1, T_1), (X_2, T_2)\} = \{\{x, -Sx\} \mid x \in \mathcal{L}\}$. The next theorem shows that this description of $\operatorname{Ker}\{(X_1, T_1), (X_2, T_2)\}$ holds in a more general context.

THEOREM 9.1. *Let (X_1, T_1) be an admissible of order p_1, and let (X_2, T_2) be a standard pair of order p_2. Then*

$$\operatorname{Ker}\{(X_1, T_1), (X_2, T_2)\} = \{\{x, -Sx\} \mid x \in \mathcal{L}\}, \tag{4}$$

where $\mathcal{L}$ and S are determined by the following statement: $\mathcal{L}$ is the largest T_1-invariant subspace $\mathbb{C}_{p_1}$ such that there exists a linear map $S : \mathcal{L} \to \mathbb{C}_{p_2}$ with the property that

$$X_1|_{\mathcal{L}} = X_2 S, \qquad ST_1|_{\mathcal{L}} = T_2 S. \tag{5}$$

Proof. Let k be the degree of the standard pair (X_2, T_2), in other words, $p_2 = nk$. Put

$$A_1 = \operatorname{col}(X_1 T_1^{i-1})_{i=1}^{k}, \qquad A_2 = \operatorname{col}(X_2 T_2^{i-1})_{i=1}^{k}.$$

Note that A_2 is invertible. So $A_2^{-1}A_1$ is a well-defined linear map from $\mathbb{C}_{p_1}$ into $\mathbb{C}_{p_2}$.

By taking together in (2) k successive equations, one sees that the system (2) can be written as

$$A_1 T_1^{\alpha} x_1 + A_2 T_2^{\alpha} x_2 = 0 \qquad (\alpha = 0, 1, 2, \ldots),$$

and hence it follows that the solution space of the system (2) is equal to the solution space of the following system:

$$\begin{cases} (A_2^{-1} A_1) x_1 + x_2 = 0, \\ (A_2^{-1} A_1) T_1^{\alpha} x_1 - T_2^{\alpha} (A_2^{-1} A_1) x_1 = 0 \qquad (\alpha = 1, 2, \ldots). \end{cases}$$

Put $\mathscr{L} = \{x \in \mathbb{C}_{p_1} \mid (A_2^{-1} A_1) T_1^{\alpha} x = T_2^{\alpha} (A_2^{-1} A_1) x, \ \alpha = 1, 2, \ldots\}$, and let $S: \mathscr{L} \to \mathbb{C}_{p_1}$ be the restriction of $A_2^{-1} A_1$ to $\mathscr{L}$. Then the observation made above shows that with this choice of $\mathscr{L}$ and S (4) holds true. It remains to prove that $\mathscr{L}$ can be characterized in the desired way.

From the definitions of $\mathscr{L}$ and S it is clear that formula (5) holds true. To see that $\mathscr{L}$ is T_1-invariant, take $x \in \mathscr{L}$ and let $\alpha \geqslant 0$. Then

$$\begin{aligned} \{(A_2^{-1} A_1) T_1^{\alpha}\} T_1 x = (A_2^{-1} A_1) T_1^{\alpha+1} x = T_2^{\alpha+1} (A_2^{-1} A_1) x = T_2^{\alpha} \{T_2 (A_2^{-1} A_1) x\} \\ = \{T_2^{\alpha} (A_2^{-1} A_1)\} T_1 x, \end{aligned}$$

and thus $T_1 x \in \mathscr{L}$. This shows that $\mathscr{L}$ is invariant under T_1.

Now let $\mathscr{L}_o$ be a T_1-invariant subspace of $\mathbb{C}_{p_1}$ and let $S_o: \mathscr{L}_o \to \mathbb{C}_{p_1}$ be a linear map such that

$$X_1|_{\mathscr{L}_o} = X_2 S_o, \qquad S_o T_1|_{\mathscr{L}_o} = T_2 S_o.$$

Take $x \in \mathscr{L}_o$. Then the T_1-invariance of $\mathscr{L}_o$ implies that $S_o T_1^{\alpha} x = T_2^{\alpha} S_o x$ for all $\alpha \geqslant 0$. But then it follows that

$$A_2 T_2^{\alpha} S_o x = A_1 T_1^{\alpha} x \qquad (\alpha = 0, 1, 2, \ldots). \tag{6}$$

By taking $\alpha = 0$ in (6), one sees that S_o is equal to the restriction of $A_2^{-1} A_1$ to $\mathscr{L}_o$. So (6) can be written as

$$(A_2^{-1} A_1) T_1^{\alpha} x = T_2^{\alpha} (A_2^{-1} A_1) x \qquad (\alpha = 0, 1, 2, \ldots),$$

and it follows that $x \in \mathscr{L}$. Thus $\mathscr{L}_o \subset \mathscr{L}$, and the proof is complete.

The characterization of $\operatorname{Ker}\{(X_1, T_1), (X_2, T_2)\}$ given in the previous theorem also holds true if (X_2, T_2) is not a standard pair but an admissible pair with $\operatorname{Ker}(X_2, T_2) = \{0\}$. If (X_2, T_2) is an arbitrary pair, then a modified version of Theorem 9.1 can be proved. These results will appear elsewhere.

COROLLARY 9.2. *Let $\mathscr{L}$ and S be as in the previous theorem. Then*

$$\operatorname{Ker} S = \operatorname{Ker}(X_1, T_1).$$

Proof. Note that $\{x, 0\} \in \operatorname{Ker}\{(X_1, T_1), (X_2, T_2)\}$ if and only if $x \in \operatorname{Ker}(X_1, T_1)$. This together with formula (4) gives the desired result.

THEOREM 9.3. *Let (X_1, T_1) be an admissible pair, and let (X_2, T_2) be a standard pair, and suppose that the spectra of T_1 and T_2 are disjoint. Then*

$$\operatorname{Ker}\{(X_1, T_1), (X_2, T_2)\} = \operatorname{Ker}(X_1, T_1) \oplus \{0\}.$$

Proof. Let $\mathscr{L}$ be as in Theorem 9.1, and let $S: \mathscr{L} \to \mathbb{C}_{p_1}$ be a linear map such that (5) holds true. Let P be a projection of $\mathscr{L}$ along Ker S, and put $\mathscr{M}_1 = \operatorname{Im} P$. The compression of T_1 to $\mathscr{M}_1$ will be denoted by $T_{\mathscr{M}_1}$, i.e.,

$$T_{\mathscr{M}_1} = PT_1 : \mathscr{M}_1 \to \mathscr{M}_1.$$

Observe that $\sigma(T_{\mathscr{M}_1})$ is a subset of $\sigma(T_1)$.

Next, we let $\mathscr{M}_2$ denote the range of S. From the second equality in (5) it is clear that $\mathscr{M}_2$ is invariant under T_2. The restriction of T_2 to $\mathscr{M}_2$, considered as a map from $\mathscr{M}_2$ into itself, will be denoted by $T_{\mathscr{M}_2}$. Note that $\sigma(T_{\mathscr{M}_2})$ is a subset of $\sigma(T_2)$.

Finally, let $S_o: \mathscr{M}_1 \to \mathscr{M}_2$ be the restriction of S to $\mathscr{M}_1$. Then S_o is invertible, and the second equality in (5) shows that

$$S_o^{-1} T_{\mathscr{M}_2} S_o = T_{\mathscr{M}_1}.$$

In other words, $T_{\mathscr{M}_1}$ and $T_{\mathscr{M}_2}$ are similar. In particular, $\sigma(T_{\mathscr{M}_1}) = \sigma(T_{\mathscr{M}_2})$. As $\sigma(T_1)$ and $\sigma(T_2)$ are disjoint, it follows that $\sigma(T_{\mathscr{M}_1}) = \sigma(T_{\mathscr{M}_2}) = \varnothing$, and hence the spaces $\mathscr{M}_1$ and $\mathscr{M}_2$ consist of the zero element only. But then $\mathscr{L} = \operatorname{Ker} S$, and formula (4) yields the desired result.

If (X_1, T_1) and (X_2, T_2) are arbitrary admissible pairs and the spectra of T_1 and T_2 are disjoint, then it can be shown that

$$\operatorname{Ker}\{(X_1, T_1), (X_2, T_2)\} = \operatorname{Ker}(X_1, T_1) \oplus \operatorname{Ker}(X_2, T_2).$$

As $\operatorname{Ker}(X_2, T_2) = \{0\}$ for a standard pair (X_2, T_2), the previous formula is the natural generalization of Theorem 9.3.

10. *Index of Stabilization*

Let (X, T) be an admissible pair of order p. By definition,

$$\operatorname{Ker}(X, T) = \{x \in \mathbb{C}_p \mid XT^\alpha x = 0, \alpha = 0, 1, 2, \ldots\}.$$

Observe that $\mathscr{N}_j = \{x \in \mathbb{C}_p \mid XT^\alpha x = 0, \alpha = 0, 1, 2, \ldots, j-1\}, j = 1, 2, \ldots,$ form a descending sequence of subspaces of $\mathbb{C}_p$, and hence there exists a positive integer $\jmath$ such that $\mathscr{N}_\jmath = \mathscr{N}_{\jmath+1} = \cdots$. The least positive integer $\jmath$

with this property will be called the *index of stabilization* of the pair (X, T). We denote this number by $\operatorname{ind}(X, T)$. As $\bigcap_j \mathcal{N}_j = \operatorname{Ker}(X, T)$, we have $\operatorname{Ker}(X, T) = \operatorname{Ker}[\operatorname{col}(XT^{i-1})_{i=1}^{\mathscr{s}}]$ if and only if $\mathscr{s} \geq \operatorname{ind}(X, T)$. Also $\operatorname{ind}(X, T)$ is the least positive integer $\mathscr{s}$ such that the operator $\operatorname{col}(XT^{i-1})_{i=1}^{\mathscr{s}}$ has maximal rank.

PROPOSITION 10.1. *The index of stabilization of the pair (X, T) is less than or equal to the degree of the minimal polynomial of T.*

Proof. Let $\mathscr{s}$ be the degree of the minimal polynomial of T, and let $A = \operatorname{col}(XT^{i-1})_{i=1}^{\mathscr{s}}$. Take x in $\operatorname{Ker} A$. We have to show that $x \in \operatorname{Ker}(X, T)$. From our choice of x it follows that $x, Tx, \ldots, T^{\mathscr{s}-1}x$ are in $\operatorname{Ker} X$. Let $\mathcal{M}$ be the linear hull of these vectors. Then $\mathcal{M} \subset \operatorname{Ker} X$. As $\mathscr{s}$ is the degree of the minimal polynomial of T, the space $\mathcal{M}$ is invariant under T. But then $T^\alpha x \in \mathcal{M} \subset \operatorname{Ker} X$ for all $\alpha \geq 0$. It follows that $x \in \operatorname{Ker}(X, T)$, and the proposition is proved.

Note that for a standard pair (X, T) the index of stabilization is precisely equal to the degree of the operator polynomial L represented by (X, T). The next theorem shows that for certain parts of (X, T) the index of stabilization still may be expressed in terms of the operator polynomial L.

THEOREM 10.2. *Let (X, T) be a standard pair representing a monic operator polynomial L of degree ℓ. Put*

$$P = (1/2\pi i) \int_\Gamma (\lambda I - T)^{-1}\, d\lambda,$$

where Γ is a closed contour consisting of regular points of L, and let

$$X_1 = X|_{\operatorname{Im} P}, \qquad T_1 = T|_{\operatorname{Im} P}.$$

Then (X_1, T_1) is an admissible pair and $\operatorname{ind}(X_1, T_1)$ is equal to the least positive integer $\mathscr{s}$ such that the operator

$$M_{\mathscr{s}} = (1/2\pi i) \int_\Gamma (\lambda^{i+j-2} L(\lambda)^{-1})_{i=1, j=1}^{\mathscr{s}, \ell}\, d\lambda$$

has maximal rank.

Proof. (In the theorem, as well as in the proof, $\operatorname{Im} P$ is identified with a space $\mathbb{C}_{p_1}$.) As $\operatorname{Im} P = \mathbb{C}_{p_1}$ is invariant under T, we see that (X_1, T_1) is an admissible pair. Further, the calculations carried out in the proof of Theorem 21 in Gohberg *et al.* [4] show that the operator $M_{\mathscr{s}}$ and

$$[\operatorname{col}(XT^{i-1})_{i=1}^{\mathscr{s}}]|_{\operatorname{Im} P}$$

have the same rank. But the latter operator is equal to $\mathrm{col}(X_1T_1^{i-1})_{i=1}^{s}$, and hence the theorem is proved.

Let $(X_1, T_1), \ldots, (X_r, T_r)$ be admissible pairs of orders $p_1, \ldots, p_r$, respectively. By definition,

(a) $\mathrm{Ker}\{(X_1, T_1), \ldots, (X_r, T_r)\} = \mathrm{Ker}(X, T)$,
(b) $\mathrm{ind}\{(X_1, T_1), \ldots, (X_r, T_r)\} = \mathrm{ind}(X, T)$,

where $X = \mathrm{row}(X_j)_{j=1}^r$ and $T = \mathrm{diag}(T_j)_{j=1}^r$. Note that (X, T) is an admissible pair of order $p_1 + \cdots + p_r$, and hence the right-hand sides of (a) and (b) are well defined. For $r = 2$ definition (a) agrees with formula (3) of Subsection 9. The number introduced in (b) will be called the *index of stabilization* of the pairs $(X_1, T_1), \ldots, (X_r, T_r)$. Clearly, this number is equal to the least positive integer s such that the solution space

$$\left\{\{x_1, \ldots, x_r\} \,\middle|\, \sum_{j=1}^{r} X_j T_j^{\alpha} x_j = 0,\ 0 \leqslant \alpha \leqslant s - 1\right\}$$

coincides with $\mathrm{Ker}\{(X_1, T_1), \ldots, (X_r, T_r)\}$. For standard pairs, representing right divisors of a monic operator polynomial, the meaning of the two entities (a) and (b) is explained by the next theorem.

THEOREM 10.3. *Let $(X_1, T_1), \ldots, (X_r, T_r)$ be standard pairs representing $L_1, \ldots, L_r$, respectively. Suppose that $L_1, \ldots, L_r$ are right divisors of a monic operator polynomial L, and let $\mathcal{M}_1, \ldots, \mathcal{M}_r$ be the corresponding supporting subspaces. Then*

(i) $\dim \mathrm{Ker}\{(X_1, T_1), \ldots, (X_r, T_r)\} = \sum_{j=1}^{r-1} \dim\{(\mathcal{M}_1 + \cdots + \mathcal{M}_j) \cap \mathcal{M}_{j+1}\}$,

(ii) $\mathrm{ind}\{(X_1, T_1), \ldots, (X_r, T_r)\} \leqslant \mathrm{degree}(L)$.

Proof. We shall use the notation introduced in the first paragraph of Subsection 4. So for any $p \geqslant 1$ we have

$$(X_j T_j^{i-1})_{i=1, j=1}^{p, r} = [\mathrm{col}(X T^{i-1})_{i=1}^{p}][\mathrm{row}(K_j)_{j=1}^{r}].$$

As (X, T) represents the polynomial L, the operator $\mathrm{col}(XT^{i-1})_{i=1}^{p}$ is injective for $p \geqslant \mathrm{degree}(L)$. Hence

$$\mathrm{Ker}(X_j T_j^{i-1})_{i=1, j=1}^{p, r} = \mathrm{Ker}[\mathrm{row}(K_j)_{j=1}^{r}]$$

whenever $p \geqslant \mathrm{degree}(L)$. But then it follows that (ii) holds. Further, we see that

$$\mathrm{Ker}\{(X_1, T_1), \ldots, (X_r, T_r)\} = \mathrm{Ker}[\mathrm{row}(K_j)_{j=1}^{r}].$$

Recall that for $1 \leqslant j \leqslant r$ the operator K_j is injective and $\operatorname{Im} K_j = \mathcal{M}_j$. By induction on r one easily sees that this implies that $\operatorname{Ker}[\operatorname{row}(K_j)_{j=1}^r]$ is given by the right-hand side of (i), and hence the theorem is proved.

11. *Common Extensions of Pairs*

Let (X, T) and (X_1, T_1) be admissible pairs with orders p and p_1, respectively. The pair (X, T) is called an *extension* of the pair (X_1, T_1), if there exist a T-invariant subspace $\mathcal{M}$ of $\mathbb{C}_p$ and a bijective operator S from $\mathbb{C}_{p_1}$ onto $\mathcal{M}$ such that

$$TS = ST_1, \qquad X_1 = XS. \tag{1}$$

For standard pairs the meaning of this notion is explained by the following theorem.

THEOREM 11.1. *Let (X, T) and (X_1, T_1) be standard pairs representing the monic operator polynomials L and L_1, respectively. Then (X, T) is an extension of (X_1, T_1) if and only if L_1 is a right divisor of L.*

Proof. Let $n\ell$ and nk be the orders of the pairs (X, T) and (X_1, T_1), respectively. In other words, we assume that $\operatorname{degree}(L) = \ell$ and $\operatorname{degree}(L_1) = k$.

Suppose that (X, T) is an extension of (X_1, T_1). Then there exist a T-invariant subspace $\mathcal{M}$ of $\mathbb{C}_{n\ell}$ and a bijective operator S from $\mathbb{C}_{nk}$ onto $\mathcal{M}$ such that formula (1) holds. As S is bijective, $\dim \mathcal{M} = nk$. Put

$$A = \operatorname{col}(XT^{i-1})_{i=1}^{k}. \tag{2}$$

From (1) we see that $AS = \operatorname{col}(X_1 T_1^{i-1})_{i=1}^{k}$. As the right-hand side of this last identity is invertible, the restriction A_o of A to $\mathcal{M}$ is a bijective operator from $\mathcal{M}$ onto $\mathbb{C}_{nk}$. Let $\operatorname{row}(W_j)_{j=1}^{k}$ be equal to A_o^{-1}. Then, by [3, Theorem 8], the operator polynomial

$$\lambda^k I - XT^k(W_1 + W_2\lambda + \cdots + W_k\lambda^{k-1}) \tag{3}$$

is a right divisor of L. As

$$[\operatorname{col}(X_1 T_1^{i-1})_{i=1}^{k}]^{-1} = S^{-1}A_o^{-1} = \operatorname{row}(S^{-1}W_j)_{j=1}^{k}$$

and $XT^kW_j = X_1 T_1{}^k(S^{-1}W_j)$ (by formula (1)), it follows that the operator polynomial (3) is equal to L_1. Thus L_1 is a right divisor of L.

Conversely, suppose that L_1 is a right divisor of L. Let A be as in formula (2). Take $\mathcal{M}$ to be the supporting subspace for L_1 (cf. [3, Section 5, Remark 2]). Then $\mathcal{M}$ is an nk-dimensional T-invariant subspace of $\mathbb{C}_{n\ell}$ and the restriction A_o of A to $\mathcal{M}$ is a bijective operator from $\mathcal{M}$ onto $\mathbb{C}_{nk}$. Put $S =$

$A_o^{-1}[\operatorname{col}(X_1 T_1^{i-1})_{i=1}^{\ell}]$. We shall show that with this choice of S formula (1) holds true. As

$$AS = [\operatorname{col}(XT^{i-1})_{i=1}^{\ell}]S = \operatorname{col}(X_1 T_1^{i-1})_{i=1}^{\ell},$$

we have $XS = X_1$. Further we see from [3, Section 5, Remark 3] that

$$[\operatorname{col}(X_1 T_1^{i-1})_{i=1}^{\ell}]T_1[\operatorname{col}(X_1 T_1^{i-1})_{i=1}^{\ell}]^{-1} = A_0 T A_0^{-1}.$$

But this is just another formulation of the first equality in formula (1).

Analogous to the similar notion for standard pairs (see Subsection 2) we define two admissible pairs (X, T) and $(\tilde{X}, \tilde{T})$ to be *similar* if the pairs (X, T) and $(\tilde{X}, \tilde{T})$ have the same order, p say, and there exists a bijective operator S on $\mathbb{C}_p$ such that

$$\tilde{X} = XS^{-1}, \qquad \tilde{T} = STS^{-1}.$$

Observe that in this case $\tilde{X}\tilde{T}^{\alpha} = XT^{\alpha}S^{-1}$ for all $\alpha \geqslant 0$. It follows that $\dim \operatorname{Ker}(X, T)$ and $\operatorname{ind}(X, T)$ do not change if the pair (X, T) is replaced by a similar one.

Note that the relation of extension is stable under similarity. That is, if (X, T) is an extension of (X_1, T_1), then $(\tilde{X}, \tilde{T})$ will be an extension of $(\tilde{X}_1, \tilde{T}_1)$ whenever $(\tilde{X}, \tilde{T})$ is similar to (X, T) and $(\tilde{X}_1, \tilde{T}_1)$ is similar to (X_1, T_1).

THEOREM 11.2. *Let (X_1, T_1) be an admissible pair of order p_1, and let (X_2, T_2) be a standard pair of order p_2. Suppose that* $\operatorname{Ker}(X_1, T_1) = \{0\}$, *and let* $q = \dim \operatorname{Ker}\{(X_1, T_1), (X_2, T_2)\}$. *Then there exists an admissible pair (X_o, T_o) of order $(p_1 + p_2) - q$ such that (X_o, T_o) is a common extension of (X_1, T_1) and (X_2, T_2) and*

(i) $\operatorname{Ker}(X_o, T_o) = \{0\}$,
(ii) $\operatorname{ind}(X_o, T_o) = \operatorname{ind}\{(X_1, T_1), (X_2, T_2)\}$.

Proof. From theorem 9.1 we know that

$$\operatorname{Ker}\{(X_1, T_1), (X_2, T_2)\} = \{\{x, -Sx\} \mid x \in \mathscr{L}\},$$

where $\mathscr{L}$ is a T_1-invariant subspace of $\mathbb{C}_{p_1}$ and $S: \mathscr{L} \to \mathbb{C}_{p_2}$ is a linear map such that

$$X_1|_{\mathscr{L}} = X_2 S, \qquad ST_1|_{\mathscr{L}} = T_2 S. \tag{4}$$

Put $S\mathscr{L} = \mathscr{M}$, and note that $\mathscr{M}$ is invariant under T_2. As $\operatorname{Ker}(X_1, T_1)$ consists of the zero element only, the map S is injective (see Corollary 9.2), and hence $\dim \mathscr{M} = \dim \mathscr{L} = q$. Furthermore, we see from the second equality in (4) that the restricted operators $T_1|_{\mathscr{L}}$ and $T_2|_{\mathscr{M}}$ are similar.

Let $e_1, \ldots, e_{p_1}$ be the canonical basis of $\mathbb{C}_{p_1}$, and let $f_1, \ldots, f_{p_2}$ be the canonical basis of $\mathbb{C}_{p_2}$. In order to prove the theorem we may replace (X_1, T_1) and (X_2, T_2) by similar pairs whenever this is convenient. Hence, replacing (X_1, T_1) and (X_2, T_2) by similar pairs, we may suppose without loss of generality that

$$\mathscr{L} = \operatorname{sp}\{e_j \mid p_1 - q + 1 \leqslant j \leqslant p_1\}, \qquad \mathscr{M} = \operatorname{sp}\{f_j \mid 1 \leqslant j \leqslant q\}.$$

Now we look at the matrices of X_1, T_1 and X_2, T_2 with respect to the canonical bases of $\mathbb{C}_{p_1}$ and $\mathbb{C}_{p_2}$, respectively. As $T_1|_{\mathscr{L}}$ and $T_2|_{\mathscr{M}}$ are similar we may suppose (after replacing the pair (X_2, T_2) by a similar pair) that the matrix of $T_1|_{\mathscr{L}}$ with respect to the basis $e_{p_1-q+1}, \ldots, e_{p_1}$ is equal to the matrix of $T_2|_{\mathscr{M}}$ with respect to the basis $f_1, \ldots, f_q$. Moreover, by the first equality in formula (4), the new pair (X_2, T_2) may be chosen such that

$$X_1 e_{p_1-q+j} = X_2 f_j \qquad (1 \leqslant j \leqslant q).$$

It follows that without loss of generality we may suppose that the matrices of X_1, T_1 and X_2, T_2 have the following form: $X_1 = (Y_1 Z)$, $X_2 = (Z Y_2)$, and

$$T_1 = \begin{bmatrix} U_1 & 0 \\ V_1 & W \end{bmatrix}, \qquad T_2 = \begin{bmatrix} W & V_2 \\ 0 & U_2 \end{bmatrix},$$

where Z is a $n \times q$ matrix and W a $q \times q$ matrix. Let $X_o : \mathbb{C}_{p_1+p_2-q} \to \mathbb{C}_n$ and $T_o : \mathbb{C}_{p_1+p_2-q} \to \mathbb{C}_{p_1+p_2-q}$ be those operators whose matrices (with respect to the canonical bases of $\mathbb{C}_{p_1+p_2-q}$ and $\mathbb{C}_n$) are given by

$$X_0 = (Y_1 Z Y_2), \qquad T_0 = \begin{bmatrix} U_1 & 0 & 0 \\ V_1 & W & V_2 \\ 0 & 0 & U_2 \end{bmatrix}.$$

Then it is not difficult to check that the pair (X_0, T_0) has the desired properties.

The previous theorem is also true if (X_2, T_2) is not a standard pair but an admissible pair with $\operatorname{Ker}(X_2, T_2) = \{0\}$. Further, it can be shown that the pair (X_o, T_o) constructed in the above proof is in a certain sense the minimal common extension with the desired properties.

THEOREM 11.3. *Let (X_1, T_1) be an admissible pair of order p_1, and let $(X_2, T_2), \ldots, (X_r, T_r)$ be standard pairs of orders $p_2, \ldots, p_r$, respectively. Suppose that* $\operatorname{Ker}(X_1, T_1) = \{0\}$, *and let*

$$q = \dim \operatorname{Ker}\{(X_1, T_1), \ldots, (X_r, T_r)\}.$$

Then there exists an admissible pair (X, T) *of order* $(p_1 + \cdots + p_r) - q$ *such that* (X, T) *is a common extension of* $(X_1, T_1), \ldots, (X_r, T_r)$ *and*

(i) $\operatorname{Ker}(X, T) = \{0\}$.
(ii) $\operatorname{ind}(X, T) = \operatorname{ind}\{(X_1, T_1), \ldots, (X_r, T_r)\}$.

Proof. We shall prove the theorem by induction on the number of standard pairs. The case $r = 2$ is covered by Theorem 11.2. Therefore, take $r > 2$ and assume that the theorem is true for $r - 1$ standard pairs. Let (X_o, T_o) be the common extension of (X_1, T_1) and (X_2, T_2) constructed in the proof of Theorem 11.2. Note that the order of (X_o, T_o) is equal to

$$p_o = (p_1 + p_2) - \dim \operatorname{Ker}\{(X_1, T_1), (X_2, T_2)\}.$$

Further, the construction of (X_o, T_o) is such that

$$q_o = \dim \operatorname{Ker}\{(X_o, T_o), (X_3, T_3), \ldots, (X_r, T_r)\}$$

is equal to $q - \dim \operatorname{Ker}\{(X_1, T_1), (X_2, T_2)\}$. It follows that

$$(p_1 + \cdots + p_r) - q = (p_o + p_3 + \cdots + p_r) - q_o. \tag{5}$$

Also, replacing (X_1, T_1) and (X_2, T_2) by (X_o, T_o) does not effect the index of stabilization. In other words

$$\operatorname{ind}\{(X_1, T_1), \ldots, (X_r, T_r)\} = \operatorname{ind}\{(X_o, T_o), (X_3, T_3), \ldots, (X_r, T_r)\}. \tag{6}$$

Now by our induction hypothesis the pairs $(X_o, T_o), (X_3, T_3), \ldots, (X_r, T_r)$ have a common extension (X, T) of order $(p_o + p_3 + \cdots + p_r) - q_o$ such that (i) holds and

$$\operatorname{ind}(X, T) = \operatorname{ind}\{(X_o, T_o), (X_3, T_3), \ldots, (X_r, T_r)\}. \tag{7}$$

From formula (5) it follows that (X, T) has the right order. Further, we see from formulas (6) and (7) that (ii) holds true for (X, T). Finally, as (X_o, T_o) is an extension of (X_1, T_1) and (X_2, T_2), the pair (X, T) is a common extension of $(X_1, T_1), \ldots, (X_r, T_r)$, and the proof is complete.

12. *Extensions to Standard Pairs*

Let (X_1, T_1) be an admissible pair of order p_1, and suppose that $\operatorname{Ker}(X_1, T_1) = \{0\}$. Let $\jmath$ be the index of stabilization of the pair (X_1, T_1). It follows that the linear map

$$\operatorname{col}(X_1 T_1^{i-1})_{i=1}^{\jmath} : \mathbb{C}_{p_1} \to \mathbb{C}_{n\jmath}$$

is injective. In particular, $p_1 \leqslant n\jmath$. Suppose that $p_2 = n\jmath - p_1 > 0$. In this subsection we deal with the question whether there exists an admissible pair (X_2, T_2) of order p_2 such that the map

$$\begin{bmatrix} X_1 & X_2 \\ X_1 T_1 & X_2 T_2 \\ \vdots & \vdots \\ X_1 T_1^{\ell-1} & X_2 T_2^{\ell-1} \end{bmatrix} : \mathbb{C}_{n\ell} \to \mathbb{C}_{n\ell}$$

is invertible. Note that in that case the operators $X = (X_1 X_2)$ and $T = T_1 \oplus T_2$ form a standard pair, which is an extension of the pair (X_1, T_1). To give an idea of what it means to extend an admissible pair to a standard pair, let us assume that T_1 is a Jordan matrix. Then the problem is to find a monic operator polynomial L such that T_1 is a part of the Jordan matrix of L and X_1 is the corresponding part of the canonical system of eigenvectors and generalized eigenvectors.

We shall show that it is possible to extend the admissible pair (X_1, T_1) to a standard pair. Further, we shall prove that the operators X_2 and T_2 may be constructed in a rather special way. We begin with a simple lemma.

LEMMA 12.1. *Let (X, T) be an admissible pair, and let $\alpha \in \mathbb{C}$. Then $(X, T + \alpha I)$ is an admissible pair and*

$$\operatorname{Ker}(X, T) = \operatorname{Ker}(X, T + \alpha I), \qquad \operatorname{ind}(X, T) = \operatorname{ind}(X, T + \alpha I).$$

Proof. For each $j \geqslant 0$, we have

$$\begin{bmatrix} I & 0 & \cdots & 0 \\ \alpha I & I & \cdots & 0 \\ \alpha^2 I & 2\alpha I & \cdots & 0 \\ \vdots & \vdots & & \vdots \\ \alpha^j I & j\alpha^{j-1} I & \cdots & I \end{bmatrix} \begin{bmatrix} X \\ XT \\ XT^2 \\ \vdots \\ XT^j \end{bmatrix} = \begin{bmatrix} X \\ X(T+\alpha I) \\ X(T+\alpha I)^2 \\ \vdots \\ X(T+\alpha I)^j \end{bmatrix}.$$

As the first factor in the left hand side of this identity is invertible, the lemma is proved.

In the sequel, whenever it is helpful, we shall identify linear maps between $\mathbb{C}_p$ and $\mathbb{C}_q$ with their matrix representations with respect to the canonical bases in these spaces. Vectors in $\mathbb{C}_n$ will be considered as $n \times 1$ matrices.

LEMMA 12.2. *Let (X, T) be an admissible pair of order p, and suppose that $\alpha \notin \sigma(T)$. Further, suppose that*

$$\operatorname{col}(XT^{i-1})_{i=1}^{\ell} \tag{1}$$

has full rank. If $p < n\ell$, *then for some vector* Z *in the canonical basis of* $\mathbb{C}_n$, *the matrix*

$$\begin{bmatrix} X & Z \\ XT & \alpha Z \\ \vdots & \vdots \\ XT^{\ell-1} & \alpha^{\ell-1}Z \end{bmatrix}$$

has full rank too.

Proof. From the proof of Lemma 12.1 it is clear that without loss of generality we may suppose that $\alpha = 0$. If the first row of X is zero, then we take Z to be the first vector in the canonical basis of $\mathbb{C}_n$ and the lemma is proved. Therefore, we shall suppose that the first row in X contains nonzero elements. As the matrix (1) has full rank and $p < n\ell$, there exists $2 \leqslant j \leqslant n\ell$ such that the first $j - 1$ rows in (1) are linearly independent and the jth row is a linear combination of the previous rows. In this unique linear combination rows of X must appear.

Suppose not. The $j \geqslant n + 1$ and the $(j - n)$th row in the matrix

$$\operatorname{col}(XT^i)_{i=1}^{\ell-1} = [\operatorname{col}(XT^{i-1})_{i=1}^{\ell-1}]T$$

is a linear combination of its predecessors. As T is invertible, this implies that the $(j - n)$th row in (1) is a linear combination of its predecessors. This is a contradiction.

So the jth row in the matrix (1) is a unique linear combination of the previous rows and in this linear combination rows of X are used. Let us assume that the kth row of X appears in this linear combination. Take Z to be the kth vector in the canonical basis of $\mathbb{C}_n$. Then, clearly, the row rank of the matrix

$$\begin{bmatrix} X & Z \\ XT & 0 \\ \vdots & \vdots \\ XT^{\ell-1} & 0 \end{bmatrix}$$

is one more than the row rank of the matrix (1). Hence the lemma is proved.

THEOREM 12.3. *Let* (X_1, T_1) *be an admissible pair of order* p_1. *Let* ℓ *be* $\operatorname{ind}(X_1, T_1)$, *and suppose that* $\operatorname{Ker}(X_1, T_1) = \{0\}$. *If* $p_1 < n\ell$, *then there exists an admissible pair* (X_2, T_2) *of order* $p_2 = n\ell - p_1$ *such that the operators* $X = (X_1X_2)$ *and* $T = T_1 \oplus T_2$ *from a standard pair of order* $n\ell$. *Further, one may choose* T_2 *such that*

$$T_2 = (\alpha_i \delta_{ij})_{i,j=1}^{p_2}, \tag{2}$$

where $\{\alpha_1, \alpha_2, \ldots, \alpha_{p_2}\}$ *is an arbitrary set of mutually different complex numbers not in* $\sigma(T_1)$.

Proof. Our conditions on (X_1, T_1) imply that the matrix $\operatorname{col}(X_1 T_1^{i-1})_{i=1}^{\ell}$ has full rank. Take p_2 different complex numbers $\alpha_1, \ldots, \alpha_{p_2}$, not in $\sigma(T_1)$. By repeatedly applying Lemma 12.2 one finds p_2 vectors $Z_1, \ldots, Z_{p_2}$ in $\mathbb{C}_n$ such that the matrix

$$\begin{bmatrix} X_1 & Z_1 & \cdots & Z_{p_2} \\ X_1 T_1 & \alpha_1 Z_1 & \cdots & \alpha_{p_2} Z_{p_2} \\ \vdots & \vdots & & \vdots \\ X_1 T_1^{\ell-1} & \alpha_1^{\ell-1} Z_1 & \cdots & \alpha_{p_2}^{\ell-1} Z_{p_2} \end{bmatrix}$$

has full rank and, hence, is invertible. Now take T_2 as in formula (2), and put $X_2 = \operatorname{row}(Z_j)_{j=1}^{p_2}$. Then the pair (X_2, T_2) has the desired properties.

THEOREM 12.4. *Let* (X_0, T_0) *be an admissible pair of order* nk. *Let* ℓ *be* $\operatorname{ind}(X_0, T_0)$, *and suppose that* $\operatorname{Ker}(X_0, T_0) = \{0\}$. *If* $q = \ell - k > 0$, *then there exist* $X_1, \ldots, X_q, T_1, \ldots, T_q$, *linear operators acting on* $\mathbb{C}_n$, *such that the operators*

$$\tilde{X} = \operatorname{row}(X_j)_{j=0}^{q}, \qquad \tilde{T} = \operatorname{diag}(T_j)_{j=0}^{q}$$

form a standard pair of order $n\ell$, $X_1, \ldots, X_q$ *are invertible, and the spectra* $\sigma(T_0), \sigma(T_1), \ldots, \sigma(T_q)$ *are mutually disjoint.*

Proof. From Theorem 12.3 it is clear that one can find $Y_1, \ldots, Y_q$, $T_1, \ldots, T_q$, linear operators acting on $\mathbb{C}_n$, such that the spectra of T_0, $T_1, \ldots, T_q$ are mutually disjoint and

$$\begin{bmatrix} X_0 & Y_1 & \cdots & Y_q \\ X_0 T_0 & Y_1 T_1 & \cdots & Y_q T_q \\ \vdots & \vdots & & \vdots \\ X_0 T_0^{\ell-1} & Y_1 T_1^{\ell-1} & \cdots & Y_q T_q^{\ell-1} \end{bmatrix} \tag{3}$$

is an invertible operator on $\mathbb{C}_{n\ell}$. Recall that the invertible operators on $\mathbb{C}_{n\ell}$ form an open set. Hence, if in (3) one replaces $Y_1, \ldots, Y_q$ by operators $\tilde{Y}_1, \ldots, \tilde{Y}_q$, then the new operator matrix will again be invertible whenever for each j the operator $\tilde{Y}_j$ is sufficiently close to Y_j. Now the invertible linear operators on $\mathbb{C}_n$ are dense in the set of all linear operators on $\mathbb{C}_n$. So by choosing invertible operators $X_1, \ldots, X_q$ on $\mathbb{C}_n$ sufficiently close to

$Y_1, \ldots, Y_q$, it may be supposed that the operator matrix

$$(X_j T_j^{i-1})_{i=1,\,j=0}^{s\ \ \,q}$$

is invertible. But then the operators $X_1, \ldots, X_q, T_1, \ldots, T_q$ have the desired properties.

Note that in the previous theorem the invertibility condition on $X_1, \ldots, X_q$ is equivalent to the requirement that the pairs $(X_1, T_1), \ldots, (X_q, T_q)$ are standard pairs which represent monic operator polynomials of degree 1.

13. *Common Monic Left Multiples*

Throughout this subsection $L_1, \ldots, L_r$ will be monic operator polynomials on $\mathbb{C}_n$ with degrees $k_1, \ldots, k_r$, respectively. Recall (see Subsection 3) that a monic operator polynomial L is called a common left multiple of $L_1 \ldots, L_r$ if for each j the polynomial L_j is a right divisor of L. In this subsection among other things we shall construct monic common left multiples of minimal degree.

Let $(X_1, T_1), \ldots, (X_r, T_r)$ be standard pairs representing $L_1, \ldots, L_r$, respectively. Put

$$\text{ind}\{L_1, \ldots, L_r\} = \text{ind}\{(X_1, T_1), \ldots, (X_r, T_r)\}.$$

We call this number the *index of stabilization* of the family $L_1, \ldots, L_r$. Note that the definition of the index does not depend on the special choice of the representing pairs $(X_1, T_1), \ldots, (X_r, T_r)$.

THEOREM 13.1. *Let $L_1, \ldots, L_r$ be monic operator polynomials, and let s be* ind$\{L_1, \ldots, L_r\}$. *Then there exists a monic common left multiple of $L_1, \ldots, L_r$ of degree s. Conversely, if L is a monic common left multiple of $L_1, \ldots, L_r$, then* degree$(L) \geqslant s$.

Proof. Let $(X_1, T_1), \ldots, (X_r, T_r)$ be standard pairs representing $L_1, \ldots, L_r$, respectively. By Theorem 11.3 the pairs $(X_1, T_1), \ldots, (X_r, T_r)$ have a common extension (X_o, T_o) such that $\text{Ker}(X_o, T_o) = \{0\}$ and

$$\text{ind}(X_o, T_o) = \text{ind}\{(X_1, T_1), \ldots, (X_r, T_r)\} = s$$

As $\text{Ker}(X_o, T_o) = \{0\}$, we can apply Theorem 12.3 to show that there exists a standard pair (X, T) of order ns such that (X, T) is an extension of (X_o, T_o). Let L be the monic operator polynomial represented by (X, T). Then degree$(L) = s$. As (X_o, T_o) is a common extension of $(X_1, T_1), \ldots, (X_r, T_r)$, the same is true for (X, T). Hence (see Theorem 11.1) $L_1, \ldots, L_r$ are right divisors of L, and the first part of the theorem is proved. The second part of the theorem is covered by Theorem 10.3(ii).

If the polynomials $L_1, \ldots, L_r$ form a commuting family, then

$$\operatorname{ind}\{L_1, \ldots, L_r\} \leqslant \sum_{j=1}^{r} \operatorname{degree}(L_j)$$

because in that case $L = \prod_{j=1}^{r} L_j$ is a common multiple of $L_1, \ldots, L_r$ and, clearly, $\operatorname{degree}(L) = \sum_{j=1}^{r} \operatorname{degree}(L_j)$.

Recall (see Subsection 6) that the family $L_1, \ldots, L_r$ is said to be minimal whenever

$$L_j(d/dx)\varphi_j = 0 \qquad (1 \leqslant j \leqslant r), \qquad \varphi_1 + \cdots + \varphi_r = 0$$

implies that $\varphi_1 = \varphi_2 = \cdots = \varphi_r = 0$. We shall now describe this property in terms of representing pairs.

PROPOSITION 13.2. *Let $L_1, \ldots, L_r$ be monic operator polynomials, and let $(X_1, T_1), \ldots, (X_r, T_r)$ be standard pairs representing $L_1, \ldots, L_r$, respectively. Then the family $L_1, \ldots, L_r$ is minimal if and only if* $\operatorname{Ker}\{(X_1, T_1), \ldots, (X_r, T_r)\} = \{0\}$.

Proof. By the previous theorem we may suppose without loss of generality that $L_1, \ldots, L_r$ are right divisors of a monic operator polynomial. But then we can apply Theorems 6.1 and 10.3(i) to get the desired result.

THEOREM 13.3. *Let $L_1, \ldots, L_r$ be a minimal family of monic operator polynomials on $\mathbb{C}_n$ with degrees $k_1, \ldots, k_r$, respectively. Put*

$$q = \operatorname{ind}\{L_1, \ldots, L_r\} - (k_1 + \cdots + k_r).$$

Then there exist monic operator polynomials $L_{r+1}, \ldots, L_{r+q}$ on $\mathbb{C}_n$ of degree 1 *such that the Vandermonde operator*

$$V(L_1, \ldots, L_r, L_{r+1}, \ldots, L_{r+q})$$

is invertible. Furthermore, $L_{r+1}, \ldots, L_{r+q}$ can be chosen such that the spectra $\sigma(L_{r+1}), \ldots, \sigma(L_{r+q})$ are mutually disjoint and have an empty intersection with $\bigcup_{j=1}^{r} \sigma(L_r)$.

Proof. Let $(X_1, T_1), \ldots, (X_r, T_r)$ be standard pairs representing $L_1, \ldots, L_r$, respectively. Put $X_0 = \operatorname{row}(X_j)_{j=1}^{r}$ and $T_0 = \operatorname{diag}(T_j)_{j=1}^{r}$. Then (X_0, T_0) is an admissible pair of order $n(k_1 + \cdots + k_r)$ and

$$s = \operatorname{ind}\{L_1, \ldots, L_r\} = \operatorname{ind}(X_0, T_0).$$

Further, as the family $L_1, \ldots, L_r$ is minimal, we have $\operatorname{Ker}(X_0, T_0) = \{0\}$. So we can apply Theorem 12.4 to show the existence of standard pairs

$(X_{r+1}, T_{r+1}), \ldots, (X_{r+q}, T_{r+q})$ of order n such that

$$(X_j T_j^{i-1})_{i=1,\, j=1}^{s \quad r+q} \tag{1}$$

is invertible. Let $L_{r+1}, \ldots, L_{r+q}$ be the monic operator polynomials represented by the pairs $(X_{r+1}, T_{r+1}), \ldots, (X_{r+q}, T_{r+q})$. Then $L_{r+1}, \ldots, L_{r+q}$ have degree 1, and

$$\sum_{j=1}^{r+q} \operatorname{degree}(L_j) = (k_1 + \cdots + k_r) + q = s.$$

So up to an invertible block diagonal operator matrix the Vandermonde operator $V(L_1, \ldots, L_{r+q})$ is equal to (1), and hence $V(L_1, \ldots, L_{r+q})$ is invertible. This proves the first part of the theorem.

To prove the second part, note that $\sigma(T_j) = \sigma(L_j)$ for $1 \leqslant j \leqslant r + q$. As $T_0 = \operatorname{diag}(T_j)_{j=1}^r$, it follows that $\sigma(T_0) = \bigcup_{j=1}^r \sigma(L_j)$. But, by Theorem 12.4, we can choose $T_{r+1}, \ldots, T_{r+q}$ in such a way that the spectra $\sigma(T_0)$, $\sigma(T_{r+1}), \ldots, \sigma(T_{r+q})$ are mutually disjoint. This gives the desired result.

Let $L(\lambda) = A_0 + A_1\lambda + \cdots + A_\ell \lambda^\ell$ be a monic operator polynomial. It is well known (e.g., see [9]) that the linear polynomial $\lambda I - T$ is a right divisor of L if and only if

$$A_0 + A_1 T + \cdots + A_\ell T^\ell = 0. \tag{2}$$

If (2) holds, then T will be called a *root* of L. In terms of roots Theorems 13.1 and 13.3 can be phrased as follows.

COROLLARY 13.4. *Let $T_1, \ldots, T_r$ be linear operators on $\mathbb{C}_n$, and let s be the least positive integer such that the rank of*

$$(T_j^{i-1})_{i=1,\, j=1}^{s \quad r}$$

is maximal as a function of s. Then $T_1, \ldots, T_r$ are roots of a monic operator polynomial of degree s. Conversely, if $T_1, \ldots, T_r$ are roots of a monic operator polynomial L, then $\operatorname{degree}(L) \geqslant s$.

COROLLARY 13.5. *Let $T_1, \ldots, T_r$ be linear operators on $\mathbb{C}_n$, and suppose that the spectra of $T_1, \ldots, T_r$ are mutually disjoint. Let s be the least positive integer such that the rank of*

$$(T_j^{i-1})_{i=1,\, j=1}^{s \quad r}$$

is maximal as a function of s. Then there exist linear operators $T_{r+1}, \ldots, T_s$ on $\mathbb{C}_n$ such that

$$(T_j^{i-1})_{i,\, j=1}^{s}$$

is invertible and the spectra of $T_1, \ldots, T_r, T_{r+1}, \ldots, T_s$ are mutually disjoint.

Note that the disjointness condition on the spectra of $T_1, \ldots, T_r$ in Corollary 13.5 implies that the family $\lambda I - T_1, \ldots, \lambda I - T_r$ is minimal (cf. Theorem 7.1).

Note. After completion of the present paper we received a reprint of V. I. Kabak, A. S. Markus, and I. V. Mereutsa, On spectral properties of divisors of an operator polynomial pencil, *Isv. Akad. Nauk Moldavian SSR, Kishinev* 2 (1976), 24–26, in which, among other things, several theorems are announced which are closely related to the results proved in Subsection 7.

REFERENCES

1. J. E. DENNIS, Jr., J. F. TRAUB, AND R. P. WEBER, The algebraic theory of matrix polynomials, *SIAM J. Numer. Anal.* **13** (1976), 831–845.
2. I. M. GEL'FAND, "Lectures on Linear Algebra," Wiley (Interscience), New York, 1963.
3. I. GOHBERG, P. LANCASTER, AND L. RODMAN, Spectral analysis of matrix polynomials, I. Canonical forms and divisors, *Linear Algebra and Appl.* **20** (1978), 1–44.
4. I. GOHBERG, P. LANCASTER, AND L. RODMAN, Spectral analysis of matrix polynomials, II. The resolvent form and spectral divisors, *Linear Algebra and Appl.*, to appear.
5. M. G. KREIN AND H. LANGER, Certain mathematical principles of the linear theory of damped vibrations of continua, *in* "Applied Theory of Functions in Continuum Mechanics" (*Proc. Internat. Symp., Tbilisi, 1963*), Vol. II, Fluid and Gas Mechanics, Math. Methods, pp. 283–322, Nauka, Moscow, 1965 [in Russian].
6. P. LANCASTER, A fundamental theorem on lambda-matrices with applications, I: Ordinary differential equations with constant coefficients, *Linear Algebra and Appl.* **18** (1977), 189–211.
7. P. LANCASTER. A fundamental theorem on lambda-matrices with applications, II: Finite difference equations, *Linear Algebra and Appl.* **18** (1977), 213–222.
8. C. C. MACDUFFEE, "The Theory of Matrices," Chelsea, New York, 1946.
9. A. S. MARKUS AND I. V. MEREUTSA, On the complete n-tuple of roots of the operator equation corresponding to a polynomial operator bundle, *Izv. Akad. Nauk SSSR Ser. Mat.* **37** (1973), 1108–1131; English transl. *Math. USSR Izv.* **7** (1973), 1105–1128.
10. I. V. MEREUTSA, On the decomposition of an operator polynomial bundle into a product of linear factors, *Mat. Issled* **8** (1973), 102–114.

AMS (MOS) subject classifications: 15A18, 15A30, 47D99.

Orbit Structure of the Möbius Transformation Semigroup Acting on H^∞ (Broadband Matching)

J. WILLIAM HELTON†

Department of Mathematics
University of California, San Diego
La Jolla, California

DEDICATED TO M. G. KREIN IN HONOR OF HIS SEVENTIETH BIRTHDAY

1. INTRODUCTION

This article concerns a certain transformation semigroup and its orbit structure. The considerations here are motivated by something called the problem of "broadband matching" in electrical network theory and Section 7 gives a brief description of that problem. The mathematical results here yield concrete engineering results and these will be presented in an engineering article. Now we describe the central problem.

Let $H^\infty(\mathbb{C}^n)$ denote the uniformly bounded analytic functions in the disk with matrix values and let $\mathscr{B}$ denote the open unit ball in that space, namely $\mathscr{B} = \{f \in H^\infty(\mathbb{C}^n): \|f(z)\| \leqslant \alpha < 1$ for some α and all $|z| < 1\}$. The closed unit ball is denoted $\bar{\mathscr{B}}$. Define a *cascade transformation* to be a map $\mathscr{F}_U$: $\bar{\mathscr{B}} \to \bar{\mathscr{B}}$ which is defined for each $S \in \bar{\mathscr{B}}$ by

$$\mathscr{F}_U(S)(z) = A(z) + B(z)S(z)(I - D(z)S(z))^{-1}C(z),$$

where the function

$$U(z) = \begin{pmatrix} A(z) & B(z) \\ C(z) & D(z) \end{pmatrix}$$

satisfies $\det(1 - B(e^{i\theta})^*B(e^{i\theta})) \not\equiv 0$ and is an *inner* function, namely, it is in $\bar{\mathscr{B}}(\mathbb{C}^{2n})$ and has unitary boundary values. The fact that a cascade transformation maps $\bar{\mathscr{B}}$ into itself is easily verified (c.f. [H–Z]), as is the fact that the set $\mathscr{G}$ of all cascade transformations is a semigroup. Good basic references are

† Partially supported by the National Science Foundation.

ISBN 0-12-287150-2

[E–P] and [A2]. Denote by $\overline{\mathscr{G}}$ the larger semigroup of all $\mathscr{F}_U$ with U inner. The subset $\{\mathscr{F}(S):\mathscr{F} \in \mathscr{G}\}$ of $\mathscr{B}$ is called the *orbit of S under* $\mathscr{G}$. Traditionally, in studying transformation groups one devotes considerable attention to their orbit structure and that is what this article addresses. In particular this article concerns two types of questions.

I. What is the orbit structure of $\mathscr{G}$ on $\overline{\mathscr{B}}$? That is, given S, describe all Σ in its orbit.

II. Given S, which point Σ in its orbit is closest to 0? What is the distance (broadband matching problem)?

The orbit $\overline{\mathscr{B}}_0$ of 0 was first computed by Arov [A1; A2] and independently by Douglas and the author [D–H] to be those functions in $\overline{\mathscr{B}}$ which are pseudomeromorphic (see also [DeW]). This must be defined: A function $F(z)$ on $\mathbb{C}$ will be called pseudomeromorphic if it is meromorphic for $|z| < 1$, meromorphic in $\{z:|z| > 1\} \cup \infty$ of bounded type, and if $\lim_{r\downarrow 1} F(re^{i\theta})$ exists and equals $\lim_{r\uparrow 1} F(re^{i\theta})$ almost everywhere. The class of all such functions in $\mathscr{B}$ will be denoted $\mathscr{B}_0$ and many other characterizations of them are given in [D–S–S]. In this article we shall never be so ambitious as to consider all functions S_0 in $\mathscr{B}$, but we shall restrict our attention to functions in $\mathscr{B}_0$ and frequently to even more restricted classes, such as the rational functions.

The paper uses repeatedly a Wiener–Hopf type of factorization which was studied very effectively by Gohberg and Krein [G–K]; also, for a recent survey see [R–R]. We say that M is a *left* (resp., *right*) *spectral factor* of an $n \times n$ matrix valued function P on the unit circle if it is in $H^\infty(\mathbb{C}^n)$ and $P(e^{i\theta}) = M(e^{i\theta})M(e^{i\theta})^*$ (resp., $M(e^{i\theta})^*M(e^{i\theta})$). Recall Proposition 5 in [D–H]. If P is a nonnegative matrix valued function on the circle which is uniformly bounded and which has a pseudomeromorphic extension, then it has both a right and left spectral factorization. A theorem [N–F], Chapter V, §4 says that if a left (resp., right) spectral factor exists it may be taken to be an outer (resp., $*$-outer) function. Here we define an *outer* function M in $H^\infty(\mathbb{C}^n)$ as having the property $\{Mf:f \in H^2(\mathbb{C}^n)\}$ is dense in the Hardy space $H^2(\mathbb{C}^n)$ of $L^2(\mathbb{C}^n)$ functions with analytic continuation onto the disk. In other words the operator multiplication by M, denoted $\mathscr{M}_M$, on $H^2(\mathbb{C}^n)$ has dense range. The function M is $*$-*outer* means that $\tilde{M}(z) \triangleq M(\bar{z})^*$ is outer. The notation M^* will be used to denote the function $M(e^{i\theta})^*$ as well as its pseudomeromorphic continuation. Denote by $\bar{H}^\infty$ (resp., $\bar{H}^2$) the Hardy space of functions f in $L^\infty(T)$ (resp., $L^2(T)$) so that f (resp., $e^{i\theta}f(e^{i\theta})$) extends to a function which is analytic outside T. We shall frequently work with subspaces of H^2 of the form $\mathscr{S}(U) = H^2(\mathbb{C}^n) \ominus \mathscr{M}_U H^2(\mathbb{C}^n)$ where U is an inner function; let $P_{\mathscr{R}}$ denote the orthogonal projection onto a subspace $\mathscr{R}$ of L^2 and denote by

$\Gamma_U(G)$ the operator

$$\Gamma_U(G) = P_{H^2} \mathcal{M}_{G(e^{i\theta})^*}\Big|_{\mathcal{S}(U)},$$

which maps $\mathcal{S}(U)$ into $H^2(\mathbb{C}^n)$.

We turn from generalities to the constructions specific to the study of cascade orbits. To a given S in $\bar{\mathscr{B}}_0$ we associate several functions. First, denote by A_S the outer right spectral factor of $1 - S(e^{i\theta})^*S(e^{i\theta})$ and by B_S' the $*$-outer left spectral factor of $1 - S(e^{i\theta})S(e^{i\theta})^*$. The function B_S' has a unique (up to unitary multiple) factorization $B_S' = B_S V$ with B_S outer and V an inner function (cf. [N–F, Chapter V, Section 4, Paragraph 3]). The function

$$\mathcal{S}_c \triangleq -A_S S^*(1 - SS^*)^{-1} B_S$$

equals $-A_S(1 - S^*S)^{-1/2} S^*(1 - SS^*)^{-1/2} B_S$ and so is in $L^\infty(\mathbb{C}^n)$; however, it is frequently not in $H^\infty(\mathbb{C}^n)$. By the Beurling theorem there is a unique inner function U_S such that $U_S\mathcal{S}_c$ is in $H^\infty(\mathbb{C}^n)$, which is minimal in the usual sense that if $V\mathcal{S}_c$ is in $H^\infty(\mathbb{C}^n)$ there is an inner W such that $V = WU_S$. Define the *complementary function* S_c of S to be $U_S\mathcal{S}_c$. Let T denote the unit circle and Δ the open unit disk. Frequently one would like to find a Σ in the orbit of S which is close to 0 on some subinterval of the circle and whose behavior off of that subinterval is irrelevant. To address this we choose a weight function $0 \leqslant \mu(e^{i\theta}) \leqslant 1$ and define the distance ρ_μ of a function f in $H^\infty(\mathbb{C}^n)$ to the origin by

$$\rho_\mu(f) = \sup_\theta \frac{1}{\mu(e^{i\theta})} \|f(e^{i\theta})\|.$$

Here $\| \ \|$ denotes the norm on $n \times n$ matrices. Typically, one would choose μ to be small on the interval of interest and near 1 off of it and thereby emphasize behavior in the desired place. Note that $\rho_u(f) \leqslant 1$ says that $\|f(e^{i\theta})\| \leqslant \mu(e^{i\theta})$ for almost all θ. Our main result on question II is

THEOREM I. *Suppose $0 \leqslant \mu(e^{i\theta}) \leqslant 1$ is pseudomeromorphic and denote by α the outer spectral factor of μ^2. Then the ρ_μ distance δ_S from the orbit under $\mathcal{G}$ of $S \in \mathscr{B}_0$ to 0 equals $\delta_S \triangleq \|\Gamma_{U_S}(S_c/\alpha)\|$. If $\delta_S < 1$ and $\delta_S < \delta' \leqslant 1$ then there is a Σ in the orbit of S of the form*

$$\Sigma = \delta' \alpha \Theta$$

where Θ is an inner function. If S is rational, then we may take $\delta_S = \delta'$.

Our analysis of question I with $S \in \mathscr{B}_0$ is absolutely complete only for scalar S; furthermore, all our results for $S \in \bar{\mathscr{B}}$ are given only for rational S.

So henceforth (in the Introduction) we present only results for scalar valued rational S and refer the reader to the body of the paper for more generality. *We always assume that S is not inner.* We begin by stating that for $S = g/q$ relatively prime polynomials our basic definitions simplify:

$$A_S = B_S \qquad \text{is the outer spectral factor of} \quad 1 - |S|^2,$$

$$S_c(z) = z^N \frac{\bar{g}(1/z)}{q(z)} \qquad \text{where} \quad N \text{ is the maximum of order of } g \text{ and order } q.$$

This is verified in Section 5. An important role will be played by the "multi-set" T_S of *transmission zeros of S*, which is defined to be the poles of

$$S^*(1 - SS^*)^{-1}$$

in $\{z: |z| \leqslant 1\}$. The term "multi-set" was introduced by D. Knuth and means that T_S consists of a set of points together with an integer assigned to each point. Note that it follows immediately from the definition that U_S is the blaske product with zeros equal precisely to $T_S \cap \Delta$. We say that two functions φ_1, φ_2 are equal on T_S if φ_1, φ_2 are analytic near each point λ of T_S and

$$\frac{d^k}{dz^k} \varphi_1(\lambda) = \frac{d^k}{dz^k} \varphi_2(\lambda), \qquad k = 0, 1, \ldots, n-1,$$

where n is the order of λ. It is easy to show (see middle of Section 5) that for T_S contained in Δ two functions φ_1, φ_2 in H^∞ are equal on T_S if and only if $\Gamma_{U_S}(\varphi_1) = \Gamma_{U_S}(\varphi_2)$. Here is our main theorem on orbit structure.

THEOREM II. *If Σ, S are rational in $\mathscr{B}$, then Σ is in the orbit of S under $\mathscr{G}$ (with rational coefficients) if and only if*

$$T_S \subset T_\Sigma \qquad \text{and} \qquad \{\text{zeros of } \Sigma^* \text{ in } \Delta\} \cap T_S = \varnothing$$

and

$$\|\Gamma_{U_S}(\varphi)\| \leqslant 1$$

for a $\varphi \in H^\infty$ which equals S_c/Σ_c on T_S.

We have endeavored in stating Theorems I and II to give the general result elegantly. This obscures their concreteness. In fact, there is a natural eigenbasis for the operator $\Gamma^*_{U_S}$ and one can use it to convert the norm conditions from Theorems I and II to a straightforward matrix text, via a procedure of Sarason [S]. In Section 6 we carry this out fully for scalar $S \in \mathscr{B}$. In the concrete statement of Theorem II the function φ never appears and one only uses S_c/Σ_c. To give an idea of what is involved now we present an illustrative special case. Suppose the transmission zeros $\{\xi_1, \ldots, \xi_n\}$ of S have order 1. Then the functions e_j in H^∞ defined by

$$e_j(z) = (1 - \bar{\xi}_j z)^{-1}$$

are eigenfunctions for Γ_{U_S} and $1 - \Gamma_{U_S}(f)^* \Gamma_{U_S}(f)$ expressed in this basis is the classical *Pick matrix*

$$\Lambda_{T_S}(f) = \{[1 - f(\xi_j)\overline{f(\xi_l)}]/(1 - \xi_j \bar{\xi}_l)\}_{j,k=1}^{n} \tag{1.1}$$

for the set T_S and values $f(\xi_j)$. Thus $\delta \geqslant \|\Gamma_{U_S}(f)\|$ if and only if $\Lambda_{T_S}(f/\delta) \geqslant 0$. So for the special case of order 1 transmission zeros we obtain a concrete form of Theorem I. Something very close to this case with $\mu \equiv 1$ appears in the engineering literature (see [Y–S, Problem 2] or [C, Theorem 4.2]) and is primarily due to Youla and Fano with Saito playing a lesser role.

We have only considered S with $\|S\|_\infty < 1$ to this point; now we turn to $S \in \mathscr{B}$. The results obtained are analogous to Theorems I and II; but, unfortunately, when $\|S\| = 1$ we have no elegant way generally to define the key operator Γ. So we shall restrict our attention to the simplest instructive case, namely, to those S whose transmission zeros inside Δ have order 1 and on T have order 2. Such an S will be said to have minimal transmission zeros. Although it is not clear what Γ_{U_S} should be in this case we can obtain an appropriate matrix Λ_{T_S} as the limit of $\Lambda_{T_{rS}}$ as $r \uparrow 1$. This limit procedure leads us to define the *Pick matrix* for a set T and a pseudomeromorphic function f to be the matrix $\Lambda_T(f)$ whose entries are as in (1.1) above for $\xi_i, \xi_j \in T$ unless $\xi_j = \xi_l$ and $|\xi_j| = 1$, in which case

$$\lambda_{ll} = \operatorname{Re} \xi_l \overline{f(\xi_l)} (df/ds)(\xi_l).$$

Now we are in a position to illustrate that Theorems I and II extend to $S \in \mathscr{B}$. (See Section 6 for stronger results.)

THEOREM III. *Suppose $0 \leqslant \mu(e^{i\theta})^2 \leqslant 1$ is pseudomeromorphic and α is its outer spectral factorization. If the rational function S is in $\bar{\mathscr{B}}$ and has minimal transmission zeros, then there is a Σ in the $\mathscr{G}$ orbit of S with $|\Sigma(e^{i\theta})| \leqslant \mu(e^{i\theta})$ if and only if $\Lambda_{T_S}(S_c/\alpha)$ is nonnegative definite.*

THEOREM IV. *Suppose the rational noninner function S in $\bar{\mathscr{B}}$ has minimal transmission zeros. Then Σ is in the orbit of S under $\bar{\mathscr{G}}$ (with rational entries) if and only if*

(1) Σ is inner, or
(2) $T_\Sigma \supset T_S$ and $\Lambda_{T_S}(S_c/\Sigma_c) \geqslant 0$ and $\{\text{zeros } \Sigma_c\} \cap T_S = \varnothing$.

The author has not worked out what the Pick matrix Λ_{T_S} is in general; however, the techniques described in Section 6 would (with perseverence) allow its computation for scalar valued rational S. It should ultimately be possible to compute the Pick matrix for 2×2 matrix valued S (and the main practical difficulty here might well be in finding spectral factorizations). Since all of this is messy, a pursuit of more mathematical interest would be to find an elegant general formulation of the positive definite Pick matrix hypothesis (as in Theorems I and II).

Although detailed historical remarks are given in Section 7 it is fitting to say here that this work depends heavily on engineering work done by Youla, Fano, Saito, and Wohlers, with Youla and Fano playing the major role. In fact most of the ingredients for Theorem III can be found among [Y1; Y2; Y–S], although oddities (such as the lemma in Section 3) occur in putting them together. However, these methods are unable to handle higher order transmission zeros or nonscalar S. The interested reader might also see the subsequent paper [H5], which gives an elegant approach to Theorems I and III, or [H3, 4], which gives some concrete engineering tests based on these results. An engineering announcement of these results is [H2].

2. Reduction to a Linear Problem

In [A; D–H] it was shown that $\overline{\mathscr{B}}_0$ equals the orbit of 0 under $\mathscr{G}$. Consequently, given S in $\overline{\mathscr{B}}_0$ there is an inner $\Delta_S \in \overline{\mathscr{B}}(\mathbb{C}^n)$ such that $S = \mathscr{F}_{\Delta_S}(0)$. Now $\mathscr{F}_V(S) = \mathscr{F}_V \circ \mathscr{F}_{\Delta_S}(0)$ and so one way to study the orbit of S would be in terms of the subsemigroup $\mathscr{G}_S = \{\mathscr{F}_V \circ \mathscr{F}_{\Delta_S}(0): V \text{ inner}\}$ of $\mathscr{G}$. That is the approach we take. Let us prepare by discussing Δ_S. Any inner function Δ in $\overline{\mathscr{B}}(\mathbb{C}^{2n})$ of the form

$$\begin{pmatrix} S & S_{12} \\ S_{21} & S_{22} \end{pmatrix}$$

is called an *inner embedding* or *dilation* for S and has the property $\mathscr{F}_\Delta(0) = S$. One such Δ is

$$\begin{pmatrix} S & B_S \\ U_S A_S & S_c \end{pmatrix}.$$

We shall henceforth denote it by Δ_S and observe that all other Δ have the form

$$\begin{pmatrix} I & 0 \\ 0 & \theta \end{pmatrix} \Delta_S \begin{pmatrix} I & 0 \\ 0 & \psi \end{pmatrix},$$

where ψ is inner and θ is a pseudomeromorphic function satisfying

$$\theta(e^{i\theta})\theta(e^{i\theta})^* = I.$$

If U, V are in $\bar{\mathscr{B}}(\mathbb{C}^{2n})$, then the W in $\bar{\mathscr{B}}(\mathbb{C}^{2n})$ which satisfies $\mathscr{F}_W = \mathscr{F}_U \circ \mathscr{F}_V$ will be called the (Redheffer) $*$-product $U * V$ of U and V. If U, V are inner then so is $U * V$. The $*$-operation has an inverse and it can be used to embed the semigroup $\mathscr{G}$ in the group $G(\mathscr{G}) \triangleq \{\mathscr{F}_U : U$ is pseudomeromorphic and $U(e^{i\theta})$ is unitary$\}$. One can check that each element of $G(\mathscr{G})$ maps $\Lambda \triangleq \{S : S$ is pseudomeromorphic and $\|S(e^{i\theta})\| \leqslant 1\}$ into itself and that $\mathscr{G}(G)$ acts transitively on the interior of this set. Thus given Σ, S in $\bar{\mathscr{B}}_0$ there is an $\mathscr{F}_U \in G(\mathscr{G})$ such that $\Sigma = \mathscr{F}_U(S)$; the problem is that U may not be in $\bar{\mathscr{B}}(\mathbb{C}^n)$. However, note that if $U \in \bar{\mathscr{B}}(\mathbb{C}^{2n})$, in other words if Σ is in the orbit of S, then $\Delta \triangleq U * \Delta_S$ is an inner embedding for Σ. Conversely if there is an inner embedding Δ for Σ so that $\Delta \div_* \Delta_S$ is in $\bar{\mathscr{B}}(\mathbb{C}^n)$, then Σ is in the orbit of S. Here $\div_*$ denotes division from the right. In other words, we shall study the orbit structure by letting the semigroup act on itself.

To study the quotients $E \div_* M$ suppose

$$E = \begin{pmatrix} A & B \\ C & D \end{pmatrix} \qquad \text{and} \qquad M = \begin{pmatrix} N & P \\ Q & R \end{pmatrix}$$

are inner functions in $H^\infty(\mathbb{C}^{2n})$ and P, Q have determinant not vanishing identically. Define

$$K = Q^{-1}[D - R]P^{-1}, \qquad H = Q^{-1}C, \qquad G = BP^{-1}.$$

Conditions guaranteeing analyticity of the quotient $E \div_* M$ are given by

LEMMA 2.1. *Suppose that E, M are inner functions as above and that P is outer. Then*:

(a) *$E \div_* M$ is inner if and only if K is analytic in the disk and $(1 + NK)^{-1}$, $G(1 + NK)^{-1}$, and $(1 + KN)^{-1}H$ are in H^∞.*

(b) *If the lower diagonal block ρ of $E \div_* M$ satisfies $\|N\rho\|$ and $\|\rho N\| < 1$, then $E \div_* M$ inner is equivalent to G, H, K belonging to H^∞.*

(c) *If all functions are rational, then $E \div_* M$ is inner if and only if G, H, K are analytic inside the disk and any pole of $-(1/z)KN$ on the unit circle has order one with residue a nonnegative matrix.*

As a consequence of this lemma we get the following theorem, which is one of our main tools in proving Theorems I–IV. First we introduce notation to the effect that two functions G, K in $L^\infty(\mathbb{C}^n)$ are pointwise isometric if and only if $\exists U$, V unitary valued functions on T so that $G = UKV$.

THEOREM 2.2. (a) *Suppose $S \in \mathscr{B}_0$ and let $\mathscr{P}_S$ denote the convex set of functions $\Omega \in \bar{\mathscr{B}}_0$ which satisfy*

$$U_S{}^*(\Omega - S_c) \text{ is in } H^\infty. \tag{2.1a}$$

One can construct a map Ξ of $\mathscr{P}_S$ into the orbit of S such that $\Xi(\Omega)$ is pointwise isometric to Ω; moreover, if Σ is in the orbit of S, there is an Ω in $\mathscr{P}_S$ with $\Xi(\Omega)$ pointwise isometric to Σ.

(b) *Suppose S is rational in $\mathscr{B}(\mathbb{C}^n\}$ and suppose* $\det(1 - S(e^{i\theta})^*S(e^{i\theta})) \not\equiv 0$. *Let e_j be those points ($|e_j| = 1$) with $\|S(e_j)\| = 1$. Let $\mathscr{P}_S$ be the convex set of those rational $\Omega \in \bar{\mathscr{B}}_0$ for which the functions*

$$U_S{}^*(\Omega - S_c) \text{ is in } H^\infty,$$

and

$$-(1/z)SA_S^{-1}U_S^{-1}(\Omega - S_c)B_S^{-1}(z) \tag{2.1b}$$

have at worst a simple pole whose residue is a nonnegative matrix. Then again the same map Ξ on $\mathscr{P}_S$ takes values in the orbit of S and has the same properties.

For $S \in \mathscr{B}_0$, Theorem 2.2 reduces our difficulties to the following mathematical problem: For given $F \in H^\infty(\mathbb{C}^n)$ find functions $\Omega \in \mathscr{B}_0(\mathbb{C}^n)$ satisfying

$$U_S{}^*(\Omega - F) \in H^\infty(\mathbb{C}^n). \tag{2.2}$$

First of all we reformulate the problem. Abbreviate $\mathscr{S} = \mathscr{S}(U_S)$, let $P_{\mathscr{S}}$ be the orthogonal projection of L^2 onto $\mathscr{S}(U_S)$, and set $M = \Gamma_{U_S}(F)$ and $T = \Gamma_{U_S}(zI)$. Since $\mathscr{M}_{U_S}$ acting on L^2 is unitary, $\mathscr{M}_{U_S}\bar{H}^2 \ominus \bar{H}^2 = H^2 \ominus \mathscr{M}_{U_S}H^2 = \mathscr{S}$. Now observe that (2.2) says $(\Omega^* - F^*)U_S g$ is in $\bar{H}^2$ if $g \in \bar{H}^2$ or, in other words, $0 = P_{H^2}(\mathscr{M}_{\Omega^*} - \mathscr{M}_{F^*})_{|\mathscr{S}}$. *That is, $\Gamma_{U_S}(\Omega) = \Gamma_{U_S}(F)$ if and only if $\Omega \in H^\infty$ satisfies* (2.2).

Fortunately, this problem is precisely one which is addressed by the modern theory of interpolation. In particular the *commutant lifting theorem* settles precisely this question. It was first discovered by D. Sarason for $n = 1$ and proved for all n by Nagy–Foias (see [S; [N–F; D–M–P]). Let S^* denote the backward shift on $H^2(\mathbb{C}^n)$, namely, $[S^*g](z) = [g(z) - g(0)]/z$.

THEOREM (*N–F*). *Suppose U is an inner function in $H^\infty(\mathbb{C}^n)$ and set $\mathscr{S} = \mathscr{S}(U)$. If the operator $M:\mathscr{S} \to H^2$ commutes with the backward shift, namely, $MT = S^*M$, where $T = \Gamma_{U_S}(zI)$, then there is a function $f \in H^\infty(\mathbb{C}^n)$ with $\|f\|_\infty = \sup_{|z|<1} \|f(z)\| = \|M\|$ so that $M = \Gamma_{U_S}(f)$.*

Remark. Although not explicitly stated in [N–F] more detail is implicit in the construction: If $\mathscr{S}$ is finite-dimensional we may take f rational; also, there is always a pseudomeromorphic f in H^∞ with $\|f\|_\infty = \|M\| + \epsilon$ and $M = \Gamma_{U_S}(f)$.

The remarkable work of Adamajan, Arov, and Krein gives much more detail:

THEOREM (A–A–K). *Under the hypothesis above on T and M for $\rho > \|M\|$ there is a cascade transformation $\mathscr{F}$ such that $M = \Gamma_{U_S}(f)$ and $\|f\| \leqslant \rho$ if and only if $f = \rho\mathscr{F}(b)$ for some b in $\mathscr{B}$. Moreover, if $\mathscr{S}$ is finite-dimensional, one may take $\rho = \|M\|$.*

The theorem just stated is not precisely what appears in the literature. For the case $n = 1$ it can be found in [A–A–K1]; for $n = n$ it is derived by taking [A–A–K3, Theorem 6.1] and performing a reduction like the one in [A–A–K2, Section 6]. In both cases construction of the cascade transformation $\mathscr{F}$ given in the theorem is described explicitly. The commutant lifting theorem plus Theorem 2.2 combine to prove Theorem I (as we now see) and the A–A–K theorem actually makes Theorem I constructive.

Proof of Theorem I. We want to see how close the orbit of S gets to 0 with respect to the μ distance. For the moment take $\mu = 1$. Theorem 2.2a reduces the problem to finding an $\Omega \subset \mathscr{B}_0$ satisfying (2.1) = (2.2) of smallest $\| \ \|_\infty$. For Ω satisfying (2.1) we have

$$\|\Omega\|_\infty \geqslant \|\Gamma_{U_S}(\Omega)\| = \|\Gamma_{U_S}(S_c)\|,$$

as desired. To get an Ω satisfying this inequality just use the N–F theorem. The refinements at the end of the theorem follow easily from the A–A–K theorem.

For $\mu \not\equiv 1$ we have α outer with $|\alpha(e^{i\theta})|^2 \equiv \mu(e^{i\theta})^2$ and we can replace S_c with $S_c = S_c/\alpha$ in (2.2) = (2.1). From the solution Ω^1 we form $\Omega = \alpha\Omega^1$, which satisfies both

$$U_S{}^*(\Omega - S_c) \in H^\infty$$

and

$$\|\Omega(e^{i\theta})\| < |\alpha(e^{i\theta})|(\|\Gamma_{U_S}(S_c/\alpha)\| + \epsilon).$$

The point $\Xi(\Omega)$ in orbit S gotten through Theorem 2.2 has ρ_μ distance to 0 less than $\|\Gamma_{U_S}(S_c/\alpha)\| + \epsilon$. Again refinements follow from the A–A–K theorem.

Proof of Theorem 2.2. Given $\Omega \in \mathscr{P}_S$ associate with it a function in the $\mathscr{G}$ orbit of S as follows: *Let β be the $*$-outer spectral factor $\beta\beta^* = 1 - \Omega\Omega^*$. Pick a pseudomeromorphic ψ with $\psi(e^{i\theta})\psi(e^{i\theta})^* = I$ so that $U_S{}^*\beta\psi$ is in H^∞.*

Now pick an inner θ so that $\theta \mathscr{Q}_c \psi$ is in H^∞. Define $\Xi(\Omega) = \theta \mathscr{Q}_c \psi$. Claim: $\Xi(\Omega)$ is in the orbit of S and is pointwise isometric to $\Omega(e^{i\theta})$.

Now we prove the claim. Let κ be the outer spectral factor $\kappa^*\kappa = 1 - \Omega^*\Omega$ and set

$$E = \begin{pmatrix} \theta \mathscr{Q}_c \psi & \theta\kappa \\ \beta\psi & \Omega \end{pmatrix}.$$

The function E is in H^∞ and one can easily compute that $E(e^{i\theta})E(e^{i\theta})^* = I$; thus E is an inner dilation for $\theta \mathscr{Q}_c \psi$ and so we can guarantee that it is in the orbit of S by establishing that $E \mathbin{\overset{*}{\underset{*}{\div}}} \Delta_S$ is inner. If $S \in \mathscr{B}_0$, then A_S and B_S have inverses in H^∞; thus by hypothesis $K \in H^\infty$ while $G = \theta\alpha(B_S)^{-1}$ and $H = A_S^{-1} U_S{}^* \beta\psi$ are in H^∞ by construction. The hypothesis of Lemma 2.1b is satisfied and so $E \mathbin{\overset{*}{\underset{*}{\div}}} \Delta_S$ is inner. This side of part (b) follows immediately from Lemma 2.1c because the argument just presented ensures that G, H, K are analytic inside the disk, and the hypothesis of Theorem 2.2b states that SK has the correct pole behavior on the boundary. Pointwise isometry of $\mathscr{Q}_c$ and Ω^* is true because $\mathscr{Q}_c = -[\alpha(1 - \Omega^*\Omega)^{1/2}]\Omega^*[(1 - \Omega\Omega^*)^{1/2}\beta]$ and the functions in brackets are inner. Pointwise isometry of Ω and Ω^* follows from the polar decomposition $\Omega = U(\Omega^*\Omega)^{1/2}$, $\Omega = (\Omega\Omega^*)^{1/2}V$ because it gives

$$\Omega^* = V^*\{U[(\Omega^*\Omega)^{1/2}]^2U^*\}^{1/2} = V^*U(\Omega^*\Omega)^{1/2}U^* = V^*\Omega U^*.$$

Finally, we must show that the range of Ξ sweeps through orbit S to the extent that it does. If Σ is in the $\mathscr{G}$ orbit of S, then some inner embedding

$$E' = \begin{pmatrix} \Sigma & \lambda \\ \gamma & \dot{\Sigma} \end{pmatrix}$$

of Σ contains Δ_S as a factor, i.e., $E' \mathbin{\overset{*}{\underset{*}{\div}}} \Delta_S$ is inner. By Lemma 2.1 we get $A_S^{-1}U_S^{-1}(\dot{\Sigma} - S_c)B_S^{-1}$ is analytic inside of the disk; thus $U_S^{-1}(\dot{\Sigma} - S_c)$ is. If $S \in \mathscr{B}_0$, then Lemma 2.1b applies and we immediately obtain $\Omega \in \mathscr{P}_S$. If S, Σ are rational, then part (c) leads us to the same conclusion. We conclude that $\dot{\Sigma}$ is in $\mathscr{P}_S$. Since E' is inner

$$\begin{aligned} \Sigma(e^{i\theta})\Sigma(e^{i\theta})^* &= I - \lambda(e^{i\theta})\lambda(e^{i\theta})^* = U(e^{i\theta})(1 - \lambda(e^{i\theta})^*\lambda(e^{i\theta}))U(e^{i\theta})^* \\ &= U(e^{i\theta})\dot{\Sigma}(e^{i\theta})^*\dot{\Sigma}(e^{i\theta})U(e^{i\theta})^*, \end{aligned}$$

where $U(e^{i\theta})$ is the unitary part of the polar decomposition $\lambda(e^{i\theta}) = U(e^{i\theta})[\lambda(e^{i\theta})^*\lambda(e^{i\theta})]^{1/2}$. Again the polar decomposition says $\Sigma = (\Sigma\Sigma^*)^{1/2}V$ and $[U\dot{\Sigma}^*U^*]^* = W(U\dot{\Sigma}^*\dot{\Sigma}U^*)^{1/2}$, which gives us

$$\dot{\Sigma}(e^{i\theta}) = [U^*W(\Sigma\Sigma^*)^{1/2}U](e^{i\theta}) = U(e^{i\theta})^*W(e^{i\theta})\Sigma(e^{i\theta})V(e^{i\theta})^*U(e^{i\theta}),$$

and so $\dot{\Sigma}$ and Σ are pointwise isometric.

Proof of Lemma 2.1. We begin by writing out the Redheffer $*$-product explicitly (cf. [R]). For

$$L = \begin{pmatrix} r & \tau \\ t & \rho \end{pmatrix} \quad \text{and} \quad V = \begin{pmatrix} r' & \tau' \\ t' & \rho' \end{pmatrix}$$

the product $L * V$ is

$$\begin{pmatrix} r + \tau r'(1-\rho r')^{-1}t & \tau(1-r'\rho)^{-1}\tau' \\ t'(1-\rho r')^{-1}t & \rho' + t\rho(1-r'\rho)^{-1}\tau' \end{pmatrix}.$$

One can use this to compute that if $L = E \mathbin{\frac{*}{*}} M$, then the coefficients r, τ, t, ρ of L are the pseudomeromorphic functions

$$\begin{aligned} \rho &= (1+KN)^{-1}K \\ \tau &= BP^{-1}(1-N\rho) = G(1-N\rho) \\ t &= (1-N)Q^{-1}C = (1-\rho N)H \\ r &= A - tN(1-\rho N)^{-1}\tau, \end{aligned}$$

where G, H, K are as defined as before.

Suppose that ρ, τ, t are in $H^\infty(\mathbb{C}^n)$. Then, since $\|L(e^{i\theta})\| \leqslant 1$, the maximum principle implies $\rho, \tau, t \in \bar{\mathscr{B}}(\mathbb{C}^n)$. Consequently, $(1-N\rho)^{-1}$ and $(1-\rho N)^{-1}$ are analytic inside the unit disk and, under the assumption $\|\rho N\|, \|N\rho\| < 1$, are in H^∞. Thus G, H, and r are analytic inside the unit disk and if $\|\rho N\|$, $\|N\rho\| < 1$, they are in H^∞; likewise for $K = \rho(1-N\rho)^{-1}$. This gives one side of (b). To finish that side of (a) observe that

$$\begin{aligned} 1 - N\rho &= 1 - NK(1+NK)^{-1} = (1+NK)^{-1}, \\ 1 - \rho N &= (1+KN)^{-1} \end{aligned}$$

and so $(1+NK)^{-1}$ is in H^∞. Since $G(1+NK)^{-1} = \tau$ and $(1-\rho N)^{-1}H = t$, they are in H^∞.

We begin to prove the converse side by noting that since the boundary values of $N\rho$ have $\| \; \| \leqslant 1$ almost everywhere there is a theorem which says that $1 - N\rho$ is analytic in the disk if and only if $(1-N\rho)^{-1}$ is too. Identity (2.3) informs us that if K is analytic inside the disk, then $(1+NK)^{-1}$ and $(1+KN)^{-1}$ are too. If, in addition, $\|N\rho\|, \|\rho N\| < 1$, we have $(1+NK)^{-1}$ and $(1+KN)^{-1} \in H^\infty$. Part (b) now follows because the expressions for ρ, τ, t, r in terms of G, H, K, and $1 + NK$ yield immediately that they are in H^∞.

With regard to part (c) observe that if $E \mathbin{\frac{*}{*}} M$ is inner, then $\operatorname{Re}(1+KN)^{-1} = \operatorname{Re}(1-\rho N) \geqslant 0$. Thus $\operatorname{Re}(1+KN) \geqslant 0$. By Theorem 2.10 [W] any pole of $-(1/z)(1+KN)$ must be first order with residue a nonnegative matrix; hence this must hold for $-(1/z)KN$. Conversely, if $1 + KN$ is analytic in the disk and satisfies this condition on its boundary poles, the fact that the boundary values satisfy $\mathrm{Re}(1 + KN)(e^{i\theta}) = \mathrm{Re}(1 - \rho N)^{-1}(e^{i\theta}) \geqslant 0$ forces

$$\mathrm{Re}(1 + KN) \geqslant 0$$

throughout $|z| \leqslant 1$. Thus $(1 + KN)^{-1}$ has no poles inside the disk; so neither does $\rho = (1 + KN)^{-1}K$, $\tau = G(1 + KN)^{-1}$, etc. Since they are rational functions they must belong to H^∞.

Remark. The reduction which Theorem 2.2a obtains can be derived from [B–D, Theorem III 4.1] for the special case that S is outer. While even for outer S that approach is not substantially shorter than this one, it is more elegant and the author feels a full proof of Theorem 2.2 along those lines might well be more pleasing than the one we have given.

3. The Boundary Case

The previous section is the core of the paper and most other results in the paper can be derived from it. In doing this we restrict our attention to special classes of scalar valued rational functions. The main ideas and principles are best illustrated at this level and extend in a straightforward fashion to the matrix case and pseudomeromorphic case. Although such generalization is straightforward it is sometimes messy so the goal here is just to illustrate the method and leave further work to someone who has a specific need for it.

Suppose S is rational in $\bar{\mathscr{B}}$ with $|S(e^{i\theta})| \not\equiv 1$. Theorem I does not apply to S unless S is in $\mathscr{B}$ and we now show how to overcome this. Suppose μ is a weight function and α is the outer spectral factor of μ. Theorem 2.2b (see also the proof of Theorem I) allows us to extend Theorem I directly to S and obtain.

Proposition 3.1. *The ρ_μ distance of orbits to 0 is $\leqslant 1$ if and only if*

$$\inf\{\|\Omega\|_\infty : U_S^{-1}A_S^{-1}(\alpha\Omega - S_c)B_S^{-1} \text{ has the property (2.1b) of Theorem 2.2b}\} \leqslant 1.$$

Whereas the $S \in \mathscr{B}$ case is settled by the commutant lifting theorem and so is intimately related to Nevanlinna–Pick–Schur–Caratheodory–Fejer type of interpolation (cf. [S]), the boundary case turns out to be related to classical Löwner interpolation. Unfortunately the Löwner theory absolutely does not fit into the context of the commutant lifting theorem and it seems to the author that extending the lifting theorem to give such a unification is a nice open problem. The Nagy–Koranyi approach fills this gap to some extent as we shall see.

The best way to begin is to describe our problem purely in terms of interpolation. When S is a scalar, $A_S = B_S$ and so an Ω satisfies the key condition on $U_S^{-1}A_S^{-2}(\alpha\Omega - S_c)$ of Theorem 2.2b if and only if

$$\frac{d^l\Omega}{dz}(\xi_k) = \frac{d^l S_c/\alpha}{dz}(\xi_k), \tag{3.1}$$

where $l = 0, 1, \ldots, n_k$,

$$\frac{d^l\Omega}{dz}(e_j) = \frac{d^l S_c/\alpha}{dz}(e_j), \qquad l = 0, 1, \ldots, n_j - 1, \tag{3.2}$$

and

$$-\left\{\operatorname{Res}\left[\frac{(z-e_j)^{n_j}}{z}\right] S(U_S^{-1}A_S^{-2}\alpha)\right\}\Bigg|_{e_j}\left[\frac{d^{n_j}\Omega}{dz}(e_j) - \frac{d^{n_j}S_c/\alpha}{dz}(e_j)\right] \geqslant 0. \tag{3.3}$$

Here ξ_k are the zeros of U_S and have order $n_k + 1$, while e_j are the zeros of A_S^{-2} and have order $n_j + 1$; in other words, they are precisely the transmission zeros of S. To interpret this, begin by supposing no e_j's exist. The numbers $(d^l/dz)(S_c/\alpha)(\xi_j)$ are prescribed and we want to find Ω with $\|\Omega\|_\infty \leqslant 1 + \epsilon$ satisfying (3.1). The case $n_k = 0$ is the classical Pick–Nevanlinna problem, and the case $\xi_1 = \xi_2 = \xi_n = 0$ is the classical Carathéodory–Fejér problem, while the general problem is a mixture. The other extreme of the problem occurs when no ξ_k's exist. The classical Löwner problem is: Given numbers w_j, γ_j, where $|w_j| = 1$, does there exist an inner function Ω satisfying

$$\Omega(e_j) = w_j, \qquad \frac{d\Omega}{dz}(e_j) = \gamma_j?$$

If so, find such an Ω. As we soon see for our orbit problem, each n_j must be odd. Consider the case $n_j \equiv 1$; set

$$w_j = \frac{S_c}{\alpha}(e_j) \qquad \text{and} \qquad \gamma_j = \frac{dS_c/\alpha}{dz}(e_j).$$

Then (3.2) requires that $\Omega(e_j) = w_j$, while (3.1–3.3) is weaker than $\Omega'(e_j) = \gamma_j$. The same phenomenon persists for higher n_j. In such cases the canonical Löwner interpolation problem is: Find an inner function Ω satisfying (3.2) and (3.3) with equality (not inequality). For $n_k = 0$ and $n_j = 1$, the combined Nevanlinna–Löwner problem is well understood. Abrahamse and Rovnyak [A–R] used a construction of Nagy–Koranyi (cf. [D, Chapter XI]) to prove that if the Pick matrix $\Lambda_{T_S}(S_c/\alpha)$ appearing in Theorem III is strictly positive definite, then there is an inner Ω satisfying (3.1)–(3.3) with equality. Conversely, if such an Ω exists, $\Lambda_{T_S}(S_c/\alpha) \geqslant 0$. However, a remarkable fact is that the problem of finding inner Ω satisfying (3.2) and (3.3) (with inequality) is in some ways more canonical than the Löwner problem itself. Namely, this problem has a solution if and only if $\Lambda_{T_S}(S_c/\alpha)$ is nonnegative. This was discovered by Krein and is described in [A–K, Chapter 3]. Thus, invoking Theorem 2.2b proves the "if" side of Theorem III.

We begin to prove the converse side by deriving a lemma.

LEMMA 3.2. *If Ω satisfies condition* (3.2) *at the point ξ, then Ω satisfies condition* (3.3) *at ξ if and only if*

$$(-1)^{(m+1)/2}\xi^m \frac{1}{\Omega(\xi)}\frac{d^m\Omega}{dz}(\xi) - (-1)^{(m+1)/2}\xi^m \frac{\alpha(\xi)}{S_c(\xi)}\frac{d^m S_c/\alpha}{dz}(\xi) \geqslant 0. \quad (3.4)$$

Here $m \overset{\Delta}{=} n_j$.

Proof. From the definition of S_c we obtain $U_S^{-1}A_S^{-2} = -S_c^{-1}(1 - S^*S)^{-1}S^*$ and substituting this into $SU_S^{-1}A_S^{-2}[\alpha\Omega - S_c]$ converts it to

$$-S^*S(1 - S^*S)^{-1}\alpha S_c^{-1}[\Omega - (S_c/\alpha)].$$

The only singular term at ξ is $(1 - S^*S)^{-1}$. Since $1 - S^*S \geqslant 0$ on $\mathbb{T}$ a zero of it has (even) order $m + 1$ and consequently satisfies

$$\frac{d^{m+1}}{dz}(1 - SS^*)(\xi)i^{m+1}\xi^{m+1} \overset{\Delta}{=} \delta^2 > 0.$$

Thus the $(-m - 1)$th Laurent term of $(1 - S^*S)^{-1}$ has coefficient

$$[(-1)^{(m+1)/2}\xi^{m+1}/\delta^2](m + 1)!$$

and we may conclude that $\operatorname{Res}(1/z)SU_S^{-1}A_S^{-1}(\alpha\Omega - S_c)B_S^{-1}$ at ξ is nonnegative if and only if

$$(-1)^{(m+3)/2}\frac{\xi^{m+1}}{\xi}\left[\frac{\alpha(\xi)}{S_c(\xi)}\frac{d^m\Omega}{dz}(\xi) - \frac{\alpha(\xi)}{S_c(\xi)}\frac{d^m S_c/\alpha}{dz}(\xi)\right] \geqslant 0.$$

Since $S_c(\xi)/\alpha(\xi) = \Omega(\xi)$, we obtain the lemma. Q.E.D.

We return to Theorem III and the case $m = 1$. Equation (3.2) holds only if $|\alpha(\xi)| = 1$; thus

$$\xi\overline{\alpha(\xi)}\frac{d\alpha}{dz}(\xi) - \xi\overline{\frac{d\alpha}{dz}}(\xi)\alpha(\xi) = 0,$$

which implies $\xi\alpha(\xi)\, d(1/\alpha)/d\xi$ is real. Likewise

$$\xi\frac{1}{S_c(\xi)}\frac{dS_c}{dz}(\xi)$$

is real. Thus (3.3) is equivalent to

$$\xi \frac{1}{\Omega(\xi)} \frac{d\Omega}{dz}(\xi) \leqslant \xi \frac{\alpha(\xi)}{S_c(\xi)} \frac{dS_c/\alpha}{dz}(\xi). \quad \text{Q.E.D.} \tag{3.5}$$

The surprising thing is how naturally this inequality meshes with the Pick matrix $\Lambda_{T_S}(S_c/\alpha)$. Namely, the right side of inequality (3.5) is just the diagonal entry of Λ corresponding to ξ. Thus we conclude that if Ω satisfies (3.1), (3.2), and the inequality (3.3), then

$$\Lambda_{T_S}(\Omega) \leqslant \Lambda_{T_S}(S_c/\alpha).$$

Moreover, if Ω is inner, then $\Lambda_{T_S}(\Omega) \geqslant 0$ and so $\Lambda_{T_S}(S_c/\alpha) \geqslant 0$. Conversely, when $\Lambda_{T_S}(S_c/\alpha) \geqslant 0$, the Krein theorem says that an inner Ω exists solving the interpolation problem (3.1)–(3.3). This finishes the proof of Theorem III.

4. Orbit Structure

In this section we prove Theorems II and IV, which give a test to determine when Σ is in the orbit of S for scalar valued functions. After that we state what happens for matrix valued functions. We call two rational functions A, B in H^∞ *coprime* if they have no common zeros.

Proof of Theorem II. Suppose Σ is in the orbit of S and is not inner. Then some dilation

$$\begin{pmatrix} \Sigma & B_\Sigma \psi \\ U_\Sigma A_\Sigma & \theta \Sigma_c \psi \end{pmatrix}, \tag{4.1}$$

of Σ has $E \mathbin{\bar{\underset{*}{+}}} \Delta_S$ inner. Here θ and ψ are inner functions. Note $B_\Sigma \not\equiv 0 \not\equiv A_\Sigma$ since Σ is not inner. Lemma 4.1b implies that $E \mathbin{\bar{\underset{*}{+}}} \Delta_S$ is inner if and only if $B_\Sigma \psi B_S^{-1}$, $A_S^{-1} U_S^{-1} \theta U_\Sigma A_\Sigma$, and $A_S^{-1} U_S^{-1} [\theta \Sigma_c \psi - S_c] B_S^{-1}$ are in H^∞. Since A_S and B_S are invertible outer, bounded analyticity of the first function is automatic while that of the last two functions is equivalent to that of

$$U_S^{-1} \theta U_S, \tag{4.2}$$

$$U_S^{-1}[\theta \Sigma_c \psi - S_c]. \tag{4.3}$$

Since U_S is the smallest inner function which makes $U_S \mathscr{S}_c \stackrel{\Delta}{=} S_c$ belong to H^∞, the function S_c cannot equal zero at a pole of U_S^{-1}. Thus the fact that $\theta \psi U_S^{-1} \Sigma_c - U_S^{-1} S_c \in H^\infty$ implies Σ_c cannot equal zero at a pole of U_S^{-1}, that is, Σ_c and U_S are coprime.

By definition, U_S is the minimal inner function so that $U_S S^*(1 - SS^*)^{-1}$ is in H^∞. Since the same holds for Σ, the statement $U_S^{-1} U_\Sigma$ is in H^∞ holds if the singularities of $\Sigma^*(1 - \Sigma\Sigma^*)^{-1}$ dominate those of $S^*(1 - SS^*)^{-1}$. To see that this is true write (4.3) as

$$-U_S^{-1}\theta\psi U_\Sigma A_\Sigma \Sigma^*(1 - \Sigma\Sigma^*)^{-1} B_\Sigma + A_S S^*(1 - SS^*)^{-1} B_S \in H^\infty;$$

observe first that the singularities of these two terms must cancel and second that A_S, B_S, A_Σ, B_Σ are outer while $U_S^{-1}\theta\psi U_\Sigma \in H^\infty$ by (4.2).

Statement (4.3) is equivalent to $(\theta\psi - \Sigma_c^{-1} S_c)^* \Sigma_c^* U_S \in \bar{H}^\infty$. Now if $\varphi \in H^\infty$ equals $\Sigma_c^{-1} S_c$ on T_S, the function

$$(\varphi - \Sigma_c^{-1} S_c)^* \Sigma_c^* U_S$$

is in $\bar{H}^\infty$. Thus $(\theta\psi - \varphi)^* \Sigma_c^* U_S \in \bar{H}^\infty$ and, since Σ_c and U_S are coprime, $(\theta\psi - \varphi)^* U_S \in \bar{H}^\infty$. Equivalently, $\Gamma_{U_S}(\theta\psi) = \Gamma_{U_S}(\varphi)$. Since $\theta\psi$ is inner, $\|\Gamma_{U_S}(\theta\psi)\| \leqslant 1$.

We initiate proof of the converse by supposing that hypothesis (2) holds. The A–A–K theorem tells us that there is an inner ψ satisfying $\Gamma_{U_S}(\psi) = \Gamma_{U_S}(\varphi)$. The argument in the preceding paragraph is reversible and implies that (4.3) with $\theta = 1$ is analytic in Δ. Analyticity of (4.2) amounts to $T_S \subset T_\Sigma$. Define E by (4.1) with $\theta = 1$. Then, as before, Lemma 2.1b implies $E \overset{*}{\underset{*}{\div}} \Delta_S \overset{\Delta}{=} W$ is inner since (4.2) and (4.3) are analytic in Δ. Clearly (see discussion beginning Section 2),

$$\Sigma = \mathscr{F}_W(S). \quad \text{Q.E.D.}$$

Proof of Theorem IV. Suppose the noninner function Σ is in the orbit of S. Then (see discussion beginning Section 2) there is an inner dilation E of the form (4.1) with $E \overset{*}{\underset{*}{\div}} \Delta_S$ in H^∞. Lemma 2.1c implies that

$$A_S^{-1} U_S^{-1} \theta U_\Sigma A_\Sigma \in H^\infty, \qquad U_S^{-1}(\Sigma_c \theta\psi - S_c) \in H^\infty \tag{4.4}$$

and

$$-(1/z)[S A_S^{-1} U_S^{-1}(\Sigma_c\theta\psi - S_c) B_S^{-1}](z) \tag{4.5}$$

has a pole of at most order 1 at each $|e_j| = 1$ transmission zero of S and the pole has nonnegative residue there. These conditions imply that inside of Δ

$$T_\Sigma \supset T_S \qquad \text{and} \qquad \{\text{zero } \Sigma_c\} \cap T_S = \varnothing,$$

via the proof of Theorem II. These conditions also hold on T because a point $e^{i\theta} \in T_S$ is a point at which $|S(e^{i\theta})| = 1$ and any $\mathscr{F}_{U(e^{i\theta})}$ maps numbers of absolute value 1 to numbers of absolute value 1. Note S_c/Σ_c is analytic near each point of T_S. Denote by $(3.1)_\Sigma$, $(3.2)_\Sigma$, $(3.3)_\Sigma$ formulas (3.1), (3.2), (3.3) with Σ_c replacing α, set $\Omega = \theta\psi$, and observe that (4.5) is equivalent to $(3.2)_\Sigma$ plus $(3.3)_\Sigma$. The second condition in (4.4) has been analyzed in the proofs of Theorem II and found to be equivalent to $(3.1)_\Sigma$ holding at the points ξ_k in $T_S \cap \Delta$. This plus application of Lemma 3.1 to (3.2) and (3.3) as in the proof of Theorem III gives the Pick matrix inequality

$$\Lambda_{T_S}(\Omega) \leqslant \Lambda_{T_S}(S_c/\Sigma_c).$$

Since Ω is inner, any Pick matrix for it is nonnegative. Thus

$$\Lambda_{T_S}(S_c/\Sigma_c) \geqslant 0.$$

Note

$$\Lambda_{T_S}(\varphi) = \Lambda_{T_S}(S_c/\Sigma_c),$$

for any rational φ in H^∞ which equals S_c/Σ_c on T_s.

Conversely,

$$\Lambda_{T_S}(S_c/\Sigma_c) \geqslant 0$$

implies $\exists\Omega$ inner satisfying the conditions $(3.2)_\Sigma$, $(3.3)_\Sigma$ above ($(3.3)_\Sigma$ can actually hold with equality). The first condition in (4.4) with $\theta = 1$ follows directly from $T_S \subset T_\Sigma$. Thus E constructed according to (4.1) with $\theta \equiv 1$ and $\Omega = \psi$ produces a function $W \stackrel{\Delta}{=} E \stackrel{*}{\underset{*}{\div}} \Delta_S$ which by Lemma 2.1c is in H^∞. Consequently, $\mathscr{F}_W(S) = \Sigma$.

We conclude this section by indicating what happens in the matrix case. First we define matrix versions of the key scalar definitions. The functions $A, B \in H^\infty(\mathbb{C}^n)$ are called *left coprime* if they have no common left inner divisor, i.e., $\nexists U$ inner $U^{-1}A \in H^\infty$ and $U^{-1}B \in H^\infty$. Suppose $\varphi, B \in H^\infty(\mathbb{C}^n)$ and B, U_S are left coprime: We say φ^* *agrees* with $(B^{-1}S_c)^*$ on the poles of B^*U_S if $[\varphi - B^{-1}S_c]^*B^*U_S \in \bar{H}^\infty(\mathbb{C}^n)$. Such a function can in practice be constructed if U_S is rational, although the construction would be tedious. tedious. If φ_1 and φ_2 are both such functions and $\mathscr{R} = \operatorname{clos} P_{H^2}B^*U_S\bar{H}^2$, then $[\varphi_1 - \varphi_2]^*B^*U_S \in \bar{H}^\infty(\mathbb{C}^n)$, which is equivalent to $P_{H^2}\mathscr{M}^*_{\varphi_1-\varphi_2}|_{\mathscr{R}} = 0$. A partial answer to the orbit problem is

Proposition 4.1. *Suppose Σ, S are in $\mathscr{B}_0(\mathbb{C}^n)$ and Σ is not inner. Then Σ is in the orbit of S if there is an inner function θ such that*

(a) $U_S{}^*\theta U_\Sigma \in H^\infty$,
(b) *$\theta\Sigma_c$ and U_S are left coprime, and*
(c) *when $\varphi \in H^\infty(\mathbb{C}^n)$ agrees with $(\theta\Sigma_c)^{-1}S_c$ we have*

$$\|P_{H^2}\mathscr{M}_\varphi{}^*|_{\mathscr{R}}\| \leqslant \delta,$$

with $\delta < 1$ and only if all this holds with $\delta = 1$. Here $\mathscr{R}$ is the subspace $\operatorname{clos} P_{H^2}(\theta\Sigma_c)^*U_S\bar{H}^2$ of H^2.

Proof. The proof follows the lines of the proof of Theorem II closely enough that one familiar with $H^p(\mathbb{C}^n)$ methods should be able to perform the extension readily.

The difficulty in applying this is that we have no control over θ and it is fortunate that θ disappears in the scalar case. We emphasize that φ as opposed to $(\theta\Sigma_c)^{-1}S_c$ is more an artifact of the way Proposition 4.1 is stated than anything essential. For example, if one had the endurance to write out the Pick matrix statement equivalent of (c), then it would not involve φ at all.

5. Simplifying the Scalar Case and Sensitivity of Orbits

This section treats two different topics. As described in the Introduction, for scalar S our definitions simplify considerably and we prove that here. Secondly, we see how sensitive the orbit structure described in Theorems I–IV is to small perturbations of S. The first part of this section is the following proposition, which describes S_c for the rational scalar case.

Proposition 5.1. *For scalar valued rational S, the function $S_c = \varphi S^*$, where φ is the smallest inner function satisfying $\varphi S^* \in H^\infty$. Write $S = g/q$ with g, q relatively prime polynomials; then*

$$S_c(z) = z^N[\bar{g}(1/z)/q(z)],$$

where N is the maximum of order of g and order q.

Proof. Let b [resp., β] be the smallest inner function which makes $b(1 - S^*S)$ [resp., $\beta(1 - S^*S)^{-1}A_S{}^2$] an H^∞ function. By construction, $(b/\beta)(1 - S^*S)A_S^{-2}$ is analytic with analytic inverse and of modulus one on $\mathbb{T}$. Since the function is rational it must equal a constant, which we may take to be 1. Thus $S_c = U_S(b/\beta)S^*$. By the definition of U_S the function $U_S(b/\beta)S^*$ has precisely the same zeros as $(b/\beta)S^*$. Any zero of b must be a pole of S^* while no zero of S^* is a pole of $(1 - S^*S)^{-1}$, and so the zeros of $(b/\beta)S^*$ precisely equal those of S^*. Thus $U_S(b/\beta) \triangleq \varphi$ is the smallest inner function which makes φS^* belong to H^∞.

To prove the last assertion in the proposition write

$$S^*(z) = \bar{g}(1/z)/\bar{q}(1/z)$$

and factor

$$q(z) = c \prod_{n=1}^{M} (z - \alpha_n)$$

with $|\alpha_n| > 1$. Set $N =$ order g; then

$$\varphi(z) = z^K \prod_{n=1}^{M} \frac{z - (1/\bar{\alpha}_n)}{1 - (1/\alpha_n)z},$$

where $K = N - M$ if $N \geqslant M$ and $K = 0$ otherwise.

$$\begin{aligned} S_c(z) &= \prod_{n=1}^{M} \frac{z - (1/\bar{\alpha}_n)}{1 - (1/\alpha_n)z} \frac{\bar{g}(1/z)}{\bar{c}\prod_{n=1}^{M}((1/z) - \bar{\alpha}_n)} z^K \\ &= \frac{z^{M+K}(-1)^M}{\bar{c}\prod_{n=1}^{M}\bar{\alpha}_n} \prod_{n=1}^{M} \frac{1}{1 - (1/\alpha_n)z} \bar{g}\left(\frac{1}{z}\right) \\ &= \frac{\bar{c}}{c}(-1)^M \prod_{n=1}^{M} \frac{\alpha_n}{\bar{\alpha}_n} \prod_{n=1}^{M} \frac{1}{c(\alpha_n - z)} z^L \bar{g}\left(\frac{1}{z}\right). \end{aligned}$$

where $L = N$ if $N \geqslant M$ and $L = N$ if $N \leqslant M$. Since S_c is determined only up to a constant with absolute value 1 we take it to be

$$z^L \frac{\bar{g}(1/z)}{q(z)}.$$

Q.E.D.

Now we analyze the sensitivity of orbits to a change in S. Let $\delta(S) =$ the ρ_1 distance of orbits to 0.

PROPOSITION 5.2 *If S is rational in $\mathscr{B}$, then δ is a continuous function of S.*

Proof. This paragraph will demonstrate for S in $\mathscr{B}$ and $\epsilon > 0$ the existence of a neighborhood $\mathscr{N}$ of S for which

$$\delta(S_1) \leqslant \delta(S) + \epsilon \tag{5.1}$$

holds whenever $S_1 \in \mathscr{N}$. By the definition of orbit there is an $\mathscr{F}$ in $\mathscr{G}$ with $|\|\mathscr{F}(S)\|_\infty - \delta(S)| < \epsilon/2$. If this $\mathscr{F} = \mathscr{F}_U$ corresponds to a $U(e^{i\theta})$ which approaches being diagonal at some point θ_0 then $\|\mathscr{F}_U(S_1)\|_{L^\infty} = 1$ for any S_1 in $\mathscr{B}$; and the demonstration is complete. So we may assume $\|\mathscr{F}(0)\|_{L^\infty} < 1$. In this case

$$\|\mathscr{F}(S) - \mathscr{F}(S_1)\|_{L^\infty} \leqslant \|B\|_{L^\infty}\|C\|_{L^\infty}\left(\frac{1}{1 - \|\mathscr{F}(0)\|_{L^\infty}}\right)^2 \|S - S_1\|_{L^\infty}$$

so for small enough $\mathscr{N}$ an S_1 in $\mathscr{N}$ satisfies $\|\mathscr{F}(S_1)\|_{L^\infty} \leqslant \|\mathscr{F}(S)\|_{L^\infty} + \epsilon/2 \leqslant \delta(S) + \epsilon$. Thus we have constructed an element of 0_{S_1} with distance $\leqslant \delta(S) + \epsilon$ to zero. This proves (5.1).

To prove the reverse inequality consider the transmission zeros of a sequence $S_n \in \mathscr{B}$ which converges to S. Since they are poles of $\bar{S}_n(1 - S_n\bar{S}_n)^{-1}$

and $\|\bar{S}_n(1 - S_n\bar{S}_n)^{-1} - \bar{S}(1 - S\bar{S})^{-1}\|_{L^\infty}$ Horowitz's theorem (cf. [T]) implies that a subset Z_n of them converges to the points of T_S. This is all that one can conclude because it is easy to construct examples where subsequences from T_{S_n} converge but not to a point in T_S. By the definition of U_S the blaske products V_{S_n} with zeros precisely at Z_n converge to it in the sense $\inf\|V_{S_n} - \lambda U_S\|_\infty \to 0$, $|\lambda| = 1$. The following argument due to D. Herrero [Hr] implies $P_{\mathscr{S}(V_{S_n})} \to P_{\mathscr{S}(U_S)}$. One can write $I - P_{\mathscr{S}(U_S)} = \mathscr{M}_{\lambda U_S} P_{H^2} \mathscr{M}_{\overline{\lambda U_S}} P_{H^2}$ and so

$$\begin{aligned} \|P_{\mathscr{S}(V_{S_n})} - P_{\mathscr{S}(U_S)}\| &\leqslant \|\mathscr{M}_{V_{S_n}} P_{H^2} \mathscr{M}_{\overline{V_{S_n}}} P_{H^2} - \mathscr{M}_{\lambda U_S} P_{H^2} \mathscr{M}_{\overline{\lambda U_S}} P_{H^2}\| \\ &\leqslant 2\|V_{S_n} - \lambda U_S\|. \end{aligned}$$

Convergence of the projections implies

$$\delta(S) \leqslant \liminf_n \delta(S_n) \tag{5.2}$$

provided that $\|(S_n)_c - S_c\|_{L^\infty} \to 0$, because by Theorem I $\delta(S) = \|P_{\mathscr{S}(U_S)} \mathscr{M}_{\overline{S_c}} P_{\mathscr{S}(U_S)}\|$ and $P_{\mathscr{S}(V_{S_n})} \leqslant P_{\mathscr{S}(U_{S_n})}$. Proposition 5.1 implies $\|(S_n)_c - S_c\|_{L^\infty} \to 0$ so we have established (5.2). Q.E.D.

We also mention but do not prove a fact which makes one feel comfortable in dealing with "ideal gains," namely, weight functions μ which are not pseudomeromorphic. Given S rational in $\mathscr{B}$, let e_j be its transmission zeros (order m_j) on $\mathbb{T}$ and let N_j be any open interval containing e_j. We introduce a norm on C^∞ functions

$$|||g||| \triangleq \|g\|_{L^2} + \sum_{j=1}^{M} \sup_{\theta \in N_j} \left[|g(\theta)| + \cdots + \left| \frac{d^{m_j+2} g}{d\theta}(\theta) \right| \right],$$

and call any $0 \leqslant \mu \leqslant 1$ in the $|||\ |||$ completion of C^∞ an admissible gain function.

THEOREM. *For fixed S the function $\delta_\mu(S)$ on the admissible gain functions is μ-continuous in $|||\ |||$.*

6. Explicit Computation

The purpose here is to compute the Pick matrix explicitly. This allows us to convert abstract conditions such as $\|\Gamma_{U_S}\| \leqslant 1$ to concrete ones. The ideas behind this computation are due to Sarason [S].

To a multiset T inside $\{z : |z| < 1\}$ we usually associate a blaske product U whose zero set is T and the corresponding space $\mathscr{S}(U)$. For $\lambda \in T$ of multiplicity n define functions

$$e_k^{\lambda}(z) = \frac{1}{1 - z\bar{\lambda}}\left(\frac{z - \lambda}{1 - z\bar{\lambda}}\right)^k, \qquad k = 0, 1, \ldots, n - 1.$$

It is well known (and ultimately we prove) that the e_k^{λ} lie in $\mathscr{S}(U)$ and together form a basis for $\mathscr{S}(U)$. Moreover, if $\psi \in H^{\infty}(\mathbb{C}^n)$, each e_k^{λ} is a generalized eigenvector for $\Gamma_U(\psi)$ corresponding to the eigenvalue $\overline{\psi(\lambda)}$. To determine whether $\|\Gamma_U(\psi)\| \leqslant 1$ let $x = \sum \alpha_i^k e_k^{\lambda_i}$ be an arbitrary vector in $\mathscr{S}(U)$ and observe that

$$\|x\|^2 - \|\Gamma_U(\psi)x\|^2 = \sum \alpha_i^k \bar{\alpha}_j^l \Lambda_T(\psi)(i, k; j, l),$$

where Λ_T is the matrix

$$\Lambda_T(\psi)(i, k; j, l) \triangleq (e_k^{\lambda_i}, e_l^{\lambda_j}) - (\Gamma_U(\psi)e_k^{\lambda_i}, \Gamma_U(\psi)e_l^{\lambda_j}).$$

Clearly $\|\Gamma_U(\psi)\| \leqslant 1$ if and only if $\Lambda_T(\psi)$ is nonnegative definite.

The heart of this section is the explicit computation of $\Lambda_T(\psi)$. First we compute $\Gamma_U(\psi)e_k^{\lambda}$. The map $\delta_{\lambda}(z) \triangleq (z - \lambda)/(1 - z\bar{\lambda})$ carries Δ conformally onto itself, carries λ into 0, and $|\delta_{\lambda}(e^{i\theta})| = 1$. Frequently, we abbreviate by suppressing repetitious subscripts; for example δ_{λ} goes to δ and $\Gamma_U(\psi)$ to Γ.

Consider the power series

$$\psi(\delta^{-1}(\omega)) = \sum_{n=0} b_n(\lambda)\omega^n; \tag{6.1}$$

then

$$\psi(z) = \sum_{n=0}^{\infty} b_n(\lambda)(\delta(z))^n.$$

To compute $\Gamma e_k^{\lambda} = P_{H^2}\psi^* e_k^{\lambda}$ first note

$$\bar{\psi}^* e_k^{\lambda}(e^{i\theta}) = \frac{1}{1 - e^{i\theta}\bar{\lambda}} \sum_{n=0}^{\infty} \bar{b}_n(\overline{\delta(e^{i\theta})})^{n-k}$$

$$= \frac{1}{1 - e^{i\theta}\bar{\lambda}} \sum_{l=0}^{k} \bar{b}_{-l+k}\delta(e^{i\theta})^l + \frac{1}{1 - e^{i\theta}\bar{\lambda}} \sum_{l=1}^{\infty} \bar{b}_{l+k}\overline{(e^{i\theta})}^l.$$

The first term is in H^2 and so is unaffected by P_{H^2}, and now we check the second term. Denote it by h and observe

$$(e^{i\theta} - \lambda)h(e^{i\theta}) = \sum_{l=1}^{\infty} \bar{b}_{l+k}\overline{\delta(e^{i\theta})}^{l-1} \triangleq \beta(e^{i\theta}).$$

The function $1/(e^{i\theta} - \lambda)$ is the boundary value of $1/(z - \lambda)$, which is analytic and uniformly bounded outside of Δ. Similarly, $\overline{\delta(e^{i\theta})} = 1/\delta(e^{i\theta})$ is the boundary of such a function having absolute value $\leqslant 1$. Thus $(1/(e^{i\theta}-\lambda))\beta(e^{i\theta})$ is in $\bar{H}^2$. We may conclude

$$\Gamma e_k{}^{\lambda} = \sum_{l=0}^{k} \overline{b_{-l+k}(\lambda)} e_l{}^{\lambda}.$$

This formula can be used to give

$$\Lambda_T(F)(i,k;j,l) \triangleq (e_k^{\lambda_i}, e_l^{\lambda_j}) - \sum_{m=0}^{k}\sum_{n=0}^{l} \overline{b_{-m+k}(\lambda_i)}(e_m^{\lambda_i}, e_n^{\lambda_j}) b_{-n+l}(\lambda_j). \quad (6.2)$$

Next we compute the entries $(e_k{}^{\lambda}, e_l{}^{\mu})$ of the Grammian matrix for $\{e_l{}^{\lambda}\}$ explicitly. The inner product

$$\frac{1}{2\pi}\int_0^{2\pi} e_k{}^{\lambda}(e^{i\theta}) e_l{}^{\mu}(e^{i\theta})\, d\theta$$

$$= \frac{1}{2\pi}\int_0^{2\pi} \frac{1}{1-e^{i\theta}\bar{\lambda}} \delta_\lambda(e^{i\theta})^k \frac{1}{1-e^{-i\theta}\mu} \overline{\delta_\mu(e^{i\theta})^l} \frac{i}{i} \frac{e^{i\theta}}{e^{i\theta}}\, d\theta$$

equals the contour integral around $\mathbb{T}$ of

$$\frac{1}{2\pi}\frac{1}{z}\frac{1}{1-z\bar{\lambda}}(\delta_\lambda(z))^k \frac{z}{z-\mu}\left(\frac{1-\bar{\mu}z}{z-\mu}\right)^l,$$

a function analytic inside the unit disk except for a pole of order $l+1$ at μ. The residue there is the order l term in the Taylor series for

$$g(z) \triangleq (z-\lambda)^k(1-z\bar{\lambda})^{-k-1}(1-\bar{\mu}z)^l.$$

We compute it using Leibnitz' rule and obtain for $l \geqslant k$

$$\frac{1}{l!}\frac{d^l g}{dz^l}\bigg|_\mu = \sum_{m=k}^{l}\sum_{n=m-k}^{m} \binom{l}{m}\binom{m}{n}\binom{k+n}{m}(-1)^{l-m-1}\delta_\lambda(\mu)^{k+n}\left(\frac{1-|\mu|^2}{\mu-\lambda}\right)^m \frac{\bar{\lambda}^n\bar{\mu}^{l-m}}{1-\mu\bar{\lambda}}.$$

The case $l<k$ follows by conjugate symmetry. Since

$$\binom{l}{r+k}\binom{r+k}{k} = \binom{l}{k}\binom{l-k}{r}$$

we can make the change of variables $r = m-k$ to obtain

$$(e_k{}^{\lambda}, e_l{}^{\mu})$$

$$= -\frac{1}{1-\mu\bar{\lambda}}\left(\frac{1-|\mu|^2}{1-\mu\bar{\lambda}}\right)^k(-\bar{\mu})^{l-k}\binom{l}{k}\sum_{r=0}^{l-k}(-1)^r\binom{l-k}{r}\sum_{n=r}^{k+r}\binom{k+n}{k+r}(\delta_\lambda(\mu)\bar{\lambda})^n. \quad (6.3)$$

Note that when $\mu = \lambda$ the formula reduces to

$$\frac{1}{l!}\frac{d^l g}{dz^l} = -\frac{1}{1 - |\lambda|^2}\,\delta_{kl},$$

where δ_{kl} is the Kronecker δ function.

In order to summarize easily introduce some matrix notation. Let $G^{\lambda\mu} = \{(e_k{}^{\lambda}, e_l{}^{\mu})\}_{k,l}$, let T_λ be the $n \times n$ Toeplitz matrix

$$T_\lambda = \begin{pmatrix} b_0(\lambda) & b_1(\lambda) & & & b_{n-1}(\lambda) \\ 0 & b_0(\lambda) & b_1(\lambda) & \cdots & \\ 0 & 0 & b_0(\lambda) & & \\ 0 & 0 & 0 & \cdots & b_0(\lambda) \end{pmatrix}, \tag{6.4}$$

and define an $n_i \times n_j$ matrix for each pair i, j:

$$M_{ij} = G^{\lambda_i\lambda_j} - T^*_{\lambda_i} G^{\lambda_i\lambda_j} T_{\lambda_j}. \tag{6.5}$$

Then the matrix whose entries are M_{ij} is $\Lambda_T(\psi)$, that is,

$$\Lambda_T(\psi) = \{M_{ij}\}_{i,j}. \tag{6.6}$$

All of this serves to put Theorems I and II into a practically applicable form since we have proved

THEOREM 6.1. *Define* Λ_{T_S} *by* (6.1), (6.4), (6.5), *and* (6.6).

(a) *The condition* $\|\Gamma_{U_S}(S_c/\alpha)\| \leqslant \delta$ *is equivalent to*

$$\Lambda_{T_S}(S_c/\alpha\delta) \geqslant 0.$$

(b) *The condition* $\|\Gamma_{U_S}(\varphi)\| \leqslant 1$ *is equivalent to*

$$\Lambda_{T_S}(S_c/\Sigma_c) \geqslant 0.$$

Having explicitly solved all problems dealing with scalar valued S and purely interior transmission zeros we turn to problems involving boundary transmission zeros, i.e., $\mathbb{T} \cap T_S$. The objective is to compute an appropriate Pick matrix in general. The obvious procedure described in the introduction of *taking* Λ_{T_S} *to be* $\lim_{r\uparrow 1} \Lambda_{T_{rS}}$ *is valid when the order of each point in* $\mathbb{T} \cap T_S$ *is* $\leqslant 2$. In fact, Λ_{T_S} can be computed very easily from (6.1), (6.4), (6.5), and (6.6) in that fashion. It is also straightforward to prove a combined Theorem I and II for rational functions (at least) which says

PROPOSITION 6.2. *Suppose all boundary transmission zeros for* S *have order* 2. *Then* $\Lambda_{T_S}(S_c/\alpha\delta_S) \geqslant 0$ *and if* $\Lambda_{T_S}(S_c/\alpha\delta) > 0$, *then* $\delta \geqslant \delta_{rS}$ *for any* r *slightly less than* 1.

Surely the last approximation theoretic statement can simply be replaced by $\delta \geqslant \delta_S$, but that would require an extension of the Nagy–Koranyi theorem to higher order interpolation. Also, *there is an obvious analog of Proposition* 6.2 *corresponding to Theorems* III *and* IV.

Unfortunately, this procedure breaks down when any boundary transmission zero has order higher than 2. In this case $\lim_{r\uparrow 1} \Lambda_{T_rS}$ always has a null vector. Consequently, the second part of Proposition 6.2 can never be used. This type of degeneracy is classical and there is a well-known way to overcome it. A matrix Λ_{T_rS} will be positive definite if and only if $M_r \Lambda_{T_rS} M_r^* \triangleq N_{rS}$ is for invertible matrices M_r. (It is universally believed that) one can always choose M_r so that $N_S \triangleq \lim_{r\uparrow 1} N_{rS}$ exists, need not be degenerate, and plays the role in Proposition 6.2 which Λ_{T_S} did. The author has not seen a systematic procedure for selecting M_r for the case studied in this paper, namely, interpolation with H^∞ functions on the disk. However, there is a classical procedure for treating the equivalent problem of interpolation on the half-plane with Pick functions, namely, functions with nonnegative imaginary part. We sketch the idea. If Ω is a matrix, then the matrix obtained from it by a particular elementary column operation followed by the same (but complex conjugated) row operation gives a new matrix Ω' of the form $M\Omega M^*$. Given f analytic on the upper half-plane (U.H.P.) define $[a, b] = (f(a) - \overline{f(b)})/(a - \bar{b})$; the Pick matrix for f and points $\lambda_1, \ldots, \lambda_n$ in U.H.P. is $\Omega = \{[\lambda_k, \lambda_l]\}_{k,l=1}^n$. Suppose, for example, we find that the limit of some sequence Ω_α of 2×2 Pick matrices

$$\begin{pmatrix} [\lambda_1, \lambda_1] & [\lambda_1, \lambda_2] \\ [\lambda_2, \lambda_1] & [\lambda_2, \lambda_2] \end{pmatrix}_\alpha$$

is degenerate and wish to remedy this. We can form

$$\Omega_\alpha' = \begin{pmatrix} [\lambda_1, \lambda_1] & \dfrac{[\lambda_1, \lambda_2] - [\lambda_1, \lambda_1]}{\lambda_2 - \lambda_1} \\ \dfrac{[\lambda_2, \lambda_1] - [\lambda_1, \lambda_1]}{\bar{\lambda}_2 - \bar{\lambda}_1} & \dfrac{\dfrac{[\lambda_2, \lambda_2] - [\lambda_2, \lambda_1]}{\lambda_2 - \lambda_1} - \dfrac{[\lambda_1, \lambda_2] - [\lambda_1, \lambda_1]}{\lambda_2 - \lambda_1}}{\bar{\lambda}_2 - \bar{\lambda}_1} \end{pmatrix}$$

through elementary column–row operations. The limit of Ω_α' will be the appropriate Pick matrix. (Everybody believes that) this procedure or a finite iteration of it works for any size matrix and any degree of degeneracy. It might be termed the "divided difference" method and a nice automation of it is described in [D, Chapter I]. Note in Ω_α' we have introduced difference quotients which frequently in the limit go to derivatives.

The method just described gives a way to obtain an alternative test to those in Proposition 6.2. One Cayley transforms S_c/α or S_c/Σ_c and inserts them into a Pick matrix derived by the "divided difference" method. This has the advantage that it can be carried out for any configuration of transmission zeros T_S and not just interior ones. The disadvantages are: it seems more difficult to perform; it does not extend to matrix valued S while the "commutant lifting" method does, while $\Omega \geqslant 0$ is rigorously proved to be a necessary condition, probably no one has ever written out a proof of sufficiency (although it is pretty clearly true).

This effort to handle boundary transmission zeros of higher order is not frivolous, since the case where T_S Cayley transforms to the multiset $\{0, \infty\}$ with order $\leqslant 4$ at 0 and $\leqslant 8$ at ∞ has current industrial interest. Although we do not give the result here, the author has worked out this case explicitly (see [H3]).

7. Motivation and History

This section gives an exposition of the electrical engineering problem underlying this paper. It is independent of the rest of the paper and so it can be read immediately after the Introduction. To set the stage let us begin with a mechanical analog of the problem. We wish to use a small fast motor to turn a big heavy fly wheel. If the two are connected directly together there will be a waste in energy, so naturally one builds a gear box which "matches" the source to the load. An electrical example is in connecting speakers to a hi-fi amplifier. The speakers can be effectively thought of as an r-ohm resistor, and the amplifier is a source of power with internal resistance R. When the two are connected together the power delivered to the load is proportional to $(r/R + r)^2$. This function achieves its maximum when $R = r$, and when R differs greatly from r direct connection of source to load is inefficient. Thus one connects them through a "matching" transformer of turns ratio $R{:}r$, thereby converting the effective resistance of the load to R, and thereby achieving a "perfect match." (This is in theory; in practice, transformers are ill behaved.)

Now we describe the general situation mathematically. The reader is referred to [H] for the basics and conventions of our setup; we use the scattering formalism. Recall that any passive network $\mathcal{N}$ with n pairs of terminals (n-port) corresponds to an $M \in \bar{\mathscr{B}}(\mathbb{C}^n)$, called its scattering matrix, and that $\mathcal{N}$ is energy conserving if and only if M is inner. Actually $M \in \bar{\mathscr{B}}$ (right half-plane) but we treat the right half-plane as being identified with $\{|z| \leqslant 1\}$ so as to be consistent with the rest of the paper. For example, a

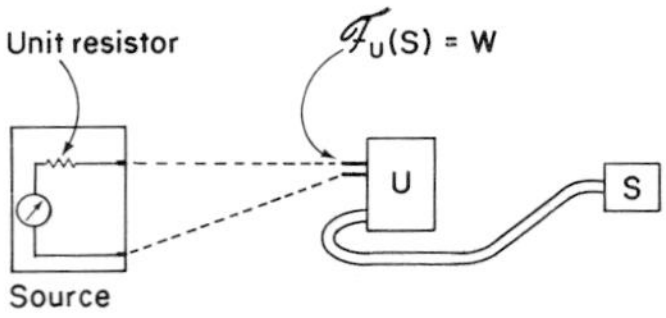

FIG. 1.

unit resistor is a 1-port with $M \equiv 0$. A typical interconnection of a 2-port and a 1-port network is to wire the two terminals of the 1-port to the second pair of terminals of the 2-port. The unconnected terminals of the 2-port can be thought of as being a new 1-port. The situation is described pictorially in Fig. 1 and mathematically as follows: the scattering matrix W for the new 1-port is

$$W(z) = \mathscr{F}_{U(z)}(S(z)), \tag{7.1}$$

where U and S are scattering matrices for the 2-port and original 1-port, respectively. Such connections are called cascade connections.

Suppose that we have a power source with unit internal resistance. If we connect it directly to a load S, then the power delivered to that load at frequency ξ is proportional to $1 - |S(\xi)|^2$. If we cascade S with a lossless 2-port U, then the power delivered to S is proportional to $1 - |\mathscr{F}_U(S)(\xi)|^2$. Thus finding the "matching circuit" U which maximizes the power transferred to the load S corresponds to question II. In the hi-fi example S is a constant and one can find a U so that $\mathscr{F}_U(S) \equiv 0$, in other words, a perfect match is possible. This can only happen if S is constant (frequency independent). Question I corresponds to determining all circuits W which can be gotten from a fixed circuit S by lossless coupling. It is frequently called the problem of compatible impedances and is mainly of interest in solving other problems such as Question II (see [Sch] for a discussion). Note that while in this exposition the load network has been a 1-port with scalar valued S, the same considerations apply to n-ports by taking S matrix valued.

Now we describe existing results on the broadband matching problem. Success has been limited to the 1-port ($n = 1$) case. The first work was Bode [B] in 1947, which studied one very simple load circuit. Fano [F] gave the first approach applicable to general load circuits, which he described through a series of examples. Fano's main idea (after extensive reinterpretation) amounts to using the old mathematical ploy of studying orbits under $\mathscr{G}$ by letting the semigroup $\mathscr{G}$ act on itself. Youla [Y1] in 1964 employed this same idea and gave a more usable systematic approach to the problem which also treats more cases. This is the accepted theoretical approach to the problem today (see [K–R]). Youla's procedure consists of reducing the matching problem to an interpolation problem studied in terms of circuit theory in the earlier paper [Y2]. Subsequently, in 1967, Youla and Saito [Y–S] discovered that a special case of the interpolation problem was classical Nevanlinna–Pick interpolation and gave circuit theory interpretations to the constructions involved. Most surely the [Y1] type of interpolation amounts to not only Nevanlinna–Pick but Carathéodory–Fejér, Löwner, etc., interpolation; however, this mathematical work is not mentioned in the engineering literature. Even [Y–S] has been virtually unpursued and little noticed (the author found it after this paper was finished). Our Lemma 2.1 and Theorem 2.2 are surely analogs of Youla's reduction, although they are so different in form that the author has never worked out the correspondence. Probably the closest relative to them in engineering literature is [W, Theorem 2.17] of Wohlers. It holds for $n = n$ as opposed to Youla's work for $n = 1$. The first work on the problem of compatible impedances was done by Schoeffler [Sch]; Wohlers subsequently gave a partial solution to the general (1-port) problem.

Some books containing treatments of broadband matching are: [W] which is terse but detailed; [K–R], a nice discussive text; and [C], which appeared a few months ago and incorporates [Y–S].

Acknowledgments

The author wishes to thank his colleagues W. Arveson and C. FitzGerald for several valuable discussions, also D. Sarason, J. Garnett, R. Horn, and D. Marshall were helpful in familiarizing the author with literature on interpolation. Also the author is most grateful to B. Abrahamse for correcting an argument in Section 3.

References

[A–A–K1] V. M. Adamajan, D. Z. Arov, and M. G. Krein, Bounded operators that commute with a contraction of class C of unit rank of non-unitary, *Functional Anal. Appl.* **3**.

[A–A–K2] V. M. Adamajan, D. Z. Arov, and M. G. Krein, Analytic properties of Schmidt pairs for a Hankel operator and the generalized Schur–Takagi problem, *Math. USSR Sb.* **15** (1971), 31–73 [in English].

[A–A–K3] V. M. Adamajan, D. Z. Arov, and M. G. Krein, Infinite block Hankel matrices and their connection with the interpolation problem, *Adad. Nauk Armenia S.S.R. Izv. Mat.* **6** (1971).

[A1] D. Z. Arov, Darlington's method for dissipative systems, *Sov. Physics Dokl.* **16**, (1972), 954–956.

[A2] D. Z. Arov, Darlington realization of matrix-valued functions, *Russian Math. Surveys, Usp.*

[A–K] N. Akiezer and M. G. Krein, "Some Questions in the Theory of Moments," Translations of Mathematics Monographs, Amer. Math. Soc., Providence, Rhode Island, 1962.

[A–R] B. Abrahamse and J. Rovnyak, An extension of the Pick interpolation theorem which includes points on the real axis (private notes).

[B–D] J. Baras and P. DeWilde, Invariant subspace methods in linear multivariable distributed systems and lumped distributed network synthesis, *Proc. IEEE* Jan. (1976), 161–178.

[B] H. Bode, "Network Analysis and Feedback Amplifier Design," Van Nostrand–Reinhold, Princeton, New Jersey, 1947.

[C] Wai-Kai Chen, "Theory and Design of Broadband Matching Networks," Pergamon, Oxford, 1976.

[DeW] P. DeWilde, Roomy scattering matrix synthesis, Stanford Univ. Technical Report, 1971.

[D] W. F. Donoghue, "Monotone Matrix Functions and Analytic Continuation," Springer, New York, 1974.

[D–H] R. G. Douglas and J. W. Helton, Inner dilations of analytic matrix functions and Darlington Synthesis, *Acta Sci. Math.* **34**, (1973), 61–67.

[D–M–P] R. G. Douglas, P. Muhly, and C. Prarey, Lifting commuting operators, *Michigan J.* **13**(1968), 385–393.

[D–S–S] R. G. Douglas, H. S. Shapiro, and A. L. Shields, Cyclic vectors and invariant subspaces for the backward shirt operator, *Ann. Inst. Fourier* **20**, (1970), 37–76.

[E–P] A. V. Efimov and V. P. Potapov, *J*-expanding functions and their role in the analytic theory of electrical circuits, *Russian Math. Surveys, Usp.* **28** (1973).

[F] R. M. Fano, Theoretical Limitations on broadband matching of arbitrary impedances, *J. Franklin Institute* **249**(Jan. 1960), 57–83; (Feb. 1960), 139–155.

[G–K] I. C. Gohberg and M. G. Krein, Systems of integral equations on a half-line with kernel depending on the difference of the arguments, *Amer. Math. Soc., Transl.* **14**(2) (1960).

[H] J. W. Helton, The characteristic functions of operator theory and electrical network realization, *Indiana J. Math.* **22**(5) (1972).

[H2] J. W. Helton, Operator theory and broadband matching, *14th Austin Conf. Continuous Systems, 1976.*

[H3] J. W. Helton, A simple test to determine gain bandwidth limitations, *Proc. IEEE Internat. Conf. Circuit Theory, Phoenix, 1977.*

[H4] J. W. Helton, Gain bandwidth limitations directly from data (to appear).

[H5] J. W. Helton, The distance of a function to H^∞ in the Poincaré metric (to appear).

[H–Z] J. W. Helton and A. Zemanian, The cascade loading of passive Hilbert parts, *SIAM J. Appl. Math.* **23**(3) (1972), 292–306.

[Hr] D. Herrero, Full range invariant subspaces of H_K.

[K–R] E. Kuh and R. A. Roher, "Theory of Linear Active Networks," Holden–Day, San Francisco, California, 1967.

[N–F] B. Szent-Nagy and C. Foias, "Harmonic Analysis of Operators on Hilbert Space," North–Holland Publ., Amsterdam, 1970.

[R] R. Redheffer, On a certain linear fractional transformation, *J. Mathematical Phys.* **39**(4) (1960), 269–286.

[R–R] M. Rosenblum and J. Rovnyak, The factorization problem for nonnegative operator valued functions, *Bull. Amer. Math. Soc.* **77**(3) (1971).

[S] D. Sarason, Generalized interpolation in H^∞, *Transl. Amer. Math. Soc.* **127**(2) (1967), 180–203.

[Sch] J. D. Schoeffler, Impedance transformation using lossless networks, *IRE Trans.* **8** (1961), 131–137.

[T] E. C. Titchmarsh, "The theory of Functions," Oxford Univ. Press, London and New York, 1939.

[W] R. Wohlers, "Lumped and Distributed Passive Networks," Academic Press, New York, 1969.

[Y1] D. C. Youla, A new theory of broadband matchings, *IEEE Trans.* **CT-11**, (1964), 30–50.

[Y2] D. C. Youla, A new theory of cascade synthesis, *IRE Trans.* **CT-8** (1961), 244–260.

[Y–S] D. C. Youla and M. Saito, Interpolation with positive real functions, *J. Franklin Institute* **284**(2) (1967), 77–108.

On the Asymptotic Number of Bound States for Certain Attractive Potentials

M. KAC

Department of Mathematics
The Rockefeller University
New York, New York

DEDICATED TO M. G. KREIN ON HIS SEVENTIETH BIRTHDAY

1.

Let $V(\mathbf{r})$ $(\mathbf{r} \in R^3)$ be a nonnegative, bounded and continuous[1] potential which is also integrable, i.e.,

$$\int V(\mathbf{r})\, d\mathbf{r} < \infty. \tag{1.1}$$

Consider the Schrödinger equation

$$\tfrac{1}{2}\nabla^2\phi + AV(\mathbf{r})\phi = -E\phi, \qquad A > 0, \tag{1.2}$$

and denote by $n(A)$ the number of bound states. The problem is to obtain an asymptotic formula for $n(A)$ as $A \to \infty$. The answer is known (see Martin [1]) to be

$$n(A) \sim (\sqrt{2}/3\pi^2)\left(\int V^{3/2}(\mathbf{r})\, d\mathbf{r}\right) A^{3/2}, \tag{1.3}$$

and, in fact, more is known (see, e.g., [2]).

The purpose of this note is to derive (1.3) and a related results by methods different from those of [1] and [2].

2.

Since we shall be interested only in bound states we set $E = -\kappa^2$ (κ real) and a rewrite (1.2) in the equivalent integral equation form

$$(1/A)\phi(\mathbf{r}) = (1/2\pi)\int \frac{\exp(-\kappa\sqrt{2}|\mathbf{r} - \boldsymbol{\rho}|)}{|\mathbf{r} - \boldsymbol{\rho}|} V(\boldsymbol{\rho})\phi(\boldsymbol{\rho})\, d\boldsymbol{\rho}, \tag{2.1}$$

[1] Continuity is assumed for the sake of simplicity only; boundedness is essential for the argument of Section 2 but can be greatly weakened for the argument of Sections 3 and 5.

ISBN 0-12-287150-2

which by setting $\psi(\mathbf{r}) = \phi(\mathbf{r})\sqrt{V(\mathbf{r})}$ becomes

$$(1/A)\psi(r) = (1/2\pi)\int \sqrt{V(\mathbf{r})}\,\frac{\exp(-r\sqrt{2}|\mathbf{r}-\boldsymbol{\rho}|)}{|\mathbf{r}-\boldsymbol{\rho}|}\,\sqrt{V(\boldsymbol{\rho})}\psi(\boldsymbol{\rho})\,d\boldsymbol{\rho}. \tag{2.2}$$

Let $\lambda_1(\kappa), \lambda_2(\kappa), \ldots$ be the eigenvalues of the integral kernel in (1.3). Then the energies of bound states of (1.2) are obtained by solving the equations

$$\lambda_1(\kappa_1) = \frac{1}{A}, \qquad \lambda_2(\kappa_2) = \frac{1}{A}, \ldots. \tag{2.3}$$

Since $\lambda_j(\kappa) \to 0$ as $\kappa \to \infty$ and since $\lambda_j(\kappa) \to \lambda_j(0)$ as $\kappa \to 0$, where the $\lambda_j(0)$ are the eigenvalues of the kernel

$$(1/2\pi)\sqrt{V(\mathbf{r})}(1/|\mathbf{r}-\boldsymbol{\rho}|)\sqrt{V(\boldsymbol{\rho})}, \tag{2.4}$$

we see that the total number of bound states $n(A)$ is equal to the number of $\lambda_j(0)$ which are greater than $1/A$.

So far there is nothing new and I have essentially repeated an argument that goes back to a 1961 paper of Schwinger [3].

The sole element of novelty comes in the form of the following theorem which is a three-dimensional extension of an old theorem of Kac *et al.* [4] (see also [5]).

Let $\tau(\mathbf{r})$ be nonnegative, even (i.e., $\tau(-\mathbf{r}) = \tau(\mathbf{r})$) and normalized by

$$\int \tau(\mathbf{r})\,d\mathbf{r} = 1. \tag{2.5}$$

Furthermore, let

$$T(\xi) = \int \tau(\mathbf{r})e^{i\zeta\cdot\mathbf{r}}\,d\mathbf{r} \tag{2.6}$$

and assume that $T(\boldsymbol{\xi}) \geqslant 0$[2] and that $T(\boldsymbol{\xi}) \in L^p$ for some $p \geqslant 1$.

Then denoting by $\omega_j(\kappa)$ the eigenvalues of the kernel

$$\kappa^3\sqrt{V(\mathbf{r})}\tau(\kappa|\mathbf{r}-\boldsymbol{\rho}|)\sqrt{V(\boldsymbol{\rho})}, \tag{2.7}$$

we have

$$\lim_{\kappa\to\infty}\frac{1}{\kappa^3}\sum_{\omega_j(\kappa)>\alpha} = \frac{1}{(2\pi)^3}\,\mu\{V(\mathbf{r})T(\boldsymbol{\xi}) > \alpha\}, \tag{2.8}$$

where $\mu\{\;\}$ denotes the six-dimensional Lebesgue measure of the set defined inside the braces provided $\mu\{V(\mathbf{r})T(\boldsymbol{\xi}) = \alpha\} = 0$. (The conditions on $V(\mathbf{r})$ are those introduced above, i.e., continuity boundedness, nonnegativity, and integrability are the whole space.)

[2] This condition is introduced for the sake of simplicity only.

The proof, except for trivial modification, is identical with the original one given in [4].

An equivalent way of stating (2.8) is

$$\lim_{\kappa \to \infty} \omega_{\kappa^3 f(\alpha)}(\kappa) = \alpha, \tag{2.9}$$

where

$$f(\alpha) = [1/(2\pi)^3]\mu\{V(\mathbf{r})T(\xi) > \alpha\} \tag{2.10}$$

and it should be understood that the subscript $\kappa^3 f(\alpha)$ in (2.9) is to be taken in the asymptotic sense, i.e., a subscript which divided by κ^3 approaches $f(\alpha)$ as $\kappa \to \infty$.

If we now set

$$\tau(r) = [1/(2\pi)^3](e^{-\sqrt{2}|\mathbf{r}|}/|\mathbf{r}|), \tag{2.11}$$

so that

$$T(\xi) = 1/(1 + |\xi|^2/2), \tag{2.12}$$

then the theorem is clearly applicable to the kernel

$$\kappa^3[1/(2\pi)^3]\sqrt{V(\mathbf{r})}[\exp(-\kappa\sqrt{2}|\mathbf{r} - \boldsymbol{\rho}|)/\kappa|\mathbf{r} - \boldsymbol{\rho}|]\sqrt{V(\boldsymbol{\rho})}, \tag{2.13}$$

and referring back to (2.2) we see that

$$\lambda_j(\kappa) = (1/\kappa^2)\omega_j(\kappa). \tag{2.14}$$

Now, for a given κ and $\alpha > 0$ set

$$A = \kappa^2/\alpha$$

and note that

$$\lambda_j(\kappa) > 1/A$$

is equivalent to

$$\omega_j(\kappa) > \alpha.$$

But for every j such that $\lambda_j(\kappa) > A^{-1}$ there is a bound state of energy $-\kappa_j^2$ ($\kappa_j > \kappa$), and it should be clear that the number of bound states with energies lower than $-\kappa^2$ is equal to the number of $\lambda_j(\kappa)$ which are greater than A^{-1}, which in turn is equal to the number of $\lambda_j(\kappa)$ which are greater than α, and that this number is asymptotically $\kappa^3 f(\alpha)$.

Combining all these remarks we see that denoting by $n(-\alpha A)$ the number of bound states with energies lower than $-\alpha A$ we have asymptotically

$$n(-\alpha A) \sim \kappa^3 f(\alpha) = \alpha^{3/2} f(\alpha) A^{3/2} \tag{2.15}$$

as $A \to \infty$.

From (2.10) and (2.12) we see that

$$f(\alpha) = \frac{4}{3}\frac{\pi 2^{3/2}}{3(2\pi)^3}\int_{V(\mathbf{r})>\alpha}\left(\frac{V(\mathbf{r})-\alpha}{\alpha}\right)^{3/2} d\mathbf{r}, \tag{2.16}$$

so that, finally,

$$n(-\alpha A) \sim \frac{\sqrt{2}}{3\pi^2}\int_{V(\mathbf{r})>\alpha}(V(\mathbf{r})-\alpha)^{3/2}\,d\mathbf{r}\,A^{3/2} \tag{2.17}$$

as $A \to \infty$.

3.

The total number $n(A)$ of bound states corresponds to $\alpha = 0$ and it is tempting to conclude from (2.17) that

$$n(A) \sim \frac{\sqrt{2}}{3\pi^2}\int V^{3/2}(\mathbf{r})\,d\mathbf{r}\,A^{3/2}, \tag{3.1}$$

which is (1.3). Unfortunately it is impossible to derive (3.1) directly from (2.17) without proving that the energies of the bound states do not cluster near 0.

However, it is quite easy to derive (3.1) by a method based on the theory of the Wiener process.

To do this we go back to the observation that $N(A)$ is equal to the number of eigenvalues $\lambda_j(0)$ of (2.4) which are greater than A^{-1}. Thus all that is required is to find the asymptotic behavior of $\lambda_j(0)$ for large j.

For the special case in which $V(\mathbf{r})$ is the indicator function of a compact region Ω I have done this in the last section of my paper [6].

Only minor modifications of the argument given there are needed to handle the more general case.

For the sake of completeness I shall sketch the argument.

Let $\mathbf{r}(\tau)$, $0 \leqslant \tau < \infty$, $\mathbf{r}(\tau) = 0$ (an arbitrarily chosen origin in R^3) be the three-dimensional Wiener process (Brownian motion), and let $\chi_A(\mathbf{r})$ be the indicator function of the set A.

One verifies by a direct calculation that

$$\sqrt{V(\mathbf{y})}\int_0^\infty E\left\{\left(\int_0^t V(\mathbf{y}+\mathbf{r}(\tau))\,d\tau\right)^k\right\}\sqrt{V(\mathbf{y}+\mathbf{r}(t))}\chi_A(\mathbf{y}+\mathbf{r}(t))\,dt$$
$$= \frac{k!}{(2\pi)^{k+1}}\int\cdots\int\frac{\sqrt{V(\mathbf{y})}}{|\mathbf{r}_1-\mathbf{y}|}\frac{V(\mathbf{r}_1)}{|\mathbf{r}_2-\mathbf{r}_1|}\cdots\frac{V(\mathbf{r}_k)}{|\mathbf{r}_K-\mathbf{r}_{K-1}|}\frac{\sqrt{V(\mathbf{r})}\chi_A(\mathbf{r})}{|\mathbf{r}-\mathbf{r}_K|}\,d\mathbf{r}_1\cdots d\mathbf{r}_k\,d\mathbf{r}$$
$$= k!\sum_{j=1}^\infty \lambda_j^{k+1}(0)\psi_j(\mathbf{y})\int_A \psi_j(\mathbf{r})\,d\mathbf{r}, \tag{3.2}$$

where the ψ's are the eigenfunctions of the kernel (2.4) (each integral on the right-hand side of (3.2) is taken over the whole of R^3).

From (3.2) by a standard argument one obtains that for $u > 0$

$$\sqrt{V(\mathbf{y})} \int_0^\infty E\left\{\exp\left(-u \int_0^t V(\mathbf{y} + \mathbf{r}(\tau))\, d\tau\right) \sqrt{V(\mathbf{y} + \mathbf{r}(t))}\chi_A(\mathbf{y} + \mathbf{r}(t))\right\} dt$$

$$= \sum_{j=1}^\infty \frac{\lambda_j(0)}{1 + u\lambda_j(0)} \psi_j(\mathbf{y}) \int_A \psi_j(\mathbf{r})\, d\mathbf{r},$$

and by a simple inversion of Laplace transform one gets

$$\sqrt{V(\mathbf{y})} \int_0^\infty E\left\{\sqrt{V(\mathbf{y} + \mathbf{r}(t))}\chi_A(\mathbf{y} + \mathbf{r}(t));\ \int_0^t V(\mathbf{y} + \mathbf{r}(\tau))\, dt > \beta\right\} dt$$

$$= \sum_{j=1}^\infty e^{-\beta/\lambda_j(0)} \lambda_j(0)\psi_j(\mathbf{y}) \int_A \psi_j(\mathbf{r})\, d\mathbf{r}. \tag{3.3}$$

By taking $A \equiv A(\delta)$ to be a sphere of radius δ with center at $\mathbf{y}$, dividing both sides of (3.3) by $4\pi\delta^3/3$ and letting $\delta \to 0$ we arrive at the formula

$$V(\mathbf{y}) \int_0^\infty \frac{dt}{(2\pi t)^{3/2}} \operatorname{Prob}\left\{\int_0^t V(\mathbf{y} + \mathbf{r}(\tau))\, d\tau > \beta \,|\, \mathbf{r}(t) = 0\right\}$$

$$= \sum_{j=1}^\infty e^{-\beta/\lambda_j(0)} \lambda_j(0)\psi_j^2(\mathbf{y}), \tag{3.4}$$

where we use conditional probabilities. Upon integration over $\mathbf{y}$, (3.4) becomes

$$\int d\mathbf{y}\, V(\mathbf{y}) \int_0^\infty \frac{dt}{(2\pi t)^{3/2}} \operatorname{Prob}\left\{\int_0^t V(\mathbf{y} + \mathbf{r}(\tau))\, d\tau > \beta \,|\, (\mathbf{r}(t)) = 0\right\}$$

$$= \sum_{j=1}^\infty e^{-\beta/\lambda_j(0)} \lambda_j(0), \tag{3.5}$$

and setting $t = \beta\xi$ and $\tau = \beta\eta$ the left-hand side of (3.5) assumes the form

$$\frac{1}{(2\pi)^{3/2}\sqrt{\beta}} \int d\mathbf{y}\, V(\mathbf{y}) \int_0^\infty \frac{d\xi}{\xi^{3/2}} \operatorname{Prob}\left\{\int_0^\xi V(\mathbf{y} + \mathbf{r}(\beta\eta))\, d\eta > 1 \,|\, (\mathbf{r}(\beta\xi)) = 0\right\}. \tag{3.6}$$

Now, as $\beta \to 0$,

$$\lim_{\beta \to 0} \operatorname{Prob}\left\{\int_0^\xi V(\mathbf{y} + \mathbf{r}(\beta\eta))\, d\eta > 1 \,|\, \mathbf{r}(\beta\xi) = 0\right\} = \begin{cases} 0 & \text{if } V(\mathbf{y})\xi < 1 \\ 1 & \text{if } V(\mathbf{y})\xi > 1 \end{cases} \tag{3.7}$$

(provided $\mu\{V(\mathbf{y})\xi = 1\} = 0$), and hence (3.6) is asymptotically

$$\frac{1}{(2\pi)^{3/2}\sqrt{\beta}} \int d\mathbf{y}\, V(\mathbf{y}) \int_{1/V(\mathbf{y})}^{\infty} \frac{d\xi}{\xi^{3/2}} = \frac{2}{(2\pi)^{3/2}} \int V^{3/2}(\mathbf{y})\, d\mathbf{y}\, \frac{1}{\sqrt{\beta}}$$

as $\beta \to 0$.

By an elementary Tauberian argument we get

$$\sum_{\lambda_j^{-1}(0) < \lambda} \lambda_j(0) \sim \frac{\sqrt{2}}{\pi^2} \int V^{3/2}(\mathbf{y})\, d\mathbf{y}\, \sqrt{\lambda}, \qquad \lambda \to \infty,$$

which implies almost at once that

$$\sum_{\lambda_j^{-1}(0) < \lambda} \sim \frac{\sqrt{2}}{3} \frac{1}{\pi^2} \left(\int V^{3/2}(\mathbf{y})\, d\mathbf{y} \right) \lambda^{3/2} \qquad \lambda \to \infty, \tag{3.8}$$

and this in turn yields (3.1).

The argument of this section has to be modified in one and two dimensions but goes through without any change for dimensions higher than three.

4.

The results (1.3) (as well as (2.17)) are what one obtains by following Planck's original quantization rule, i.e., the allowable discrete energy levels $-\epsilon_n$ are obtained by setting

$$\mu\{\tfrac{1}{2}|\mathbf{p}|^2 - AV(\mathbf{r}) < -\epsilon_n\} = n, \qquad n = 1, 2, \ldots \tag{4.1}$$

(the units are so chosen that $m = 1$, $h^3 = 1$).

This is analogous to the more familiar situation in which $V(\mathbf{r}) \to \infty$ as $\mathbf{r} \to \infty$, and the asymptotic behavior of high eigenvalues of the Schrödinger equation

$$\tfrac{1}{2}\nabla^2\phi - V(\mathbf{r})\phi = -E\phi \tag{4.2}$$

is again obtained by Planck's quantization rule

$$\mu\{\tfrac{1}{2}|\mathbf{p}|^2 + V(\mathbf{r}) < E_n\} = n, \qquad n = 1, 2, \ldots. \tag{4.3}$$

The justification of this fundamental result in the one-dimensional case goes back to the early days of quantum mechanics and is associated with the names of Wentzel, Kramers, and Brillouin (WKB). The rigorous justification requires additional assumptions on the rate at which $V(x)$ approaches infinity as $x \to \infty$ and was first given by Titchmarsh [7].

A proof based on quite different principles was suggested by me in [6] and executed by Ray [8].

This proof again uses the theory of the Wiener process and is based on the formula

$$\sum_{n=1}^{\infty} e^{-E_n t} = \frac{1}{(\sqrt{2\pi t})^3} \int d\mathbf{r}\, E\left\{\exp\left(-\int_0^t V(\mathbf{r} + \mathbf{r}(\tau))\, dt\right) \middle| \mathbf{r}(t) = 0\right\}. \tag{4.4}$$

To obtain the behavior of E_n for large n one needs to study the right-hand side of (4.4) for small t ($t \to 0$). This is the "trivial" limit because for small τ, $\mathbf{r}(\tau)$ $(0 \leqslant \tau \leqslant t,\ \mathbf{r}(t) = 0)$ is replaced by 0, obtaining

$$\sum_{n=1}^{\infty} e^{-E_n t} \sim \frac{1}{(\sqrt{2\pi t})^3} \int d\mathbf{r}\, e^{-tV(\mathbf{r})} = \frac{1}{(2\pi)^3} \iint d\mathbf{r}\, dp \exp\left(-t\left(\frac{|p|^2}{2} + V(\mathbf{r})\right)\right), \tag{4.5}$$

which at once yields a formal justification of Planck's quantization rule embodied in (4.3).

The rigorous justification of (4.5), however, is not at all trivial and requires imposition of a variety of technical conditions on the potential V.

It is now evident that the proof of Section 3 is in essence the same as the one discussed in this section, for it is also based on replacing $r(\beta\eta)$ $(0 \leqslant \eta \leqslant \xi$, $\mathbf{r}(\beta\xi) = 0)$ by 0 for small β.

The justification is much easier, however, for

$$\begin{aligned} \text{Prob}&\left\{\int_0^\xi V(\mathbf{y} + \mathbf{r}(\beta\eta))\, d\eta > 1 \,\middle|\, \mathbf{r}(\beta\xi) = 0\right\} \\ &= \text{Prob}\left\{\int_0^\xi V(\mathbf{y} + \sqrt{\beta}\mathbf{r}(\eta))\, d\eta > 1 \,\middle|\, \mathbf{r}(\xi) = 0\right\}. \end{aligned}$$

In the space of paths $\mathbf{r}(\eta)$, $0 \leqslant \eta \leqslant \xi$, tied at ξ by the condition $\mathbf{r}(\xi) = 0$, one clearly has

$$\lim_{\beta \to 0} \int_0^\xi V(\mathbf{y} + \sqrt{\beta}\mathbf{r}(\eta))\, d\eta = \xi V(\mathbf{y}) \tag{4.6}$$

for every path. Thus (3.7) follows almost at once.

It is gratifying that although the problem of the asymptotic number of bound states appears to be quite different from the problem of the asymptotic behavior of high energy levels they turn out to be closely related when approached by the methodology of the Wiener process.

5.

To round out the discussion, I shall show how one can also obtain the asymptotic result for $n(-\alpha A)$ by the method based on the theory of the Wiener process.

The clue is the following extension of (3.5)

$$\int d\mathbf{y}\, V(\mathbf{y}) \int_0^\infty \frac{e^{-\kappa^2 t}}{(2\pi t)^{3/2}} \operatorname{Prob}\left\{\int_0^t V(\mathbf{y}+\mathbf{r}(\tau))\,dt > \beta \,\middle|\, \mathbf{r}(t)=0\right\} dt$$

$$= \sum_{j=1}^\infty e^{-\beta/\lambda_j(\kappa)}\lambda_j(\kappa), \tag{5.1}$$

which can be derived by trivially modifying the derivation of (3.5).

Once again setting $t = \beta\xi$, $\tau = \beta\eta$, we obtain

$$\frac{1}{(2\pi)^{3/2}\sqrt{\beta}} \int d\mathbf{y}\, V(r) \int_0^\infty \frac{e^{-\kappa^2\beta\xi}}{\xi^{3/2}} \operatorname{Prob}\left\{\int_0^\xi V(\mathbf{y}+\mathbf{r}(\beta\eta))\,d\eta > 1 \,\middle|\, \mathbf{r}(\beta\xi)=0\right\} d\xi$$

$$= \sum_{j=1}^\infty e^{-\beta/\lambda_j(\kappa)}\lambda_j(\kappa), \tag{5.2}$$

and setting

$$\beta = a/\kappa^2 \tag{5.3}$$

(5.2) becomes

$$\frac{1}{\sqrt{a}}\frac{1}{(2\pi)^{3/2}} \int d\mathbf{y}\, V(\mathbf{y}) \int_0^\infty \frac{e^{-a\xi}}{\xi^{3/2}} \operatorname{Prob}\left\{\int_0^\xi V(\mathbf{y}+r(\beta\eta))\,dn > 1 \,\middle|\, \mathbf{r}(\beta\xi)\,d\xi = 0\right\}$$

$$= \frac{1}{\kappa^3}\sum_{j=1}^\infty e^{-a/\kappa^2\lambda_j(\kappa)}(\kappa^2\lambda_j(\kappa)),$$

or, letting $\kappa \to \infty$ (i.e., $\beta \to 0$),

$$\frac{1}{\sqrt{a}}\frac{1}{(2\pi)^{3/2}} \int d\mathbf{y}\, V(\mathbf{y}) \int_{1/V(\mathbf{y})}^\infty \frac{e^{-a\xi}}{\xi^{3/2}}\,d\xi = \lim_{\kappa\to\infty} \frac{1}{\kappa^3}\sum_{j=1}^\infty e^{-a/\kappa^2\lambda_j(\kappa)}(\kappa^2\lambda_j(\kappa)). \tag{5.3}$$

Using the elementary identity

$$\frac{1}{(2\pi)^{3/2}}\frac{1}{\sqrt{a}} \int_0^t \frac{e^{-a/\tau}\,d\tau}{\tau^{1/2}} = \frac{1}{\sqrt{2\pi^2}} \int_0^t \frac{e^{-a/\tau}}{\tau^{3/2}}\sqrt{t-\tau}\,d\tau, \qquad a > 0, \tag{5.4}$$

we arrive after a few elementary transformations at the formula

$$\lim_{\kappa\to\infty} \frac{1}{\kappa^3}\sum_{j=1}^\infty e^{-a/\kappa^2\lambda_j(\kappa)}(\kappa^2\lambda_j(\kappa)) = \frac{\sqrt{2}}{3\pi^2}\int^\infty e^{-a/\tau}\tau\,d\tau\left\{\int_{V(\mathbf{y})>\tau}\left(\frac{V(\mathbf{y})}{\tau}-1\right)^{3/2} d\mathbf{y}\right\}, \tag{5.5}$$

which can be used as a basis for an alternate proof of (2.17). Of course, the proof based on the theory of the Wiener process is much less elementary than the proof sketched in Section 2, but it is always satisfying to see how different approaches hang together.

REFERENCES

1. A. MARTIN, Bound states in the strong coupling limit, *Helv. Phys. Acta* **45** (1972), 140–148.
2. E. H. LIEB AND W. E. THIRRING, Inequalities for the moments of eigenvalues of the Schrödinger Hamiltonian and their relation of Sobolev inequalities, *in* "Studies in Mathematical Physics, Essays in Honor of Valentine Bargmann," Princeton Series in Physics, pp. 269–304, Princeton Univ. Press, Princeton, New Jersey 1976; BARRY SIMON, On the number of bound states of two body Schrödinger operators–A review, *ibid.*, pp. 327–350; E. LIEB, Bounds on the eigenvalues of the Laplace and Schrödinger operators, *Bull. Amer. Math. Soc.*, **82**(5) (1976), 751–753.
3. J. SCHWINGER, On the bound states of a given potential, *Proc. Nat. Acad. Sci. USA* **47** (1961), 122–129.
4. M. KAC, W. L. MURDOCK, AND G. SZEGÖ, On the eigenvalues of certain Hermitian forms, *J. Rational Mech. Anal.* **2** (1952), 767–800.
5. U. GRENADER AND G. SZEGÖ, "Toeplitz Forms and Their Applications," Section 11.8(b), Univ. of California Press, Berkely, California, 1958.
6. M. KAC, On some connections between probability theory and differential and integral equations, *Proc. Second Berkeley Symp. Math. Statistics and Probability*, pp. 189–215, Univ. of California Press, Berkeley, California, 1951.
7. E. C. TITCHMARSH, "Eigenfunction Expansions," Vols. I and II, Oxford Univ. Press, London and New York, 1953 and 1958.
8. D. B. RAY, On spectra of second order differential operators, *Trans. Amer. Math. Soc.* **77** (1954), 299 321.

Generalized Krein–Levinson Equations for Efficient Calculation of Fredholm Resolvents of Nondisplacement Kernels†

THOMAS KAILATH, LENNART LJUNG,‡ AND MARTIN MORF

Department of Electrical Engineering
Stanford University
Stanford, California

DEDICATED TO THE MEMORY OF NORMAN LEVINSON AND, WITH ADMIRATION, TO THE CONTINUATION OF THE REMARKABLY PROLIFIC AND MULTIFACETED WORK OF MARK GRIGOR'EVICH KREIN

The assumptions of stationarity or homogeneity (or its equivalents) are often imposed on many physical problems because they lead to tractable integral equations with displacement (also called convolution or Toeplitz) kernels. Such equations have many nice properties, and it seems unreasonable that they should all be lost if the kernel is not exactly Toeplitz. We shall associate with any kernel an index α that will measure how non-Toeplitz it is and we shall prove that it takes α times as much computation to find the Fredholm resolvent of such a kernel than of a Toeplitz kernel. This new classification of kernels in terms of their degree of nontoeplitzness seems to have several interesting ramifications.

I. INTRODUCTION

Problems in many fields lead to integral equations of the form

$$f(t) + \int_0^T K(t, u)f(u)\,du = m(t), \qquad 0 \leqslant t \leqslant T \leqslant \infty, \tag{1}$$

where $m(.)$ and $K(.,.)$ are given functions. It is well known that the solution of such so-called Fredholm equations of the second kind can be written as

$$f(t) = m(t) - \int_0^T H(t, u; T)m(u)\,du, \tag{2}$$

† This work was supported by the Air Force Office of Scientific Research, Air Force Systems Command, under Contract AF44-600-74-C-0068, and in part by the National Science Foundation under Contract NSF-Eng-75-18952 and the Joint Services Electronics Program under Contract N00014-75-C-0601.

‡ Present address: Linköping University, Sweden.

ISBN 0-12-287150-2

where the so-called Fredholm resolvent $H(.,.;T)$ is defined via the integral equation

$$H(t, s; T) + \int_0^T H(t, u; T)K(u, s)\, du = K(t, s), \qquad 0 \leqslant t, s \leqslant T. \tag{3}$$

Actually, the resolvent also depends on the lower limit of integration in (1) and is therefore a function of four variables. However, the dependence on the lower limit is very similar to that on the upper limit. Therefore, for simplicity of notation, we have chosen the lower limit to be zero and written H as a function of three variables. Smoothness assumptions on $K(.,.)$ are required in order to ensure that the equation (3) has a unique solution, for example, that $K(.,.)$ be square integrable on $[0, T] \times [0, T]$ (and obviously also that no eigenvalue be equal to -1). In this paper we shall assume that $K(t, s)$ is differentiable in t and s, but many of our results also hold for square-integrable kernels. For simplicity of presentation, we shall also assume that $K(t, s)$ is an $r \times r$, real and symmetrix matrix kernel, so that

$$K(t, s) = K'(s, t),$$

where the prime denotes (Hermitian) transpose. Extensions to the nonsymmetric case are straightforward but notationally more burdensome and are therefore not considered here.

Now, explicit formulas for the resolvent are available only for a very small number of kernels. Therefore recourse has to be had to various computational and approximation schemes. Many of these schemes amount to replacing (3) by some approximating set of N linear equations in N unknowns, where N may be quite large for a good approximation. Since it takes $O(N^3)$ operations (multiplications and additions) to solve such equations, it is worthwhile to try to exploit any special structure in the original problem (as reflected in the kernel $K(.,.)$) that can reduce the computational burden.

An especially important and widely occurring type of special structure is the so-called *shift-invariant* or *displacement* or *stationarity* structure in which the kernel $K(.,.)$ has the form (with an obvious abuse of notation)

$$K(t, s) = K(t - s). \tag{4}$$

Such kernels are often called *Toeplitz* or *displacement* or *convolution* kernels.

In this case, a number of special results are available. In particular, when (4) holds and $T = \infty$ (1) becomes the celebrated Wiener–Hopf equation extensively studied in the classic papers of Krein [1] and Gohberg and Krein [2]. For finite T, and kernels with rational spectral densities, explicit but cumbersome expressions were derived by several authors, e.g., [3, 4].

Rather less known, at least to communication and control engineers, is a more general result of the Soviet astronomer Sobolev, who showed [5] that though the resolvent of a scalar (symmetric) Toeplitz kernel is not Toeplitz, it obeys a special relation (which we shall call the Sobolev identity),

$$\left(\frac{\partial}{\partial t}+\frac{\partial}{\partial s}\right)H(t, s; T) = a(T; t)a(T; s) - a(T; T-t)a(T; T-s), \tag{5}$$

where the function $a(T; s)$ can be found by solving the integral equation

$$a(T; t) + \int_0^T a(T; u)K(u-t)\,du = K(T-t), \qquad 0 \leqslant t \leqslant T. \tag{6}$$

The point is that the function of *two* variables $H(t, s; T)$ is now determined by a function of *one* variable $a(T; .)$ obeying the so-called [6] truncated Wiener–Hopf equation (6), which is easier to approximate and solve than the resolvent equation (3). In fact, $a(T; .)$ can be recursively computed via the differential equations

$$\left(\frac{\partial}{\partial t}+\frac{\partial}{\partial s}\right)a(t; s) = -a(t; t-s)a(t; 0), \tag{7a}$$

or

$$\frac{\partial}{\partial t}a(t; t-s) = -a(t; s)a(t; 0), \qquad 0 \leqslant s \leqslant t \leqslant T, \tag{7b}$$

where the boundary condition $a(t; 0)$ can be found using the relation

$$a(t; 0) = K(t, 0) - \int_0^t a(t; u)K(u, 0)\,du. \tag{8}$$

It is not hard (cf. Section III) to obtain simple discretization schemes for solving (7) and (8) and to check that if $[0, T]$ is divided into N subintervals, it takes only a number of computations proportional to N^2 to determine a from (7) and (8) and H from (5). For large N, this can mean a very substantial reduction over the $O(N^3)$ operations required to solve equations with non-Toeplitz kernels and is a powerful argument for the use of the stationarity assumption (4) in physical problems. We shall call (7) and (8) the Krein–Levinson equations. Their discrete-time analogs were first introduced by Levinson [7] in studying a least-squares estimation problem and were soon noted to be the recursions for the Szego orthogonal polynomials on the unit circle [8, 9]. Later, Krein [10] presented a continuous analog of these polynomials, which are exactly our functions $a(t; .)$; it also transpires [11] that the Sobolev identity (5) is just the Christoffel–Darboux formula for these continuous analogs. We should also note that Gohberg and Semencul [12] and Gohberg and Heinig [13] independently obtained the Sobolev identity for nonsymmetric matrix Toeplitz kernels.

Now, there are many problems in which there is some time- or shift-invariance feature that may not always lead to Toeplitz kernels $K(.,.)$, but should yield kernels that are in some sense close to Toeplitz. For example, the resolvent of a Toeplitz kernel is not Toeplitz, but in a very definite sense it must be close to a Toeplitz kernel. Similarly, the composition of two Toeplitz kernels yields a non-Toeplitz kernel that should somehow also be close to Toeplitz. So also for kernels that are asymptotically (as $t, s \to \infty$) Toeplitz, and so on.

Nevertheless, presently known solution methods do not provide any computational reductions for these "almost Toeplitz" kernels. This sharp distinction between Toeplitz and non-Toeplitz kernels is unreasonable, but it does not seem to have been closely studied before. In this paper we shall show that we can associate with any kernel $K(t, s)$ a number α that measures in a certain sense its "distance" from a Toeplitz kernel. Moreover, this notion will be given an operational significance by showing that it takes α times more operations to determine the resolvent of such a kernel than of a Toeplitz kernel.

Briefly, we first define an operator (of the "divergence" type)

$$\lrcorner F(t, s) = \left(\frac{\partial}{\partial x} + \frac{\partial}{\partial y}\right) F(x, y)\Big|_{\substack{x=t \\ y=s}}. \tag{9}$$

Then we note that for a Toeplitz kernel

$$\lrcorner K(t - s) = 0 \tag{10a}$$

but its resolvent is not Toeplitz, i.e.,

$$\lrcorner H(t, s; T) \neq 0, \qquad 0 \leqslant t, s \leqslant T < \infty.$$

However, the resolvent is not arbitrary but has "displacement rank" 2 in the sense that, according to the Sobolev formula (5) (for scalar kernels)

$$\lrcorner H(t, s; T) = \sum_1^2 \tilde{\lambda}_i q_i(t) q_i(s), \tag{10b}$$

where

$$q_1(t) = a(T; t), \qquad q_2(t) = a(T; T - t); \qquad \tilde{\lambda}_1 = 1, \quad \tilde{\lambda}_2 = -1.$$

In general, i.e., for an arbitrary kernel $K(t, s)$, suppose that we know that

$$\lrcorner K(t, s) = \sum_1^{\tilde{\alpha}} \tilde{\lambda}_i p_i(t) p_i(s) \tag{11a}$$

for some functions $p_i(.)$ and some $\tilde{\lambda}_i$ and some $\tilde{\alpha}$, possibly infinite (though we shall not consider any convergence problems for infinite $\tilde{\alpha}$). We shall show that we have a "closure property" in the sense that

$$\lrcorner H(t, s; T) = \sum_{1}^{\tilde{\alpha}+2r} \tilde{\lambda}_i q_i(t) q_i(s) \tag{11b}$$

for some functions $q_i(.)$ and the same $\tilde{\alpha}$; moreover, the $q_i(.)$ can be determined from the $K(.,.)$ and the $p_i(.)$ by solving not more than $\tilde{\alpha} + 2r$ truncated Wiener–Hopf integral equations or Krein–Levinson-type differential equations.

The smallest number $\tilde{\alpha}$ such that we can write $K(t, s)$ in the form (11a) for some $p_i(.)$ will be called *a* first-order displacement rank of $K(.,.)$.

We say *a* displacement rank because slightly modified definitions might be more appropriate in different applications (cf. Section II and [14, 15]). We say first order because the operation can clearly be repeated and we could, for example, consider the rank of $\lrcorner\, ^3K(t, s)$ and so on.

SOME EXAMPLES. Let us consider some simple kernels arising often in statistical signal detection and estimation problems.

The covariance function of any stationary process is Toeplitz and therefore has displacement rank $\tilde{\alpha} = 0$. Perhaps the simplest nonstationary process is the Wiener or Brownian motion process with covariance function $\min(t, s)$. It is easy to see that

$$\lrcorner \min(t, s) = 1, \qquad \text{so that} \quad \tilde{\alpha} = 1. \tag{12}$$

For covariance kernels, the name 'index of nonstationarity' for $\tilde{\alpha}$ is more appropriate than displacement rank, and we see that the Wiener process has index 1. Another simple process is the so-called Brownian-bridge or pinned Wiener process with covariance function $\min(t, s) - ts$. We see that

$$\lrcorner(\min(t, s) - ts) = 1 - t - s, \qquad \text{and} \qquad \tilde{\alpha} = 2. \tag{13}$$

Other examples can be given (e.g., by using Lemma 1 below), but here we shall only note that in control theory the kernel $K(t, s)$ is often the covariance of a process generated by passing white noise through a system describable by an nth order constant coefficient differential equation. An exact formula for the $\tilde{\alpha}$ of such a kernel can be derived (and was in fact at the origin of the results of this paper—(cf. [14, 16]), but here it suffices to note that the index $\tilde{\alpha}$ for such a process is always upper bounded by the order of the differential equation, a fact that will be helpful in the actual application of our results (cf. the remarks in Section III).

We should also mention that discrete-time (or matrix) analogs of the above results have been obtained (cf. [17, 18]). Briefly, it is well known that it takes a number of operations proportional to N^3 to invert a general

$N \times N$ matrix, while it takes only a number of operations proportional to N^2 to invert an $N \times N$ Toeplitz matrix. We have shown that, roughly speaking, with any $N \times N$ matrix we can associate a displacement rank $\tilde{\alpha}$, $1 \leqslant \tilde{\alpha} \leqslant N$, such that it takes $N^2\tilde{\alpha}$ operations to invert a matrix of rank $\tilde{\alpha}$; more details and examples can be found in [17, 18].

Some further results and possible extensions are briefly noted in Section IV.

II. Generalized Sobolev Identities

To present and derive our generalizations of the Sobolev and Krein–Levinson equations, it will be helpful to introduce some operator notation. Thus, as is traditional, we shall write

$$KL \triangleq \int_0^T K(t, u)L(u, s)\, du. \tag{14}$$

Then the resolvent equation (1) can be written as

$$H + HK = K. \tag{15}$$

The operator notation leads immediately to the following useful identities:

$$H = K(I + K)^{-1} = (I + K)^{-1}K, \tag{16a}$$

$$I - H = (I + K)^{-1}, \qquad (I - H)(I + K) = I = (I + K)(I - H). \tag{16b}$$

We shall also write the Dirac delta function at q as

$$\delta_q = \delta(. - q) \tag{17}$$

and note that

$$K\delta_q L \triangleq \int_0^T K(t, u)\delta(u - q)L(u, s)\, du = K(t, q)L(q, s). \tag{18}$$

As in (18), we shall sometimes abuse the notation by switching freely between operator and function notation when we feel this might be helpful.

Lemma 1. *With*

$$\lrcorner \triangleq \frac{\partial}{\partial t} + \frac{\partial}{\partial s}, \tag{19}$$

we have

$$\lrcorner(K_1 + K_2) = \lrcorner K_1 + \lrcorner K_2 \tag{20}$$

and

$$\lrcorner(K_1 K_2) = (\lrcorner K_1)K_2 + K_1(\lrcorner K_2) + K_1(\delta_0 - \delta_T)K_2. \tag{21}$$

Proof. The proof of (20) is trivial. For (21), note that

$$\begin{aligned}\lrcorner(K_1K_2) &= \int_0^T \left[\frac{\partial}{\partial t} K_1(t,u)\right] K_2(u,s)\,du + \int_0^T K_1(t,u)\left[\frac{\partial}{\partial s} K_2(u,s)\right] du \\ &= \int_0^T \left[\lrcorner K_1 - \frac{\partial}{\partial u} K_1(t,u)\right] K_2\,du + \int_0^T K_1\left[\lrcorner K_2 - \frac{\partial}{\partial u} K_2(u,s)\right] du \\ &= (\lrcorner K_1)K_2 + K_1(\lrcorner K_2) - \int_0^T \frac{\partial}{\partial u}[K_1(t,u)K_2(u,s)]\,du,\end{aligned}$$

which immediately yields (21). ∎

We use this result to study the resolvent. Thus note that

$$\begin{aligned}\lrcorner H &= \lrcorner(K - HK) \\ &= \lrcorner K - (\lrcorner H.K + H.\lrcorner K + H(\delta_0 - \delta_T)K)\end{aligned} \tag{22}$$

so that

$$\lrcorner H.(I + K) = (I - H)\lrcorner K - H(\delta_0 - \delta_T)K.$$

Postmultiplying by $(1 + K)^{-1}$ and using (16) gives

$$\lrcorner H = (I - H)\lrcorner K(1 - H) - H(\delta_0 - \delta_T)H. \tag{23}$$

Suppose now that

$$K \text{ is Toeplitz,} \qquad \text{so that } \lrcorner K \equiv 0. \tag{24a}$$

Then we have the identity

$$\lrcorner H = H\delta_T H - H\delta_0 H, \tag{24b}$$

or, more explicitly,

$$\left(\frac{\partial}{\partial t} + \frac{\partial}{\partial s}\right) H(t,s;T) = H(t,T;T)H(T,s;T) - H(t,0;T)H(0,s;T). \tag{24c}$$

A significant thing about this formula is that it shows that the resolvent of a Toeplitz kernel is completely determined by knowledge of its values along the four boundaries $s = T$, $t = T$, $s = 0$, $t = 0$. Therefore the problem of determining the 2-variable resolvent (for fixed T) is reduced to that of determining four 1-variable functions. In the *scalar* case, these are related to each other since the Toeplitz nature of K implies (use (1)) that

$$H(t,s;T) = H(T - s, T - t; T).$$

Therefore, the values along the $s = 0$ boundary are determined by those along the $t = T$ boundary, and similarly for the $s = T$ and $t = 0$ boundaries.

For stochastic applications, the kernel K is symmetric, so that

$$K(t, s) = K'(s, t) \qquad \text{and} \qquad H(t, s; T) = H'(s, t; T). \tag{25}$$

Then, if we define

$$a(T, t) = H(T, t; T), \qquad b(T, t) = H(0, t; T), \tag{26a}$$

we can note that

$$a'(T, t) = H'(T, t; T) = H(t, T; T), \tag{26b}$$

$$b'(T, 0) = H'(0, t; T) = H(t, 0; T), \tag{26c}$$

so that Sobolev's identity for symmetric Toeplitz kernels can be written

$$\lrcorner H(t, s; T) = a'(T, t)a(T, s) - b'(T, t)b(T, s). \tag{27}$$

For *scalar* ($r = 1$) and symmetric Toeplitz kernels, we have furthermore that

$$b(T, t) = H(0, t; T) = H(t, 0; T) = H(T, T - t; T) = a(T, T - t), \tag{28}$$

which gives the formula (5) first derived by Sobolev.

However, our goal is the study of non-Toeplitz kernels, for which (23) shows that the significant quantity is $\lrcorner K$. Now, $\lrcorner K$ is some (symmetric) function of two variables and therefore it has a representation of the form

$$\lrcorner K(t, s) = \sum_{0}^{\tilde{\alpha}} \tilde{\lambda}_i P_i(t) P_i(s) \tag{29}$$

for some $\tilde{\alpha}$, possibly infinite. While representations as in (29) are quite general, in particular applications slight variations can be useful. Thus in linear least-squares estimation, we have found it convenient to consider a slightly modified representation for $\lrcorner K$, say

$$\lrcorner K(t, s) = K(t, 0)K(0, s) + \sum_{1}^{\alpha} \lambda_i d_i(t) d_i'(s) \tag{30a}$$

$$= K\delta_0 K + d(t) \Lambda d'(s), \tag{30b}$$

where $d(.)$ is an $r \times \alpha$ matrix function and Λ is an $\alpha \times \alpha$ matrix. We shall work with this representation here. First, let $c'(T, .)$ be the solution of the integral equation

$$c'(T, t) + \int_0^T K(t, s) c'(T, s)\, ds = d(t), \qquad 0 \leqslant t \leqslant T, \tag{31a}$$

so that also

$$c'(T, t) = d(t) - \int_0^T H(t, u; T) d(u)\, du \tag{31b}$$

Then we can establish the following result.

THEOREM 1 (*Generalized Sobolev identity*). *Let the $r \times r$ symmetrix matrix kernel $K(.,.)$ have the representation*

$$\lrcorner K(t, s) = K(t, 0)K(0, s) + d(t)\Lambda d'(s) \tag{32a}$$

for some $r \times \alpha$ matrix $d(.)$ and some $\alpha \times \alpha$ matrix Λ. Then the resolvent of K obeys the relation

$$\lrcorner H(t, s) = a'(T, t)a(T, s) + c'(T, t)\Lambda c(T, s), \tag{32b}$$

where $a'(T, t) \triangleq H(t, T; T)$ and $c'(T, .)$ is defined by (31), *viz.*, $c' = (I - H)d$. *Note that* $a' = H\delta_T = (I - H)K\delta_T$.

Proof. We use (30)–(31) in (23) to get

$$\begin{aligned}\lrcorner H &= (I - H)K\delta_0 K(I - H) + (I - H)d\Lambda d'(I - H) - H(\delta_0 - \delta_T)H \\ &= H\delta_0 H + c'\Lambda c - H\delta_0 H + H\delta_T H,\end{aligned}$$

which is (32b). ∎

Note that when K is Toeplitz, we must choose

$$\alpha = r, \qquad \Lambda = -I, \qquad d(t) = K(t, 0), \tag{33a}$$

so as to make

$$0 = K\delta_0 K + d\Lambda d' = \lrcorner K.$$

Then

$$c' = (I - H)d = (I - H)K\delta_0 = H\delta_0 = b', \tag{33b}$$

so that (32) reduces correctly to the previously derived formulas (27) or (24).

In fact, reference to the discussion below (24) suggests that the functions $a(T, .)$ and $c(T, .)$ may appropriately be called *generalized boundary functions* of $H(t, s; T)$. The Sobolev identity shows that knowledge of these $(\alpha + 2r)r$ boundary functions suffices to determine the resolvent, and this fact can be (theoretically and) computationally significant, as we shall show in the next section.

III. GENERALIZED KREIN–LEVINSON EQUATIONS

The generalized Sobolev identity of Theorem 1 shows that the resolvent H is completely determined by knowledge of the generalized boundary functions $a(T; .)$ and $c(T; .)$. Like H, these functions are themselves determined in terms of certain integral equations, albeit simpler ones than the equation for H. However, the major advantage of the specification in terms of $a(T; .)$

and $c(T;.)$ is that these functions can be efficiently computed via certain differential equations.

THEOREM 2 (*Generalized Krein–Levinson equations*). *The functions $a(T;.)$ and $c(T;.)$ satisfy the differential equations*

$$\left(\frac{\partial}{\partial T}+\frac{\partial}{\partial t}\right)a(T;t)=c'(T;T)\Lambda\, c(T;t) \tag{34a}$$

and

$$\frac{\partial}{\partial T}c(T;t)=-c(T;T)a(T;t). \tag{34b}$$

Proof. For (34b) we note from the defining integral equation (31) that

$$\frac{\partial c'}{\partial T}(T;t)+\int_0^T K(t,u)\frac{\partial c'(T;u)}{\partial T}\,du=-K(t,T)c'(T;T)$$

or, in operator notation,

$$(I+K)\frac{\partial c'}{\partial T}=-K\delta_T c'.$$

Therefore,

$$\begin{aligned}\frac{\partial c'}{\partial T}&=-(I-H)K\delta_T c'\\&=-H\delta_T c'=-H(t,T;T)c'(T;T),\end{aligned}$$

from which (34b) follows by recalling (cf. (26)) that $a'(T;t)=H(t,T;T)$. For (34a), we start with the defining equation (cf. (26) and (1))

$$a(T;t)+\int_0^T a(T;u)K(u,t)\,du=K(T,t), \tag{34c}$$

and calculate $\lrcorner a$ by the method used in the proof of Lemma 1—everything will be the same as for $\lrcorner H$ except for an additional term arising from the upper limit of the integral. In other words, we shall get the expression

$$\lrcorner a=a'a+c'\Lambda c-a'a=c'\Lambda c,$$

which is (34a). We omit the detailed calculations, which can be carried out in several different ways. ∎

For scalar symmetric Toeplitz kernels, we have (cf. (33) and (28)) that

$$\Lambda=-1,\quad c(T;t)=a(T;T-t),$$

so that (34a, b) reduce to the Krein–Levinson equations (7a, b).

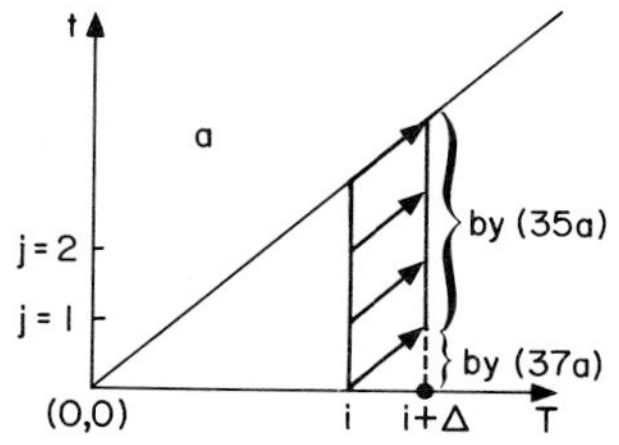

FIG. 1.

The necessary boundary conditions for the generalized Krein–Levinson equations will become apparent by seeing how we might actually solve (34). The numerical aspects of this question have not been studied yet, but as a first step we can use a simple discretization scheme to replace (34) by the recursions

$$a(i + \Delta; j + \Delta) = a(i; j) + \Delta c'(i; i)\Lambda c(i; j) \tag{35a}$$

$$c(i + \Delta; j) = c(i; j) - \Delta c(i; i)a(i; j). \tag{35b}$$

The way in which these equations are propagated is shown in the figures. We assume that $a(i; j)$, $c(i; j)$ have been determined for all $j \leqslant i$, and now have to compute $a(i + \Delta; j + \Delta)$, $c(i + \Delta; j + \Delta)$ for all $j \leqslant i + \Delta$. From (35a) and Fig. 1 we see how to obtain all values of $a(i + \Delta; j + \Delta)$ except $a(i + \Delta; 0)$. However, this value can be obtained by returning to the defining integral equation (34c) for a, which for $t = 0$ gives

$$a(T; 0) = K(T, 0) - \int_0^T a(T; u)K(u, 0)\, du. \tag{36a}$$

The discretized version

$$a(i + \Delta; 0) = K(i + \Delta; 0) - \sum_{\Delta}^{i+\Delta} a(i + \Delta; j)K(j, 0)\Delta \tag{37a}$$

then allows us to complete the calculation of $a(i + \Delta; j + \Delta)$.

Similarly, (35b) and Fig. 2 show how to obtain all values of $c(i + \Delta; j + \Delta)$ except for $c(i + \Delta; i + \Delta)$. This can be obtained by returning to the definition

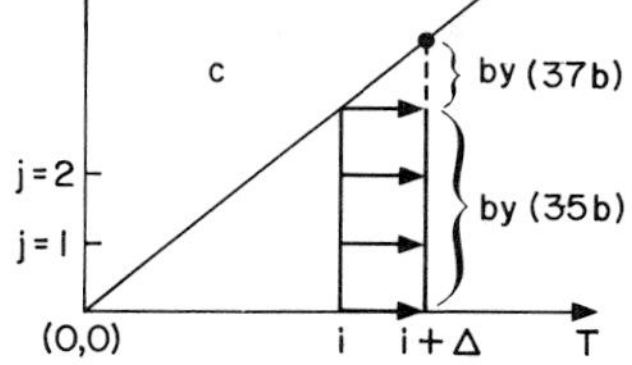

FIG. 2.

(31) for c,

$$c'(T; t) = d(t) - \int_0^T H(t, u; T)\, d(u)\, du$$

which at $t = T$ gives

$$c'(T; T) = d(T) - \int_0^T a(T, u)\, d(u)\, du. \tag{36b}$$

The discretized version

$$c'(i + \Delta; i + \Delta) = d(i + \Delta) - \sum_{\Delta}^{i+\Delta} a(i + \Delta; j)\, d(j)\, \Delta \tag{37b}$$

then allows us to complete the calculation of $c(i + \Delta; j + \Delta)$. We should note that the recursions (37a) and (37b) are essentially the same, a useful fact that might not have been apparent if, as we did for a, we had used the expression $c' = d - Kc'$ instead of $c' = d - H\,d$. This is a generalization of a trick used by Burg in the Toeplitz case (cf. [8]).

III.1. The Krein–Siegert–Bellman Identity

This is a good place to introduce a well-known identity for the resolvent of non-Toeplitz kernels,

$$\frac{\partial H(t, s; T)}{\partial T} = -H(t, T; T) H(T, s; T), \tag{38}$$

which was independently derived by Krein [19], Siegert [20], and Bellman [21]. The formula (38), which can be derived by a method similar to that used above to obtain (34b), also yields a method of recursively calculating $H(t, s; T)$, but it will generally take more effort than the procedure we have described above. The reason is that to determine $H(t, s; T)$ via (38) we shall have to calculate $H(t, s; T')$ for all $T' \leqslant T$; on the other hand, via the generalized Krein–Levinson and Sobolev equations, we need only to calculate the generalized boundary values of $H(t, s; T')$ for $T' \leqslant T$. In fact, if $T = N\Delta$, an operation count will show that the latter method will require only a number of operations proportional to $N^2\alpha$, as opposed to N^3 for methods based on (38). We should mention that Kalaba and Zagustin [22] have shown that the calculations via (38) can be carried out via a certain integro-differential equation, which we may call a "slow" Krein–Levinson equation because it requires N^3 computations and cannot exploit any 'closeness to Toeplitz' property (as reflected in a low value of α) that may be present in the problem.

It is important to note that the number α that appears in the generalized Sobolev identity and the Krein–Levinson equations is not necessarily the displacement rank of K, i.e., it is not the smallest number that will yield the representation (32a) for $\lrcorner K$. This is useful because in applications it may often be easier to obtain upper bounds on the displacement rank and non-minimal representations for $\lrcorner K$.

III.2. Operator Factorizations

We may note that the integrated form of the Krein–Siegert–Bellman identity,

$$H(t, s; T) = H(t, s; t) + H(s, t; s) - \int_0^T H(t, u; u)H(u, s; u)\, du, \qquad (39)$$

has been very extensively and effectively used by Gohberg and Krein (see, e.g., [23, Chapter IV]) to study problems of operator factorization. The integrated form of our generalized Sobolev identity (32) shows that every linear integral operator can be written as a sum of products of upper and lower triangular (i.e., Volterra) Toeplitz operators. This representation has many intriguing features. Its implications for operator factorization will be described elsewhere; an application to the calculation of likelihood ratios (Radon–Nikodym derivatives) of Gaussian processes is presented in [24].

We may remark also that since the integrated forms do not contain derivatives, they may be expected to hold for square-integrable kernels as well, and this is in fact true.

IV. Further Results and Extensions

The generalized Krein–Levinson equations use no structure of the non-Toeplitz kernel except that displayed by the representation (32a) for $\lrcorner K$. In particular physical problems more may be known, and this knowledge may be used to simplify the recursions even further.

Thus, in radiative transfer theory even the Toeplitz kernel is assumed to have the further structure

$$K(t - s) = \int_0^1 e^{-\alpha|t-s|} w(\alpha)\, d\alpha.$$

and in this case the X and Y functions of Ambartsumian and Chandrasekhar [25, 26] were shown to give an alternative solution method for the finite interval Wiener–Hopf equation and, through identities like (38), for the

Fredholm equation as well. We have shown that for nonstationary kernels generated by passing white noise through a system describable by a vector differential equation, leading to kernels of the form

$$\begin{aligned} K(t, s) &= Me^{F(t-s)}N(s), \qquad && t \geqslant s, \\ &= N'(t)e^{F'(s-t)}M', && t \leqslant s, \end{aligned}$$

the generalized Krein–Levinson differential equations can be reduced to generalized Chandrasekhar X and Y equations and to generalized Riccati equations (cf. [14, 15, 27]). The displacement rank for such kernels never exceeds the order of the differential equation (which is the size of the matrix F in the above formula).

This brings up the important point that we do not need to know the displacement rank of K to use our results. In view of the order-of-magnitude reduction in complexity, from $O(N^3)$ to $O(N^2\alpha)$, we could envisage several complete trial solutions with different "guessed" values of α (and conveniently chosen functions $d(.)$—cf. (32a)), with less computational effort than for a single direct $O(N^3)$ solution of the original integral equation. In other words, for physical problems where there generally exists some uncertainty as to the exact nature of the kernel $K(., .)$, our fast solution methods make feasible an interactive "trial-and-error" method of obtaining a satisfactory mathematical model (and solution).

In this connection, it is worth stressing that our method should not be confused with the well-known method of approximating the given kernel $K(t, s)$ by a degenerate or separable kernel of the form $\sum_1^N \lambda_i p_i(t) p_i'(s)$, $N < \infty$. Except in very degenerate cases, such an approximation will *destroy* any Toeplitz or shift-invariance properties of the original kernel. What we do instead is to first remove any Toeplitz part of the kernel by applying the $\lrcorner$ operation and then approximate the remainder by a degenerate kernel. In other words, we try to preserve as much of the Toeplitz structure as possible.

Notice also that if $\lrcorner K$ does not have low enough rank, we could compute $\lrcorner^2 K$, then $\lrcorner^3 K$, and so on as necessary. Such continued differentiation would not be numerically very desirable, but it is worth remarking that what we have will be a kind of general Taylor series representation about Toeplitz kernels as the "origin."

Furthermore, the methods need not be restricted to Toeplitz-type structure. For example, we know that Hankel operators, which correspond to kernels of the form $K(t + s)$, also have nice Fredholm resolvents. Thus, in certain applications it should be of interest to classify operators according to their degree of "non-Hankelness," and this can be done by using the operator

$$\left(\frac{\partial}{\partial t} - \frac{\partial}{\partial s}\right) K(t, s).$$

Various other special types of kernels can also be considered. In fact, it seems to us that the results we have described above are only a special case of some more general algebraic framework, but we have not been able to pursue this thought.

Even in the restricted framework of operators close to Toeplitz, there seem to be many directions worth pursuing. For example, it is known that Toeplitz kernels are closely related to classical objects in analysis like orthogonal polynomials, harmonic functions, moment problems, etc. There should be some generalizations of these objects corresponding to non-Toeplitz kernels that have a finite displacement rank. The theory of poly-harmonic functions (see, e.g., [28]) seems to have some relevance in this connection.

References

1. M. G. Krein, Integral equations on a half-line with kernel depending upon the difference of the arguments, *Usp. Mat. Nauk* (*N.S.*), **13** (1958), 3–120; English transl. *Amer. Math. Soc. Transl. Ser. 2* **22** (1962), 163–288.
2. I. C. Gohberg and M. G. Krein, Systems of integral equations on the half-line with kernels depending on the difference of the arguments, *Usp. Mat. Nauk* **20** (1958), 3–72; English transl. *Amer. Math. Soc. Transl. Ser. 2* **14** (1960), 217–287.
3. L. A. Zadeh and J. R. Ragazzini, An extension of Wiener's theory of prediction, *J. Appl. Phys.* **21** (1950), 645–655.
4. D. Slepian and T. T. Kadota, Four integral equations of detection theory, *SIAM J. Appl. Math.* **17** (1969), 1102–1117.
5. V. V. Sobolev, "A Treatise on Radiative Transfer," Van Nostrand Reinhold, Princeton, New Jersey, 1963 [Russian original 1956].
6. I. G. Gohberg and I. A. Fel'dman, "Convolution Equations and Projection Methods for their Solutions," Math. Monographs No. 41, Amer. Math. Soc., 1974.
7. N. Levinson, The Wiener RMS (Root-Mean-Square) error criterion in filter design and prediction, *J. Math. Phys.* **25** (1947), 261–278.
8. R. A. Wiggins and E. A. Robinson, Recursive solution to the multichannel filtering problem, *J. Geophys. Res.* **70** (1965), 1885–1891.
9. U. Grenander and G. Szego, "Toeplitz Forms and Their Applications," Univ. of California Press, Berkeley, California, 1958.
10. M. G. Krein, The continuous analogues of theorems on polynomials orthogonal on the unit circle, *Doklady Akad. Nauk SSSR* **104** (1955), 637–640.
11. T. Kailath, A. Vieira, and M. Morf, Inverses of Toeplitz operators, innovations, and orthogonal polynomials, *SIAM Rev.* **20** (1978), 106–119.
12. I. C. Gohberg and A. A. Semencul, On the inversion of finite Toeplitz matrices and their continuous analogs, *Mat. Issled* **2** (1972), 201–233 [in Russian].
13. I. C. Gohberg and G. Heinig, On Toeplitz matrix integral operators over a finite interval, *Rev. Roumaine Math. Pures Appl.* **20**, no. 1 (1975), 55–73 [in Russian].
14. T. Kailath, L. Ljung, and M. Morf, Recursive input–output and state-space solutions for continuous-time linear estimation problems, *Proc. IEEE Decision and Control Conf. Florida, Dec. 1976*, pp. 182–185, IEEE, New York, 1976.

15. B. Friedlander, T. Kailath, and M. More, A modified displacement rank and some applications, *Proc. IEEE Decision and Control Conf., New Orleans, Dec. 1977*, pp. 958–961, IEEE, New York, 1977.
16. T. Kailath, Some new results and insights in linear least-squares estimation, Proc. *First Joint IEEE–USSR Workshop Information Theory, Moscow, December 1975* [rept. with corrections in T. Kailath, "Lectures in Linear-Least-Squares Estimation," Springer-Verlag, Berlinand, New York, 1978.]
17. T. Kailath, S.-Y. Kung, and M. More, Displacement ranks of matrices and linear equations, *J. Math. Anal. Appl.*, to appear.
18. B. Friedlander, T. Kailath, M. Morf, and L. Ljung, Levinson- and Chandrasekhar-type equations for a general discrete-time linear estimation problem, *IEEE Trans. Automat. Contr.* **AC-23** (1978).
19. M. G. Krein, On integral equations leading to second-order differential equations *Doklady Akad. Nauk SSSR* **97** (1954), 21–24.
20. A. J. F. Siegert, A systematic approach to a class of problems in the theory of noise—Part II, *IRE Trans. Information Theory* **IT-3** (1957), 38–43.
21. R. E. Bellman, Functional equations in the theory of dynamic programming—VII: A partial differential equation for the Fredholm resolvent, *Proc. Amer. Math. Soc.* **8** (1957), 435–440.
22. R. E. Kalaba and E. A. Zagustin, Krein–Bellman boundary formulas for Fredholm resolvents, preprint, March 1976.
23. I. C. Gohberg and M. G. Krein, "Theory of Volterra Operators in Hilbert Space," Transl. of Mathematical Monographs, Vol. 28, Amer. Math. Soc. Providence, Rhode Island, 1970.
24. T. Kailath, B. Lévy, L. Ljung, and M. Morf, Fast time-invariant implementations of Gaussian signal detectors, *IEEE Trans. Information Theory* **IT-24** (1978).
25. V. A. Ambartzumian, Diffuse reflection of light by a foggy medium, *Doklady Akad. Sci. SSSR* **38** (1943), 229–232.
26. S. Chandrasekhar, On the radiative equilibrium of a stellar atmosphere, Pt. XXI, *Astrophys. J.* **106** (1947), 152–216; **107** (1948), 48–72.
27. T. Kailath, Some new algorithms for recursive estimation in constant linear systems, *IEEE Trans. Information Theory* **IT-19** (1973), 750–760.
28. D. D. Gakhov, "Boundary Value Problems," Pergamon, Oxford, 1962 [Russian original 1958].

AMS (MOS) subject classifications: 45B05, F05, L05; 47B30, B35, G05.

TOPICS IN FUNCTIONAL ANALYSIS
ADVANCES IN MATHEMATICS SUPPLEMENTARY STUDIES, VOL. 3

Trotter's Product Formula for an Arbitrary Pair of Self-Adjoint Contraction Semigroups

TOSIO KATO†

Department of Mathematics,
University of California, Berkeley
Berkeley, California

DEDICATED TO PROFESSOR M. G. KREIN ON HIS SEVENTIETH BIRTHDAY

1. INTRODUCTION

The purpose of this paper is to show that

$$\operatorname*{s-lim}_{n\to\infty} (e^{-tA/n}e^{-tB/n})^n \tag{1.1}$$

exists whenever A, B are nonnegative self-adjoint operators in a Hilbert space, to describe the limit, and to give some applications of the results.

The existence of (1.1) is known under various additional conditions (see Chernoff [1] for comprehensive discussion of the problem, and Belyi and Semenov [2] for some recent results). But it seems to have remained an open question for an arbitrary pair $A, B \geqslant 0$.

Let us state our results more precisely. Let A, B be nonnegative self-adjoint operators in a Hilbert space H. Let $D' = D(A^{1/2}) \cap D(B^{1/2})$ and let H' be the closure of D', where $D(T)$ denotes the domain of T. We define the *form sum*

$$C' = A \dotplus B \tag{1.2}$$

of A, B as the self-adjoint operator in H' associated with the nonnegative, closed quadratic form

$$u \mapsto \|A^{1/2}u\|^2 + \|B^{1/2}u\|^2, \qquad u \in D', \tag{1.3}$$

which is densely defined in H'. (For quadratic forms and the associated self-adjoint operators see, e.g., Kato [3, Chapter 6].) We assert that (1.1) exists and equals $e^{-tC'}P'$, where P' is the orthogonal projection of H onto H'.

Actually we shall prove these results for more general products of the form $[f(tA/n)g(tB/n)]^n$, where f and g are real valued, Borel measurable

† This work was partially supported by NSF Grant MCS 76-04655.

ISBN 0-12-287150-2

functions on $[0, \infty)$ such that

$$0 \leqslant f(s) \leqslant 1, \qquad f(0) = 1, \qquad f'(0) = -1, \tag{1.4}$$

and similarly for g. Convergence of such general products is useful in applications to numerical analysis. Simple examples of functions satisfying (1.4) are

$$f(s) = e^{-s}, \qquad f(s) = (1 + k^{-1}s)^{-k}, \qquad k > 0. \tag{1.5}$$

Similar product formulas were considered in an earlier paper [4], but the functions f, g were more restricted than (1.4) and could not cover the important case (1.1).

THEOREM. *Let H, A, B, D', H', C', P', f, and g be as defined above. Then*

$$s\text{-}\lim_{n\to\infty} [f(tA/n)g(tB/n)]^n = e^{-tC'}P', \qquad t > 0, \tag{1.6}$$

the convergence being uniform in $t \in [0, T]$ for any $T > 0$ when applied to $u \in H'$ and in $t \in [T_0, T]$ for any $0 < T_0 < T$ when applied to $u \perp H'$.

The proof of the theorem will be given in Sections 2 and 3, and some simple applications will be discussed in Section 4. Here we state a lemma on the properties of $f(tA)$, $g(tB)$ that follow directly from (1.4).

LEMMA 1.1. *We have*

(a) $f(tA) \in B(H), \qquad 0 \leqslant f(tA) \leqslant 1, \qquad t > 0,$

(b) $0 \leqslant [1 - f(tA)]^{1/2} \to 0, \qquad t \downarrow 0$

(c) $t^{-1/2}[1 - f(tA)]^{1/2}u \to A^{1/2}u, \qquad t \downarrow 0, \quad u \in D(A^{1/2}),$

and similarly for $g(tB)$.

Here and in what follows $B(H)$ denotes the set of all bounded linear operators on H to H, and $\to$ denotes strong convergence in $B(H)$ or in H, unless otherwise stated. The lemma is an easy consequence of the spectral theorem, and the proof will be omitted. It should be noted that $s^{-1}[1 - f(s)]$ is a bounded function by virtue of (1.4).

2. PROOF OF THE THEOREM, PART 1

For $t > 0$ set

$$F_t = f(tA)g(tB) \in B(H), \qquad S_t = t^{-1}(1 - F_t). \tag{2.1}$$

The first step in the proof of the theorem is to establish the convergence

$$(1 + S_t)^{-1} \to (1 + C')^{-1}P', \qquad t \downarrow 0. \tag{2.2}$$

This will already prove the theorem if D' is dense in H so that $H' = H, P' = 1$, by a well-known theorem of Chernoff [1, Theorem 1.1]. In the general case we need some supplementary arguments, which will be given in the next section.

It is convenient to introduce the following bounded self-adjoint operators for $t > 0$.

$$\begin{aligned} &A_t = t^{-1}[1 - f(tA)] \geqslant 0, \qquad B_t = t^{-1}[1 - g(tB)] \geqslant 0, \\ &K_t = 1 + A_t + B_t \geqslant 1. \end{aligned} \tag{2.3}$$

A simple calculation then gives

$$1 + S_t = 1 + A_t + B_t - tA_tB_t = K_t^{1/2}(1 - Q_t)K_t^{1/2}, \tag{2.4}$$

$$Q_t = tK_t^{-1/2}A_tB_tK_t^{-1/2}. \tag{2.5}$$

Note that F_t, S_t, and Q_t are in general not self-adjoint.

LEMMA 2.1.

$$(1 - Q_t)^{-1} \in B(H), \qquad \|(1 - Q_t)^{-1}\| \leqslant 2.$$

Proof. Since $Q_t \in B(H)$, it suffices to show that the numerical range of Q_t is contained in the disk about the origin with radius $\frac{1}{2}$ (see, e.g., [3, Chapter 5, Theorem 3.2]).

For any $w \in H$, we have

$$|(Q_t w, w)| \leqslant t\|A_t^{1/2}B_t^{1/2}\| \, \|B_t^{1/2}K_t^{-1/2}w\| \, \|A_t^{1/2}K_t^{-1/2}w\|.$$

Since $tA_t^{1/2}B_t^{1/2} = [1 - f(tA)]^{1/2}[1 - g(tB)]^{1/2}$ has norm $\leqslant 1$, we obtain

$$|(Q_t w, w)| \leqslant \tfrac{1}{2}(\|B_t^{1/2}K_t^{-1/2}w\|^2 + \|A_t^{1/2}K_t^{-1/2}w\|^2) \leqslant \|w\|^2/2$$

by (2.3). This proves the required result.

LEMMA 2.2. *Fix $u \in H$ and set $w_t = K_t^{-1}u$. Given any sequence $t_n \downarrow 0$, we can extract a subsequence such that*

$$w_t \rightharpoonup w \in D', \qquad A_t^{1/2}w_t \rightharpoonup A^{1/2}w, \qquad B_t^{1/2}w_t \rightharpoonup B^{1/2}w \tag{2.6}$$

along this subsequence, where $\rightharpoonup$ denotes weak convergence.

Proof. Taking the inner product of $u = K_t w_t = (1 + A_t + B_t)w_t$ with w_t, we obtain

$$\|w_t\|^2 + \|A_t^{1/2}w_t\|^2 + \|B_t^{1/2}w_t\|^2 = (u, w_t). \tag{2.7}$$

Hence $\|w_t\|^2 \leqslant \|u\| \, \|w_t\|$, $\|w_t\| \leqslant \|u\|$, so that (2.7) does not exceed $\|u\|^2$. Thus we can extract a subsequence of $\{t_n\}$ such that w_t, $A_t^{1/2}w_t$, and $B_t^{1/2}w_t$ all converge weakly along this subsequence. Let w, w', and w'' be their weak limits, respectively. We have to show that $w \in D' = D(A^{1/2}) \cap D(B^{1/2})$ and $w' = A^{1/2}w$, $w'' = B^{1/2}w$.

To this end, let $x \in D(A^{1/2})$. Then $A_t^{1/2}x \to A^{1/2}x$ (strongly) by Lemma 1.1, so that

$$(w', x) = \lim(A_t^{1/2}w_t, x) = \lim(w_t, A_t^{1/2}x) = (w, A^{1/2}x).$$

Since this is true for every $x \in D(A^{1/2})$, it follows that $w \in D(A^{1/2})$ with $A^{1/2}w = w'$. Similarly, one proves that $w \in D(B^{1/2})$ with $B^{1/2}w = w''$.

LEMMA 2.3. $w = (1 + C')^{-1}P'u$.

Proof. Let $z \in D'$. If $t \downarrow 0$ along the subsequence stated in Lemma 2.2, we have

$$\begin{aligned}(u, z) &= ((1 + A_t + B_t)w_t, z)\\ &= (w_t, z) + (A_t^{1/2}w_t, A_t^{1/2}z) + (B_t^{1/2}w_t, B_t^{1/2}z)\\ &\to (w, z) + (A^{1/2}w, A^{1/2}z) + (B^{1/2}w, B^{1/2}z)\\ &= (w, z) + (C'^{1/2}w, C'^{1/2}z),\end{aligned}$$

where we have again used Lemma 1.1; recall that C' is the self-adjoint operator associated with the form (1.3). Since the last equality is true for every $z \in D' = D(C'^{1/2})$, we conclude that $w \in D(C')$ and $(1 + C')w = P'u$, which is equivalent to the desired result.

LEMMA 2.4. *We have as $t \downarrow 0$*

$$w_t \to w, \qquad A_t^{1/2}w_t \to A^{1/2}w, \qquad B_t^{1/2}w_t \to B^{1/2}w. \tag{2.8}$$

Proof. Lemma 2.3 shows that the weak limits given in (2.6) are independent of the particular subsequence chosen. Thus a standard argument shows that the weak convergence (2.6) takes place as $t \downarrow 0$, without any restriction on subsequences.

We have to show that actually we have strong convergence here. To this end we come back to (2.7) and let $t \downarrow 0$, obtaining

$$\begin{aligned}\|w_t\|^2 + \|A_t^{1/2}w_t\|^2 + \|B_t^{1/2}w_t\|^2 &\to (u, w) = (P'u, w)\\ &= ((1 + C')w, w) = \|w\|^2 + \|A^{1/2}w\|^2 + \|B^{1/2}w\|^2,\end{aligned}$$

where we have used Lemma 2.3, noting that $w \in D' \subset H'$. Another standard argument based on (2.6) then gives

$$\|w_t - w\|^2 + \|A_t^{1/2}w_t - A^{1/2}w\|^2 + \|B_t^{1/2}w_t - B^{1/2}w\|^2 \to 0,$$

which is equivalent to (2.8).

Proof of (2.2). Let $u \in H$ and set $w_t = K_t^{-1}u$ as above, so that Lemmas 2.3 and 2.4 hold. Thus we have by (2.4)

$$(1 + S_t)^{-1}u = K_t^{-1/2}(1 - Q_t)^{-1}K_t^{-1/2}u$$
$$= K_t^{-1}u + K_t^{-1/2}(1 - Q_t)^{-1}Q_tK_t^{-1/2}u.$$

Since $K_t^{-1}u = w_t \to w = (1 + C')^{-1}P'u$ and since $\|K_t^{-1/2}(1 - Q_t)^{-1}\| \leqslant 2$ by (2.3) and Lemma 2.1, (2.2) will follow if we show that

$$Q_tK_t^{-1/2}u = tK_t^{-1/2}A_tB_tw_t \to 0, \qquad t \downarrow 0.$$

Since $\|K_t^{-1/2}A_t^{1/2}\| = \|A_t^{1/2}K_t^{-1/2}\| \leqslant 1$ by (2.3), it suffices to show that

$$tA_t^{1/2}B_tw_t \to 0.$$

But this follows from the facts that $B_t^{1/2}w_t \to B^{1/2}w$ by (2.8) and $tA_t^{1/2}B_t^{1/2} = [1 - f(tA)]^{1/2}[1 - g(tB)]^{1/2} \to 0$ by Lemma 1.1. This completes the proof of (2.2).

3. Proof of the Theorem, Part 2

For the proof of the theorem, it suffices to prove the following two results separately: as $n \to \infty$ we have

$$F_{t/n}^n u \to e^{-tC'}u \qquad \text{if} \quad u \in H', \tag{3.1}$$

$$F_{t/n}^n u \to 0 \qquad \text{if} \quad u \perp H', \tag{3.2}$$

together with the uniformity of convergence in t specified in the theorem.

Proof of (3.1). This is easily done by using the existing approximation theory for a C_0-semigroup by a family of discrete semigroups (see [3, Chapter 9, §3]). There are two things to be noticed, however. First, our C_0-semigroup $U(t) = e^{-tC'}$ to be approximated acts on H' and not on H, while the approximating discrete semigroups $U_n(t) = F_{\tau_n}^{[t/\tau_n]}$ act on H, where $\tau_n \downarrow 0$ is the sequence of "time units" to be chosen appropriately. Actually, this does not cause any difficulty. Indeed, starting from the "resolvent convergence"

$$(1 + S_t)^{-1}u \to (1 + C')^{-1}u, \qquad t \downarrow 0, \qquad u \in H', \tag{3.3}$$

implied by (2.2), all the results of [3, Chapter 9, §3] can be justified easily.

Second, it is convenient to extend the results of [3] by admitting a continuous family of discrete semigroups $U_\tau(t) = F_\tau^{[t/\tau]}$, where the time unit parameter τ varies over all positive real numbers, rather than a sequence of such semigroups. Again, all the results of [3] can be extended to this case.

With these generalizations, the approximation theorem [3, Theorem 3.6] shows that

$$F_\tau^{[t/\tau]}u \to e^{-tC'}u, \qquad \tau \downarrow 0, \quad u \in H', \tag{3.4}$$

uniformly in $t \in [0, T]$. Note that (3.3) is exactly the "resolvent convergence" for the generator family $\tau^{-1}(1 - F_\tau)$ required in this theorem.

(3.4) gives the required result (3.1) on restricting τ to $\tau = t/n$, $n = 1, 2, \ldots,$ for a fixed t. In view of the uniformity of convergence (3.4) stated above, the convergence (3.1) is also uniform in t, as required in the theorem.

Proof of (3.2). This is quite different from the case of (3.1) and cannot be reduced to a general theorem valid for any C_0-semigroups. The basis for the proof is again given by (2.2):

$$(1 + S_t)^{-1}u \to 0, \qquad t \downarrow 0, \quad u \perp H', \tag{3.5}$$

but the fact that S_t is in general not self-adjoint causes a little trouble. Thus we first "symmetrize" S_t in (3.5) by introducing new operators (cf. (2.1))

$$F_t' = f(tA)^{1/2}g(tB)f(tA)^{1/2} \geqslant 0, \qquad S_t' = t^{-1}(1 - F_t') \geqslant 0, \tag{3.6}$$

which are self-adjoint.

A simple computation then gives

$$(1 + S_t')^{-1} = (1 + t)^{-1}[t + f(tA)^{1/2}g(tB)(1 + S_t)^{-1}f(tA)^{1/2}]. \tag{3.7}$$

Since $f(tA)^{1/2}u = u - [1 - f(tA)^{1/2}]u$, with the second term tending to 0 as $t \downarrow 0$, and since $f(tA)$, $g(tB)$, and $(1 + S_t)^{-1}$ are all uniformly bounded in t, (3.7) and (3.5) give

$$(1 + S_t')^{-1}u \to 0, \qquad t \downarrow 0, \quad u \perp H'. \tag{3.8}$$

From this it is easy to deduce, using the resolvent equation, that

$$(\lambda + S_t')^{-1}u \to 0, \qquad t \downarrow 0, \tag{3.9}$$

uniformly in $\lambda \in [T^{-1}, T]$.

(3.6) implies that $0 \leqslant F_t' = 1 - tS_t' \leqslant (1 + tS_t')^{-1}$ and hence

$$0 \leqslant F_{t/n}'^{2n} \leqslant [1 + (t/n)S_{t/n}']^{-2n} \leqslant (1 + 2tS_{t/n}')^{-1} \tag{3.10}$$

because $F_{t/n}'$ and $S_{t/n}'$ commute. Hence

$$\begin{aligned} \|F_{t/n}'^{n}u\|^2 &= (F_{t/n}'^{2n}u, u) \leqslant ((1 + 2tS_{t/n}')^{-1}u, u) \\ &= (2t)^{-1}([(2t)^{-1} + S_{t/n}']^{-1}u, u) \to 0, \qquad n \to \infty, \end{aligned} \tag{3.11}$$

by (3.9), uniformly in $t \in [T^{-1}, T]$.

To deduce (3.2) from (3.11), we need only a slight adjustment. Since

$$F_{t/n}^{n+1} = f(tA/n)^{1/2} F_{t/n}^{\prime n} f(tA/n)^{1/2} g(tB/n)$$

and

$$f(tA/n)^{1/2} g(tB/n)u = u - [1 - f(tA/n)^{1/2} g(tB/n)]u$$

with the last term tending to 0 as $n \to \infty$, (3.11) implies that $F_{t/n}^{n+1} u \to 0$. Then

$$F_{t/n}^{n} u = F_{t/n}^{n+1} u + F_{t/n}^{n}[1 - f(tA/n)g(tB/n)]u \to 0,$$

completing the proof of (3.2).

4. Applications

A. *Schrödinger Operators*

Consider the formal Schrödinger operator

$$L = -\Delta + q(x)\cdot, \qquad x \in \Omega \subset R^m, \tag{4.1}$$

where Ω is an arbitrary open set in R^m and q is assumed to be a nonnegative measurable function on Ω. Set

$$a[u] = \int_\Omega |\mathrm{grad}\, u|^2\, dx, \qquad b[u] = \int_\Omega q|u|^2\, dx, \tag{4.2}$$

with $D(a) = H_0{}^1(\Omega)$ (Sobolev space) and with $D(b)$ consisting of all $u \in L^2(\Omega)$ with $b[u] < \infty$. The forms a, b are densely defined and closed in $H = L^2(\Omega)$. Let A, B be the associated self-adjoint operators in H. Then $A \subset -\Delta$ with the Dirichlet boundary condition (in a generalized sense) and B is the maximal multiplication operator with q.

For these A, B we have the convergence (1.6) with $C' = A \dotplus B$, a self-adjoint operator in H', the closure of $D' = D(a) \cap D(b)$. C' may be regarded as a realization of L, but its concrete meaning is in general not clear.

The size of H' depends on how well behaved q is. If $q \in L^1_{\mathrm{loc}}(\Omega)$, D' is dense in H. Hence $H' = H$, $P' = 1$, and the limit in (1.6) is a C_0-semigroup $e^{-tC'}$. In this case one can show that C' is exactly the realization of L in H.

If q is ill behaved, it may happen that $D' = \{0\} = H'$, so that the limit in (1.6) is zero. This is the case if, for example, $m = 1$ and q is nowhere locally integrable in Ω; note that $D(a)$ for $m = 1$ consists of rather smooth functions. A similar example was given by [1, Chapter 7] using the Wiener integral.

It should be noted that if $\partial\Omega$ is small (so that $R^m \backslash \Omega$ has capacity zero), then $H = L^2(\Omega) = L^2(R^m)$ and $H_0{}^1(\Omega) = H^1(R^m)$ so that $-A$ is the usual "free" Laplacian. Construction of $C' = A \dotplus B$ in the case $\partial\Omega = \{0\}$ has been extensively studied in the literature.

B. *Segal's Lemma*

It is known that

$$\|e^{-(A \dotplus B)}\| \leqslant \|e^{-A}e^{-B}\| \tag{4.3}$$

holds for nonnegative self-adjoint operators A, B under certain additional conditions which ensure, among other things, that $C' = A \dotplus B$ is self-adjoint in H (see Segal [5], Reed and Simon [6, Chapter 10]). Since the proof of (4.3) depends only on Trotter's product formula (and some interpolation theorems), it follows from our theorem that (4.3) is valid for any pair $A, B \geqslant 0$; note that $\|e^{-tC'}P'\| = \|e^{-tC'}\|$. In the extreme case, $H' = \{0\}$, the left side of (4.3) should be interpreted to be zero.

The left side of (4.3) is equal to $e^{-\gamma}$, where γ is the lower bound of $C' = A \dotplus B$, which is also the lower bound of the form (1.3). Again one should set $\gamma = +\infty$ if $D' = \{0\}$.

Another inequality of a similar nature is

$$\|e^{-(A \dotplus B)}\|_p \leqslant \|e^{-A}e^{-B}\|_p, \qquad 0 < p < \infty, \tag{4.4}$$

where $\| \;\|_p$ denotes the p-norm for compact operators, defined as the pth root of the sum of the pth powers of the singular values of the operator (see, e.g., Gohberg and Krein [7, p. 121]). Since (4.4) can be proved by using Trotter's product formula (cf. [6, p. 334]), it is again valid for any pair $A, B \geqslant 0$. As before note that $\|e^{-tC'}P'\|_p = \|e^{-tC'}\|_p$, and the left side of (4.4) should be taken to be zero if $D' = \{0\}$.

Remark. It is trivial to extend (4.3) and (4.4) to the case in which A, B are bounded from below but not necessarily nonnegative. Moreover, it is possible, under certain conditions, to extend them to the case in which A is not bounded from below. In such a case one has the result that whenever the right side is finite, the left side is also finite and the inequality holds (so that $A \dotplus B$ is bounded from below).

5. Generalizations to Improper Operators

The appearance of the operator C', which is not densely defined in H, in (1.6) suggests the following generalization, not only of the product formula but also of nonnegative operators A, B in general.

Consider the set $S(H)$ of all linear operators A in H with the following properties. A has its domain $D(A)$ and range $R(A)$ in a common (closed) subspace M_A of H and A is a nonnegative self-adjoint operator in M_A. M_A will be called the *domain subspace* of A.

$A \in S(H)$ is said to be *proper* if $M_A = H$ (so that A is self-adjoint in H), and *improper* otherwise. A is (improperly) bounded if A is bounded on M_A in the usual sense. A is properly bounded if A is proper and bounded (so that $A \in B(H)$).

We imagine that $A \in S(H)$ takes the value $+\infty$ in $M_A{}^{\perp}$. To make this statement precise, we associate with A a spectral measure E_A on $[0, \infty]$ (more precisely, on Borel subsets of $[0, \infty]$ with an obvious topology) with values in the set $P(H)$ of all orthogonal projections on H. For each Borel set $\Delta \subset [0, \infty)$, $E_A(\Delta)$ is by definition the value of the usual spectral measure of A as a self-adjoint operator in M_A, regarded as an element of $P(H)$ in an obvious way. On the other hand, we set $E_A\{\infty\}H = M_A{}^{\perp}$. It is easy to see that this defines a spectral measure on $[0, \infty]$. Conversely, a spectral measure on $[0, \infty]$ with values in $P(H)$ corresponds to a unique operator A in $S(H)$ in the manner described.

$A \in S(H)$ is proper if and only if $E_A\{\infty\} = 0$. A is bounded if and only if $E_A[\mu, \infty) = 0$ for some finite $\mu > 0$. A is properly bounded if and only if $E_A[\mu, \infty] = 0$ for some finite $\mu > 0$.

A simple example of an improper operator is O_M, which has domain M and range $\{0\}$, where M is a (closed) subspace of H. O_M is improperly bounded, with its spectral measure concentrated on $\{0\}$ and $\{\infty\}$. We shall denote by ∞ the special case of O_M for $M = \{0\}$. The spectral measure for ∞ is concentrated on $\{\infty\}$, with $E_\infty\{\infty\} = 1$.

For each Borel measurable function ϕ on $[0, \infty]$ to itself and for each $A \in S(H)$, the operator $B = \phi(A) \in S(H)$ is defined by $E_B(\Delta) = E_A(\phi^{-1}(\Delta))$. An important example is $\phi(s) = e^{-s}$ with $\phi(\infty) = 0$; this defines e^{-A} for every $A \in S(H)$. e^{-A} is always properly bounded ($e^{-A} \in B(H)$) and equals zero on $M_A{}^{\perp}$. In particular, $e^{-\infty} = 0$. Another example of similar nature is $\phi(s) = (1 + s)^{-1}$ with $\phi(\infty) = 0$. This gives $(1 + A)^{-1} \in B(H)$ for each $A \in S(H)$. Note that $(1 + A)^{-1}$ is not necessarily equal to the inverse of $1 + A$ in the usual sense.

For any pair $A, B \in S(H)$, their form sum $A \dotplus B \in S(H)$ can be defined via the associated quadratic forms, exactly as before. Thus $S(H)$ is closed under the form addition $\dotplus$.

Our theorem can now be generalized to any pair $A, B \in S(H)$:

$$\text{s-}\lim_{n \to \infty} [f(tA/n)g(tB/n)]^n = e^{-t(A \dotplus B)}, \tag{5.1}$$

where f, g should be defined on $[0, \infty]$ satisfying (1.4) and $0 \leqslant f(\infty) < 1$, and similarly for g. The proof of this generalized convergence theorem is essentially the same as before, although some additional considerations are required.

As an application of (5.1), choose for B the operator O_M considered above, where M is an arbitrary subspace of H, and assume $g(\infty) = 0$. Then $g(tB) = P$ for all $t > 0$, where P is the projection onto M, and (5.1) gives

$$\operatorname*{s-lim}_{n\to\infty} [f(tA/n)P]^n = e^{-tA_P}, \tag{5.2}$$

where $A_P = A \dotplus O_M \in S(H)$. The quadratic form a_P for A_P is equal to the one a for A restricted on $D' = D(a) \cap M$. (5.2) is zero on $D'^{\perp}$.

If we further specialize $A = O_N$, where N is another subspace with the projection Q, we have $A_P = O_N \dotplus O_M = O_{N \cap M}$ so that $e^{-tA_P} = R$, where R is the projection onto $N \cap M$. If we assume $f(\infty) = 0$, (5.2) thus reduces to the well-known formula $\lim (QP)^n = R$.

Remarks. 1. Inequalities (4.3) and (4.4) may be extended to the case $A, B \in S(H)$.

2. The class $S(H)$ has other properties, with useful applications. For example, one can extend not only the usual semiorder $\leqslant$ to $S(H)$ but also the *spectral order*, under which $S(H)$ becomes a lattice (for spectral order see Olson [8], Arveson [9]). As an application, one can prove the convergence formula

$$(A^n \dotplus B^n)^{1/n} \to A \vee B, \qquad n \to \infty, \quad A, B \in S(H);$$

here $\to$ denotes generalized strong convergence (or resolvent convergence), that is, $T_n \to T$ means $(1 + T_n)^{-1} \to (1 + T)^{-1}$ (strongly), where $(1 + T)^{-1} \in B(H)$ was defined above. These results will be published elsewhere.

Addendum

In a private communication to the author, Barry Simon has given the following generalization of the theorem.

Theorem (Simon). *Let A, B be any m-sectorial operators in H, with the associated closed sectorial forms a, b, respectively. Let P' be the orthogonal projection onto the closure H' of $D' = D(a) \cap D(b)$.* Then the conclusion of the theorem in text is true for $f(s) = g(s) = e^{-s}$. (For the terminology see [3, Chapters 6 and 7]).

Sketch of Simon's proof. One may assume that the forms a, b have vertices zero. Consider the families $a(k) = \operatorname{Re} a + k \operatorname{Im} a$, $b(k) = \operatorname{Re} b + k \operatorname{Im} b$, with $D(a(k)) = D(a)$, $D(b(k)) = D(b)$. It is easy to see that $a(k)$, $b(k)$ are holomorphic of type (a) in the strip $-\epsilon < \operatorname{Re} k < \epsilon$ with ϵ sufficiently small. The associated m-sectorial operators $A(k)$, $B(k)$ are holomorphic of type (B), and the operators $e^{-tA(k)}$, $e^{-tB(k)}$ are contractive for $t > 0$ and holomorphic in k (see [3, Chapter 9, Theorem 2.6 and the attached footnote]). Now the theorem in the text shows that

$$s\text{-}\lim_{n\to\infty} [e^{-tA(k)/n}e^{-tB(k)/n}]^n = e^{-tC'(k)}P'$$

with $C'(k) = A(k) \dotplus B(k)$, if k is real with $-\epsilon < k < \epsilon$. It follows from Vitali's theorem (see Hille and Phillips [10, Theorem 3.14.1]) that the same is true for any k in the strip. The desired result follows on setting $k = i$.

REFERENCES

1. P. R. CHERNOFF, Product formulas, nonlinear semigroups, and addition of unbounded operators, *Mem. Amer. Math. Soc.* **140** (1974), 1–121.
2. A. C. BELYI AND YU. A. SEMENOV, preprints, Kiev (1974).
3. T. KATO, "Perturbation Theory for Linear Operators," 2nd ed., Springer-Verlag, Berlin and New York, 1976.
4. T. KATO, On the Trotter-Lie product formula, *Proc. Japan Acad.* **50** (1974), 694–698.
5. I. SEGAL, Notes towards the construction of nonlinear relativistic quantum fields. III: Properties of the C*-dynamics for a certain class of interactions, *Bull. Amer. Math. Soc.* **75** (1969), 1390–1395.
6. M. REED AND B. SIMON, "Methods of Modern Mathematical Physics," Vol. II, Academic Press, New York, 1975.
7. I. C. GOHBERG AND M. G. KREIN, "Introduction to the Theory of Linear Non-Selfadjoint Operators in Hilbert Space," Izdat. Nauka, Moscow, 1965.
8. M. P. OLSON, The selfadjoint operators of a von Neumann algebra form a conditionally complete lattice, *Proc. Amer. Math. Soc.* **28** (1971), 537–544.
9. W. ARVESON, On groups of automorphisms of operator algebras, *J. Functional Anal.* **15** (1974), 217–243.
10. E. HILLE AND R. S. PHILLIPS, "Functional Analysis and Semigroups," *Amer. Math. Soc. Colloq. Publ. Vol. 31*, Providence, Rhode Island, 1957.

AMS (MOS) subject classifications: 47B25, 47D05.

The Time Delay Operator and a Related Trace Formula†

PETER D. LAX

*Courant Institute of Mathematical Sciences,
New York University,
New York, New York*

AND

RALPH S. PHILLIPS

*Department of Mathematics
Stanford University
Stanford California*

DEDICATED TO MARK G. KREIN ON THE OCCASION OF HIS SEVENTIETH BIRTHDAY

The time delay operator was introduced into quantum mechanics by Eisenbud [2] and Wigner [8]; see also Smith [7] and Jauch and Marchand [3]. In this paper we define the time delay operator for classical wave equations in terms of the Lax–Phillips scattering theory and show its relation to the scattering matrix.

In Section 3 we derive a trace formula which appears to be related to the time delay operator. This formula is an extension to the classical case of one derived by the authors in [5] for the automorphic wave equation.

1. INTRODUCTION

In this note we study one-parameter groups $U(t)$ of unitary operators in a separable Hilbert space $\mathscr{H}$. We think of $U(t)$ as describing the evolution of a physical system in time; that is, if the initial state of the system is f, its state at time t is $U(t)f$. We take the classical rather then the quantum mechanical view, regarding $\|f\|^2$ as the total energy of the state f.

Let $\mathscr{K}$ be some subspace of $\mathscr{H}$ and K the orthogonal projection of $\mathscr{H}$ onto $\mathscr{K}$. The component of f in $\mathscr{K}$ is then Kf, and the energy contained in this component is $\|Kf\|^2$.

The quantity

$$\int_{-\infty}^{\infty} \|KU(t)f\|^2\, dt \tag{1.1}$$

† This work was supported in part by the National Science Foundation under grant numbers MCS76–07039 and MCS76–07289.

ISBN 0-12-287150-2

measures the energy of the $\mathscr{K}$-component of $U(t)f$ throughout its whole history. Since K is an orthogonal projection,

$$K^* = K, \qquad K^2 = K,$$

and since $U(t)$ is a unitary group,

$$U(-t) = U^*(t).$$

Using these facts we deduce that

$$\|KU(t)f\|^2 = (KU(t)f, KU(t)f) = (U(-t)KU(t)f, f).$$

Substituting this above we can rewrite (1.1) as

$$(Tf, f) \tag{1.2}$$

where

$$T = \int_{-\infty}^{\infty} U(-t)KU(t)\,dt; \tag{1.3}$$

the integral is to be taken in the strong sense. T is called the *time delay operator of the subspace* $\mathscr{K}$.

It is convenient to interpret the integral (1.3) in the sense of some summability such as

$$Tf = \lim_{\lambda \to 0} T_\lambda f, \tag{1.4}$$

where

$$T_\lambda = \int_{-\infty}^{\infty} e^{-\lambda|t|} U(-t)KU(t)\,dt. \tag{1.5}$$

The domain of T consists of those f for which the limit in (1.4) exists. If this domain is dense, then T commutes with $U(s)$. For

$$T_\lambda U(s) = \int_{-\infty}^{\infty} e^{-\lambda|t|} U(-t)KU(t+s)\,dt = U(s)T_\lambda',$$

where

$$T_\lambda' = e^{\lambda s} \int_s^\infty e^{-\lambda r} U(-r)KU(r)\,ds + e^{-\lambda s} \int_{-\infty}^{s} e^{\lambda r} U(-r)KU(r)\,ds.$$

If f is in the domain of T, then $T_\lambda' f$ converges with $T_\lambda f$. Using the fact that the integral in (1.1) converges it is easy to prove that $\lim(T_\lambda' f, f) = (Tf, f)$ and by polarization that weak $\lim T_\lambda' f = Tf$. It then follows that

$$U(s)T = TU(s). \tag{1.6}$$

It is helpful to look at T in the spectral representation for U; that is, take $\mathscr{H}$ to be $L_2(\mathbb{R}, \mathscr{N})$, consisting of functions $f(\sigma)$ which take their values in some auxiliary Hilbert space $\mathscr{N}$, where the action of $U(t)$ is multiplication by $e^{i\sigma t}$.

If we suppose that in this representation K can be written as an integral operator with a continuous Hermitian symmetric kernel $q(\sigma, \rho)$,

$$(Kf)(\sigma) = \int q(\sigma, \rho) f(\rho)\, d\rho, \tag{1.7}$$

then

$$(U(-t)KU(t)f)(\sigma) = \int q(\sigma, \rho) e^{i(\rho - \sigma)t} f(\rho)\, d\rho. \tag{1.8}$$

We take f to have bounded support. Multiplying (1.8) by $e^{-\lambda|t|}$ and integrating with respect to t we get

$$T_\lambda f = 2 \int q(\sigma, \rho) \frac{\lambda}{\lambda^2 + (\rho - \sigma)^2} f(\sigma)\, d\sigma.$$

Clearly,

$$\lim_{\lambda \to 0} (T_\lambda f)(\sigma) = 2\pi q(\sigma, \sigma) f(\sigma). \tag{1.9}$$

So, in this representation T is multiplication by $2\pi q(\sigma, \sigma)$ and every f with bounded support belongs to the domain of T. Thus T is defined on a dense domain and since it corresponds to a multiplicative operator in the spectral representation it is essentially self-adjoint.

The choice of the space $\mathscr{K}$ is dictated by physical considerations. In the quantum mechanical case, where $U(t) = \exp(itH)$, H the Hamiltonian, $|f(x)|^2$ is interpreted, for $\|f\| = 1$, as a probability density, and (Kf, f) is the probability of finding the system in $\mathscr{K}$; so (Tf, f) can be interpreted as the expected time that the system spends in the subspace $\mathscr{K}$. Eisenbud [2] and Wigner [8] have chosen $\mathscr{K}$ to consist of functions $f(x)$ whose support lies in some ball $|x| \leqslant R$. In this paper we study the classical analog of the quantum mechanical case. Here the appropriate choice for $\mathscr{K}$ is suggested by the scattering theory developed by the authors [4]. In this theory, which will be outlined below, the Hilbert space is decomposed into three orthogonal subspaces: an incoming subspace $\mathscr{D}_-$, an outgoing subspace $\mathscr{D}_+$, and the interacting subspace $\mathscr{K} = \mathscr{H} \ominus (\mathscr{D}_- \oplus \mathscr{D}_+)$. K is orthogonal projection onto $\mathscr{K}$, and the quantity (1.9) is the total energy of interaction of the wave with the scatterer.

In the Lax–Phillips scattering theory the incoming and outgoing subspaces satisfy the following properties:

$$\begin{aligned} U(t)\mathscr{D}_- &\subset \mathscr{D}_- \quad \text{for} \quad t \leqslant 0, \\ U(t)\mathscr{D}_+ &\subset \mathscr{D}_+ \quad \text{for} \quad t \geqslant 0; \end{aligned} \tag{1.10i}$$

$$\wedge\, U(t)\mathscr{D}_- = 0 = \wedge\, U(t)\mathscr{D}_+; \tag{1.10ii}$$

$$\overline{\vee\, U(t)\mathscr{D}_-} = \mathscr{H} = \overline{\vee\, U(t)\mathscr{D}_+}; \tag{1.10iii}$$

$$\mathscr{D}_- \perp \mathscr{D}_+. \tag{1.10iv}$$

Associated with $\mathscr{D}_-$ and $\mathscr{D}_+$ are two unitary translation representations T_- and T_+ of U onto $L_2(\mathbb{R}, \mathscr{N})$,

$$T_\pm : f \in \mathscr{H} \to k_\pm(s) \in L_2(\mathbb{R}, \mathscr{N}), \tag{1.11}$$

taking $\mathscr{D}_-, \mathscr{D}_+$ onto $L_2(\mathbb{R}_-, \mathscr{N})$, $L_2(\mathbb{R}_+, \mathscr{N})$, respectively, with the action of $U(t)$ going into right translation:

$$U_t : k(s) \to k(s - t) \tag{1.12}$$

(here $\mathscr{N}$ is an auxiliary Hilbert space). The incoming and outgoing spectral representations of U are obtained by Fourier transforming the translation representations:

$$\tilde{k}_\pm(\sigma) = (1/\sqrt{2\pi}) \int e^{i\sigma s} k_\pm(s)\, ds. \tag{1.13}$$

The scattering operator S is defined as the mapping

$$S : k_- = T_- f \to k_+ = T_+ f. \tag{1.14}$$

It is easy to show that S is unitary on $L_2(\mathbb{R}, \mathscr{N})$, commutes with translation U_t, and, in the presence of (iv), is causal, that is,

$$SL_2(\mathbb{R}_-, \mathscr{N}) \subset L_2(\mathbb{R}_-, \mathscr{N}). \tag{1.15}$$

Thus S can be realized as a distribution-valued convolution operator with support in $\mathbb{R}_-$. The spectral representer of S, called the scattering matrix and denoted by $\mathscr{S}$, is a multiplicative operator, that is,

$$\mathscr{S} : \tilde{k}_-(\sigma) \to \tilde{k}_+(\sigma) = \mathscr{S}(\sigma)\tilde{k}_-(\sigma). \tag{1.16}$$

Moreover, $\mathscr{S}(\sigma)$ is unitary on $\mathscr{N}$ for each σ in $\mathbb{R}$ and is the boundary value for an operator-valued function $\mathscr{S}(z)$, holomorphic in the lower half-plane, $\operatorname{Im} z < 0$, where $\mathscr{S}(z)$ is a contraction operator on $\mathscr{N}$.

We denote the orthogonal projections of $\mathscr{H}$ onto $\mathscr{D}_-{}^\perp$ and $\mathscr{D}_+{}^\perp$ by P_- and P_+, respectively. The operators

$$Z(t) = P_+ U(t) P_-, \qquad t \geqslant 0, \tag{1.17}$$

annihilate $\mathscr{D}_- \oplus \mathscr{D}_+$ and form a semigroup of contraction operators on $\mathscr{K}$. This semigroup of operators plays an important role in the theory. For instance, there is a one-to-one correspondence between the points λ in the spectrum of the infinitesimal generator B of Z and the points z in the lower half-plane at which $\mathscr{S}(z)$ is not invertible; this correspondence is $\lambda = -iz$.

The time delay operator T for $K = P_- P_+$ and the operators Z, being built of the same ingredients, are related. One of these relations can be seen by taking f to belong to $\mathscr{K}$; then

$$f = P_- f = P_+ f.$$

Furthermore, by (1.10i) $U(t)$ maps $\mathscr{D}_-^{\perp}$ into itself for $t \geqslant 0$ and $\mathscr{D}_+^{\perp}$ into itself for $t \leqslant 0$. Hence, for f in $\mathscr{K}$

$$\begin{aligned}(Tf, f) &= \int_{-\infty}^{\infty} \|P_-P_+U(t)f\|^2\,dt = \int_{-\infty}^{0} [\|P_-U(t)f\|^2 + \int_0^{\infty} \|P_+U(t)f\|^2]\,dt \\ &= \int_0^{\infty} \|Z^*(t)f\|^2\,dt + \int_0^{\infty} \|Z(t)f\|^2\,dt. \end{aligned} \tag{1.18}$$

Another relation is exhibited by some explicit formulas for T and Z. In Section 2 we shall show that in the outgoing spectral representation T is represented by

$$T \sim -i\mathscr{S}(\sigma)\,\partial_\sigma \mathscr{S}^*(\sigma). \tag{1.19}$$

In Section 3 we shall show that for any ψ of class $C_0^{\infty}(\mathbb{R}_+)$ for which

$$C = \int_0^{\infty} \psi(t)Z(t)\,dt \tag{1.20}$$

is of trace class,

$$\operatorname{tr} C = (-i/\sqrt{2\pi}) \int \Psi(\sigma) \operatorname{tr}[\mathscr{S}(\sigma)\partial_\sigma \mathscr{S}^*(\sigma)]\,d\sigma; \tag{1.21}$$

here Ψ is the Fourier transform of ψ. Formula (1.21) is reminiscent of the well-known trace formula for self-adjoint operators due to Birman and Krein [1].

In Section 3 we also show how to apply this theory to the wave equation in the exterior of a star-shaped obstacle with Dirichlet boundary conditions.

2. The Time Delay Operator

We assume—and this is satisfied in all cases of physical interest—that the scattering matrix $\mathscr{S}(\sigma)$ introduced in (1.16) has the following additional properties:

$$\begin{aligned}&\text{(i)}\quad \mathscr{S}(\sigma) \text{ is analytic at every point of the real axis.}\\ &\text{(ii)}\quad |\mathscr{S}(z)| \leqslant \text{const}/|\operatorname{Im} z| \text{ for } z \text{ in the lower half-plane.}\end{aligned} \tag{2.1}$$

With $\mathscr{D}_-$ and $\mathscr{D}_+$ defined as in Section 1, we set

$$\mathscr{K} = \mathscr{H} \ominus (\mathscr{D}_- \ominus \mathscr{D}_+).$$

Since by (1.10iv) the subspaces $\mathscr{D}_-$ and $\mathscr{D}_+$ are orthogonal, the projection K onto $\mathscr{K}$ is

$$K = P_-P_+, \tag{2.2}$$

where P_- and P_+ are the orthogonal projections onto $\mathscr{D}_-^{\perp}$ and $\mathscr{D}_+^{\perp}$.

We shall determine the time-delay operator T associated with $\mathscr{K}$. To be able to apply formula (1.9) we shall represent K as an integral operator in the, say, outgoing spectral representation. Let f be any element of $\mathscr{H}$, $f(\sigma)$ its outgoing spectral representer:

$$f \underset{+}{\leftrightarrow} f(\sigma).$$

In the outgoing translation representation P_+ acts as multiplication by h_+:

$$h_+(s) = \begin{cases} 1 & \text{for} \quad s < 0 \\ 0 & \text{for} \quad s > 0. \end{cases} \tag{2.3$_+$}$$

Therefore, in the outgoing spectral representation P_+ acts as convolution with H_+, the Fourier transform of h_+:

$$H_+(\sigma) = (-i/2\pi\sigma) + \tfrac{1}{2}\delta(\sigma), \tag{2.4$_+$}$$

$$P_+ f \underset{+}{\leftrightarrow} \int H_+(\rho - \sigma) f(\sigma)\, d\sigma.$$

To determine the action of P_- we first have to switch to the incoming representation; this is accomplished by multiplying with $\mathscr{S}^*(\rho)$:

$$P_+ f \underset{-}{\leftrightarrow} \mathscr{S}^*(\rho) \int H_+(\rho - \sigma) f(\sigma)\, d\sigma.$$

The action of P_- in this representation is convolution with H_-, the Fourier transform of h_-:

$$h_-(s) = \begin{cases} 0 & \text{for} \quad s < 0 \\ 1 & \text{for} \quad s > 0 \end{cases} \tag{2.3$_-$}$$

$$H_-(\sigma) = (i/2\pi\sigma) + \tfrac{1}{2}\delta(\sigma), \tag{2.4$_-$}$$

$$P_- P_+ f \underset{-}{\leftrightarrow} \int H_-(\tau - \rho) \mathscr{S}^*(\rho) \int H_+(\rho - \sigma) f(\sigma)\, d\sigma\, d\rho.$$

Finally, we switch back to the outgoing representation through multiplication by $\mathscr{S}(\tau)$:

$$P_- P_+ f \leftrightarrow \mathscr{S}(\tau) \iint H_-(\tau - \rho) \mathscr{S}^*(\rho) H_+(\rho - \sigma) f(\sigma)\, d\sigma\, d\rho. \tag{2.5}$$

(2.5) represents $P_- P_+$ as an integral operator, with kernel

$$q(\tau, \sigma) = \mathscr{S}(\tau) \int H_-(\tau - \rho) H_+(\rho - \sigma) \mathscr{S}^*(\rho)\, d\rho. \tag{2.6}$$

Using formulas (2.4)$_\pm$ for $H_\pm$ we get

$$q(\tau, \sigma) = \mathscr{S}(\tau) \int \left[\frac{i}{2\pi(\tau - \rho)} + \tfrac{1}{2}\delta(\tau - \rho) \right] \left[\frac{-i}{2\pi(\rho - \sigma)} + \tfrac{1}{2}\delta(\rho - \sigma) \right] \mathscr{S}^*(\rho)\, d\rho. \tag{2.7}$$

We take $\tau \neq \sigma$; then $\delta(\tau - \rho)\delta(\rho - \sigma) = 0$. Using this and the partial fraction decomposition

$$\frac{1}{(\tau - \rho)}\frac{1}{(\rho - \sigma)} = \frac{1}{\tau - \sigma}\left(\frac{1}{\tau - \rho} + \frac{1}{\rho - \sigma}\right) \tag{2.8}$$

in (2.7) we get

$$q(\tau, \sigma) = \mathscr{S}(\tau) \int \left[\frac{1}{(2\pi)^2(\tau - \sigma)}\left(\frac{1}{\tau - \rho} + \frac{1}{\rho - \sigma}\right) + \frac{i}{4\pi}\frac{\delta(\rho - \sigma)}{\tau - \rho} - \frac{i}{4\pi}\frac{\delta(\rho - \tau)}{\rho - \sigma}\right] \mathscr{S}^*(\rho)\, d\rho. \tag{2.9}$$

Next we use the following consequence of Cauchy's formula:

Let $g_+(z)$ be analytic in the upper half-plane, and

$$|g_+(z)| \leqslant \frac{\text{const}}{\text{Im } z}, \qquad \text{Im } z > 0;$$

then

$$\int \frac{g_+(\rho)}{\rho - \sigma}\, d\rho = \pi i g_+(\sigma). \tag{2.10$_+$}$$

Similarly, for $g_-(z)$ analytic in the lower half-plane and satisfying there

$$|g_-(z)| \leqslant \frac{\text{const}}{|\text{Im } z|}, \qquad \text{Im } z < 0,$$

$$\int \frac{g_-(\rho)}{\rho - \sigma}\, d\rho = -\pi i g_-(\sigma). \tag{2.10$_-$}$$

Using $(2.10)_+$ in (2.9) with $g_+(\rho) = \mathscr{S}^*(\rho)$ we get

$$\begin{aligned} q(\tau, \sigma) &= \frac{i\mathscr{S}(\tau)}{4\pi}\left[\frac{1}{\tau - \sigma}(\mathscr{S}^*(\sigma) - \mathscr{S}^*(\tau)) + \frac{\mathscr{S}^*(\sigma)}{\tau - \sigma} - \frac{\mathscr{S}^*(\tau)}{\tau - \sigma}\right] \\ &= \frac{-i}{2\pi}\mathscr{S}(\tau)\frac{\mathscr{S}^*(\tau) - \mathscr{S}^*(\sigma)}{\tau - \sigma}. \end{aligned} \tag{2.11}$$

We claim that the limiting form of (2.11) as σ tends to τ is valid on the diagonal; in other words, that

$$q(\sigma, \sigma) = (-i/2\pi)\, \mathscr{S}(\sigma)\, \partial_\sigma \mathscr{S}^*(\sigma). \tag{2.12}$$

To prove this we shall verify that the integral operator whose kernel g is given by (2.11) and (2.12) is indeed the projection onto $\mathscr{K}$. We have to verify that

$$\int q(\tau, \sigma) f(\sigma)\, d\sigma = 0$$

when f represents an element of $\mathscr{D}_+$ or $\mathscr{D}_-$, and that

$$\int q(\tau, \sigma) f(\sigma)\, d\sigma = f(\tau)$$

when f represents an element of $\mathscr{K}$.

From (2.11),

$$\int q(\tau, \sigma) f(\sigma)\, d\sigma = \frac{i}{2\pi} \mathscr{S}(\tau) \int \frac{\mathscr{S}^*(\sigma) - \mathscr{S}^*(\tau)}{\tau - \sigma} f(\sigma)\, d\sigma. \tag{2.13}$$

An $f(\sigma)$ representing an element of $\mathscr{D}_+$ is analytic in the upper half-plane. The integrand in (2.13) is an analytic function of σ in the upper half-plane, so by the Cauchy integral theorem we can shift the line of integration to $z = \sigma + i\mu$, $\mu > 0$. Estimating the resulting integral by Schwarz's inequality shows that the integral (2.13) is zero.

Because of the unitarity of $\mathscr{S}$, we can rewrite (2.13) in the form

$$\int q(\tau, \sigma) f(\sigma)\, d\sigma = \frac{i}{2\pi} \int \frac{\mathscr{S}(\tau) - \mathscr{S}(\sigma)}{\tau - \sigma} \mathscr{S}^*(\sigma) f(\sigma)\, d\sigma. \tag{2.13'}$$

Now if $f(\sigma)$ represents an element of $\mathscr{D}_-$, $\mathscr{S}^*(\sigma)f(\sigma)$ is analytic in the lower half-plane. In this case the integrand is analytic in the lower half-plane, so again the same argument shows that the integral is zero.

We rewrite (2.13) as

$$\int q(\tau, \sigma) f(\sigma)\, d\sigma = \frac{i}{2\pi} \mathscr{S}(\tau) \int \frac{\mathscr{S}^*(\sigma) f(\sigma)\, d\sigma}{\tau - \sigma} - \frac{i}{2\pi} \int \frac{f(\sigma)\, d\sigma}{\tau - \sigma}. \tag{2.13''}$$

Now when $f(\sigma)$ represents an element of $\mathscr{K}$, then $f(\sigma)$ is analytic in the lower half-plane and $\mathscr{S}^*(\sigma)f(\sigma)$ is analytic in the upper half-plane. Applying $(2.10)_+$ to the first term in (2.13″) with $g_+ = \mathscr{S}^* f$ and $(2.10)_-$ to the second term with $g_- = f$ we get that

$$\int q(\tau, \sigma) f(\sigma)\, d\sigma = \tfrac{1}{2} \mathscr{S}(\tau) \mathscr{S}^*(\tau) f(\tau) + \tfrac{1}{2} f(\tau) = f(\tau).$$

This completes the verification that the integral operator whose kernel q is given by formulas (2.11) and (2.12) is indeed the projection onto $\mathscr{K}$.

We now apply formula (1.9) to our operator K to conclude

THEOREM 2.1. *Suppose $\mathscr{S}(\sigma)$ satisfies conditions* (2.1). *Then the time-delay operator T as defined by* (1.5) *is essentially self-adjoint and has in the* *outgoing spectral representation the form*

$$-i\mathscr{S}(\sigma)\,\partial_\sigma \mathscr{S}^*(\sigma). \tag{2.14}$$

From its definition, the operator T is nonnegative; therefore, the operator (2.14) is nonnegative for all σ. This is easy to deduce directly from the fact that $\mathscr{S}(z)$ is unitary for z real and a contraction for $\operatorname{Im} z \leqslant 0$; for then

$$|\mathscr{S}(\sigma - i\mu)f|^2 \leqslant |\mathscr{S}(\sigma)f|^2$$

for $\mu < 0$, which implies that

$$\frac{d}{d\mu}|\mathscr{S}(\sigma - i\mu)f|^2 \leqslant 0.$$

From this the nonnegativity of (2.14) is easily deduced.

The operator T represented as multiplication by (2.14) is a bounded operator if and only if $\partial_\sigma \mathscr{S}^*(\sigma)$ is uniformly bounded on the whole σ axis. Whether this is the case or not depends on the location of the poles of $\mathscr{S}(z)$ in the upper half-plane. When these poles approach the real axis, $\partial_\sigma \mathscr{S}(\sigma)$ cannot be uniformly bounded; on the other hand, if $\mathscr{S}(\sigma)$ is holomorphic and bounded in a strip $\operatorname{Im} z < \beta$, $\beta > 0$, then $\partial_\sigma \mathscr{S}(\sigma)$ is bounded on the whole real axis, and so the time delay operator T is bounded.

It is worth remarking that in the latter case the boundedness of T can be deduced directly from definition (1.5) or (1.3), without appealing to formula (2.14). We have shown in [4, p. 84] that $\mathscr{S}(z)$ is holomorphic and bounded in a strip above the real axis if and only if $Z(t)$ decays exponentially:

$$\|Z(t)\| \leqslant \text{const}\, e^{-\alpha t}, \tag{2.15}$$

α any positive number. To deduce the boundedness of T from (2.15), we decompose $\mathscr{D}_-$ and $\mathscr{D}_+$ as follows:

$$\mathscr{D}_- = \bigcup_{j<0} \mathscr{D}_-^{(j)}, \qquad \mathscr{D}_+ = \bigcup_{j>0} \mathscr{D}_+^{(j)},$$

where

$$\mathscr{D}_+^{(j)} = U(j-1)\mathscr{D}_+ \ominus U(j)\mathscr{D}_+, \qquad j = 1, 2, \ldots,$$
$$D_-^{(j)} = U(1+j)\mathscr{D}_- \ominus U(j)\mathscr{D}_-, \qquad j = -1, -2, \ldots.$$

Every f in $\mathscr{H}$ can be decomposed as

$$f = \sum_{-\infty}^{\infty} f_j, \qquad f_0 \in \mathscr{K}, \qquad f_j \in \begin{cases} \mathscr{D}_+^{(j)} & \text{for } j > 0 \\ \mathscr{D}_-^{(j)} & \text{for } j < 0. \end{cases}$$

Since the decomposition is orthogonal,

$$\|f\|^2 = \sum \|f_j\|^2. \tag{2.16}$$

Since for $j < 0$, $f_j \in U(1 + j)\mathscr{D}_-$, it follows from (1.10i) that for $t < -j - 1$

$$U(t)f_j \in \mathscr{D}_-.$$

Hence,

$$P_- U(t)f_j = 0 \qquad \text{for} \quad t < -j - 1. \tag{2.17}$$

On the other hand, $U(-j)f_j$ is orthogonal to $\mathscr{D}_-$; so,

$$\begin{aligned} P_- U(-j)f_j &= U(-j)f_j, \\ U(t)f_j &= U(t + j)P_- U(-j)f_j \qquad \text{for} \quad t > -j. \end{aligned} \tag{2.17$'$}$$

Finally, $U(t)f_j \in \mathscr{D}_+$ for $t > 0$ and $j > 0$, so

$$P_+ U(t)f_j = 0 \qquad \text{for} \quad j > 0, \quad t > 0. \tag{2.17$''$}$$

Combining (2.17), (2.17′), and (2.17″) we deduce the following for $t > 0$:

$$\begin{aligned} P_+ P_- U(t)f &= \sum_{-\infty}^{\infty} P_+ P_- U(t)f_j \\ &= P_+ P_- U(t)f_{-[t]} + \sum_{-[t] < j \leqslant 0} Z(t + j)U(-j)f_j; \end{aligned} \tag{2.18}$$

here $[t]$ denotes the integer part of t. By the triangle inequality and using estimate (2.15) we get

$$\|P_+ P_- U(t)f\| \leqslant \|f_{-[t]}\| + \text{const} \sum_{-[t] < j \leqslant 0} e^{-\alpha(t + j)}\|f_j\|.$$

Using the Schwarz inequality yields

$$\|P_+ P_- U(t)f\|^2 \leqslant k\left(\|f_{-[t]}\|^2 + \text{const} \sum_{-[t] < j \leqslant 0} e^{-\alpha(t + j)}\|f_j\|^2\right), \tag{2.19}$$

where

$$k = (1 + \text{const} \sum_{[-t] < j} e^{-\alpha(t + j)});$$

so,

$$\int_0^{\infty} \|P_+ P_- U(t)f\|^2 \, dt \leqslant k \sum_{j < 0} \|f_j\|^2 + \frac{\text{const}}{\alpha} \sum_{j \leqslant 0} \|f_j\|^2 \leqslant k'\|f\|^2.$$

The integral from $-\infty$ to 0 is estimated similarly, using Z^* in place of Z. This shows that the quadratic form

$$(Tf, f) = \int_{-\infty}^{\infty} \|P_+ P_- U(t)f\|^2 \, dt$$

is bounded, and, hence, that T is a bounded operator.

3. A Related Trace Formula

We turn now to the proof of the trace formula (1.21) for operators of the type

$$C = \int_0^\infty \psi(t) Z(t)\, dt, \tag{3.1}$$

where ψ lies in $C_0^\infty(\mathbb{R}_+)$. We shall denote the trace class norm of an operator by the subscript 1 as in $\|\cdot\|_1$.

THEOREM 3.1. *Suppose that the scattering matrix $\mathscr{S}$ is of the form*

$$\mathscr{S}(\sigma) = e^{-i\sigma\rho} \mathscr{S}_0(\sigma), \tag{3.2}$$

where

$$\mathscr{S}_0(\sigma) = I + K(\sigma) \tag{3.3}$$

and $\partial_\sigma K(\sigma)$ is an operator of trace class on $\mathscr{N}$ with $\|\partial_\sigma K(\sigma)\|_1$ being of polynomial growth. If, in addition, C is of trace class, then

$$\operatorname{tr} C = (-i/\sqrt{2\pi}) \int_{-\infty}^{\infty} \Psi(\sigma) \operatorname{tr}[\mathscr{S}_0(\sigma) \partial_\sigma \mathscr{S}_0{}^*(\sigma)]\, d\sigma. \tag{3.4}$$

Proof. We first compute (Cf, f) in the outgoing translation representation. It will be convenient to work with all f in $\mathscr{D}_- \oplus \mathscr{K} = \mathscr{D}_+{}^\perp$ rather than in $\mathscr{K}$ alone; in other words, we consider all f with $T_+ f = k$ having support in $\mathbb{R}_-$. For such f and $t \geqslant 0$,

$$Z(t) f = P_+ U(t) P_- f.$$

As in the proof of Theorem 2.1, P_+ can be characterized in the outgoing translation representation as multiplication by h_+ and P_- as Sh_-S^*. We can therefore write

$$T_+ Z(t) f = h_+ U_t S h_- S^* T_+ f, \tag{3.5}$$

where U_t denotes right translation by t units. Setting $k = T_+ f$ we therefore have

$$(Z(t)f, f) = (h_+ U_t S h_- S^* k, k)_{L_2} = (h_- S^* k, S^* U_{-t} h_+ k)_{L_2}.$$

Since we are considering only k with support in $\mathbb{R}_-$, we can write this, at least formally, as

$$(Z(t)f, f) = \int_0^\infty du \left(\int_{-\infty}^{\infty} ds\, S^*(u - s) k(s), \int_{-\infty}^{\infty} dr\, S^*(u - r) k(r + t) \right)_{\mathscr{N}}; \tag{3.6}$$

the action of h_- is taken care of by the u range of integration. Finally, if we include ψ and the t integration this becomes

$$(Cf, f) = \int_0^\infty dt \int_0^\infty du \int_{-\infty}^{\infty} ds \int_{-\infty}^{\infty} dr \left(S^*(u - s)k(s), \overline{\psi}(t) S^*(u + t - r)k(r) \right)_{\mathcal{N}}. \tag{3.7}$$

The expression (3.7) can be rigorized if we replace the distribution-valued S^* by a mollified version. In this respect we treat the above two S^*'s differently. The one on the right in (3.7) can be thought of as already mollified by $\overline{\psi(-t)}$ and the resulting operator is

$$S_\psi{}^*(s) = \int_{-\infty}^{\infty} e^{-i\sigma s} \overline{\Psi(\sigma)} \mathcal{S}^*(\sigma)\, d\sigma. \tag{3.8}$$

Note that this requires an interchange in the order of the t and r integrations, but since ψ has compact support this is easily justified. We obtain a mollification for the S^* on the left by replacing C in the outgoing translation representation by

$$C_\epsilon k = C(\varphi_\epsilon * k), \tag{3.9}$$

where

$$\varphi_\epsilon(s) = (1/\epsilon)\varphi(s/\epsilon)$$

and φ in $C_0{}^\infty$ is $\geqslant 0$ with $\int \varphi\, ds = 1$. Since we are treating C as an operator on $\mathbb{R}_-$ we should really write C_ϵ as $C(h_+(\varphi_\epsilon * k))$. However, this h_+ is superfluous since its action is nullified by that of P_+ on the left in $Z(t) = P_+U(t)P_-$. Hence, convolution with φ_ϵ in (3.9) can be shifted over from k to the left S^* in (3.7). Denoting the so-mollified S^* by $S_\epsilon{}^*(s)$ and the Fourier transform of φ_ϵ by Φ_ϵ, we have

$$S_\epsilon{}^*(s) = \int e^{-i\sigma s} \Phi_\epsilon(\sigma) \mathcal{S}^*(\sigma)\, d\sigma. \tag{3.10}$$

We can now write

$$(C_\epsilon f, f) = \int_0^\infty du \int_{-\infty}^{\infty} ds \int_{-\infty}^{\infty} dr \left(S_\epsilon{}^*(u - s)k(s), S_\psi{}^*(u - r)k(r) \right)_{\mathcal{N}}. \tag{3.11}$$

Because of causality (1.15), the support of $S^*(s)$ is contained in $\mathbb{R}_+$. By choosing φ so that its support also lies in $\mathbb{R}_+$, the mollified $S_\epsilon{}^*(s)$ will inherit this property.

We note that

$$is\, S_\epsilon{}^*(s) = \int e^{-i\sigma s} \partial_\sigma (\Phi_\epsilon(\sigma) \mathcal{S}^*(\sigma))\, d\sigma.$$

For any integer n we have the estimate

$$|\Phi_\epsilon(\sigma)|, |\partial_\sigma \Phi_\epsilon(\sigma)| \leqslant \text{const}\,(1 + \sigma^2)^{-n},$$

and since we have assumed that $\|\partial_\sigma \mathscr{S}^*(\sigma)\|_{\mathscr{N}}$ is of polynomial growth, it follows that $\|sS_\epsilon^*(s)\|_{\mathscr{N}}$ is bounded and square integrable. The same is trivially true of $\|S_\epsilon^*(s)\|_{\mathscr{N}}$. It follows from this that $\|S_\epsilon^*(s)\|_{\mathscr{N}}$ is integrable, and the same is true of $\|S_\psi^*(s)\|_{\mathscr{N}}$. The integrand in (3.11) is therefore absolutely integrable.

We note that C_ϵ is the product of a trace class operator with the bounded operator defined as convolution with φ_ϵ and, hence, C_ϵ is itself of trace class. Moreover, since convolution with φ_ϵ converges strongly to the identity it follows that

$$\lim_{\epsilon \to 0} \|C - C_\epsilon\|_1 = 0 \tag{3.12}$$

and, hence, that

$$\operatorname{tr} C = \lim_{\epsilon \to 0} \operatorname{tr} C_\epsilon . \tag{3.13}$$

Next we choose an orthonormal basis $\{n_k\}$ for $\mathscr{N}$ and denote by P_k the orthogonal projection of $L_2(\mathbb{R}_-, \mathscr{N})$ onto $L_2(\mathbb{R}_-)n_k$. Then

$$\operatorname{tr} C = \sum_k \operatorname{tr}[P_k C P_k]. \tag{3.14}$$

The analogs of (3.12) and (3.13) hold for $P_k C P_k$, so that

$$\operatorname{tr}[P_k C P_k] = \lim_{\epsilon \to 0} \operatorname{tr}[P_k C_\epsilon P_k].$$

It follows that

$$\operatorname{tr} C = \sum_k \lim_{\epsilon \to 0} \operatorname{tr}[P_k C_\epsilon P_k]. \tag{3.15}$$

It is clear from (3.11) that $P_k C_\epsilon P_k$ is an integral operator on $L_2(\mathbb{R}_-)$ with kernel

$$\int_0^\infty du\, (S_\epsilon^*(u - s)n_k, S_\psi^*(u - r)n_k)_{\mathscr{N}} . \tag{3.16}$$

This is obviously continuous and integrable along the diagonal. The trace of $P_k C_\epsilon P_k$ can, therefore, be evaluated as the integral of this kernel along the diagonal:

$$\operatorname{tr}[P_k C_\epsilon P_k] = \int_{-\infty}^0 ds \int_0^\infty du\, (S_\epsilon^*(u - s)n_k, S_\psi^*(u - s)n_k)_{\mathscr{N}} .$$

Making the substitution

$$v = u - s \qquad \text{and} \qquad w = u + s$$

and performing the w integration, this becomes

$$\operatorname{tr}[P_k C_\epsilon P_k] = \int_0^\infty dv\, (vS_\epsilon^*(v)n_k, S_\psi^*(v)n_k)_{\mathscr{N}} . \tag{3.17}$$

Because of causality the lower limit of integration can be replaced by $-\infty$. Since the Fourier transform of $ivS_\epsilon^*(v)$ is $\partial_\sigma \mathscr{S}_\epsilon^*(\sigma)$, the spectral version of (3.17) is

$$\operatorname{tr}[P_k C_\epsilon P_k] = -i \int_{-\infty}^{\infty} (\partial_\sigma \mathscr{S}_\epsilon^*(\sigma) n_k, \mathscr{S}_\psi^*(\sigma) n_k)_{\mathscr{N}}\, d\sigma$$

$$= -i \int_{-\infty}^{\infty} (\partial_\sigma(\Phi_\epsilon(\sigma)\mathscr{S}^*(\sigma) n_k), \overline{\Psi(\sigma)}\mathscr{S}^*(\sigma) n_k)_{\mathscr{N}}\, d\sigma. \quad (3.18)$$

Passing to the limit as $\epsilon \to 0$, it is easy to see that

$$\lim_{\epsilon \to 0} \operatorname{tr}[P_k C_\epsilon P_k] = \frac{-i}{\sqrt{2\pi}} \int_{-\infty}^{\infty} \Psi(\sigma)(\partial_\sigma \mathscr{S}^*(\sigma) n_k, \mathscr{S}^*(\sigma) n_k)_{\mathscr{N}}\, d\sigma. \quad (3.19)$$

We now make use of the relations (3.2) and (3.3):

$$\mathscr{S}(\sigma) = e^{-i\sigma\rho}\mathscr{S}_0(\sigma), \qquad \mathscr{S}_0(\sigma) = I + K(\sigma)$$

(by assumption, $\partial_\sigma \mathscr{S}_0(\sigma) = \partial_\sigma K(\sigma)$ is of trace class). The integrand in (3.19) can be written in terms of $\mathscr{S}_0$ as

$$i\rho\Psi(\sigma) + \Psi(\sigma)(\partial_\sigma \mathscr{S}_0^*(\sigma) n_k, \mathscr{S}_0^*(\sigma) n_k)_{\mathscr{N}}.$$

Using this together with the fact that

$$\int \Psi(\sigma)\, d\sigma = (2\pi)^{1/2}\psi(0) = 0$$

and combining the resulting form of (3.19) with (3.15), we get

$$\operatorname{tr} C = (-i/\sqrt{2\pi}) \sum \int_{-\infty}^{\infty} \Psi(\sigma)(\partial \mathscr{S}_0^*(\sigma) n_k, \mathscr{S}_0^*(\sigma) n_k)_{\mathscr{N}}\, d\sigma. \quad (3.20)$$

By assumption, $\|\mathscr{S}_0(\sigma)\,\partial_\sigma \mathscr{S}_0^*(\sigma)\|_1$ is of polynomial growth. Hence,

$$\sum |\Psi(\sigma)|\, |(\partial_\sigma \mathscr{S}_0^*(\sigma) n_k, \mathscr{S}_0^*(\sigma) n_k)_{\mathscr{N}}|$$

is integrable and we can interchange the order of integration and summation in (3.20). Since

$$\operatorname{tr}[\mathscr{S}_0(\sigma)\,\partial_\sigma \mathscr{S}_0^*(\sigma)] = \sum (\partial_\sigma \mathscr{S}_0^*(\sigma) n_k, \mathscr{S}_0^*(\sigma) n_k)_{\mathscr{N}}$$

this gives us the desired relation (3.4) and completes the proof of Theorem 3.1.

We conclude this section by showing that Theorem 3.1 is applicable to the scattering problem for the wave equation with Dirichlet boundary conditions in an exterior domain G in $\mathbb{R}^3$:

$$\begin{aligned} u_{tt} &= \Delta u && \text{in} \quad G, \\ u &= 0 && \text{on } \partial G. \\ u(x,0) &= f_1, && u_t(x,0) = f_2. \end{aligned} \quad (3.21)$$

In this case $\mathscr{H}$ consists of all data $f = \{f_1, f_2\}$ obtained by completion from $C_0(G)$ data with respect to the energy norm:

$$\|f\|^2 = \int_G (|\nabla f_1|^2 + |f_2|^2)\, dx. \tag{3.22}$$

The solution operator $U(t)$ taking initial data f into $\{u(\cdot, t), u_t(\cdot, t)\}$ is generated by

$$A = \begin{bmatrix} 0 & I \\ \Delta & 0 \end{bmatrix};$$

here Δ is defined in the weak sense on functions vanishing on ∂G. We shall suppose that the scattering object is contained in a sphere of radius ρ about the origin. For further details on how this problem is set up see [4, Chapter 5].

For φ in $C_0{}^\infty(\mathbb{R}_+)$ we define

$$Z(\varphi) = \int_0^\infty \varphi(t) Z(t)\, dt. \tag{3.23}$$

It is easy to prove that

$$Z(\varphi * \theta) = Z(\varphi) Z(\theta). \tag{3.24}$$

LEMMA 3.2. *If ∂G is of class C^3 and φ in $C_0{}^\infty(\mathbb{R})$ has support in $(2\rho, \infty)$, then $Z(\varphi)$ is Hilbert–Schmidt.*

Since the product of two Hilbert–Schmidt operators is of trace class, it follows from Lemma 3.2 and the relation (3.24) that

COROLLARY 3.3. *If φ and θ belong to $C_0{}^\infty(\mathbb{R})$ and both have their support in $(2\rho, \infty)$, then $Z(\varphi * \theta)$ is of trace class.*

Proof of Lemma 3.2. Let $U_0(t)$ denote the solution operator for the wave equation in free space and set

$$M = U(2\rho) - U_0(2\rho). \tag{3.25}$$

If we treat data f in $\mathscr{H}$ as extended to be zero in the complement of G, then it makes sense to consider the action of M on $\mathscr{H}$. As shown in [4, pp. 152–153], M maps $\mathscr{H}$ into $\mathscr{H}$ with the following properties:

(a) $\|M\| \leqslant 2$,
(b) Mf depends only on the behavior of f in the sphere $|x| < 5\rho$,
(c) $Z(2\rho) = P_+ M P_-$.

For φ in $C_0{}^\infty(\mathbb{R}_+)$ with support in $(2\rho, \infty)$, $\varphi_0(t) = \varphi(t + 2\rho)$ lies in $C_0{}^\infty(\mathbb{R}_+)$ and

$$Z(\varphi) = Z(2\rho) Z(\varphi_0) = P_+ M P_- P_+ U(\varphi_0) P_- = P_+ M U(\varphi_0) P_-; \tag{3.26}$$

the P_-P_+ in the next-to-last member is redundant because (1) $U(\varphi_0)\mathscr{D}_-^{\perp} \subset \mathscr{D}_-^{\perp}$ and (2) both $U(2\rho)$ and $U_0(2\rho)$ take $\mathscr{D}_+$ into itself and therefore the $\mathscr{D}_+$ component of $U(\varphi_0)P_-$ will be annihilated by the P_+ on the left. We now choose ζ in $C_0^{\infty}(\mathbb{R}^3)$ so that

$$\zeta(x) = \begin{cases} 1 & \text{for } |x| < 5\rho \\ 0 & \text{for } |x| > 6\rho. \end{cases}$$

Then by property (b) above

$$MU(\varphi_0) = M\zeta U(\varphi_0). \tag{3.27}$$

In view of (3.26) and (3.27) it suffices to prove that $\zeta U(\varphi_0)$ is Hilbert–Schmidt. Now for any g in $\mathscr{H}$, $\{u_1, u_2\} = U(\varphi_0)g$ belongs to the domain of all powers of the generator A of U. Hence if ∂G is of class C^3 it follows from elliptic theory and the Sobolev lemma that

$$|u_1(x)|, |\Delta u_1(x)|, |u_2(x)| \leqslant \text{const} \|g\| \tag{3.28}$$

for all x in G. By the Riesz representation theorem there exist functions $k_{ij}(x, y)$ $(i, j = 1, 2)$ with

$$\{k_{i1}(x, \cdot), k_{i2}(x, \cdot)\} \qquad \text{in} \quad \mathscr{H},$$

of norm uniformly bounded for all x in G, and such that

$$u_i(x) = (\{k_{i1}(x, \cdot), k_{i2}(x, \cdot)\}, g).$$

It follows that $\zeta U(\varphi_0)$ is an integral operator with kernel matrix

$$(\zeta(x)k_{ij}(x, y))$$

and by (3.28) that

$$\iint_{G\times G} (|\nabla_x \nabla_y \zeta k_{11}|^2 + |\nabla_x \zeta k_{12}|^2 + |\nabla_y \zeta k_{21}|^2 + |\zeta k_{22}|^2)\, dx\, dy \leqslant \text{const.}$$

This proves that $\zeta U(\varphi_0)$ is Hilbert–Schmidt and hence, by (3.26) and (3.27), so is $Z(\varphi)$. This completes the proof of Lemma 3.2.

LEMMA 3.4. *If*

$$\|[i\sigma - B]^{-1}\| \leqslant \text{const}$$

for all real σ, then for the wave equation (3.21) *in an exterior domain, $\partial_\sigma \mathscr{S}_0(\sigma)$ is of trace class and*

$$\|\partial_\sigma \mathscr{S}_0(\sigma)\|_1 \leqslant \text{const}\, \sigma^2. \tag{3.29}$$

Remark. The hypothesis of Lemma 3.4 is satisfied for the wave equation with Dirichlet boundary conditions when the scattering object is star shaped (see [6]).

Proof. The operator $K(\sigma)$ in (3.3) can be constructed as follows (see [4, pp. 167–171]): Let v denote the outgoing scattered wave; that is, v is the solution of the differential equation

$$\Delta v + \sigma^2 v = 0 \tag{3.30}$$

satisfying for a given unit vector ω the boundary condition

$$v = -e^{-i\sigma x \cdot \omega} \quad \text{on} \quad \partial G \tag{3.31}$$

and having asymptotic behavior

$$v(x) \sim s(\theta, \omega; \sigma) e^{i\sigma r}/r \qquad \text{for} \quad \text{large } r, \tag{3.32}$$

where $x = r\theta$ and $\theta = x/|x|$. The kernel of $K(\sigma)$ is given by

$$k(\omega, \theta; \sigma) = -\overline{s(-\theta, \omega; \sigma)}. \tag{3.33}$$

Next we choose $\zeta \in C^\infty(\mathbb{R}^3)$ so that

$$\zeta(x) = \begin{cases} 1 & \text{on the scattering object} \\ 0 & \text{for} \quad |x| \geqslant \rho \end{cases}$$

and set

$$w = v + \zeta e^{-i\sigma \cdot \omega}. \tag{3.34}$$

Then w satisfies the differential equation

$$\Delta w + \sigma^2 w = 2\nabla\zeta \cdot \nabla e^{-i\sigma x \cdot \omega} + (\Delta\zeta) e^{-i\sigma x \cdot \omega} \equiv h, \qquad w = 0 \qquad \text{on} \quad \partial G. \tag{3.35}$$

We note that

$$\|\{0, h\}\|, \|\{0, \partial_\sigma h\}\| = O(|\sigma|) \tag{3.36}$$

uniformly in ω. Rewriting (3.35) in vector form we get

$$[i\sigma - A]\begin{bmatrix} w \\ i\sigma w \end{bmatrix} = \begin{bmatrix} 0 \\ -h \end{bmatrix}. \tag{3.35$'$}$$

It can be shown (see [4, pp. 162–163]) that for $|x| < \rho$,

$$[i\sigma - B]\begin{bmatrix} w \\ i\sigma w \end{bmatrix} = \begin{bmatrix} 0 \\ -h \end{bmatrix}. \tag{3.37}$$

It now follows from our assumption on the resolvent of B and the estimate (3.36) that

$$\int_{G_\rho} (|\nabla w|^2 + \sigma^2 |w|^2)\, dx = O(\sigma^2), \tag{3.38}$$

where $G_\rho = G \cap \{|x| < \rho\}$. Recall that

$$\partial_\sigma[i\sigma - B]^{-1} = -i[i\sigma - B]^{-2}.$$

Hence, inverting (3.37) and differentiating with respect to σ gives

$$\begin{bmatrix} \partial_\sigma w \\ iw + i\sigma\,\partial_\sigma w \end{bmatrix} = -i[i\sigma - B]^{-2}\begin{bmatrix} 0 \\ -h \end{bmatrix} + [i\sigma - B]^{-1}\begin{bmatrix} 0 \\ -\partial_\sigma h \end{bmatrix}.$$

Again invoking our assumption on the resolvent of B and using the estimates (3.36) and (3.38), we obtain

$$\int_{G_\rho} (|\nabla \partial_\sigma w|^2 + \sigma^2 |\partial_\sigma w|^2)\, dx = O(\sigma^2). \tag{3.39}$$

We now choose η in $C^\infty(\mathbb{R}^3)$ so that

$$\eta(x) = \begin{cases} 0 & \text{on the support of } \zeta \\ 1 & \text{for } |x| > \rho. \end{cases}$$

Then ηw is smooth over all of $\mathbb{R}^3$ where it satisfies the differential equation

$$\Delta(\eta w) + \sigma^2(\eta w) = 2\nabla\eta \cdot \nabla w + (\Delta\eta) w \equiv g, \tag{3.40}$$

and it follows from (3.38) and (3.39) that

$$\int |g|^2\, dx, \int |\partial_\sigma g|^2\, dx = O(\sigma^2). \tag{3.41}$$

The free space solution of (3.40) with the required asymptotic behavior is given by

$$\eta w = \int \gamma(x - y; \sigma) g(y; \omega, \sigma)\, dy, \tag{3.42}$$

where

$$\gamma(x; \sigma) = -e^{i\sigma|x|}/4\pi|x|.$$

It is clear from (3.34) that $\eta w = v$ for $|x| > \rho$. From the relation (3.42) one easily shows that

$$k(\omega, \theta; \sigma) = \overline{s(-\theta, \omega; \sigma)} = -(i\sigma/8\pi^2) \int e^{i\sigma y\cdot\theta}\, \overline{g(y; \omega, \sigma)}\, dy. \tag{3.43}$$

Finally the kernel for the integral operator $\partial_\sigma \mathscr{S}_0 = \partial_\sigma K$ is

$$\begin{aligned} \partial_\sigma k(\omega, \theta; \sigma) = {} & (\sigma/8\pi^2) \int e^{i\sigma y\cdot\theta}\, y\cdot\theta\, \overline{g(y; \omega, \sigma)}\, dy \\ & - (i\sigma/8\pi^2) \int e^{i\sigma y\cdot\theta}\, \partial_\sigma \overline{g(y; \omega, \sigma)}\, dy - (i/8\pi^2) \int e^{i\sigma y\cdot\theta}\, \overline{g(y; \omega, \sigma)}\, dy. \end{aligned} \tag{3.44}$$

We treat each term on the right side of (3.44) as the kernel of the composition of two operators: the g factor being the kernel of an operator on $L_2(S_2)$ into $L_2(G_\rho)$, and the exponential being the kernel of an operator on $L_2(G_\rho)$ into $L_2(S_2)$ (here S_2 denotes the unit sphere in $\mathbb{R}^3$). Each of these factors correspond to Hilbert–Schmidt kernels; the Hilbert–Schmidt norm of the g factors are by (3.41) of order $|\sigma|$, as are those of the second factors. It follows that $\partial_\sigma \mathcal{S}_0$ is of trace class with trace norm of order σ^2. This concludes the proof of Lemma 3.4.

Lemmas 3.2 and 3.4 show that the trace formula (3.4) applies to the wave equation (3.21) in the exterior of a star-shaped obstacle for functions $\psi = \varphi * \theta$ where φ, θ belong to $C_0^\infty(R)$ with support in $(2\rho, \infty)$.

References

1. M. Sh. Birman and M. G. Krein, On the theory of wave operators and scattering operators, *Dokl. Akad. Nauk. SSSR* **144** (1962), 475–478.
2. L. Eisenbud, Ph.D. Thesis, Princeton Univ., 1948.
3. J. M. Jauch and J. P. Marchand, The time delay operator for simple scattering systems, *Helv. Phys. Acta* **40** (1967), 217–229.
4. P. D. Lax and R. S. Phillips, "Scattering Theory," Academic Press, New York, 1967.
5. P. D. Lax and R. S. Phillips, "Scattering Theory for Automorphic Functions," Annals of Mathematical Studies, Princeton Univ. Press, Princeton, New Jersey, 1976.
6. P. D. Lax, C. S. Morawetz, and R. S. Phillips, Exponential decay of solutions of the wave equation in the exterior of a star-shaped obstacle, *Commun. Pure and Appl. Math.* **16** (1963), 427–455.
7. F. T. Smith, Lifetime matrix in collision theory, *Phys. Rev.* **118** (1960), 349–356.
8. E. P. Wigner, Lower limit for the energy derivative of the scattering phase shift, *Phys. Rev.* **98** (1955), 145–147.

Boussinesq's Equation as a Hamiltonian System[†]

H. P. McKEAN

Courant Institute of Mathematical Sciences
New York University,
New York, New York

DEDICATED TO M. G. KREIN ON THE OCCASION OF HIS SEVENTIETH BIRTHDAY

1. INTRODUCTION

Zakharov (1973) has initiated the study of the equation of Boussinesq (1872)

$$\frac{\partial^2 q}{\partial t^2} = 3\frac{\partial^2}{\partial x^2}\left(\frac{\partial^2 q}{\partial x^2} + 4q^2\right) \tag{1}$$

from the standpoint of Hamiltonian mechanics.[1] Let C_1^∞ be the class of real infinitely differentiable functions of $0 \leqslant x < 1$ of period 1 and assume that (1) is (locally) well posed in that space. The appropriate Hamiltonian is

$$H_2 = \int_0^1 [(3/2)p^2 + (4/3)q^3 - (1/2)(q')^2]\,dx \tag{2}$$

and (1) is equivalent to[2]

$$\begin{pmatrix} q^\cdot \\ p^\cdot \end{pmatrix} = \begin{pmatrix} 3p' \\ q''' + 8qq' \end{pmatrix} = \begin{pmatrix} 0 & D \\ D & 0 \end{pmatrix}\begin{pmatrix} \partial H_2/\partial q \\ \partial H_2/\partial p \end{pmatrix}. \tag{3}$$

Zakharov (1973) proved the existence of infinitely many integrals of (3); for example,

$$\begin{aligned} H_{-1} &= \int_0^1 q\,dx, \qquad H_0 = \int_0^1 (3/2)p\,dx, \\ H_1 &= \int_0^1 pq\,dx, \qquad H_2 = (2), \\ H_3 &= \int_0^1 [(-1/2)(p')^2 + 2p^2q + (1/6)(q'')^2 + (8/9)q^4 - 2q(q')^2]\,dx, \\ H_4 &= \int_0^1 [(5/2)p^3 + (20/3)pq^3 + (15/2)p(q')^2 + 10pqq'' + p''q'']\,dx. \end{aligned} \tag{4}$$

[†] The support of the National Science Foundation under Grant No. NSF-GP-37069X is gratefully acknowledged. Reproduction in whole or in part is permitted for any purpose of the United States Government.

[1] (1) differs from Zakharov (1973) by choice of units. Berryman (1976) lists the recent literature.

[2] $' = D = \partial/\partial x$. $\partial H_2/\partial q$ and $\partial H_2/\partial p$ signify the gradients of H_2 by q and by p.

ISBN 0-12-287150-2

Further integrals are provided by the (simple) eigenvalues of

$$L = \sqrt{-1}\,(D^3 + qD + Dq) + p \tag{5}$$

arising from the eigenfunctions of period 1, say, and Zakharov (1973) proved that the latter as well as the former integrals are all in involution. The connection between (3) and (5) is provided by the fact that (3) is equivalent to $\dot{L} = [Q, L]$ with $Q = \sqrt{-1}\,(3D^2 + 4q)$.

The purpose of this paper is to derive these results from a more systematic viewpoint, paralleling the known development for the equation of Korteweg–de Vries (1895), $\partial q/\partial t = 3q\,\partial q/\partial x - (1/2)\,\partial^3 q/\partial x^3$, as explained, for instance, in McKean–Trubowitz (1976). M. Adler [private communication] has proved the crucial Proposition 3 and justified the recursion $\mathbf{X}_{n+2}(q, p) = \mathbf{K}\,\mathrm{grad}\,H_n$ of Theorem 3, independently; a full account of his results will appear presently.

The paper is dedicated to M. G. Krein on the occasion of his seventieth, birthday as a token of my admiration of his deep and elegant contributions to hard analysis, generally, and to spectral theory in particular.

2. Eigenfunctions of L

The eigenfunctions of L play a central role. Let y_1, y_2, y_3 be the solutions of $Ly = \lambda y$ with $D^{i-1}y_j(0, \lambda) = 1$ if $i = j$ and vanishing otherwise $(1 \leqslant i, j \leqslant 3)$; for example, $y_1(x, \lambda) = (1/3)[\exp(\sqrt{-1}kx) + \exp(\sqrt{-1}k\omega x) + \exp(\sqrt{-1}k\omega^2 x)]$ for $q = p = 0$, $k^3 = \lambda$, and $\omega = \exp(2\pi\sqrt{-1}/3)$. The notations $e^{\times}(x, \lambda) = [e(x, \lambda^*)]^*$ and $[e_1, e_2] = e_1'e_2 - e_1e_2'$ are employed below.

Proposition 1. *Let $e_1(x, \lambda)$ and $e_2(x, \lambda)$ be any two solutions of $Le = \lambda e$. Then $[e_1, e_2]^{\times} = e_3$ is another such; in particular,*

$$\begin{aligned} [y_1, y_2]^{\times} &= -y_1 + 2q(0)y_3 \\ [y_2, y_3]^{\times} &= -y_3 \\ [y_3, y_1]^{\times} &= y_2. \end{aligned} \tag{6}$$

Proof. The computation is best performed in two stages: first, verify that $e_3'' - [e_1', e_2']^{\times} + 2qe_3 = 0$ and then use $Le = \lambda e$ for $e = e_1, e_2$ to compute

$$[e_1', e_2']^{\times} = (e_1'''e_2' - e_1'e_2''')^{\times} = (q' + \sqrt{-1}p - \sqrt{-1}\lambda)e_3.$$

The evaluations (6) will be self-evident.

PROPOSITION 2. *Let the discriminant $\Delta(\lambda)$ be the trace of the matrix M with entries $D^{i-1}y_j(1, \lambda)$ $(1 \leqslant i, j \leqslant 3)$, i.e., $\Delta(\lambda) = y_1(1, \lambda) + y_2'(1, \lambda) + y_3''(1, \lambda)$. Then the eigenvalues of M (= multipliers of L) are determined by the vanishing of the cubic*

$$\epsilon^3 - \epsilon^2 \Delta + \epsilon \Delta^\times - 1 = 0.$$

Proof. The computation makes use of (6):

$$\det(\epsilon - M) = \epsilon^3 - \sigma_1 \epsilon^2 + \sigma_2 \epsilon - \sigma_3$$

with $\sigma_1 = \operatorname{sp} M = \Delta$, $\sigma_3 = \det M = 1$, and

$$\begin{aligned}\sigma_2 &= y_2' y_3'' + y_1 y_2' + y_1 y_3'' - y_2 y_1' - y_3 y_1'' - y_3' y_2'' \\ &= [y_3, y_2]'' - y_3''' y_2 + y_3 y_2''' + [y_2, y_1] + [y_3, y_1]' \\ &= (y_3'')^\times + (y_2')^\times + y_1{}^\times - 2q(0) y_3{}^\times + 2q(1)[y_3, y_2] \\ &= \Delta^\times,\end{aligned}$$

$Ly = \lambda y$ being used to compute $y_3''' y_2 - y_3 y_2'''$.

PROPOSITION 3. *Let e_1 and e_2 be as in Proposition 1. Then $f = (\sqrt{-1}\,[e_1, e_2{}^\times], e_1 e_2{}^\times)$ satisfies $\mathbf{K}f = 3\lambda \mathbf{D}f$, in which $\mathbf{D}$ and $\mathbf{K}$ are the skew-symmetrical 2×2 operators*

$$\mathbf{D} = \begin{pmatrix} 0 & D \\ D & 0 \end{pmatrix}$$

and

$$\mathbf{K} = \begin{pmatrix} D^3 + qD + Dq & 3pD + 2p' \\ 3pD + p' & \frac{1}{3}D^5 + \frac{5}{3}(qD^3 + D^3 q) - (q''D + Dq'') + \frac{16}{3} qDq \end{pmatrix}.$$

Proof. The computation is tiresome but perfectly elementary. $Le = \lambda e$ is used to eliminate e''', e'''', etc., in favor of e, e', e'', and the stated result is verified, with tears.

PROPOSITION 4. *Let ϵ be a fixed number of modulus 1, let λ be a (necessarily real) simple eigenvalue of the (self-adjoint) restriction of L to the class C_ϵ^∞ of functions e satisfying $e(x+1) = \epsilon e(x)$, and let the eigenfunction satisfy $\|e\|_2 = 1$. Then*

$$\operatorname{grad} \lambda = \begin{pmatrix} \partial\lambda/\partial q \\ \partial\lambda/\partial p \end{pmatrix} = \begin{pmatrix} \sqrt{-1}[e, e^*] \\ ee^* \end{pmatrix};$$

in particular, $f = \operatorname{grad}\lambda$ is a solution of $\mathbf{K}f = 3\lambda\mathbf{D}f$.

Proof. This is a standard perturbation calculation. Let $q^\cdot$ and $p^\cdot$ be infinitesimal variations of q and p. Then $(\mathbf{K} - \lambda)e^\cdot + \mathbf{K}^\cdot e - \lambda^\cdot e = 0$, so that

$$\lambda^\cdot = \lambda^\cdot \int_0^1 |e|^2\,dx = \int_0^1 e^*[\sqrt{-1}(q^\cdot D + Dq^\cdot) + p^\cdot]e\,dx.$$

The rest will be plain, keeping in mind the reality of λ.

PROPOSITION 5. grad $\Delta = (\partial\Delta/\partial q, \partial\Delta/\partial p)$ *is also a solution of* $\mathbf{K}f = 3\lambda\mathbf{D}f$; *naturally, the same is true of* $\Delta^\times$.

Proof. Let $y = y_1, y_2$, or y_3 and let $q^\cdot$ and $p^\cdot$ be infinitesimal variations of q and p. Then, for fixed $0 \leqslant \xi \leqslant 1$,

$$-\sqrt{-1}\,y^\cdot(\xi) = \int_0^\xi D(\xi,\eta)[\sqrt{-1}(q^\cdot D + Dq^\cdot) + p^\cdot]y(\eta)\,d\eta$$

with

$$D(\xi,\eta) = \begin{vmatrix} y_1(\xi) & y_2(\xi) & y_3(\xi) \\ y_1(\eta) & y_2(\eta) & y_3(\eta) \\ y_1'(\eta) & y_2'(\eta) & y_3'(\eta) \end{vmatrix}$$
$$= y_1(\xi)y_3^\times(\eta) - y_2(\xi)y_2^\times(\eta) + y_3(\xi)[y_1^\times(\eta) - 2q(0)y_3^\times(\eta)],$$

by (6); in particular, if $\xi > \eta$, then[3]

$$-\sqrt{-1}\,\partial y(\xi)/\partial q(\eta) = \sqrt{-1}[y'(\eta)D(\xi,\eta) - y(\eta)D'(\xi,\eta)]$$
$$-\sqrt{-1}\,\partial y(\xi)/\partial p(\eta) = y(\eta)D(\xi,\eta).$$

The expansion of $D(\xi,\eta)$ makes plain that grad Δ = grad$[y_1(1,\lambda) + y_2'(1,\lambda) + y_3''(1,\lambda)]$ is a sum of terms $(\sqrt{-1}[e_1, e_2^\times], e_1e_2^\times)$. This observation permits the application of Proposition 3 to finish the proof.

3. HAMILTONIAN FLOWS

Let $\Delta(\lambda)$ be the discriminant $y_1(1,\lambda) + y_2'(1,\lambda) + y_3''(1,\lambda)$ and, for fixed λ, define a vector field $\mathbf{X}$ on $C_1^\infty \times C_1^\infty$ by the rule:

$$\mathbf{X}\begin{pmatrix} q \\ p \end{pmatrix} = \mathbf{D}\,\text{grad}\,\Delta(\lambda) = \mathbf{D}\begin{pmatrix} \partial\Delta/\partial q \\ \partial\Delta/\partial p \end{pmatrix}.$$

[3] $' = \partial/\partial\eta$.

$\mathbf{X}$ may be viewed as a Hamiltonian field, $\mathbf{D}$ being skew-symmetric; in conformity with this attitude, the Poisson bracket

$$[A, B] = \int_0^1 \operatorname{grad} \mathrm{A}\, \mathbf{D} \operatorname{grad} \mathrm{B}\, dx$$

$$= \int_0^1 \left[\frac{\partial A}{\partial q} D \frac{\partial B}{\partial p} - \frac{\partial A}{\partial p} D \frac{\partial B}{\partial q} \right] dx$$

may be regarded as inducing a natural symplectic structure into the phase space of points $(q, p) \in C_1^\infty \times C_1^\infty$.

THEOREM 1. *The values of Δ for different arguments are in involution relative to the bracket; in particular, the associated flows $(q^{\cdot}, p^{\cdot}) = \mathbf{X}(q, p)$ commute.*[4]

Proof. Let $f_1 = \operatorname{grad} \Delta(\lambda_1)$ and $f_2 = \operatorname{grad} \Delta(\lambda_2)$ for fixed $\lambda_1 \neq \lambda_2$. Then, by Proposition 5,

$$\begin{aligned} 3\lambda_2[\Delta(\lambda_1), \Delta(\lambda_2)] &= 3\lambda_2 \int_0^1 f_1 \mathbf{D} f_2 \\ &= \int_0^1 f_1 \mathbf{K} f_2 \\ &= -\int_0^1 f_2 \mathbf{K} f_1 \\ &= -3\lambda_1 \int_0^1 f_2 \mathbf{D} f_1 \\ &= 3\lambda_1 \int_0^1 f_1 \mathbf{D} f_2 \\ &= 3\lambda_1 [\Delta(\lambda_1), \Delta(\lambda_2)]. \end{aligned}$$

The vanishing of the bracket follows. The commuting of the associated flows is a standard fact of Hamiltonian mechanics.

THEOREM 2. *The simple eigenvalues of L figuring in Proposition 4 are in involution not only with themselves but also with the values of Δ.*

Proof. $(\mathbf{K} - 3\lambda\mathbf{D}) \operatorname{grad} \lambda = 0$ for such eigenvalues, by Proposition 4. The rest of the proof is a variant upon that of Theorem 1.

THEOREM 3. *The recipe $\mathbf{X}_{n+2}(q, p) = \mathbf{K} \operatorname{grad} H_n$, beginning from $H_{-1} = \int_0^1 q$ or from $H_0 = \int_0^1 (3/2)p$, produces two series of Hamiltonian vector*

[4] The (local) existence of the flow $\exp(t\mathbf{X})$ is easily proved, Δ being a smooth function of qp.

fields: $\mathbf{X}_n(q, p) = D \operatorname{grad} H_n$; *for example,*[5]

$$\mathbf{X}_{-1}\begin{pmatrix} q \\ p \end{pmatrix} = \mathbf{D} \operatorname{grad} H_{-1} = 0$$

$$\mathbf{X}_1\begin{pmatrix} q \\ p \end{pmatrix} = \mathbf{K} \operatorname{grad} H_{-1} = \begin{pmatrix} q' \\ p' \end{pmatrix} = \mathbf{D} \operatorname{grad} H_1$$

$$\mathbf{X}_0\begin{pmatrix} q \\ p \end{pmatrix} = \mathbf{D} \operatorname{grad} H_0 = 0$$

$$\mathbf{X}_2\begin{pmatrix} q \\ p \end{pmatrix} = \mathbf{K} \operatorname{grad} H_0 = \begin{pmatrix} 3q' \\ q''' + 8qq' \end{pmatrix} = \mathbf{D} \operatorname{grad} H_2.$$

The Hamiltonians $H_{-1}, H_0, H_1, H_2, \ldots$ *so defined are in involution not only with themselves but also with* Δ *and the eigenvalues of Theorem* 2.

Proof. The proof is self-evident except for one item: It must be proved that the rule $\mathbf{X}_n(q, p) = \mathbf{K} \operatorname{grad} H_{n-2}$ produces out of H_{n-2} a vector field $\mathbf{X}_n$ of the required form $\mathbf{X}_n(q, p) = \mathbf{D} \operatorname{grad} H_n$ with a bona fide Hamiltonian H_n expressible as an integral over $0 \leqslant x \leqslant 1$ of a polynomial in q, p, q', p', q'', etc., as in the list (4) of $H_{-1}, \ldots, H_4$. The proof is presented at the end of the next section; an alternative proof might be modeled upon Lax (1976, pp. 356–358).

4. ESTIMATING THE PERIODIC SPECTRUM OF L

The starting point[6] is the fact that for large real k, $Le = k^3 e$ has an eigenfunction $e(\xi, k)$ of the form

$$e(\xi, k) = \exp\left[\sqrt{-1}k\xi + \int_0^\xi C(\eta, k)\, d\eta\right] \qquad (0 \leqslant \xi \leqslant 1),$$

$$C(\eta, k) \sim \sum_{n=1}^{\infty} k^{-n} C_n(\eta) \qquad (k \to \pm\infty);$$

$C(\eta, k)$ is of class C_1^{∞}, and the development may be differentiated as many times as you please. The evaluations $C_1 = (2\sqrt{-1/3})q$, $C_2 = (1/9)q' - (\sqrt{-1/3})p$, etc., follow by substitution of $e(x, k)$ into $Le = k^3 e$; clearly, C_3, C_4, etc., are polynomials in q, p, q', p', q'', etc.

The construction of $e(x, k)$ will be outlined: Let q and p be modified outside $0 \leqslant x \leqslant 1$ so as to vanish smoothly in the vicinity of $x = \pm\infty$. The

[5] See (4).

[6] Zakharov (1973); compare Erdélyi (1956) for such matters.

operator $A = D^2 + \sqrt{-1}kD - k^2 + 2q$ is invertible in the class of summable functions if $k^2 > 2\|q\|_\infty$, and $\|A^{-1}\|_1 = O(k^{-2})$ in that space. Now $Le = k^3 e$ is equivalent to

$$A(D - \sqrt{-1}k)e = (\sqrt{-1}p - q' - 2\sqrt{-1}kq)e \equiv Be,$$

or, what is the same if $\|e\|_\infty < \infty$,

$$e(\xi) = e(0)e\sqrt{-1}k\xi + \int_0^\xi e^{\sqrt{-1}k(\xi-\eta)}A^{-1}Be(\eta)\,d\eta.$$

Let $e(0) = 1$, $e_0(\xi) = \exp(\sqrt{-1}k\xi)$, and

$$e_{n+1}(\xi) = \int_0^\xi e^{\sqrt{-1}k(\xi-\eta)}A^{-1}Be_n(\eta)\,d\eta.$$

Then

$$\|e_{n+1}\|_\infty < \|A^{-1}\|_1 \big|\|p\|_1 + \|q'\|_1 + 2k\|q\|_1)\|e_n\|_\infty\big| = O(1/k)\|e_n\|_\infty,$$

so $e = e_0 + e_1 + e_2 + \cdots$ is nicely convergent if k is so large that $\delta = O(1/k\}$ is <1, and $|e| < \sum_{n=0}^{\infty} \delta^n = (1 - \delta)^{-1}$.

The reason for outlining the construction of $e(x, k)$ is to make it plain that if now $\lambda = k^3$ is a periodic eigenvalue of L with eigenfunction e of period 1, say, then $e(x)$ *is proportional to* $e(x, k)$, so that the large periodic eigenvalues $\lambda = k^3$ of L are determined by requiring that $e(x, k)$, $e'(x, k)$, and $e''(x, k)$ match at $x = 0$ and $x = 1$. The requirement is equivalent to

$$\exp\left[\sqrt{-1}k + \int_0^1 C(\eta, k)\,d\eta\right] = 1,$$

by the periodicity of C, so the (necessarily simple) distant periodic spectrum is determined by

$$k = 2\pi n + \sqrt{-1}\int_0^1 C(\eta, k)\,d\eta = 2\pi n + \sqrt{-1}\int_0^1 \left[\frac{C_1(\eta)}{k} + \cdots\right] d\eta,$$

and this relation may be solved for k:

$$k = 2\pi n - \frac{1}{3\pi n}\int_0^1 q + \frac{1}{12\pi^2 n^2}\int_0^1 p + \cdots;$$

or, what is more to the point,

$$k^3 = 8\pi^3 n^3 - 4\pi n\int_0^1 q + \int_0^1 p + \cdots.$$

Now you are in a position to prove the existence of the Hamiltonians proposed in Theorem 3. Let λ_n $(-\infty < n < \infty)$ be the eigenvalues of L arising from eigenfunctions of period 1. The preceding expansion of $\lambda_n = k_n{}^3$

implies (1) that $\sum(\lambda - \lambda_n)^{-1}$ may be expanded for large imaginary $\lambda = k^3$ as

$$\frac{d_2}{k^2} + \frac{d_3}{k^3} + \frac{d_4}{k^4} + \frac{d_5}{k^5} + \cdots,$$

in which

$$d_2 = \sqrt{-1}/3, \qquad d_4 = (2/9)\int_0^1 q, \qquad d_5 = (2/9)\sqrt{-1}\int_0^1 p,$$

and (2) that the near periodic spectrum influences only the numbers d_3, d_6, d_9, etc., attached to whole powers of k^{-3}. Now, by Proposition 2, $\sum(\lambda - \lambda_n)^{-1}$ is just the logarithmic derivative of[7] $\Delta_1 = \Delta^\times - \Delta$, so by Proposition 5[8]

$$f = \operatorname{grad}\sum(\lambda - \lambda_n)^{-1} = \operatorname{grad}(\log \Delta_1)^{\cdot} = (\operatorname{grad}\log \Delta_1)^{\cdot}$$

satisfies

$$\begin{aligned}\mathbf{K}f &= (\mathbf{K}\operatorname{grad}\log\Delta_1)^{\cdot} = (3\lambda\mathbf{D}\operatorname{grad}\log\Delta_1)^{\cdot}\\ &= 3\mathbf{D}\operatorname{grad}\log\Delta_1 + 3\lambda\mathbf{D}\operatorname{grad}(\log\Delta_1)^{\cdot},\end{aligned}$$

or, what is more to the point,

$$\mathbf{K}f^{\cdot} = 6\mathbf{D}f + 3\lambda\mathbf{D}f^{\cdot}.$$

The expansion $f = \sum k^{-n}\operatorname{grad} d_n$ is now substituted, the upshot being that

$$\mathbf{K}\operatorname{grad} d_n = (3/n)(n-3)\mathbf{D}\operatorname{grad} d_{n+3} \qquad (n \geqslant 2);$$

in particular, starting from

$$d_4 = (2/9)H_{-1} \qquad \text{or} \qquad d_5 = (4\sqrt{-1}/27)H_0,$$

it appears that at every stage of the proposed recursion $\mathbf{K}\operatorname{grad} H_{n-2} = \mathbf{D}\operatorname{grad} H_n$ $(n = 1, 2, 3, \ldots)$, the existence of a bona fide Hamiltonian is assured; for example, $\mathbf{K}\operatorname{grad} H_{-1}$ is proportional to $D\operatorname{grad} d_7$, so $H_1 =$ constant $\times\, d_7$; $\mathbf{K}\operatorname{grad} H_1$ is proportional to $D\operatorname{grad} d_{10}$, so $H_3 =$ constant $\times\, d_{10}$; and so forth. The proof is finished.

5. The Solitary Wave

The flows $\mathbf{X}_n(q, p) = \mathbf{D}\operatorname{grad} H_n$ $(n = 1, 2, 3, \ldots)$ preserve a closed 1-dimensional manifold in qp-space corresponding to the solitary wave for (1); presumably, they have many finite- and infinite-dimensional invariant manifolds, but this is not yet proved. The solitary wave arises by making

[7] Δ_1 is an integral function of λ of order $\frac{1}{3}$.
[8] $^{\cdot} = \partial/\partial\lambda$.

$\mathbf{X}_1(q, p)$ proportional to $\mathbf{X}_2(p, q)$. Let

$$3\mathbf{X}_1\begin{pmatrix} q \\ p \end{pmatrix} = 3\begin{pmatrix} q' \\ p' \end{pmatrix} = \mathbf{X}_2\begin{pmatrix} q \\ p \end{pmatrix} = \begin{pmatrix} 3p' \\ q''' + 8qq' \end{pmatrix},$$

i.e., let $(q(x + 3t), p(x + 3t))$ be a solution of (3). Then $q' = p'$, i.e., $q = p + a$; also $3p' = (q'' + 4q^2)'$, i.e., $3p = q'' + 4q^2 + b$; and $(q')^2/2 = (3/2)q^2 - (4/3)q^3 - (3a + b)q + c$, a, b, c being constants of integration. Let $a = b = c = 0$ for simplicity. Then the general solution $q \in C_1{}^\infty$ is a translate of $(-3/2)p(x + \sqrt{-1}\omega'/2)$, in which p is a Weierstrassian elliptic function with real period 1 and a certain imaginary period $\sqrt{-1}\omega'$. Let M be the simple closed curve in $C_1{}^\infty \times C_1{}^\infty$ comprising the translations of (q, p) by $0 \leqslant x < T$, the latter being the primitive real period of p.

PROPOSITION 6. *$\mathbf{X}_3 = 3\mathbf{X}_1$ on M and all the local flows* $\exp(t\mathbf{X}_n)$ *preserve M, acting by mere translation at speeds* 3, 9, 27, ...

Proof. The final statement is immediate from the rule $\mathbf{K}\,\text{grad}\, H_n = \mathbf{X}_{n+2}(q, p)$ and the evaluation of

$$\mathbf{X}_3\begin{pmatrix} q \\ p \end{pmatrix} = \mathbf{D}\,\text{grad}\, H_3 = \mathbf{K}\,\text{grad}\, H_1 = \mathbf{K}\begin{pmatrix} p \\ q \end{pmatrix}$$

as $3\mathbf{X}_1(q, p)$. The computation is elementary; by use of $(1/2)(q')^2 = (3/2)q^2 - (4/3)q^3$ and $q = p$, one finds

$$(D^3 + qD + Dq)p + (3pD + p')q = q''' + 8qq' = 3q'$$

and

$$\begin{aligned}(3pD + p')p + [(1/3)D^5 + (5/3)(qD^3 + D^3q) - (q''D + Dq'') + (16/3)qDq]q \\ = 4qq' + (1/3)q''''' + 4qq''' + 8q'q'' + (32/3)q^2q' = 3q' = 3p'.\end{aligned}$$

PROPOSITION 7. *The (nonlocal) flows based upon Δ and upon the simple periodic spectrum of L also preserve M.*

Proof. The present proof could also have been employed for Proposition 6. Let $q_0 = p_0 = (-3/2)p(x + \sqrt{-1}\omega'/2)$. Then $f_0 = (q_0, p_0)$ is a solution of $3\mathbf{X}_1 f = \mathbf{X}_2 f$. Let $\mathbf{X}$ be a constant multiple of any of the cited nonlocal fields. Then $e^{\mathbf{X}}$ commutes with $\mathbf{X}_1$ and $\mathbf{X}_2$, so $e^{\mathbf{X}}f_0 = f = (q, p)$ is also a solution of $3\mathbf{X}_1 f = \mathbf{X}_3 f$ with the same integrals H_{-1}, H_0, H_1, H_2 as $f_0 = (q_0, p_0)$; in particular, $q' = p'$ and $3p' = q''' + 8qq'$ imply, as before,

$$q = p + a \tag{7}$$

$$3p = q'' + 4q^2 + b, \tag{8}$$

and

$$(1/2)(q')^2 + (4/3)q^3 - (3/2)q^2 + (3a + b)q = c, \tag{9}$$

the numbers a, b, c being constants of integration. To prove that $f = (q, p)$ is a translate of $f_0 = (q_0, p_0)$ it suffices to check that $a = b = c = 0$. This follows from the invariance of these qualities under $e^{\mathbf{X}}$: in fact, by (7)

$$a = \int_0^1 q - \int_0^1 p$$

is conserved, as is

$$\int_0^1 q^2 = \int_0^1 pq + a \int_0^1 q,$$

so (8) implies the conservation of

$$b = 3 \int_0^1 p - 4 \int_0^1 q^2;$$

similarly, by (7)

$$\int_0^1 p^2 = \int_0^1 pq - a \int_0^1 p,$$

so that

$$(4/3) \int_0^1 q^3 - (1/2) \int_0^1 (q')^2 = -(3/2) \int_0^1 p^2 + H_2$$

is conserved, while by (8)

$$-\int_0^1 (q')^2 + 4 \int_0^1 q^3 = -b \int_0^1 q + 3 \int_0^1 pq$$

is, too, and the invariance of c is inferred by integrating (9) over a period. The proof is finished.

References

J. Berryman, Stability of solitary waves in shallow water, *Phys. Fluids* **19** (1976), 771–777.

J. Boussinesq, Théorie des ondes et des remous qui se propagent le long d'un canal rectangulaire horizontal, *J. Math. Pures Appl.* **7** (1872), 55–108.

A. Erdélyi, "Asymptotic Expansions," Dover, New York, 1956.

D. J. Korteweg and G. De Vries, On the change of form of long waves advancing in a rectangular canal, and on a new type of long stationary wave, *Phil. Mag.* **39** (1895), 422–443.

P. Lax, Almost periodic solutions of the *KdV* equation, *SIAM Rev.* **18** (1976), 351–375.

H. P. McKean and E. Trubowitz, Hill's operator and hyperelliptic function theory in the presence of infinitely many branch points, *Commun. Pure Appl. Math.* **29** (1976), 143–226.

V. Zakharov, On stochastization of one-dimensional chains of non-linear oscillators, *Zh. Eksp. Teor. Fiz.* **65** (1973), 219–225.

TOPICS IN FUNCTIONAL ANALYSIS
ADVANCES IN MATHEMATICS SUPPLEMENTARY STUDIES, VOL. 3

On the Set of Minimal Extensions of a Subadditive Functional

D. P. MILMAN

Department of Mathematical Sciences
Tel Aviv University
Tel Aviv, Israel

DEDICATED TO M. G. KREIN ON THE OCCASION OF HIS SEVENTIETH BIRTHDAY

1

Let E be a Fréchet space with quasi-norm P that defines the topology in E,[1] let K_0 be a vector subspace of E, and let p_0 be a subadditive functional on K_0.

We say that p_0 is compatible (with the topology in E) if from

$$\{x_n\}_1^\infty \subset K_0, \qquad \lim_{n\to\infty} P(x_n - x_0) = 0, \qquad \lim_{m,\, n\to\infty} p_0(x_n - x_m) = 0$$

it follows that

$$x_0 \in K_0, \qquad \lim_{n\to\infty} p_0(x_n - x_0) = 0 = \lim_{n\to\infty} p_0(x_0 - x_n).$$

For example, let A be a closed operator defined on K_0. If E_1 is a Fréchet space with quasi-norm P_1 (which defines the topology in E_1) and A sends K_0 into E_1, then the functional $p_0(x) = P_1(Ax)$ is subadditive and compatible. It is clear that p_0 is continuous at zero if and only if A is continuous at zero (in the subspace K_0). Thus, Banach's closed-graph theorem may be considered a special case of the following theorem:

THEOREM A. *If K_0 is of the second category in itself, p_0 is compatible, and* $\lim p_0(x/n) = 0$, *then p_0 is continuous at zero.*

[1] Recall that a Fréchet space is a locally convex separated and complete metric space in which the metric is defined by a sequence of seminorms $\{p_n(x)\}_{n=1}^\infty$ by the formula

$$P(x-y) = \sum_{n=1}^\infty \frac{1}{2^n} \frac{p_n(x-y)}{1 + p_n(x-y)};$$

the functional P is the quasi-norm that defines the topology.

ISBN 0-12-287150-2

The Banach–Steinhaus theorem about a family $\{A_\alpha\}$ of bounded linear operators that send E into E_1 is a special case of Theorem A. To verify this, set $p_0(x) = \sup_\alpha p_1(A_\alpha x)$ and apply Theorem A.

The main result of this paper is a theorem that includes as special cases Theorem A, the Hahn–Banach theorem, the Krein–Milman theorem, and a few others.

2

In the following E will denote either a vector space or a Fréchet space.

A set K, $\varnothing \neq K \subset E$, is called a wedge in E if from $x, y \in K$, $\lambda \geqslant 0$, it follows that $x + y$, $\lambda x \in K$. A functional p on K is said to be subadditive if $p(x + y) \leqslant p(x) + p(y)$, $x, y \in K$, $p(0) = 0$; if, moreover, $p(\lambda x) = \lambda p(x)$, p is called sublinear on K.

A functional p is called a subadditive extension of p_0 from K_0 to E if p is subadditive on E and coincides with p_0 on K_0.

Let K_1, K_2 be two wedges in E, and let p_1, p_2 be sublinear functionals on K_1, K_2, respectively. We denote by $\hat{K}_{p_0}$ the set of all subadditive extensions p of p_0 from K_0 to E that satisfy the following condition:

$$p(x_2 - x_1) \leqslant p_1(x_1) + p_2(x_2), \qquad x_1 \in K_1, \quad x_2 \in K_2. \tag{α}$$

A functional $p \in \hat{K}_{p_0}$ is called a minimal extension of p_0, if for any $q \in \hat{K}_{p_0}$ the inequality $q(x) \leqslant p(x)$ for all $x \in E$ implies that $q = p$. We denote by K_{p_0} the set of all such minimal extensions of p_0.

3

The main result of this paper describes the set K_{p_0} in the following theorem:

THEOREM B. (a) *Let E be a vector space. If*

$$\begin{gathered} K_2 - K_1 + K_0 = E, \qquad p_0(x_2 - x_1) \leqslant p_1(x_1) + p_2(x_2), \\ x_1 \in K_1, \quad x_2 \in K_2, \quad x_2 - x_1 \in K_0, \end{gathered} \tag{1}$$

then $K_{p_0} \neq \varnothing$ and when p_0 is linear all p from K_{p_0} are linear, K_{p_0} is convex and $\operatorname{supp}_{p \in K_{p_0}}(x) = \operatorname{supp}_{p \in \mathscr{E}_{p_0}}(x)$ where $\mathscr{E}_{p_0}$ denotes the set of all extreme points of K_{p_0}.

(b) *Let E be a Fréchet space. Let K_0 and p_0 satisfy these conditions: K_0 is of the second category in itself, p_0 is compatible and has the right derivative at zero,*

$$p'_{0+}(x) \overset{\text{def}}{=} \lim_{\lambda \to +0} \frac{p_0(\lambda x)}{\lambda}, \qquad x \in K_0;$$

when $\bar{K}_0 = E$, *one can replace the last condition by a weaker*: $\lim p_0(x/n) = 0$, $x \in K_0$.

By these conditions $K_{p_0} \neq \varnothing$ and every $p \in K_{p_0}$ is continuous at zero. When $\bar{K}_0 \neq E$, there exists a continuous seminorm p_2 of E that majorizes p_0 ($p_0 \leqslant p_2$ on K_0). If we choose $K_2 = E$, $K_1 = \{0\}$, $p_1 = 0$, then we have functionals p_1, p_2 as in case (a) of this theorem. Thus, all from case (a) is fulfilled and each $p \in K_{p_0}$ is p_2-majorized, in accordance with (α) (here, $p_1 = 0$), and because p is continuous at zero. By case (b), if p_0 is linear then K_{p_0} is compact in the weak*-topology, each convex and continuous functional on K_{p_0} achieves its maximum on a facial subset of K_{p_0}, and every facial subset contains an extreme point of K_{p_0}.[2] The convex closed hull of $\mathscr{E}_{p_0}$ coincides with K_{p_0} ($\overline{\text{Conv}\, \mathscr{E}_{p_0}} = K_{p_0}$).

For a proof of Theorem B we use only Zorn's lemma, transfinite induction, and Baire's category theorem.

4. Sketch of the Proof of Theorem B

We show that in case (b) there is a continuous seminorm in E that majorizes p_0 in K_0; thus, case (b) reduces to case (a). In case (a) we introduce

$$\hat{p}_0(x) = \inf_{x = x_2 - x_1 + x_0} [p_0(x_0) + p_1(x_1) + p_2(x_2)], \quad \text{where} \quad x_i \in K_i, \quad i = 0, 1, 2. \tag{2}$$

One shows that $\hat{p}_0 \in \hat{K}_{p_0} = K_{\hat{p}_0}$, where $K_{\hat{p}_0}$ is the set of all subadditive and p_0-majorizing extensions of p_0 from K_0 to E, and that this set is inductive in the natural order, in the decreasing direction. From Zorn's lemma it follows that $K_{p_0} \neq \varnothing$. One shows that if p_0 is linear, then each $p \in K_{p_0}$ is linear and K_{p_0} is convex. The equality $\text{supp}_{p \in K_{p_0}}(x) = \text{supp}_{p \in \mathscr{E}_{p_0}}(x)$, $x \in E$, is derived by means of transfinite induction (the proof is in Section 6).

5. The Special Cases

It is clear that Theorem B contains as special cases Theorem A (case (b) of Theorem B when $\bar{K}_0 = E$) and the Hahn–Banach theorem (in two forms). Thus, Theorem B contains the three principal theorems of linear analysis. If E is a Banach space, $K_0 = 0$, $p_0 = 0$, we can use case (b) of Theorem B by taking p_2 to be the norm in E; then K_{p_0} is the unit ball in E^*, so that Theorem B contains Alaoglu's theorem about weak* compactness of the unit ball of E^*. If E is a Fréchet space, $K \subset E^*$ is a weakly* compact and

[2] γ is called a facial subset of K_{p_0} if $0 \neq \gamma \neq K_{p_0}$, γ is closed, and from $(f_1 + f_2)/2 \in \gamma$, $f_1, f_2 \in K_{p_0}$ it follows that $f_1, f_2 \in \gamma$.

convex set, and $p_2(x) = \sup p_{p \in K}(x)$, $x \in E$, then from Theorem A it follows that p_2 is a continuous (sublinear) functional and, taking $p_0 = 0$ ($K_0 = \{0\}$), we can apply Theorem B, case (b); here $K_{p_0} = \{p \in E^* : p(x) \leqslant p_2(x),\ x \in E\}$ and, using the geometrical form of the Hahn–Banach theorem, we see that $K = K_{p_0}$ (if $q \notin K$, then there exists $x_1 \in E$ such that $q(x_1) > \sup p_{p \in K}(x_1) = p_2(x_1)$). This gives the Krein–Milman theorem (for functionals) and also Bauer's theorem (on achieving the maximum of a convex functional in an extreme point) as special cases of Theorem B.

Now we want to verify that the Mazur–Orlicz theorem and monotone extension theorem are special cases of Theorem B.

The Mazur–Orlicz theorem states [4] that if E is a vector space, $p_2(x)$ is a sublinear functional on E, $a_1(x)$ is a real finite functional on a set $X_1 \subset E$ ($X_1 \neq \varnothing$), and for any finite sequence of elements $\{x_k\} \subset X$ and nonnegative numbers $\{\epsilon_k\}$ we have $\sum_k \epsilon_k a_1(x_k) \leqslant p_2(\sum_k \epsilon_k x_k)$, then there exists a linear functional "p" majorized by "p_2" and majorizing "a_1." Note that the functional $p_1(x) = -\sup \sum_k \epsilon_k a_1(x_k)$, where $x = \sum_k \epsilon_k x_k$, is sublinear on the smallest wedge K_1 containing X_1; moreover, $0 \leqslant p_1(x_1) + p_2(x_1)$, $x_1 \in K_1$. Setting $K_2 = E$, $K_0 = \{0\}$, $p_0 = 0$, we see that the assumptions of case (a) in Theorem B are satisfied. This implies the existence of the required functional p (it will be linear, since p_0 is linear).

The monotone extension theorem states that if E, K_1, K_0, and p_0 are the same as in case (a) of Theorem B, but $K_2 = \{0\}$, $p_2 = 0$, and p_0 is nonnegative on $K_0 \cap K_1$, then p_0 has a linear extension on E nonnegative on K_1. Note that the functional $p_1(x) = -\sup p_0(y)$ where $y \in K_0 \cap K_1 \cap (x - K_1)$, $x \in K_1$ is finite, sublinear, and nonpositive on K_1,[3] and for $x_1 \in K_1 \cap K_0$ we have $p_0(-x_1) = p_1(x_1)$. Thus, the assumptions of case (a) of Theorem B are satisfied, and therefore there exists a linear extension p of p_0 such that $p(x_2 - x_1) \leqslant p_1(x_1) + p_2(x_2)$, $x_1 \in K_1$, $x_2 \in K_2$; but $K_2 = \{0\}$, so that $p(-x_1) \leqslant p_1(x_1) \leqslant 0$ and $p(x_1) \geqslant 0$.

6. Details of Proof of Theorem B: Case (a)—E is a Vector Space

(1) First of all we prove that

$$\hat{p}_0 \in \hat{K}_{p_0} = K_{\hat{p}_0} \tag{3}$$

in the notations of Section 4 ($\hat{K}_{p_0}$ is defined in Section 2). After this we shall prove $K_{p_0} \neq \varnothing$.

[3] For $x \in E$, $y \in K_0 \cap K_1 \cap (x - K_1)$, we have $x = x_0 - x_1$, where $x_0 \in K_0$, $x_1 \in K_1$, and $y = x - x'$ where $x' \in K_1$; so $x' + y = x = x_0 - x_1$, $x_1 + x' = x_0 - y \in K_0 \cap K_1$, $p_0(y) \leqslant p_0(x_0)$. For fixed x_0 we obtain $\sup p_0(y) \leqslant p_0(x_0)$, $p_1(x) \geqslant -p_0(x_0)$. Thus, p_1 is finite; one checks directly that it is sublinear.

By the first property in (1), $-x = y_2 - y_1 + y_0$, where $y_i \in K_i$, $i = 0, 1, 2$. We have $x = x_2 - x_1 + x_0$, $(x_2 + y_2) - (x_1 + y_1) = -(x_0 + y_0) \in K_0$, $x_2 + y_2 \in K_2$, $x_1 + y_1 \in K_1$.

By the second property in (1)

$$p_0[-(x_0 + y_0)] \leqslant p_1(x_1 + y_1) + p_2(x_2 + y_2).$$

Using the subadditivity property, we obtain

$$\begin{aligned} -p_0(x_0) - p_0(y_0) &\leqslant -p_0(x_0 + y_0) \leqslant p_0[-(x_0 + y_0)] \\ &\leqslant p_1(x_1) + p_1(y_1) + p_2(x_2) + p_2(y_2) \\ &\quad - [p_0(y_0) + p_1(y_1) + p_2(y_2)] \\ &\leqslant p_0(x_0) + p_1(x_1) + p_2(x_2). \end{aligned}$$

We consider y_0, y_1, y_2 in this inequality as fixed, and x_0, x_1, x_2 as varying for fixed x; it follows from the definition of $\hat{p}_0$ that

$$-\infty < -[p_0(y_0) + p_1(y_1) + p_2(y_2)] \leqslant \hat{p}_0(x).$$

This proves that the functional $\hat{p}_0(x)$ is finite.

It follows directly from the definition of $\hat{p}_0$ that

$$\begin{gathered} \hat{p}_0(x_0) \leqslant p_0(x_0), \qquad \hat{p}_0(-x_1) \leqslant p_1(x_1), \qquad \hat{p}_0(x_2) \leqslant p_2(x_2), \\ x_i \in K_i, \quad i = 0, 1, 2. \end{gathered} \tag{4}$$

That $\hat{p}_0$ is subadditive follows directly from the definition (via the defining property of the infimum).

In order to show that $\hat{p}_0$ is an extension of p_0 it will suffice to verify that $p_0(x) \leqslant \hat{p}_0(x)$ for $x \in K_0$. We have $p_0(x) \leqslant p_0(x - x_0) + p_0(x_0)$ for $x_0 \in K_0$. Taking the decomposition $x = x_2 - x_1 + x_0$, we see that $x - x_0 = x_2 - x_1$. It follows from (1) that $p_0(x - x_0) \leqslant p_1(x_1) + p_2(x_2)$. Therefore, $p_0(x) \leqslant p_0(x_0) + p_1(x_1) + p_2(x_2)$. Hence it follows that $p_0(x) \leqslant \hat{p}_0(x)$. This shows that $\hat{p}_0(x) = p_0(x)$ for $x \in K_0$.

We have thus proved that $\hat{p}_0$ is a subadditive extension of p_0.

It follows from (4) that for $x_1 \in K_1$, $x_2 \in K_2$,

$$\hat{p}_0(x_2 - x_1) \leqslant \hat{p}_0(-x_1) + \hat{p}_0(x_2) \leqslant p_1(x_1) + p_2(x_2).$$

Hence $\hat{p}_0$ is a subadditive extension of p_0 from K_0 to E, satisfying condition (α), that is, $\hat{p}_0 \in \hat{K}_{p_0}$.

If $p \in K_{\hat{p}_0}$, then from the definition of $K_{\hat{p}_0}$ and because $\hat{p}_0$ satisfies (α) we have $p(x_2 - x_1) \leqslant \hat{p}_0(x_2 - x_1) \leqslant p_1(x_1) + p_2(x_2)$, $x_1 \in K_1$, $x_2 \in K_2$, so $p \in \hat{K}_{p_0}$. Thus, $K_{\hat{p}_0} \subset \hat{K}_{p_0}$. If $p \in \hat{K}_{p_0}$, then for $x_i \in K_i$, $i = 0, 1, 2$, and $x = x_2 - x_1 + x_0$ we have

$$p(x) \leqslant p(x_0) + p(x_2 - x_1) = p_0(x_0) + p(x_2 - x_1) \leqslant p_0(x_0) + p_1(x_1) + p_2(x_2),$$

and from the definition of $\hat{p}_0$ follows that $p(x) \leqslant \hat{p}_0(x)$, $x \in E$, so $p \in K_{\hat{p}_0}$. Thus, we have proved that $\hat{p}_0 \in \hat{K}_{p_0} = K_{\hat{p}_0}$. Note that $\hat{p}_0$ is the greatest functional in $\hat{K}_{p_0}$.

Now, we shall prove that $\hat{K}_{p_0}$ is inductively ordered, in the decreasing sense, by the ordinary order for real functions; that is, any linearly ordered subset $\{q_t\}_{t \in T} \subset \hat{K}_{p_0}$ has a minimum in $\hat{K}_{p_0}$.

For given $x \in E$, fixed $t_1 \in T$, and any $t \geqslant t_1$ we have $-q_{t_1}(-x) \leqslant -q_t(-x) \leqslant q_t(x)$, whence it follows that $-q_{t_1}(-x) \leqslant \inf_{t \in T} q_t(x)$. Thus the functional $q_T(x)$ defined as $\inf_{t \in T} q_t(x)$, $x \in E$, is finite. If x, $y \in E$, we have $q_T(x + y) \leqslant q_t(x + y) \leqslant q_t(x) + q_t(y)$; hence, taking the infimum of the expression on the right, we see that p_T is subadditive. It is clear that p_T is an extension of p_0 from K_0 to E and that $q_T(x) \leqslant q_t(x) \leqslant \hat{p}_0(x)$, $x \in E$. Thus $q_T \in \hat{K}_{p_0}$.

We have thus shown that $\hat{K}_{p_0}$ is inductively ordered. By Zorn's lemma there exists a minimal $p \in \hat{K}_{p_0}$, that is, $p \in K_{p_0}$.

We have proved $K_{p_0} \neq \varnothing$.

Note that in case $K_1 = \{0\}$, $p_1 = 0$, our result means that if p_0 is p_2-majorized, then it has a p_2-majorized minimal subadditive extension from K_0 to L.[4] Note that from the definition of $\hat{p}_0$ it follows directly that if p_0 is sublinear, then $\hat{p}_0$ is also sublinear, and if p_0 is linear, then

$$\hat{p}_0(x + y) = p(x) + p(y) \qquad \text{for} \quad y \in K_0, \quad x \in E. \tag{5}$$

Note for $p \in \hat{K}_{p_0}$ that from $-p(x) \leqslant p(-x)$ it follows that $|p(x)| \leqslant \max[\hat{p}_0(x), \hat{p}_0(-x)]$, $x \in E$. So, when p_0 is sublinear we have $|p(\lambda x)| \leqslant |\lambda| \max[\hat{p}_0(x), \hat{p}_0(-x)]$, $x \in E$; from this it follows that in this case every $p \in \hat{K}_{p_0}$ is continuous on each one-dimensional subspace of E. In particular, if p_0 is sublinear, then each $p \in \hat{K}_{p_0}$ that is additive must be linear.

(2) Now we consider the case when p_0 is linear. To prove that $p \in K_{p_0}$ is linear it suffices to verify that $p(-x) = -p(x)$, $x \in E$, since in this case we have $p(x + y) \leqslant p(x) + p(y) = -[p(-x) + p(-y)] \leqslant -p(-x - y) = p(x + y)$.

If we suppose that p is not linear, then there exists $x_1 \in E \setminus K$ for which $-p(x_1) < p(-x_1)$. Denote $q_1(\lambda x) = \lambda p(x_1)$, $-\infty < \lambda < \infty$. This q_1 is a linear functional on the one-dimensional space $\{\lambda x_1\}_{-\infty < \lambda < \infty}$, and $q_1(\lambda x_1) \leqslant p(\lambda x_1)$ since $-p(x_1) < p(-x_1)$.

By the first note in part (1) of this proof, the functional q_1 has a subadditive p-majorized extension $\hat{p}$ from $\{\lambda x_1\}_{-\infty < \lambda < \infty}$ to E. From $p(x) = p_0(x)$ for

[4] In this case we name $\hat{p}_0$ the envelope of p_0 and p_2. Here

$$\hat{p}_0(x) = \inf_{y \in K_{p_0}} [p_2(x - y) + p_0(y)], \qquad x \in E;$$

in essence, I introduced this concept in [1] and used it in [2].

$x \in K_0$, and since p_0 is linear, we have $p_0(x) = -p_0(-x) \leqslant -\hat{p}(-x) \leqslant \hat{p}(x) \leqslant p_0(x)$ for $x \in K_1$; and besides, $\hat{p}(x) \leqslant p(x) \leqslant \hat{p}_0(x)$ for all $x \in E$.

Thus $\hat{p} \in K_{\hat{p}_0}$, and since p is minimal ($p \in K_{p_0}$), from $\hat{p} \leqslant p$ it follows that $\hat{p} = p$. But $\hat{p}(-x_1) = -p(x_1) < p(-x_1)$. This contradiction shows that p is linear. So, we have proved that every $p \in K_{p_0}$ is linear, if p_0 is linear.

If $p \in \hat{K}_{p_0}$ is linear and $q(x) \leqslant p(x)$, $x \in E$, then $p(x) = -p(-x) \leqslant -q(-x) \leqslant q(x) \leqslant p(x)$, so that $q = p$; thus, each linear p from $\hat{K}_{p_0}$ belongs to K_{p_0}. Hence it follows that K_{p_0} is convex.

Thus, if p_0 is linear, every $p \in K_{p_0}$ is linear and K_{p_0} is convex. Note, that

$$K_{p_0} = \{f \in E^{\#} : f(x) \leqslant \hat{p}_0(x),\ x \in E\} \tag{6}$$

where $E^{\#}$ is the set of all linear functionals on E.

(3) Here p_0 is also linear. We shall prove $\operatorname{supp}_{p \in K_{p_0}}(x)\ \operatorname{supp}_{p \in \mathscr{E}_{p_0}}(x)$, where $\mathscr{E}_{p_0}$ denotes the set of all extreme points of K_{p_0}. We shall choose $\{q_t\}_{t \in T} \subset \hat{K}_{p_0}$ where $q_T \in \hat{K}_{p_0}$, with special properties.

The case when p_0 is linear is trivial; in this case $K_{p_0} = \{p_0\} = K_{p_0}$.

If $\hat{p}_0$ is not linear, then there exists an element x_1 for which $-\hat{p}_0(x_1) < p(-x_1)$. We denote by ζ_1 one of the maximal wedges from E in which $\hat{p}_0$ is additive and which contains K_0 and x_1. Such a ζ_1 exists by Zorn's lemma since (see (5)) for $0 \leqslant \lambda < \infty$ and $y \in K_0$ we have $\hat{p}_0(\lambda x_1 + y) = \lambda \hat{p}_0(x_1) + \hat{p}_0(y)$.

For $x', y', x'', y'' \in \zeta_1$, $x'' - x' = x = y'' - y'$, we have $x' + y'' = y' + x''$, $\hat{p}_0(x') + \hat{p}_0(y'') = \hat{p}_0(y') + \hat{p}_0(x'')$, $\hat{p}_0(x'') - \hat{p}_0(x') = \hat{p}_0(y'') - \hat{p}_0(y') \overset{\text{def}}{=} q_0(x) \leqslant p_0(x)$, so that q_0 is a univalued p_0-majorized linear functional on the vector space $\zeta_1 - \zeta_1 \overset{\text{def}}{=} E_1$. As we saw before, the envelope q_1 of q_0 and $\hat{p}_0$ is maximal among the $\hat{p}_0$-majorized sublinear extensions of q_0 from E_1 to E, and since $q_0|_{K_0} = p_0$, we have $q_1 \in \hat{K}_{p_0}$. Note that $q_1(y) = \hat{p}_0(y) = q_0(y)$ for $y \in \zeta_1$ and $q_1 \neq \hat{p}_0$ because $q_1(-x_1) = q_0(-x_1) = -\hat{p}_0(x_1) < \hat{p}_0(-x_1)$.

We prove now that $q_1 \in K_{p_0}$ implies $q_1 \in \mathscr{E}_{p_0}$.

Suppose $\frac{1}{2}(\varphi_1 + \varphi_2) = q_1 \in K_{p_0}$, $\varphi_1, \varphi_2 \in K_{p_0}$. Then $\frac{1}{2}(\varphi_1(y) + \varphi_2(y)) = q_1(y) = \hat{p}_0(y) = q_0(y)$ for $y \in \zeta_1$, so that $[\hat{p}_0(y) - \varphi_1(y)] + [\hat{p}_0(y) - \varphi_2(y)] = 0$, and since $\varphi_1 \leqslant \hat{p}_0$, $\varphi_2 \leqslant \hat{p}_0$, we obtain on $y \in \zeta_1$ $\varphi_1(y) = \varphi_2(y) = \hat{p}_0(y) = q_0(y)$. φ_1, φ_2 are linear on E and q_0 is linear on E_1, so that φ_1, φ_2 are $\hat{p}_0$-majorized extensions of q_0 from E_1 to E. But q_1 is the maximal sublinear $\hat{p}_0$-majorized extension of q_0, so that $\varphi_1 \leqslant q_1$, $\varphi_2 \leqslant q_1$, and, because $(q_1 - \varphi_1) + (q_1 - \varphi_2) = 0$, we have $\varphi_1 = q_1 = \varphi_2$. Thus, $q_1 \in \mathscr{E}_{p_0}$.

On the base of the last consideration we construct transfinite consequences $\{x_t, E_t, q_{0t}, q_t\}$. If t is not a limiting transfinite number and q_{t-1} is not linear, then there exists x_t for which $-q_{t-1}(x_t) < q_{t-1}(-x_t)$ and we choose: ζ_t, the maximal wedge in which q_{t-1} is additive and that contains E_{t-1} and x_t, $E_t = \zeta_t - \zeta_t$, q_{0t}, the linear functional on E_t that coincides with

q_{t-1} on ζ_t; and q_t, the envelope of q_{0t} and q_{t-1}. q_t is the maximal q_{t-1}-majorized sublinear extension of q_{0t} from E_t to E. Note that $q_{0t}(x) = q_{t-1}(x)$ for $x \in \zeta_\tau$, $\tau \leqslant t$.

If t is a limiting transfinite number, we choose $E_t = \bigcup_{\tau<t} E_\tau$, $q_{0t}(x) = q_{0\tau}(x)$ for $x \in E_\tau$ and $\tau < t$, $q_t(x) = \inf_{\tau<t} q_\tau(x)$, $x \in E$. Note that if $x \in E_t$, there exists $\tau < t$ for which $x \in E_\tau$, and then $q_\tau(x) = q_{0\tau}(x) = q_{0t}(x)$, so that $q_t(x) = q_{0t}(x)$. Thus, q_t is a q_τ-majorized extension of q_{0t} for every $\tau < t$.

In both cases (1) $q_{0t}(x) = q_{0\tau}(x)$ for $x \in E_\tau$, $\tau \leqslant t$, and (2) any sublinear extension of q_{0t} from E_t to E that is q_τ-majorized for every $\tau < t$ is also q_t-majorized; thus, q_t is maximal among the sublinear extensions of q_{0t} that are q_τ-majorized for any $\tau < t$.

The process of constructing $\{E_t\}$ must end, because the E_t are different and the set $E \backslash K_0$ has a definite power. Thus, there exists an increasing (in the set-theoretical sense) transfinite sequence $\{E_t\}_{1 \leqslant t \leqslant \gamma}$ for which either γ is not limiting and $E_{\gamma-1} \neq E = E_\gamma$ or γ is limiting, $E_t \neq E$ for $t < \gamma$, and $E = E_\gamma$ $(E_\gamma = \bigcup_{t<\gamma} E_t)$.

In both cases $E_\gamma = E$ and we construct $q_{0\gamma}$, q_γ as before, but now $q_{0\gamma}$ is given over E, $q_\gamma = q_{0\gamma} \in K_{p_0}$, and q_γ is unique among the sublinear functionals on E that are q_τ-majorized for $\tau < \gamma$.

We shall prove that $q_\gamma \in \mathscr{E}_{p_0}$, that is, if $q_\gamma = \frac{1}{2}(\varphi_1 + \varphi_2)$, $\varphi_1, \varphi_2 \in K_{p_0}$, then $\varphi_1 = \varphi_2$. It suffices to verify that the statement that φ_1 and φ_2 are both q_t-majorized linear extensions of q_{0t} is borne out by transfinite induction on t.[5]

For $t = 1$, $x \in \zeta_1$, we have $\hat{p}_0(x) = q_{01}(x) = q_{0\gamma}(x) = \frac{1}{2}(\varphi_1(x) + \varphi_2(x))$, $\varphi_1, \varphi_2 \leqslant \hat{p}_0$, so that $\varphi_1(x) = q_{01}(x) = \varphi_2(x)$ and, since φ_1, φ_2 are linear on E and q_{01} is linear on E_1, we see that φ_1, φ_2 are $\hat{p}_0$-majorized linear extensions of q_{01}; but q_1 is the maximal sublinear extension to be $\hat{p}_0$-majorized, so that $\varphi_1, \varphi_2 \leqslant q_1$.

Let it be assumed that for $1 \leqslant \tau < t$, φ_1 and φ_2 are both q_τ-majorized extensions of $q_{0\tau}$; we must show that they are q_t-majorized extensions of q_{0t}.

When t is limiting this follows from the definitions of q_{0t} and q_t. If t is not limiting we have for $x \in \zeta_t$, $q_{t-1}(x) = q_{0t}(x)$, $q_\gamma(x) = \frac{1}{2}(\varphi_1(x) + \varphi_2(x))$, $\varphi_1, \varphi_2 \leqslant q_{t-1}$, so that $\varphi_1(x) = q_{0t}(x) = \varphi_2(x)$. φ_1, φ_2 are linear on E and q_{0t} is linear on E_t. Thus φ_1, φ_2 are q_{t-1}-majorized linear extensions of q_{0t}. But q_t is the maximal sublinear q_{t-1}-majorized extension of q_{0t}, so that $\varphi_1, \varphi_2 \leqslant q_t$.

We see that transfinite induction shows that φ_1 and φ_2 are q_t-majorized extensions of q_t for $t \leqslant \gamma$. For $t = \gamma$ we obtain $\varphi_1 = q_\gamma = \varphi_2$. Thus, we have proved that $q_\gamma \in \mathscr{E}_{p_0}$. Recall that x_1 was an arbitrary element from E for which $-\hat{p}_0(x_1) < \hat{p}_0(-x_1)$. So $\operatorname{supp}_{p \in \mathscr{E}_{p_0}}(x) = \hat{p}_0(x)$ when $-\hat{p}_0(x) < \hat{p}_0(-x)$.

[5] From $q_\gamma \geqslant \varphi_1, \varphi_2$, $q_\gamma = \frac{1}{2}(\varphi_1 + \varphi_2)$, it follows that $\varphi_1 = q_\gamma = \varphi_2$.

If $-\hat{p}_0(-x) = \hat{p}_0(x)$, then $-\hat{p}_0(x) \leqslant -p(x) \leqslant p(-x) \leqslant \hat{p}_0(-x)$ for $p \in K_{p_0}$, so that $p(x) = \hat{p}_0(x)$. Thus, $\operatorname{supp}_{p \in \mathscr{E}_{p_0}}(x) = \hat{p}_0(x)$ for every $x \in E$. But $\operatorname{supp}_{p \in \mathscr{E}_{p_0}}(x) \leqslant \operatorname{supp}_{p \in K_{p_0}}(x) \leqslant \hat{p}_0(x)$. Thus,

$$\sup_{p \in \mathscr{E}_{p_0}} p(x) = \hat{p}_0(x) = \sup_{p \in K_{p_0}} p(x), \qquad x \in E. \tag{7}$$

Here we also have that for each $x \in E$,

$$\sup_{p \in K_{p_0}} p(x) = \max_{p \in K_{p_0}} p(x). \tag{8}$$

7. Details of Proof of Theorem B: Case (b)—E is a Fréchet Space

(1) We shall first prove that p_0 is continuous in a neighborhood of zero in K_0. Instead of p_0 we can take $p_s(x) = \max[p_0(x), p_0(-x)]$, since $-p_0(x) < p_0(-x)$ and therefore $|p_0(x)| \leqslant p_s(x)$, $x \in K_0$. Note that since p_0 is compatible, the same is true for p_s.

Set $\mathscr{D}(\epsilon) = \{x \in K_0 : p_s(x) \leqslant \epsilon\}$, $0 < \epsilon < \infty$. Since $\lim_{n\to\infty} p_s(x/n) = 0$, it follows that $K_0 = \bigcup_{n=1}^{\infty} n\mathscr{D}(\epsilon)$ and so, *a fortiori*, $K_0 = \bigcup_{n=1}^{\infty} n\overline{\mathscr{D}(\epsilon)} \cap K_0$. Since K_0 is of the second category in itself, there exists n_0 for which $n_0\overline{\mathscr{D}(\epsilon)} \cap K_0 \supset S(x_0, r) \cap K_0$.[6] If $x_0 \in n_0\mathscr{D}(\epsilon) \cap K_0$, there exists $y_0 \in \mathscr{D}(\epsilon)$ such that $P(n_0 y_0 - x_0) < r$. Set $r_1 = r - P(n_0 y_0 - x_0)$, $\delta(\epsilon) = \min(r_1/n_0, \epsilon)$.

Let $x \in S(0, \delta(\epsilon)) \cap K_0$. Then $P[n_0(x + y_0) - x_0] \leqslant n_0 P(x) + P(n_0 y_0 - x_0) < r_1 + P(ny - x) = r$. Thus $n_0(x + y_0) \in S(x_0, r) \subset n_0\overline{\mathscr{D}(\epsilon)}$, $y_0 + x \in \overline{\mathscr{D}(\epsilon)} \cap K_0$ (since $x, y_0 \in K_0$). Any neighborhood of $y_0 + x$ contains an element $y_1 \in \mathscr{D}(\epsilon)$, and so any neighborhood of x contains an element $y_1 - y_0$, where $y_0, y_1 \in \mathscr{D}(\epsilon)$. Since $p_s(y_1 - y_0) \leqslant 2\epsilon$, it follows that $x \in \overline{\mathscr{D}(2\epsilon)} \cap K$.

We have proved $S(0, \mathscr{D}(\epsilon)) \cap K_0 \subset \overline{\mathscr{D}(2\epsilon)} \cap K_0$. Set $A(\epsilon) = \mathscr{D}(2\epsilon) \cap S(0, \delta(\epsilon))$.

Let $x \in S(0, \delta(\epsilon)) \cap \bar{K}_0$. In any neighborhood of x there is an element from $S(0, \delta(\epsilon) \cap K_0$, hence also an element of $\mathscr{D}(2\epsilon)$, i.e., an element from $A(\epsilon)$. Thus $x \in \overline{A(\epsilon)}$.

We have proved that $S(0, \delta(\epsilon)) \cap \bar{K}_0 \subset \overline{A(\epsilon)}$. Let $z_0 \in S(0, \delta(\epsilon)) \cap \bar{K}_0$.

There is an element $z_1 \in A(\epsilon)$ in the neighborhood $S(z_0, \delta(\epsilon/2))$. We have $z_0 - z_1 \in S(0, \delta(\epsilon/2))$. Since $z_0 - z_1 \in \bar{K}_0 - K_0 \subset \bar{K}_0$, it follows that $z_0 - z_1 \in S(0, \delta(\epsilon/2)) \cap \bar{K}_0$.

Thus, if $z_0 \in S(0, \delta(\epsilon)) \cap \bar{K}_0$ there exists $z_1 \in A(\epsilon)$ such that $z_0 - z_1 \in S(0, \delta(\epsilon/2)) \cap \bar{K}_0$. We may therefore construct by induction a sequence $\{z_n\}_{n=1}^{\infty}$ with properties: $z_n \in A(\epsilon/2^{n-1})$, $z_0 - \sum_{k=1}^{n} z_k \in S(0, \delta(\epsilon/2^n)) \cap \bar{K}_0$, $n = 1, 2, 3, \ldots$.

[6] $S(x_0, r) = \{x \in E : P(x - x_0) < r\}$, where P is the quasi-norm in E that defines its metric.

Set $x_n = \sum_{k=1}^{n} z_k$, $n = 1, 2, \ldots$. We have

$$\{x_n\}_1^\infty \subset K_0, \qquad P(z_0 - x_n) \leqslant \delta\left(\frac{\epsilon}{2^n}\right) \leqslant \frac{\epsilon}{2^n}, \qquad x_n - x_m = \sum_{k=m+1}^{n} z_k,$$

$$p_s(x_n - x_m) \leqslant \sum_{k=m+1}^{n} p_s(z_k) \leqslant \sum_{k=m+1}^{n} \frac{\epsilon}{2^{k-2}}, \qquad n > m.$$

Hence it follows that $\lim_{n\to\infty} P(z_0 - x_n) = 0$, $\lim_{n,m\to\infty} p_s(x_n - x_m) = 0$, where $\{x_n\}_{n=1}^\infty \subset K_0$.

Since the functional p_s is compatible, we have $z_0 \in K_0$, $\lim p_s(z_0 - x_n) = 0$. Note that $p_s(x_n) \leqslant \sum_{k=1}^{n} p_s(z_k) \leqslant \sum_{k=1}^{n} \epsilon/2^{k-2} < 4\epsilon$. Hence $p_s(z_0) \leqslant p(z_0 - x_n) + p_s(x_n) \leqslant p_s(z_0 - x_n) + 4\epsilon$. Letting $n \to \infty$ we get $p_s(z_0) \leqslant 4\epsilon$.

We have thus shown that if $z_0 \in S(0, \delta(\epsilon)) \cap \bar{K}_0$, then $z_0 \in K_0$ and $p_s(z_0) \leqslant 4\epsilon$. This means that p_s and hence also p_0 are continuous at zero and $E = K_0$.

This completes the proof of case (b) under assumption that K_0 is dense in E.

(2) If K_0 is not dense in E we assume that $p'_{0+}(x)$ exists for $x \in K_0$.

We shall now prove that $p'_{0+}(x)$ is compatible, and hence continuous, in K_0.

Let $\lim_{n,m\to\infty} p'_{0+}(x_n - x_m) = 0 = \lim_{n\to\infty} P(x_n - x_0)$, where $\{x_n\}_{n=1}^\infty \subset K_0$. Since $p_0(x_n - x_m) \leqslant p'_{0+}(x_n - x_m)$ and $p_0(x)$ is compatible, it follows that $x_0 \in K_0$. Since $\lim_{n,m\to\infty} p'_{0+}(x_n - x_m) = 0$, it follows that for any $\epsilon > 0$ there exists N such that, if $n, m > N$, then $|p'_{0+}(x_n - x_m)| < \epsilon$. Since

$$p'_{0+}(x_n - x_m) = \sup_k \left[k p_0\left(\frac{x_n - x_m}{k}\right)\right]$$

it follows that

$$k p_0\left(\frac{x_n - x_m}{k}\right) < \epsilon \qquad \text{for} \quad m, n > N, \quad k = 1, 2, \ldots.$$

Since p_0 is continuous and

$$p_0\left(\frac{x_0 - x_m}{k}\right) \leqslant p_0\left(\frac{x_0 - x_n}{k}\right) + p_0\left(\frac{x_n - x_m}{k}\right),$$

it follows that

$$k p_0\left(\frac{x_0 - x_m}{k}\right) \leqslant \epsilon$$

and, similarly, $k p'_{0+}(x_0 - x_n) \leqslant \epsilon$, $n > N$, $k = 1, 2, \ldots$. Taking suprema over k, we see that $p'_{0+}(x_n - x_0) \leqslant \epsilon$, $p'_{0+}(x_0 - x_n) \leqslant \epsilon$, $n > N$.

Thus, the functional p'_{0+} is compatible and therefore also continuous.

On this basis we can prove that there is a seminorm in E majorizing p'_{0+}. We shall use the sublinearity of p'_{0+}.

Let $\{P_k\}_{k=1}^{\infty}$ be the sequence of seminorms defining the topology in E. It follows from the continuity of p'_{0+1} that for any $\epsilon > 0$ there exists $\delta > 0$ and a number N such that if $\max_{1 \leqslant k \leqslant N} P_k(x) \leqslant \delta$, then $|p'_{0+}(x)| < \epsilon$.

Set $\tilde{P}(x) = (\epsilon/\delta) \max_{1 \leqslant k \leqslant N} P_k(x)$. For the case $\tilde{P}(x_1) = \lambda_1 > 0$ we have $\max_{1 \leqslant k \leqslant N} P_k(x_1) = (\lambda_1 \delta/\epsilon)$ or $\max_{1 \leqslant k \leqslant N} P_k((\epsilon/\lambda_1)x_1) = \delta$. Hence it follows that $|p'_{0+}((\epsilon/\lambda_1)x_1)| < \epsilon$ and therefore $|p'_{0+1}(x_1)| < \lambda_1 = \tilde{P}(x_1)$. Thus, if $\tilde{P}(x_1) > 0$, then $|p'_{0+}(x_1)| < \tilde{P}(x_1)$.

Now let $\tilde{P}(x_1) = 0$. Then $(\delta/\epsilon)\tilde{P}(nx_1) = 0 < \delta$ and so $|p'_{0+}(nx_1)| < \epsilon$, $p'_{0+}(x_1) < \epsilon/n$, $n = 1, 2, \ldots.$ But this implies $p'_{0+}(x_1) = 0 = P(x_1)$. Thus, $|p'_{0+}(x)| \leqslant \tilde{P}(x)$, $x \in E$.

We have shown that p'_{0+} has a majorizing continuous seminorm in E; hence the same is true for $p_0(x)$. Thus everything proved for case (a) is applicable here.

(3) Here p_0 is linear. As we proved, K_{p_0} is convex and every $p \in K_{p_0}$ is linear. In our case (E is a Fréchet space) from (6) follows $K_{p_0} = \{f \in E^*: f(x) \leqslant \hat{p}_0(x),\ x \in E\}$, where E^* denotes the space of all linear continuous functionals. By definition, the weak*-topology in E^* is generated by the sets $\{f \in E^*: f(x) > c\}$, where $x \in E$, as open sets; so, the sets $\{f \in E^*: f(x) \leqslant c\}$ are closed sets in this topology. But K_{p_0} is an intersection of such sets, so that K_{p_0} is closed in the weak*-topology.

We have seen ((7) and (8)) that for each $x \in E$ $\sup_{f \in K_{p_0}} f(x) = \max_{f \in K_{p_0}} f(x) = \hat{p}_0(x)$, and from this it follows that $\min_{f \in K_{p_0}} f(x) = -\hat{p}_0(-x)$. Since K_{p_0} is convex, the set $K_{p_0}(x) = \{f(x)\}_{f \in K_{p_0}}$ coincides with the segment $[-\hat{p}_0(-x), \hat{p}_0(x)]$.

Let $\prod_0$ denote the topological product of the segments $\{-\hat{p}_0(-x), \hat{p}_0(x)\}_{x \in E}$. Then for each $f \in K_{p_0}$,

$$\{f(x)\}_{x \in E} \in \prod{}_0 \quad \text{and} \quad \tilde{K}_{p_0} \stackrel{\text{def}}{=} \{\{f(x)\}_{x \in E}\}_{f \in K_{p_0}} \subset \prod{}_0.$$

It is clear that by these definitions $\tilde{K}_{p_0}$ is homeomorphic to K_{p_0}, if we consider K_{p_0} in the weak*-topology.

We shall prove that K_{p_0} is compact in the weak*-topology because $\prod_0$ is compact.

Let γ be either K_{p_0} or a facial subset of K_{p_0}, and let $r(f)$ be a convex continuous functional on K_{p_0} (in the weak*-topology) that is not constant on γ. Since γ is compact, we have $c \stackrel{\text{def}}{=} \sup_{f \in \gamma} r(f) = \max_{f \in \gamma} r(f)$, the set $\gamma_1 \stackrel{\text{def}}{=} \{f \in \gamma: r(f) = c\}$ is closed, and $\varnothing \neq \gamma_1 \neq \gamma$.

If $\frac{1}{2}(\varphi_1 + \varphi_2) = f \in \gamma_1$, $\varphi_1, \varphi_2 \in K_{p_0}$, then $f \in \gamma$ and from the definition of γ it follows that $\varphi_1, \varphi_2 \in \gamma$. We have $c = r(f) \leqslant \frac{1}{2}(r(\varphi_1) + r(\varphi_2))$, $r(\varphi_1) \leqslant c$, $r(\varphi_2) \leqslant c$, so that $r(\varphi_1) = c = r(\varphi_2)$ and $\varphi_1, \varphi_2 \in \gamma_1$. Thus, γ_1 is also a facial set of K_{p_0}. One can also take $r = x \in E$, which is not constant on γ (if $f_1, f_2 \in \gamma$, $f_2 - f_1 \neq 0$, there exists $x_0 \in E$ for which $(f_2 - f_1)(x_0) \neq 0$).

The set of all facial subsets of γ is inductively ordered, in the set-theoretical sense, in the decreasing direction. By Zorn's lemma there exists a minimal element in them, and this is a facial set containing one point from K_{p_0}. Such a point must be an extreme point of K_{p_0}.

Thus we have proved that every facial subset of K_{p_0} contains an extreme point of K_{p_0}.

Now we consider the convex closed hull $K(\mathscr{E}_{p_0})$ of $\mathscr{E}_{p_0}$. By the Banach–Steinhaus theorem (which is contained in Theorem A) the functional $\max_{p \in K(\mathscr{E}_{p_0})} p(x)$ is continuous (and sublinear). As we have seen, in such a case $K(\mathscr{E}_{p_0}) = \{f \in E^* : f(x) \leqslant \max_{p \in K(\mathscr{E}_{p_0})} p(x), x \in E\}$, so that $K(\mathscr{E}_{p_0})$ is an intersection of closed sets. If we suppose $f_0 \in K_{p_0} \backslash K(\mathscr{E}_{p_0})$, then there exists x_0 for which $\max_{p \in K(\mathscr{E}_{p_0})} p(x_0) < f_0(x_0)$; but this contradicts (7).

Thus we have proved $K_{p_0} = K(\mathscr{E}_{p_0})$, and the proof of Theorem B is ended.

8. Addition and Conclusion

(1) Set $\Gamma^x = \{f \in K_{p_0} : f(x) = \max_{p \in K_{p_0}} p(x)\}$; if $\varnothing \neq \Gamma^x \neq K_{p_0}$, Γ^x is called a face of K_{p_0}. As we have seen, if $-\hat{p}_0(x_1) \neq \hat{p}_0(-x_1)$, then there exists $f_1 \in K_{p_0}$ for which $f_1(x_{x_1}) = \hat{p}_0(x_1)$, and since $\max_{p \in K_{p_0}} p(x) = \hat{p}_0(x)$, $\min_{p \in K_{p_0}} p(x) = -\hat{p}_0(-x)$, we have $f_1 \in \Gamma^{x_1} \neq K_{p_0}$, so that Γ^{x_1} is a face of K_{p_0}. Of course, if $-\hat{p}_0(x_1) = \hat{p}_0(-x_1)$, then Γ^{x_1} is not a face ($\Gamma^{x_1} = K_{p_0}$).

If f_1 from K_{p_0} belongs to a face, we say that the set $M(f_1) = \{x \in E : f_1 \in \Gamma^x\}$ is a support bundle of K_{p_0} in point f_1; this $M(f_1)$ is a wedge in which $\hat{p}_0(x)$ is additive since for $x \in M(f_1)$ $f_1(x) = \hat{p}_0(x)$. Set $\zeta_0 = \{x \in E : \hat{p}_0(-x) = -\hat{p}_0(x)\}$.[7]

[7] Note that ζ_0 is a vector space that equals the intersection of all maximal wedges in which $\hat{p}_0(x)$ is additive. Indeed, if $y_1, y_2 \subset \zeta_0$, then $\hat{p}_0(y_1 + y_2) \leqslant -[\hat{p}_0(-y_1) + \hat{p}_0(-y_2)] \leqslant -\hat{p}_0(-y_1 - y_2) \leqslant \hat{p}_0(y_1 + y_2)$, $\hat{p}_0(-y_1 - y_2) = -\hat{p}_0(y_1 + y_2)$, $y_1 + y_2 \in \zeta_0$.

If $x \in \zeta_1$, $y \in \zeta_0$, and ζ_1 is maximal, then $\hat{p}_0(x) + \hat{p}_0(y) = \hat{p}_0(x) - \hat{p}_0(-y) \leqslant \hat{p}_0(x + y) \leqslant \hat{p}_0(x) + \hat{p}_0(y)$. For $x_1, x_2 \in \zeta_1$, $y_1, y_2 \in \zeta_0$, we have

$$\begin{aligned}\hat{p}_0&[(x_1 + y_1) + (x_2 + y_2)] \\ &= \hat{p}_0(x_1 + x_2) + \hat{p}_0(y_1 + y_2) \\ &= [\hat{p}_0(x_1) + \hat{p}_0(y_1)] + [\hat{p}_0(x_2) + \hat{p}_0(y_2)] \\ &= \hat{p}_0(x_1 + y_1) + \hat{p}_0(x_2 + y_2),\end{aligned}$$

so that p_0 is additive in $\zeta_1 + \zeta_0$. But ζ_1 is maximal and so $\zeta_1 + \zeta_0 = \zeta_1$, $\zeta_0 \subset \zeta_1$. As we have seen in part (3) of Section 6, if $x \in E \backslash \zeta_0$, then there exists a maximal ζ_1 that contains $-x_1$, $x_1 \notin \zeta_1$.

Our considerations in part (3) of Section 6 have shown that if $x_1 \in \zeta_0$, there exists a maximal wedge ζ_1 in which $\hat{p}_0$ is additive and that contains x_1, and there exists $f_1 \in \mathscr{E}_{p_0}$ (in part (3) of Section 6 it was q_γ) for which $f_1(x) = \hat{p}_0(x)$, $x \in \zeta_1$. Set $K_{p_0}(\zeta_1) = \{f \in K_{p_0} : f(x) = \max_{p \in K_{p_0}} p(x), x \in \zeta_1\}$. It is clear that $f_1 \in K_{p_0}(\zeta_1) = \bigcap_{x \in \zeta_1} \Gamma^x = \bigcap_{x \in \zeta_1 \backslash \zeta_0} \Gamma^x \subset \Gamma^{x_1}$, so that $K_{p_0}(\zeta_1)$ is an intersection of faces (but not necessarily a face) that contains f_1 and $\zeta_1 \subset M(f_1)$. But since ζ_1 is a maximal wedge in which $\hat{p}_0$ is additive, we have $M(f_1) = \zeta_1$. For this reason we call $K_{p_0}(\zeta_1)$ an extreme face of K_{p_0}.

Thus, we have proved that for each $x_1 \in \zeta_0$ the set Γ^{x_1} is a face of K_{p_0} and that there exists an extreme face $K_{p_0}(\zeta_1) \subset \Gamma^{x_1}$ containing an extreme point f_1 of K_{p_0}.

Note that for two maximal wedges ζ_1, ζ_2 in which $\hat{p}_0$ is additive either $\zeta_1 = \zeta_2$ and $K_{p_0}(\zeta_1) = K_{p_0}(\zeta_2)$ or $\zeta_1 \neq \zeta_2$ and $K_{p_0}(\zeta_1) \cap K_{p_0}(\zeta_2) = \varnothing$.

These considerations show that in part (3) of Section 6 we obtained extreme points of a special type, i.e., those which belong to extreme faces of K_{p_0}. We could choose there the maximal set ζ_1 arbitrarily (and after this take $x_1 \in \zeta_1 \backslash \zeta_0$). Thus, each extreme face contains an extreme point of K_{p_0}. We can choose from each extreme face one (only one) point and then we get a set $X \subset K_{p_0}$ for which (as we see in part (3) of Section 6) $\operatorname{supp}_{p \in X}(x) = \max_{p \in K_{p_0}} p(x)$, $x \in E$. We have seen that this X can be taken from $\mathscr{E}_{p_0}$.

We have seen that in case (b) (when E is a Fréchet space) it follows that $K_{p_0} = K(X)$, where $K(X)$ is the closed convex hull of X (in the weak*-topology).

This last results, a theorem of Milman and Rutman, is, in essence, an intensification of the Krein–Milman theorem [3].

(2) I think that the range of problems considered here comprises the first stage of functional analysis. I believe that the next (completed) stages may be organized in a similar way, i.e., in such a way that the basic material is contained in one theorem as special cases of the latter.

Milman published [2] a special case of the above theorem, in which it was shown that it implies the three fundamental Banach theorems;[8] in that paper minimal extensions of a *sublinear* functional were considered, whereas here minimal extensions of a *subadditive* functional are also considered. The difference between "sublinear" and "subadditive" becomes clear if we observe that on R_1 the set of all symmetric sublinear functions is one dimensional ($p(t) = \lambda|t|$, $\lambda \geqslant 0$), while the set of all symmetric subadditive functions has infinite dimension (for example, each of the functions $p(t) = |\sin \omega t|$, $\omega > 0$, is symmetric and subadditive).

[8] At the same time, Zabreiko published [5] a theorem implying two of the three fundamental Banach theorems (closed-graph and Banach–Steinhaus).

References

1. D. P. Milman, Separators of nonlinear functionals and their linear extensions, *Izv. Akad. Nauk SSR, Ser. Mat.* **27** (1963), 1189–1210 [in Russian].
2. D. P. Milman, On sublinear continuations of functionals, *Dokl. Akad. Nauk SSSR* **186** (1969), 257–260; English transl. *Soviet Math. Dokl.* **10** (1969), 582–585.
3. D. P. Milman and M. A. Rutman, On a more precise theorem about the completeness of the system of extremal points of a regularly convex set, *Dokl. Akad. Nauk SSSR* **60** (1948), 25–27 [in Russian].
4. R. Sikorski, On a theorem of Mazur and Orlicz, *Studia Math.* **13** (1973), 180–182.
5. P. P. Zabreiko, On a theorem for semi-additive functionals, *Funkcsional. Anal. i Prilozen.* **3** (1969), 86–89 [in Russian].

AMS (MOS) subject classification: 46A30.

Symmetric Singular Integral Operators with Arbitrary Deficiency

JOEL D. PINCUS

Department of Mathematics
State University of New York
Stony Brook, New York

THIS PAPER IS DEDICATED TO M. G. KREIN ON THE OCCASION OF HIS SEVENTIETH BIRTHDAY

INTRODUCTION

The primary goal of this paper is the construction of an explicit solution theory for a class of singular integral equations with coefficients that are badly discontinuous and unbounded.

We are interested in singular integral equations of the form $\mathscr{L}x(\lambda) = lx(\lambda)$ on $L_2(E)$, where E is a real set,

$$\mathscr{L}x(\lambda) = A(\lambda)x(\lambda) + \frac{1}{\pi i}\int_E \frac{\bar{k}(\lambda)k(t)}{t-\lambda}\,x(t)\,dt.$$

We assume only that $A(\lambda)$ is a real-valued function that defines a self-adjoint multiplication operator on $L_2(E)$ and that $k(\lambda) \in L_2(E)$. There is no loss of generality if we take also $k(\lambda) \neq 0$ a.e. on E, and we do this. Small modifications make the present theory applicable also to the case where $|k(\lambda)|^2/1 + \lambda^2$ is in $L_1(E)$.

The present theory applies to symmetric Wiener–Hopf and Toeplitz operators, and we will also describe these results.

From an abstract point of view our procedure may be understood in the following way. First we associate a symmetric operator with the formal singular integral operator $\mathscr{L}$. This is not entirely easy to do because it is not *a priori* obvious how to associate a domain of symmetry in $L_2(E)$ with $\mathscr{L}$. We do this by approximating the symmetric operator we seek by a sequence of bounded self-adjoint operators obtained from $\mathscr{L}$ by truncation in such a way that the truncated sequence of resolvents $(L_n - l)^{-1}$ converges weakly for $\operatorname{Im} l \neq 0$. It is quite easy to see that the standard modes of convergence such as graph convergence or strong resolvent convergence fail in general under our conditions and it is therefore necessary for us to define a new mode of convergence for sequences of self-adjoint (or unitary)

ISBN 0-12-287150-2

operators to symmetric (or partially isometric) limits. This is done in Section I.

Let $T_l = 1 + (l - \bar{l})R(l)$, where $R(l)$ is the weak limit of the sequence $(L_n - l)^{-1}$. The minimal power dilation W_l of T_l satisfies an intertwining relation on the dilation space; namely, $W_l H_2 = (H_2 + \mathscr{D}(l))W_l$, where H_2 and $\mathscr{D}(l)$ are self-adjoint operators defined on the dilation space and $\mathscr{D}(l)$ is in trace class. The wave operator $W_-(l)$ associated with the perturbation problem $H_2 \to H_2 + \mathscr{D}(l)$ also intertwines the perturbed and unperturbed operators, and we show that $W(l)$ is the minimal power dilation of $W_-(l)S_-(H_2 \colon T_l)$. Here $S_-(H_2 \colon T_l)$ is the symbol of T_l with respect to H_2 defined as the strong limit $s\text{-lim}_{t\to-\infty} e^{itH_2}T_l e^{-itH_2}P_{\mathrm{ac}}(H_2)$. The defect spaces of the contraction T_l may now be characterized: $1 - T_l{}^*T_l = PP_{\mathrm{s}}(l)P$ and $1 - T_lT_l{}^* = PP_{\mathrm{s}}(\bar{l})P$ where $P_{\mathrm{s}}(l)$ is the projection onto the singular space of $H_2 + \mathscr{D}(l)$ and P is the projection on the base space.

The geometry associated with these relations is completely characterized by the principal function of $W(l)$ and H_2; this is a function $G_l(\lambda, e^{i\theta})$ defined on the cartesian product of the line and unit circle, which we exhibit explicitly. In turn this function is obtained by a change of variables from the principal function $g(\nu, \mu)$ of $H_1 = lW(l) - \bar{l}/W(l) - 1$ and H_2. We regard this function as the main invariant. Indeed the major point of this paper is that T_l can be regarded as being a function of the two self-adjoint operators H_1 and H_2, namely, $T_l = P_{\mathrm{ac}}(H_2)H_1 - \bar{l}/H_1 - l$. The index of T_l can be obtained from this equation because *the principal function of* T_l (which is equal to the index of T_l on components of the complement of the essential spectrum of T_l) can be obtained from $g(\nu, \mu)$ by a simple mapping. If $\chi_{\mathrm{ac}}(\lambda)$ denotes the characteristic function of the absolutely continuous spectrum of H_2, then the essential discontinuity set of $g(\nu, \mu)$ maps under $\chi_{\mathrm{ac}}(\lambda)\nu - \bar{l}/\nu - l$ to the unit circle (which is the essential spectrum of T_l) and the index of T_l is simply "the number of times" that this image set "surrounds the origin." The meaning of these phrases will be explained later. The appropriate definitions are determined by the way T_l is embedded in the algebra generated by H_1 and H_2.

The solutions of the singular integral equation $\mathscr{L}x(\lambda) = lx(\lambda)$ turn out simply to be the projections back to the original Hilbert space of the vectors in the singular space of $H_2 + \mathscr{D}(l)$.

Previous work on "determining functions" and their associated "principal functions" (see [1–5]) makes it possible to construct these singular spaces explicitly. The contraction T_l is a quasi-orthogonal extension of the Cayley transform of the symmetric operator of interest, and by including the results on symmetric Toeplitz operators at the end of the paper we indicate that the class of symmetric operators for which our description has validity is not only the singular integral operators $\mathscr{L}$. Indeed, the abstract structure illustrated here includes a great many operators, and we will discuss elsewhere the application of our methods to differential operators arising in potential scattering problems. Similarly, we postpone to another place a discussion of the relation between the present results and certain invariant subspace results associated with canonical models.

Here our main goal is to show that we are dealing with a new method for solving singular integral equations, and our exposition will be restricted to this end. Thus we concentrate on the relationship between the classical Riemann–Hilbert problem associated with the equation $\mathscr{L}x(\lambda) = \lambda x(\lambda)$ and the underlying geometry which we have found.

The Riemann–Hilbert problem of which we speak has from the very beginning of the theory of singular integral equations been used as the major tool for actually finding solutions. These factorization problems have also played the major role in the associated index theory of singular integral operators. However, it is always assumed that the coefficients in the singular integral equation are either continuous or else that they have a few simple discontinuities. In this case it is not difficult to construct solutions; indeed, there are several ways in which one can proceed and the associated index theory (approached simply as a homotopy theory) is not difficult. The situation which we study is the case in which the "localization" results, due originally to Sinonenko [6], and since developed by several other authors, e.g., Douglas [7], do not apply. Thus the Riemann–Hilbert barrier function $S(l, \lambda)$ defined in equation (1.5) below *does not* satisfy conditions of "local sectorality" which make it possible to assign to it an index or winding number about the origin in a simple way.

Our treatment of the Riemann–Hilbert problem associated with $S(l, \lambda)$ *is not classical.* We *do not* seek a sectionally holomorphic function $\Omega_l(\omega)$ such that

$$S(l, \lambda) = \Omega_l(\lambda + i0)^{-1}(\lambda + i/\lambda - i)^{\text{index}\, S(l,\lambda)}\Omega_l(\lambda - i0),$$

where index $S(l, \lambda)$ is the "number of times" the curve traced out by $S(l, \lambda)$ wraps about the origin as λ traverses the real axis. Rather, we make a more *intrinsic* factorization: $S(l, \lambda) = \varphi_l(\lambda + i0, 0)^{-1}\varphi_l(\lambda - i0, 0)$ in terms of a function of two complex variables

$$\varphi_l(\omega, z) = 1 + (1/\pi)d_l{}^* W(l)(W(l) - z)^{-1}(H_2 - \omega)^{-1}\, d_l$$

with $\mathscr{D}(l) = d_l \otimes d_l$. We consider the classical method (which seems to have been universally followed since the days of Plemelj–Carleman) to be quite misleading; the essential geometry is masked by a procedure which works in smooth cases but is not generally defined. This geometry is present also in the smooth cases; it is not introduced by any pathology in the coefficients, and we believe that its occurrence is fundamental to the study of Riemann–Hilbert and associated index problems.

I. Riemann–Hilbert Problems

We begin with a formal derivation. Suppose $\mathscr{L}x = \bar{l}x$, $x \in L_2$. That is,

$$(A(\lambda) - \bar{l})x(\lambda) + \frac{1}{\pi i} \int \frac{\bar{k}(\lambda)k(\mu)}{\mu - \lambda} x(\mu)\, d\mu = 0. \tag{1.1}$$

Let

$$\phi(l, z) = \frac{1}{2\pi i} \int \frac{k(\mu)x(\mu)}{\mu - z} d\mu \qquad \text{for} \quad \operatorname{Im} z \neq 0.$$

Under a variety of conditions we have the Plemelj relations:

$$\phi(l, \lambda + i0) - \phi(l, \lambda) - i0) = k(\lambda)x(\lambda),$$

$$\phi(l, \lambda + i0) + \phi(l, \lambda) - i0) = \frac{1}{\pi i} \int \frac{k(\mu)x(\mu)}{\mu - \lambda} d\mu.$$

On the assumption that these relations are valid we could derive

$$\frac{A(\lambda) - \bar{l} - |k(\lambda)|^2}{A(\lambda) - \bar{l} + |k(\lambda)|^2} \phi(l, \lambda - i0) = \phi(l, \lambda + i0). \tag{1.2}$$

We would conclude that $\phi(l, z)$ is a sectionally holomorphic function of z which satisfies the barrier relation (1.2).

It is not at all difficult to find one solution of the barrier relation. Let $g(v, \lambda)$ be the characteristic function of the set of pairs $\{(v, \lambda) : A(\lambda) - |k(\lambda)|^2 < v < A(\lambda) + |k(\lambda)|^2\}$. Define

$$E(l, z) = \exp \frac{1}{2\pi i} \iint g(v, \mu) \frac{dv}{v - l} \frac{d\mu}{\mu - z}. \tag{1.3}$$

Since

$$\log \frac{A(\lambda) - l + |k(\lambda)|^2}{A(\lambda) + l - |k(\lambda)|^2} = \int g(v, \lambda) \frac{dv}{v - l},$$

the Plemelj relations show that $E(l, z)^{-1}$ is a sectionally holomorphic solution of the barrier relation. However, $[E(\bar{l}, \lambda + i0)^{-1} - E(\bar{l}, \lambda - i0)^{-1}]/k(\lambda)$ is not a solution of (1.1). To see that, we note that $\int |k(\lambda)|^2\, d\lambda < \infty$ implies that $\iint g(v, \lambda)\, dv\, d\lambda < \infty$. This in turn implies that $\lim_{z \to \infty} E(l, z) = 1$. Yet $\phi(l, z) \to 0$ as $z \to \infty$.

A more delicate analysis is necessary. Suppose that $E^{-1}(\bar{l}, z)$ has zeros in z. Call these zeros $\lambda_j(l)$. Then the functions $\phi_j(l, z) = E^{-1}(\bar{l}, z)/[z - \lambda_j(l)]$ will satisfy (1.2) and it is at least conceivable that

$$w_j(l, \lambda) = \frac{\phi_j(l, \lambda + i0) - \phi_j(l, \lambda - i0)}{k(\lambda)} \tag{1.4}$$

may be a square integrable solution of (1.1). To make sense of this formal method for finding solutions we have to be able to study the z zeros of $E^{-1}(\bar{l}, z)$—and (this is the difficulty) obtain sharp enough estimates to prove the square integrability of our putative solution. Standard estimates of the Hilbert transform of nonsmooth unbounded functions (even quite refined ones) are ill adapted for this purpose, and we are forced to proceed in a more indirect way.

We will form a new barrier. Since

$$\frac{A(\lambda) - l + |k(\lambda)|^2}{A(\lambda) - l - |k(\lambda)|^2} E(l, \lambda - i0) = E(l, \lambda + i0)$$

and

$$\frac{A(\lambda) - \bar{l} + |k(\lambda)|^2}{A(\lambda) - \bar{l} - |k(\lambda)|^2} E(\bar{l}, \lambda - i0) = E(\bar{l}, \lambda + i0),$$

we have

$$\begin{aligned} &\frac{A(\lambda) - l + |k(\lambda)|^2}{A(\lambda) - \bar{l} + |k(\lambda)|^2} \frac{A(\lambda) - \bar{l} - |k(\lambda)|^2}{A(\lambda) - l - |k(\lambda)|^2} E(l, \lambda - i0) E^{-1}(\bar{l}, \lambda - i0) \\ &\quad = E(l, \lambda + i0) E^{-1}(\bar{l}, \lambda + i0). \end{aligned}$$

The new barrier

$$S(l, \lambda) = \frac{A(\lambda) - l + |k(\lambda)|^2}{A(\lambda) - \bar{l} + |k(\lambda)|^2} \frac{A(\lambda) - \bar{l} - |k(\lambda)|^2}{A(\lambda) - l - |k(\lambda)|^2} \tag{1.5}$$

is unimodular, and we will see that it is the *scattering operator* corresponding to a self-adjoint perturbation problem $H_2 \to H_2 + \mathscr{D}(l)$ where H_2 is a self-adjoint extension to a bigger Hilbert space of multiplication by λ (i.e., M) and $\mathscr{D}(l)$ is an operator with one-dimensional range on the extension Hilbert space.

In fact, we will show that $E(l, z)E^{-1}(\bar{l}, z) = \det(1 + (H_2 - z)^{-1}\mathscr{D}(l))$ and that the zeros of $E^{-1}(l, z)$ correspond to points in the singular spectrum of $H_2 + \mathscr{D}(l)$.

It will even turn out that the projection back to L_2 of the eigenvectors of $H_2 + \mathscr{D}(l)$ (which lie in the extension Hilbert space) gives us the solutions we seek of $\mathscr{L}x = \bar{l}x$.

We note here that index $S(l, \lambda)$ is in general not zero although it is formed by taking a quotient of two barriers each of which has zero index (both barriers have positive imaginary parts in the upper half-plane). The preceding facts illustrate the sense in which the classical $(\lambda + i/\lambda - i)^{\text{index } S(l,\lambda)}$ term is unnatural and ad hoc. Of course, because $S(l, \lambda)$ is not locally sectorial, the classical method fails in our case anyway since index $S(l, \lambda)$ is not defined by the usual procedures. However, we do give a meaning later in this paper to index $S(l, \lambda)$ as a kind of winding number and our comments have force with respect to this extended meaning.

Construction of the Symmetric Operator and Statement of Results about Deficiency Spaces

We approach the problem of associating a symmetric operator with the formal operator $\mathscr{L}$ in the following way:

Let

$$A_n(t) = \begin{cases} A(t) & \text{when } \quad |A(t)| \leqslant n \\ 0 & \text{when } \quad |A(t)| > n, \end{cases}$$

$$k_n(t) = \begin{cases} k(t) & \text{when } \quad |k(t)| \leqslant n \\ 0 & \text{when } \quad |k(t)| > n. \end{cases}$$

Let

$$L_n x(\lambda) = A_n(\lambda)x(\lambda) + \frac{1}{\pi i}\int_E \frac{\overline{k}_n(\lambda)k_n(t)}{t - \lambda}\, x(t)\, dt.$$

The sequence of resolvents $(L_n - l)^{-1}$ converges weakly to an operator $R(l)$ (for the proof, see [2, Theorem 1]). We will show that $1 + (l - \overline{l})R(l)$ is a contraction operator in $L_2(E)$ which is the extension of an isometric operator V_l in $L_2(E)$, and that V_l is the Cayley transform of a symmetric operator L which is obtained by restricting the formal singular integral operator $\mathscr{L}$ to a domain which we describe.

Our analysis is given entirely in terms of the so-called principal function of the pair of operators H_1 and H_2. This is the function $g(t, x)$ introduced above as the characteristic function of the set of points (t, x) for which $A(t) - k(t)|^2 < x < A(t) + |k(t)|^2$. In terms of the principal function we define

$$h_l(t) = \begin{cases} \dfrac{\operatorname{Im} l}{\pi} \displaystyle\int \frac{g(x, t)}{|x - l|^2}\, dx & \text{if } \quad \operatorname{Im} l > 0 \\[2ex] \dfrac{|\operatorname{Im} l|}{\pi} \displaystyle\int \frac{1 - g(x, t)}{|x - l|^2}\, dx & \text{if } \quad \operatorname{Im} l < 0. \end{cases}$$

It follows that $0 \leqslant h_l(t) \leqslant 1$, and we have

$$h_i(t) = \frac{1}{\pi}\tan^{-1}(A(t) + |k(t)|^2) - \frac{1}{\pi}\tan^{-1}(A(t) - |k(t)|^2).$$

There exist positive Borel measures on the real line μ_l so that

$$\exp \int \frac{h_l(t)}{t - z}\,dt = 1 + \frac{1}{\pi}\int \frac{d\mu_l(t)}{t - z}.$$

Let $\mu_{i,s}$ and $\mu_{-i,s}$ be the singular part of μ_i and μ_{-i} entering into the Lebesgue decomposition of these measures.

Let $n =$ dimension $L_2(\mu_{i,s})$, $m =$ dimension $L_2(\mu_{-i,s})$.

THEOREM. *The deficiency indices of the symmetric operator L described above are (m, n). If the measures above are not atomic then the corresponding deficiency indices are infinite.*

Let

$$E(l, z) = \exp\frac{1}{2\pi i}\iint g(v, \lambda)\frac{dv}{v - l}\frac{d\lambda}{\lambda - z}$$

and define $H(l, \lambda) = E(l, \lambda + i0) - E(l, \lambda - i0)$. Consider the case when the two singular measures defined above are purely atomic. Suppose that the atoms of $\mu_{-i,s}$ are $\{\lambda_1^-, \lambda_2^-, \ldots, \lambda_m^-\}$ while those of $\mu_{i,s}$ are $\{\lambda_1^+, \lambda_2^+, \ldots, \lambda_n^+\}$.

THEOREM. *The functions*

$$w_j^-(\lambda) = \frac{\bar{H}(i, \lambda)}{k(\lambda)}\frac{1}{\lambda - \lambda_j^-}, \qquad j = 1, 2, \ldots, m,$$

$$w_r^+(\lambda) = \frac{\bar{H}(-i, \lambda)}{k(\lambda)}\frac{1}{\lambda - \lambda_r^+}, \qquad r = 1, 2, \ldots, n,$$

are in $L_2(E)$. They satisfy the formal eigenvalue equations $(\mathscr{L} + i)w_j^-(\lambda) = 0$ and $(\mathscr{L} - i)w_r^+(\lambda) = 0$ and span the deficiency spaces of the operator L.

When (m, n) are finite the operator $1 + 2iR(i)$ is the extension of a closed isometric operator V_i acting in $L_2(E)$ and $1 + 2iR(i) = V_i \oplus U_i$, where U_i is a linear contraction operator with domain of definition in $L_2(E) \ominus D_{V_i}$ and range in $L_2(E) \ominus VD_{V_i}$. The operator U_i is constructed explicity in Theorem 4.3 below.

The proper context for all these results is the study of two unitary operators satisfying a commutator identity. The present results are simply a study of certain compressions of such operators. For transformation properties and index results in associated algebras see [4].

We now turn to the task of defining the "limit" of a sequence of self-adjoint operators.

Let L_n be a sequence of self-adjoint operators defined on a Hilbert space H. We assume:

(1) $R(l) = \lim (L_n - l)^{-1}$ exists in the weak operator topology for every nonreal number l, and

(2) $-\lim_{\lambda \to \infty} i\lambda R(i\lambda) = 1$ in the weak operator topology.

These two conditions are enough to establish the existence of a generalized resolution of the identity F_λ such that $R(l) = \int dF_\lambda/(\lambda - l)$. By the Naimark dilation theorem there is an essentially unique orthogonal resolution of the identity E_λ on a Hilbert space $\bar{H} \supset H$ such that $PE_\lambda = F_\lambda$, where P is the projection from $\bar{H}$ to H. Now define the self-adjoint operator $\hat{H}_1$ in $\bar{H}$ by setting $\hat{H}_1 = \int \lambda \, dE_\lambda$. Then $P(\hat{H}_1 - l)^{-1} = R(l)$.

Let $N_l = P(\hat{H}_1 - \bar{l})^{-1}(\bar{H}_1 - l)^{-1}P - P(\hat{H}_1 - \bar{l})^{-1}P(\hat{H}_1 - l)^{-1}P$. For arbitrary f and g in H we have $((\hat{H}_1 - l)^{-1}f, (\hat{H}_1 - l)^{-1}g) = (R(l)f, R(l)g) + (N_1 f, g)$.

Thus, if f is in the null space of N_l we will have $\|H_1 - l)^{-1}f\| = \|R(l)f\|$ and $(1 - P)(\hat{H}_1 - l)^{-1}f = 0$. Now let $T_l = 1 + (l - \bar{l})R(l)$. Since $(\hat{H}_1 - l)^{-1}f = R(l)f$, we can conclude that $T_l f = (\hat{H}_1 - \bar{l})(\hat{H}_1 - l)^{-1}f$. But $(\hat{H}_1 - \bar{l})(\hat{H}_1 - l)^1 = W_l$ is unitary, hence $\|T_l f\| = \|f\|$. This shows that T_l restricted to the orthogonal complement of the range of N_l is an isometry. Call this restriction V_l. If we now assume that

(3) the set of all $V_l y - y$, y in the null space of N_l, is dense in H, then, by the classical results of von Neumann, $L = (l1 - \bar{l}V_l)(1 - V_l)^{-1}$ is a symmetric operator defined on the dense range of $1 - V_l$. Furthermore, neither the range of $1 - V_l$ nor the symmetric operator L is l dependent. We show first the range of $1 - V_l$ is not l dependent.

Suppose that f_1 is in the null space of N_{l_1}. Then $(\hat{H}_1 - l_1)^{-1}f_1$ is in the domain of H_1 and we can define $f_2 = (\hat{H}_1 - l_2)(\hat{H}_1 - l_1)^{-1}f_1$. We claim that f_2 is in the null space of N_{l_2}. To see this let us note that $(\hat{H}_1 - l_2)^{-1}f_2 = (\hat{H}_1 - l_1)^{-1}f_1$. But $(\hat{H}_1 - l_1)^{-1}f_1 = P(\hat{H}_1 - l_1)^{-1}f_1$, hence $P(\hat{H}_1 - l_2)^{-1}f_2 = (\hat{H}_1 - l_2)^{-1}f_2$, and for any g in H $((\hat{H}_1 - l_2)^{-1}f_2, (\hat{H}_1 - l_2)^{-1}g) = (R(l_2)f_2, R(l_2)g)$. Thus $(N_{l_2}f_2, g) = 0$ for all g in H, and we conclude that $N_{l_2}f_2 = 0$. It follows that any vector $R(l_1)f_1$ in the range of $V_{l_1} - 1$ is also in the range of $V_{l_2} - 1$.

We show that L is not l dependent. This is a consequence of the fact that $(\hat{H}_1 - \bar{l})(\hat{H}_1 - l)^{-1}$ is an extension of V_l. Thus, suppose that $g = (1 - V_{l_1})f_1 = (1 - V_{l_2})f_2$; then $\hat{H}_1 g = (l_1 - \bar{l}_1 V_{l_1})(1 - V_{l_1})g = (l_2 - \bar{l}_2 V_{l_2})(1 - V_{l_2})g$ because

$(l_1 - \bar{l}_1 V_{l_1})(1 - V_{l_1})g = (l_1 - \bar{l}_1 V_{l_1})f_1 = (l_1 - \bar{l}_1 V_{l_1})f_1 = (l_1 - \bar{l}_1 W_{l_1})f_1 = \hat{H}_1(1 - W_{l_1})f_1 = \hat{H}_1(1 - V_{l_1})f_1 = \hat{H}_1 g$. This shows that L is well defined. Of course if we do not make assumption (3) then L is a nondensely defined symmetric operator. (In our application below L will be densely defined. But there are situations in which it is necessary to drop this assumption). We call L the "limit" of the sequence $\{L_n\}$. Note that L does not depend upon $\{L_n\}$ except through F_λ.

Since $PW_lP = T_l$, a calculation shows that $N_l = -(l - \bar{l})^{-2}(1 - T_l{}^*T_l)$. We know that $T_l(1 - T_l{}^*T_l)^{1/2} = (1 - T_{\bar{l}}{}^*T_{\bar{l}})^{1/2}T_l$ and that $(1 - T_l{}^*T_l)^{1/2}T_l{}^* = T_l{}^*(1 - T_l{}^*T_l)^{1/2}$. We can thus deduce that T_l maps the range of N_l into the range of $N_{\bar{l}}$. Theorem 4.3 below gives more precise information—which is needed for a spectral analysis of the limiting operator.

The symmetric singular integral operators which we introduce in this paper are interesting perhaps because they can have arbitrary deficiency indices and because all of the calculations of characteristic operator functions and associated considerations can be explicitly carried through in terms of the principal function alone (see [2]).

It will be convenient for us to work with another representation of the operator $\hat{H}_1$. To this end we introduce another self-adjoint operator H_1; it will be seen later that $\hat{H}_1$ is unitarily equivalent to H_1.

In the following pages we make use of the easily verified identity $\bar{E}(\bar{l}, \bar{z}) = E^{-1}(l, z)$. We also summarize certain facts previously established by the author which are crucial for our present purpose.

The author has shown (see [4]) that there exists a Hilbert space $\mathscr{H}$ and self-adjoint operators H_1 and H_2 on $\mathscr{H}$ so that

$$E(l, z) = 1 + (1/\pi i)k(H_1 - l)^{-1}(H_2 - z)^{-1}K^*,$$

where k^* is a map of the complex numbers into a one-dimensional subspace of $\mathscr{H}$ and k is the adjoint of k^* which maps $\mathscr{H}$ back into the complex numbers. Furthermore, it is known that

$$(H_1 - l)^{-1}(H_2 - z)^{-1} - (H_2 - z)^{-1}(H_1 - l)^{-1}$$
$$= (H_2 - z)^{-1}(H_1 - l)^{-1}\frac{k^*k}{\pi i}(H_1 - l)^{-1}(H_2 - z)^{-1} \qquad (1.6)$$

and the pair (H_1, H_2) is determined up to unitary equivalence by $g(v, \lambda)$, the principal function of (H_1, H_2).

The following theorem is also known (see [9]):

THEOREM. *The absolutely continuous part of H_2 is the smallest invariant subspace of H_2 which contains the range of k^*. Call this space H. There is a unitary map of H to $L_2(E)$ which takes H_2/H to M, carries k^* into the* *multiplication operator* $\overline{k}(\lambda)$, *and carries* k *to the integral operator* $\int k(\lambda)\cdot d\lambda$. *Furthermore*, $P(H_1 - l)^{-1} = R(l)$, *where* P *is the projection to* H.

This theorem identifies the multiplication operator M as the absolutely continuous part of the abstractly constructed operator H_2.

The resolvant commutator identity (1.6) above implies immediately that

$$E(l, z)\overline{E}(l, \overline{z}) = 1 + \frac{1}{\pi}\frac{(l - \overline{l})}{i} k(H_1 - l)^{-1}(H_2 - z)^{-1}(H_1 - \overline{l})^{-1}k^*. \quad (1.7)$$

Thus, if we set $d_l = [(l - \overline{l})/i]^{1/2}(H_1 - \overline{l})^{-1}k^*$, and think of k^* as a vector in $\mathscr{H}$, we will have $E(l, z)\overline{E}(l, \overline{z}) = 1 + (1/\pi)((H_2 - z)^{-1} d_l, d_l)$.

With the notation of the Introduction, we see that

$$E(i, z)\overline{E}(i, \overline{z}) = \exp \int h_i(t)\frac{dt}{t - z} = 1 + \frac{1}{\pi}\int \frac{d\mu_i(t)}{t - z},$$

where $\mu_i(\Delta)$ is the spectral measure $\|E^2(\Delta)\, d_i\|^2$ associated with the resolution of the identity $E^2(\Delta)$ of H_2.

We remind the reader of the definition of a perturbation determinant, and some basic facts about trace class perturbation theory.

If A and B are self-adjoint operators with $B - A$ in trace class, then $D(z) = \det(1 + (B - A)(A - z)^{-1})$ exists and is called the perturbation determinant corresponding to the perturbation $A \to B$.

It is known that the wave operators $\Omega_{\pm} = s\text{-}\lim_{t \to \pm\infty} e^{itB}e^{-itA}P_{\mathrm{ac}}(A)$ and that $S = \Omega_+{}^*\Omega_-$ is a partial isometry having $H_{\mathrm{ac}}(A)$ as both initial and final space. Further, the restriction of S to $H_{\mathrm{ac}}(A)$ is unitary and commutes with A. $S = \int \oplus S(\lambda)\, d\lambda$ on the direct integral space $= \int \oplus H_\lambda\, d\lambda$, which diagonalizes A_{ac}, where $S(\lambda) = 1 + 0(\lambda)$ and $0(\lambda)$ is trace class in H_λ. This last fact implies that det $S(\lambda)$ exists. It is known that

$$\frac{D(\lambda - i0)}{D(\lambda + i0)} = \det S(\lambda) = e^{-2\pi i\delta(\lambda)}.$$

where $\delta(\lambda)$ is a real-valued summable function, the so-called spectral displacement function.

The main property of the spectral displacement function is expressed by a trace relation, $\mathrm{Tr}[f(B) - f(A)] = \int f'(t)\delta(t)\, dt$, for a certain class of functions $f(t)$. This relation is the extension by M. G. Krein of results of the physicist I. M. Lifshitz. (A generalization to the case of von Neumann algebras is now also known and there are certain two-variable analogs as well. See [4]).

For our purpose $A = H_2$, and $B = H_2 + d_l \otimes d_l$. Since the perturbation here has one-dimensional range the scattering operator is simply a scalar multiplication operator, and (1.7) implies that it is the function $S(l, \lambda)$ introduced above.

It is easy to show that

$$(1 + (1/\pi)(H_2 - z)^{-1} d_i, d_i))^{-1} = 1 - (1/\pi)((H_2 + d_i \otimes d_i - z)^{-1} d_i, d_i).$$

Thus, if we define μ_{-i} as the unique positive Borel measure such that

$$1 - \frac{1}{\pi} \int \frac{d\mu_{-i}(t)}{t - z} = \exp - \int h_i(t) \frac{dt}{t - z},$$

it follows at once that

$$E(-i, z)\bar{E}(-i, \bar{z}) = 1 - \frac{1}{\pi} \int \frac{d\mu_{-i}(t)}{t - z},$$

and

$$\int \frac{d\mu_{-i}(t)}{t - z} = ((H_2 + d_i \otimes d_i - z)^{-1} d_i, d_i).$$

We wish to compute the Radon–Nikodym derivative of μ_i with respect to Lebesgue measure.

For this purpose we note that

$$\lim_{YO} \frac{y}{\pi} \int \frac{d\mu_i(t)}{(t - x)^2 + y^2} = \frac{d}{dt} \mu_i(t)\big|_{t=x} \qquad \text{a.e.}$$

However, a simple calculation shows that

$$\frac{E(l, t + i0)\bar{E}(l, t - i0) - E(l, t - i0)\bar{E}(l, t + i0)}{l - \bar{l}} = \begin{cases} \frac{1}{2}|H(l, t)/k(t)|^2, & t \in E, \\ 0, & t \notin E, \end{cases}$$

where E is the support of $k(t)$.

We can conclude then that

$$\frac{i}{z - \bar{z}} \operatorname{Im} E(i, z)\bar{E}(i, \bar{z}) = \frac{1}{2} \int \frac{d\mu_i(t)}{|t - z|^2}$$

$$= \frac{1}{4} \int \left| \frac{H(i, t)}{k(t)} \right|^2 \frac{dt}{|t - z|^2} + \sum \frac{\mu_i(j)}{|\lambda_j{}^+ - z|^2},$$

where $\mu_i(j)$ is the mass of μ_i concentrated at the point $\lambda_j{}^+$. We take $j = 1, 2, \ldots, n$.

Define

$$F(w) = E(i, w)\bar{E}(i, \bar{w}).$$

Then, as we have already seen,

$$\frac{1}{F(w)} = 1 - \frac{1}{\pi} \int \frac{d\mu_{-i}(t)}{t - w}.$$

But,

$$\operatorname{Im}\left[-\frac{1}{F(w)}\right]=\frac{1}{2i}\left[\frac{1}{F(\bar{w})}-\frac{1}{F(w)}\right]=\frac{\operatorname{Im}F(w)}{|F(w)|^2}.$$

However, for x real,

$$\operatorname{Im}\left[-\frac{1}{F(x+i\epsilon)}\right]=\int\frac{2\epsilon\,d\mu_{-i}(t)}{(t-x)^2+\epsilon^2}\geqslant\int_{x-\epsilon}^{x+\epsilon}\frac{2\epsilon}{(t-x)^2+\epsilon^2}\,d\mu_i(t)$$

$$\geqslant\int_{x-\epsilon}^{x+\epsilon}\frac{d\mu_{-i}(t)}{\epsilon}=\frac{\delta_{\mu_{-i}}(x+\epsilon)-\delta_{\mu_{-i}}(x-\epsilon)}{\epsilon},$$

where $\delta_{-i}(t)=\mu_{-i}(-\infty,t)$.

Thus,

$$\operatorname{Im}\left[\frac{-\epsilon}{F(x+i\epsilon)}\right]\geqslant\delta_{\mu_{-i}}(x+\epsilon)-\delta_{\mu_{-i}}(x-\epsilon)$$

and so

$$\frac{1}{\operatorname{Im}\left[\dfrac{-\epsilon}{F(x+i\epsilon)}\right]}\leqslant\frac{1}{\delta_{\mu_{-i}}(x+\epsilon)-\delta_{\mu_{-i}}(x-\epsilon)}.$$

But, setting $w=x+i\epsilon$, we get

$$\frac{1}{4\pi}\int\left|\frac{H(i,t)}{k(t)}\right|^2\frac{dt}{|t-(x+i\epsilon)|^2}+\frac{1}{2\pi}\sum_j\frac{\mu_i(j)}{|\lambda_j^+-(x+i\epsilon)|^2}\leqslant\frac{1}{\delta_{\mu_{-i}}(x+\epsilon)-\delta_{\mu_{-i}}(x-\epsilon)}.$$

Let the set of atoms of μ_{-i} be denoted by $\{\lambda_j^-\}$, $j=1,2,\ldots,m$. It is clear that the sets $\{\lambda_j^-\}$ and $\{\lambda_j^+\}$ are disjoint. Indeed, the above inequality gives us

$$\frac{1}{4\pi}\left|\frac{H(i,t)}{k(t)}\right|^2\frac{dt}{(t-\lambda_r^-)+\epsilon^2}+\frac{1}{2\pi}\sum_j\frac{\mu_i(j)}{(\lambda_j^+-\lambda_r^-)^2+\epsilon^2}$$

$$\leqslant\frac{1}{\delta_{\mu_{-i}}(\lambda_r^-+\epsilon)-\delta_{\mu_{-i}}(\lambda_r^--\epsilon)}.$$

The right-hand side remains finite as ϵ approaches zero.

Thus, since the sequence of measurable functions

$$\left\{\left|\frac{H(i,t)}{k(t)}\right|^2\frac{1}{(t-\lambda_r^-)^2+\epsilon_n^2}\right\}\quad\text{converges to}\quad\left|\frac{H(i,t)}{k(t)}\right|^2\frac{1}{(t-\lambda_r^-)^2}$$

almost everywhere dt, the Fatou theorem tells us that

$$\int \left| \frac{H(i, t)}{k(t)} \right|^2 \frac{dt}{(t - \lambda_r^-)^2} < \infty, \qquad r = 1, 2, \ldots, m.$$

It is quite important for us that somewhat more is true. Our sequence is positive and converges in measure to an L_1 dominant. Using in turn the Lebesgue dominated convergence theorem, a theorem of Fichtenholz, and the Vitali convergence theorem, we can conclude that the functions

$$\left\{ \left| \frac{H(i, t)}{k(t)} \right|^2 \frac{1}{(t - \lambda_r^-)^2 + \epsilon^2} \right\}$$

are uniformly integrable.

We can conclude therefore, that for any continuous function $f(t)$,

$$\lim_{n \to \infty} \int \left| \frac{H(i, t)}{k(t)} \right|^2 \frac{1}{(t - \lambda_r^-)^2 + \epsilon_n^2} f(t)\, dt = \int \left| \frac{H(i, t)}{k(t)} \right|^2 \frac{1}{(t - \lambda_r^-)^2} f(t)\, dt$$

We can repeat the foregoing reasoning with only small variations to show that the functions

$$\left\{ \left| \frac{H(-i, t)}{k(t)} \right|^2 \frac{1}{(t - \lambda_s^+)^2 + \epsilon^2} \right\}$$

also are uniformly integrable.

II

We will show now that the L_2 functions $w_j^-(\lambda)$ and $w_r^+(\lambda)$ defined by

$$w_j^-(\lambda) = \frac{\bar{H}(i, \lambda)}{k(\lambda)} \frac{1}{(\lambda - \lambda_j^-)}, \qquad j = 1, 2, \ldots, m,$$

and

$$w_r^+(\lambda) = \frac{\bar{H}(-i, \lambda)}{k(\lambda)} \frac{1}{(\lambda - \lambda_r^+)} \qquad r = 1, 2, \ldots, n,$$

satisfy the formal eigenvalue equations

$$\begin{aligned} (\mathscr{L} + i) w_j^-(\lambda) &= 0, \qquad j = 1, 2, \ldots, m, \\ (\mathscr{L} - i) w_j^+(\lambda) &= 0, \qquad r = 1, 2, \ldots, n. \end{aligned} \tag{2.1}$$

Of course the corresponding statement for arbitrary values of l is true as well.

LEMMA 2.1. *If $A(\lambda)$ is real and measurable finite a.e. and $k(\lambda) \in L_2(E)$, then* $\lim_{y \to 0} E(i, \lambda_j^- + iy) = 0$.

Let

$$\theta_j(z) = \frac{1}{2\pi i} \int \frac{k(\lambda) w_j^-(\lambda)}{\lambda - z} \, d\lambda.$$

We also have the following result:

LEMMA 2.2. $\theta_j(z) = -(1/z - \lambda_j^-) \, \bar{E}(i, \bar{z})$. $\operatorname{Im} z \neq 0$.

Proof of Lemma 2.1. The following inequality is basic: Let $\bar{\lambda}$ be one of the λ_j^-. Then

$$\int_{\lambda < \bar{\lambda}} 1 - h_i(\lambda) \frac{d\lambda}{\bar{\lambda} - \lambda} + \int_{\lambda > \bar{\lambda}} h_i(\lambda) \frac{d\lambda}{\lambda - \bar{\lambda}} < \infty$$

(see [10]). Two conclusions follow from this inequality:

(1) $A(\lambda) + |k(\lambda)|^2$ has left Lebesgue value $+\infty$ at $\bar{\lambda}$ with density one, i.e., for any $\delta > 0$,

$$\lim_{\epsilon \to 0} \frac{\text{measure}[\lambda\colon A(\lambda) + |k(\lambda)|^2 \geqslant \delta \cap \bar{\lambda} - \epsilon < \lambda < \bar{\lambda}]}{\epsilon} = 1;$$

(2) $A(\lambda) - |k(\lambda)|^2$ has left Lebesgue value $-\infty$ at $\bar{\lambda}$ with density one.

To prove these facts we simply note that

$$\int_{\lambda < \bar{\lambda}} \frac{1}{2} \mp \frac{1}{\pi} (\arctan A(\lambda) \pm |k(\lambda)|^2) \frac{d\lambda}{\bar{\lambda} - \lambda} < +\infty,$$

where we have taken $\tan^{-1}(\infty) = \frac{1}{2}\pi$, $\tan^{-1}(-\infty) = -\frac{1}{2}\pi$. We also note

(3) $|k(\lambda)|^2$ has left (right) Lebesgue value $+\infty$ (0).

The left value assertion comes from (1) by subtraction; the right value assertion comes from

$$\int_{\lambda > \lambda} \tan^{-1}(A(\lambda) + |k(\lambda)|^2) - \tan^{-1}(A(\lambda) - |k(\lambda)|^2) \frac{d\lambda}{\lambda - \bar{\lambda}} < \infty.$$

Finally, we will need to observe that

(4) the set of Lebesgue values of $A(\lambda)/|k(\lambda)|^2$ at $\bar{\lambda}$ is contained in $[-1, 1]$.

For if $A(\lambda)/|k(\lambda)|^2$ had a Lebesgue value greater than one at $\bar{\lambda}$, then $|k(\lambda)|^2[A(\lambda)/|k(\lambda)|^2 - 1]$ would have a Lebesgue value of $+\infty$, but this contradicts (2) above. Similarly, a Lebesgue value less than -1 at $\bar{\lambda}$ is ruled out by (1).

We now begin our proof that

$$\lim_{\substack{\operatorname{Im} z \downarrow 0 \\ \operatorname{Re} z = \bar{\lambda}}} E(i, z) = 0.$$

It is clear that we have

$$\begin{aligned}
|E(i, z)| &= \exp \frac{1}{2\pi} \iint g(v, \lambda) \frac{(\lambda - \lambda) + yv}{(\lambda - \bar{\lambda})^2 + y^2} \frac{dv\, d\lambda}{v^2 + 1} \\
&= \exp\left[\frac{1}{2\pi} \iint_{\lambda < \bar{\lambda}} g(v, \lambda) \frac{(\lambda - \bar{\lambda})}{(\lambda - \bar{\lambda})^2 + y^2} \frac{dv}{v^2 + 1} d\lambda \right. \\
&\quad \left. + \frac{1}{2\pi} \iint_{\lambda > \bar{\lambda}} g(v, \lambda) \frac{dv}{v^2 + 1} \frac{(\lambda - \bar{\lambda})}{(\lambda - \bar{\lambda})^2 + y^2} d\lambda \right] \\
&\quad \cdot \exp\left[\frac{1}{2\pi} \iint_{\lambda < \bar{\lambda}} g(v, \lambda) \frac{yv}{v^2 + 1} \frac{d\lambda\, dv}{(\lambda - \bar{\lambda})^2 + y^2} \right. \\
&\quad \left. + \frac{1}{2\pi} \iint_{\lambda > \bar{\lambda}} g(v, \lambda) \frac{yv}{v^2 + 1} dv \frac{d\lambda}{(\lambda - \bar{\lambda})^2 + y^2}\right].
\end{aligned}$$

Now, we note that

$$\begin{aligned}
&\frac{1}{2\pi} \iint_{a < \lambda < \bar{\lambda}} g(v, \lambda) \frac{(\lambda - \bar{\lambda})}{(\lambda - \bar{\lambda})^2 + y^2} \frac{dv}{v^2 + 1} d\lambda \\
&= \frac{1}{2\pi} \iint_{a < \lambda < \bar{\lambda}} (g(v, \lambda) - 1) \frac{(\lambda - \bar{\lambda})}{(\lambda - \lambda)^2 + y^2} \frac{dv}{v^2 + 1} d\lambda \\
&+ \iint_{a < \lambda < \bar{\lambda}} \frac{1}{2\pi} \frac{(\lambda - \bar{\lambda})}{(\lambda - \bar{\lambda})^2 + y^2} \frac{dv}{v^2 + 1} d\lambda \cong \frac{1}{2} \ln|y| \qquad \text{as} \quad y \to 0.
\end{aligned}$$

For, the first term on the right remains bounded as $y \to 0$ by the basic inequality. The second term is $-\frac{1}{4} \ln y^2 + \ln((a - \bar{\lambda})^2 + y^2)$.

Consider next

$$\begin{aligned}
&\frac{1}{2\pi} \iint_{\lambda > \bar{\lambda}} g(v, \lambda) \frac{v\, dv}{v^2 + 1} \frac{y\, dv}{(\lambda - \bar{\lambda})^2 + y^2} \\
&\quad = \frac{1}{4\pi} \int_{\lambda > \bar{\lambda}} \ln \frac{1 + (A(\lambda) + |k(\lambda)|^2)^2}{1 + (A(\lambda) - |k(\lambda)|^2)^2} \frac{y\, d\lambda}{(\lambda - \bar{\lambda})^2 + y^2}.
\end{aligned}$$

Since

$$\ln \frac{1 + (A(\lambda) + |k(\lambda)|^2)^2}{1 + (A(\lambda) - |k(\lambda)|^2)^2}$$

is in $L_1(R)$ and $k(\lambda)$ has right Lebesgue value zero with density one, the Fatou theorem shows that the integral on the right has limit zero as y approaches zero.

It remains to consider

$$\frac{1}{2\pi} \iint_{\lambda > \bar{\lambda}} g(v, \lambda) \frac{dv}{v^2 + 1} \frac{(\lambda - \bar{\lambda})}{(\lambda - \bar{\lambda})^2 + y^2} d\lambda$$

and

$$\frac{1}{2\pi} \iint_{\lambda < \bar{\lambda}} g(v, \lambda) \frac{yv\, dv}{v^2 + 1} \frac{d\lambda}{(\lambda - \bar{\lambda})^2 + y^2}.$$

The first integral remains bounded by the basic inequality, and some results connecting Hilbert transforms with Poisson integrals.[1]

To show that $|E(i, \bar{\lambda} + iy)| \to 0$ as $y \to 0^+$ it suffices to show that the second integral approaches infinity no faster than $-\frac{1}{4}\ln|y|$.

But

$$\frac{1 + (A(\lambda) + |k(\lambda)|^2)^2}{1 - (A(\lambda) - |k(\lambda)|^2)} \leqslant 1 + |k(\lambda)|^4 \left(\frac{A(\lambda)}{|k(\lambda)|^2} + 1 \right)^2,$$

and using (4) it can be seen that it suffices for us to examine how quickly

$$y \int_{\lambda < \bar{\lambda}} \frac{\ln[1 + 4|k(\lambda)|]^4}{(\lambda - \bar{\lambda})^2 + y^2} d\lambda \tag{*}$$

approaches ∞ as $y \to 0$.

First we must show that we can replace

$$1 + |k(\lambda)|^4 \left(\frac{A(\lambda)}{|k(\lambda)|^2} + 1 \right)^2$$

by $1 + 4|k(\lambda)|^2$ for the purpose of obtaining an upper estimate for the Poisson integral.

For this purpose recall that if

$$P_y(x, f) \equiv \frac{1}{\pi} \int \frac{y}{(t - x)^2 + y^2} f(t)\, dt,$$

[1] See, for example, Garsia [11, p. 103]; in particular the inequality $|H_y(x, f) - Q_y(x, f)| \leqslant P_y(x, |f|)$ is valid for f in L_∞.

where f is integrable, then

$$|P_y(x, f)| \leqslant \sup_{h>0} \frac{1}{2h} \int_{x-h}^{x+h} |f(u)|\, du$$

and the Hardy–Littlewood function on the right in turn can be estimated by

$$\int_E |f(u)|\, du = \int_0^\infty \text{measure}(E \cap \lambda_f(t))\, dt,$$

where $\lambda_f(t) = \{u : |f(u)| > t\}$.

For our purpose we set

$$f(\lambda) = \ln \max\left(|k(\lambda)|^4 \left(\frac{A(\lambda)}{|k(\lambda)|^2} + 1 \right)^2 - 4|k(\lambda)|^4, 0 \right)$$

for $\lambda < \bar{\lambda}$, while $f(\lambda) = 0$ for $\lambda > \bar{\lambda}$, and take

$$E_h = \left\{ \lambda : |k(\lambda)|^4 \left(\frac{A(\lambda)}{|k(\lambda)|^2} + 1 \right)^2 > 4|k(\lambda)|^4 \text{ and } \bar{\lambda} - h < \lambda < \bar{\lambda} \right\}.$$

Then

$$\int_{E_h} |f(\lambda)|\, d\lambda = \int_0^\infty \text{measure}(E_h \cap \lambda_f(t))\, dt.$$

But from (4) we can see that

$$\overline{\lim_{h\to 0}} \frac{1}{h} \text{measure}(E_h \cap \lambda_f(t)) = 0.$$

By the monotone convergence theorem of B. Levi we can now conclude that

$$\overline{\lim_{h\to 0}} \frac{1}{2h} \int_{E_h} |f(\lambda)| d\lambda = 0.$$

Now we estimate (*). First we remark that $(\lambda - \bar{\lambda})(1 + 2|k(\lambda)|^2)$ has Lebesgue value 0 on the left with density one. To see this let $F(\lambda) = (\lambda - \bar{\lambda})(1 + 2|k(\lambda)|^2)$. Then $F(\lambda)$ is locally integrable and

$$(1/h) \int_0^h F(\bar{\lambda} - t)\, dt = (1/h) \int_0^h t(1 + 2|k(\bar{\lambda} - t)|^2)\, dt \leqslant \int_0^h (1 + 2|k(\bar{\lambda} - t)|^2)\, dt.$$

The last integral approaches zero as h approaches zero because $|k(\cdot)|^2$ is integrable.

Thus $(\lambda - \bar{\lambda})^2(1 + 4|k(\lambda)|^4) \leqslant (\lambda - \bar{\lambda})^2(1 + 2|k(\lambda)|^2)^2$ has left Lebesgue value zero at $\bar{\lambda}$ with density one.

Hence, for $a < \bar{\lambda}$, a finite,

$$\frac{1}{4\pi} \lim_{y\to 0} y \int_{a<\lambda<\bar{\lambda}} \ln(\lambda - \bar{\lambda})^2(1 + 4|k(\lambda)|^4) \frac{d\lambda}{(\lambda - \bar{\lambda})^2 + y^2} = -\infty.$$

Thus the term

$$-\frac{1}{4\pi}\int_{a<\lambda<\bar{\lambda}} \ln\frac{1+(A(\lambda)+|k(\lambda)|^2)^2}{1+(A(\lambda)-|k(\lambda)|^2)^2}\frac{y\,d\lambda}{(\lambda-\bar{\lambda})^2+y^2}$$

grows at most like $-(1/4\pi)y\int_{a<\lambda<\bar{\lambda}}\ln(\lambda-\bar{\lambda})^2/(\lambda-\bar{\lambda})^2+y^2\,d\lambda$.

If we integrate by parts we see that this last integral is equal to

$$\frac{1}{2\pi}\ln(\bar{\lambda}-a)\tan^{-1}((\bar{\lambda}-a)/y)+\frac{1}{2\pi}\int_a^{\bar{\lambda}}\tan^{-1}((\bar{\lambda}-\lambda)/y)\frac{d\lambda}{\bar{\lambda}-\lambda}.$$

(We have chosen our branch of $\tan^{-1}$ so that $\tan^{-1}(0)=0$.)

The first term in the above expression is bounded as y approaches zero; we thus look at the integral. Let $t=(\bar{\lambda}-\lambda)/y$; then our integral becomes

$$\begin{aligned}\frac{1}{2\pi}\int_0^{(\bar{\lambda}-a)/y}\frac{1}{t}\tan^{-1}t\,dt &= \frac{1}{2\pi}\int_0^1\frac{1}{t}\tan^{-1}t\,dt\\ &\quad+\frac{1}{2\pi}\int_1^{(\bar{\lambda}-a)/y}\frac{1}{t}\left(\tan^{-1}t-\frac{\pi}{2}\right)dt\\ &\quad+\frac{1}{2\pi}\int_1^{(\bar{\lambda}-a)/y}\frac{1}{t}\frac{\pi}{2}\,dt.\end{aligned}$$

The first two integrals are bounded as y approaches zero. The last integral is

$$\frac{1}{4}\ln\frac{\bar{\lambda}-a}{y}\cong-\frac{1}{4}\ln y.$$

Thus we can conclude finally that

$$\frac{1}{4\pi}\int\ln(1+4|k(\lambda)|^4)\frac{y}{(\lambda-\bar{\lambda})^2+y^2}\,d\lambda$$

blows up at most like $-\frac{1}{4}\ln y$. This is what we wished to show, and the proof of the lemma is complete.

Although we have restricted ourselves to the case where $|k(\lambda)|^2$ is integrable, the reader will see that essentially the present lemma remains valid if it is assumed that $|k(\lambda)|^2/(1+\lambda^2)$ is integrable. The lemma is valid also if we take $y\to 0^-$.

We turn to the proof of Lemma 2.2.

Note that if we define

$$J(z,w)=\frac{1}{2\pi i}\int\frac{\bar{E}(i,\lambda+i0)-\bar{E}(i,\lambda-i0)}{\lambda-w}\frac{d\lambda}{\lambda-z},\qquad \operatorname{Im}w\neq 0,\quad \operatorname{Im}z\neq 0,$$

then we have already shown that $\lim_{w\to\lambda_j^-} J(z, w) = \theta_j(z)$. This follows from the uniform integrability result established in Section I.

But the integral in $J(z, w)$ can be rewritten as a contour integral, using the fact that $\bar{E}(\bar{l}, \bar{z}) = E^{-1}(l, z)$.

$$J(z, w) = \frac{1}{2\pi i} \int_\Gamma \frac{E^{-1}(-i, \gamma)}{(\gamma - w)(\gamma - z)} d\gamma,$$

where Γ is a contour which consists of two equally spaced horizontal lines—one above the real axis and one below the real axis.

By the residue theorem, we have

$$J(z, w) = -\frac{1}{z - w}[E^{-1}(-i, z) - E^{-1}(i, w)],$$

and by virtue of Lemma 2.1 the conclusion follows.

THEOREM 2.1. *The eigenvalue equations* (2.1) *are satisfied.*

By Lemma 2.2 we have

$$\theta_j(\lambda + i0) - \theta_j(\lambda - i0) = \frac{1}{\lambda - \lambda_j^-} \bar{H}(i, \lambda),$$

$$\theta_j(\lambda + i0) + \theta_j(\lambda - i0) = -\frac{\bar{E}(i, \lambda + i0) + \bar{E}(i, \lambda - i0)}{\lambda - \lambda_j^-}.$$

However, it is also true that

$$\theta_j(\lambda - i0) + \theta_j(\lambda - i0) = \frac{1}{\pi i} \int \frac{k(t) w_j^-(t)}{t - \lambda} dt.$$

Thus if

$$T = (A(\lambda) + i) w_j^-(\lambda) + \frac{\bar{k}(\lambda)}{\pi i} \int \frac{k(t) w_j^-(t)}{t - \lambda} dt,$$

we get

$$T = \frac{A(\lambda) + i}{\lambda - \lambda_j^-} \left[\frac{\bar{E}(i, \lambda + i0) - \bar{E}(i, \lambda - i0)}{k(\lambda)} \right] - k(\lambda) \left[\frac{\bar{E}(i, \lambda + i0) + \bar{E}(i, \lambda - i0)}{\lambda - \lambda_j^-} \right].$$

Since $[A(\lambda) + i + |k(\lambda)|^2] E^{-1}(-i, \lambda - i0) = [A(\lambda) + i - |k(\lambda)|^2] E^{-1}(i, \lambda + i0)$, we conclude that $T = 0$. This proves the theorem.

III

In this section we construct the operator N_l and determine its range. Since the dimension of the range N_l depends only upon the half-plane in which l is situated, we take $l = \pm i$.

THEOREM 3.1

$$(H_1 - i)^{-1}f, (H_1 - i)^{-1}g) = (R(i)f, R(i)g) + \sum_{j=1}^{m} (f, \tilde{\tilde{w}}_j^-)(\tilde{\tilde{w}}_j^-, g),$$

$$(H_1 + i)^{-1}f, (H_1 + i)^{-1}g) = (R(\bar{i})f, R(\bar{i})g) + \sum_{j=1}^{n} (f, w_j^+)(w_j^+, g),$$

where

$$\tilde{\tilde{w}}_j^- = \mu_{-i}(\lambda_j)(2\pi)^{1/2}w_j^-, \qquad j = 1, 2, 3, \ldots, m,$$
$$\tilde{\tilde{w}}_r^+ = \mu_{+i}(\lambda_r)(2\pi)^{1/2}w_r^+, \qquad r = 1, 2, 3, \ldots, n.$$

The proof depends upon the basic commutator identity:

$$((H_1 - i)^{-1}(H_2 - z_1)^{-1}k^*, (H_1 - i)^{-1}(H_2 - z_2)^{-1}k^*)$$
$$= k(H_2 - \bar{z}_2)^{-1}(H_1 + i)^{-1}(H_1 - i)^{-1}(H_2 - z_1)^{-1}k.$$

But

$$(H_2 - z_1)^{-1}(H_1 - i)^{-1}k^*E(i, z_1) = (H_1 - i)^{-1}(H_z - z_1)^{-1}k^*.$$

Thus,

$$((H_1 - i)^{-1}(H_2 - z_1)^{-1}k^*, (H_1 - i)^{-1}(H_2 - z_2)^{-1}k^*)$$
$$= E(i, z_1)\bar{E}(i, z_2)((H_2 - z_1)^{-1}(H_1 - i)^{-1}k^*, (H_2 - z_2)^{-1}(H_1 - i)^{-1}k^*)$$
$$= E(i, z_1)\bar{E}(i, z_2)\left[k(H_1 + i)^{-1}\frac{[(H_2 - \bar{z}_2)^{-1} - (H_2 - z_1)^{-1}]}{\bar{z}_2 - z_1}(H_1 - i)^{-1}k^*\right].$$

But

$$E(i, z)\bar{E}(i, z) = 1 + \frac{2}{\pi}k(H_1 - i)^{-1}(H_2 - z)^{-1}(H_1 + i)^{-1}k^*,$$

$$E(-i, z)\bar{E}(-i, z) = 1 - \frac{2}{\pi}k(H_1 + i)^{-1}(H_2 - z)^{-1}(H_1 - i)^{-1}k^*.$$

Hence,

$$((H_1 - i)^{-1}(H_2 - z_1)^{-1}k^*, (H_1 - i)^{-1}(H_2 - z_2)^{-1}k^*$$

$$= \frac{2}{\pi}\frac{E(i, z_1)\bar{E}(i, z_2)}{\bar{z}_2 - z_1}[E(-i, z_1)\bar{E}(-i, \bar{z}_1) - E(-i, \bar{z}_2)\bar{E}(-i, z_2)]$$

$$= \frac{2}{\pi}\frac{E(i, z_1)\bar{E}(i, z_2)}{\bar{z}_2 - z_1}\left[\left(1 - \int \frac{d\mu_{-i}(t)}{t - z_1}\right) - \left(1 - \int \frac{d\mu_{-i}(t)}{t - \bar{z}_2}\right)\right]$$

$$= \frac{2}{\pi} E(i, z_1)\bar{E}(i, z_2) \int \frac{d\mu_{-i}(t)}{(t - z_1)(t - \bar{z}_2)}$$

$$= E(i, z_1)E(i, z_2)\frac{1}{2\pi}\int \left|\frac{H(-i, t)}{k(t)}\right|^2 \frac{dt}{(t - z_1)(t - \bar{z}_2)}$$

$$+ \frac{1}{\pi}\sum_{j=1}^{m} \frac{\mu_{-i}^{(j)}}{(\lambda_j^- - z_1)(\lambda_j^- - z_2)}$$

But

$$R(l)\frac{\bar{k}(t)}{t - z} = \begin{cases} \dfrac{1}{2}\dfrac{\bar{H}(\bar{l}, t)}{k(t)}\dfrac{E(l, z)}{t - z}, & t \text{ in } E \\ 0 & t \text{ not in } E, \end{cases}$$

where E is the support of $k(t)$.

We will indicate the proof of this identity. Let $E_n(l, z)$ be the determining function of the pair of operators L_n and M. Then

$$E_n(l, z) = 1 + (1/\pi)k_n(L_n - l)^{-1}(M - z)^{-1}k_n^*.$$

The commutator identities then give quite simply that

$$\left((L_n - l)^{-1}\frac{\bar{k}_n(t)}{t - w}, \frac{\bar{k}_n(t)}{t - \bar{z}}\right) = \frac{\pi i}{z - w}(1 - E_n(l, z)\bar{E}_n(\bar{l}, \bar{z})).$$

The inner product in this equation can be rewritten as

$$\int k_n(t)\left[(L_n - l)^{-1}\frac{\bar{k}_n(t)}{t - w}\right]\frac{dt}{t - z} = \frac{\pi i}{z - w}(1 - E_n(l, z)\bar{E}_n(\bar{l}, \bar{z})).$$

Now use the Plemelj formulas to solve for $k_n(t)[(L_n - l)^{-1}\bar{k}_n(t)/(t - w)]$. We also use the fact that

$$E_n(l, z) = \exp\frac{1}{2\pi i}\iint g_n(x, t)\frac{dx}{x - l}\frac{dt}{t - z},$$

where $g_n(x, t)$ is the characteristic function of the set of points (t, x) for which $A_n(t) - |k_n(t)|^2 < x < A_n(t) + |k_n(t)|^2$. Since $g_n(x, t)$ converges to the integrable function $g(x, t)$ the result follows.

Thus we have

$$\left(R(i)\frac{\bar{k}(t)}{t - z_1}, R(i)\frac{\bar{k}(t)}{t - z_2}\right) = \frac{1}{4}\int \left|\frac{\bar{H}(-i, t)}{k(t)}\right|^2 \frac{dt}{(t - z_1)(t - \bar{z}_2)} E(i, z_1)\bar{E}(i, z_2).$$

Accordingly,

$$((H_1 - i)^{-1}(H_2 - z_1)^{-1}k^*, (H_1 - i)^{-1}(H_2 - z_2)^{-1}k^*)$$
$$= \left(R(i)\frac{\bar{k}(t)}{t - z_1}, R(i)\frac{\bar{k}(t)}{t - \bar{z}_2}\right) + \sum_{j=1}^{m} \frac{\mu_{-i}(j)}{(\lambda_j^- - z_1)(\lambda_j^- - \bar{z}_2)}. \tag{3.1}$$

But, we have already seen that

$$\frac{E(i, z)}{z - \lambda_j^-} = \frac{1}{2\pi i}\int \frac{H(i, t)}{t - \lambda_j^-}\frac{dt}{t - z} = \lim_{w \to \lambda_j^-} \int_\Gamma \frac{E(i, \gamma)}{(\gamma - w)(\gamma - z)}\, d\gamma,$$

where Γ is a pair of horizontal straight lines—one above the real axis and one below it.

Thus,

$$\frac{E(i, z)}{z - \lambda_j^-} = 2\pi i \left(\frac{\bar{k}(t)}{t - z}, w_j^-(t)\right).$$

Substituting this into (3.1) gives us

$$\left((H_1 - i)^{-1}\frac{\bar{k}(t)}{t - z_1}, (H_1 - i)^{-1}\frac{k(t)}{t - z_2}\right)$$
$$= \left(R(i)\frac{\bar{k}(t)}{t - z_1}, R(i)\frac{k(t)}{t - z_2}\right) + \sum_{j=1}^{m}\left(\frac{\bar{k}(t)}{t - z_1}, \tilde{\tilde{w}}_j^-\right)\left(\tilde{\tilde{w}}_j^-, \frac{k(t)}{t - z_2}\right).$$

Since sums of the form $\sum k(t)/(t - z_s)$ are dense in $L_2(E)$, we have proved the theorem.

IV

THEOREM 4.1. *$1 + 2iR(i)$ restricted to the orthogonal complement of the span $\{w_j^-\}$, $j = 1, \ldots, m$, is an isometry.*

Proof. By Theorem 3.1 we can see that if $(f, w_j^-) = 0$, $j = 1, \ldots, m$, then $(H_1 - i)^{-1}f = P(H_1 - i)^{-1}f = R(i)f$. To see this it suffices to note that $\|(H_1 - i)^{-1}f\| = \|P(H_1 - i)^{-1}f\| = \|R(i)f\|$ and therefore

$$(1 - P)(H_1 - i)^{-1}f = 0.$$

But $1 + 2i(H_1 - i)^{-1}$ is an isometry. Thus,

$$\|f\| = \|(H_1 + i)(H_1 - i)^{-1}f\| = \|(1 + 2iR(i)f\|.$$

THEOREM 4.2. *$V_i = 1 + 2iR(i)$ restricted to the orthogonal complement of the span of w_j^- is the Cayley transform of a symmetric operator L.*

Proof. We must show that the range of $1 - V_i$ is dense in $L_2(E)$, i.e., that hypothesis (3) of the introduction is satisfied.

In order to do this we will produce a dense set of functions in the range of $1 - V_i$. Let $f(t)$ be any rational function with simple poles at some finite set z_r of nonreal points and zeros at the points λ_j^- with a partial fraction expansion of the form $f(t) = \sum_r [\beta_r/(t - z_r)]$. Then define

$$B(t) = \bar{k}(t) \sum \beta_r \frac{E^{-1}(i, z_r)}{t - z_r}.$$

We claim that $B(t)$ is orthogonal to all the functions $w_j^-(t)$. To see this we simply note that

$$\left(\frac{\bar{k}(t)}{t - z}, w_j^-(t)\right) = \frac{2\pi i}{\lambda_j - z} E(i, z),$$

and thus $(B(t), w_j^-(t)) = 2\pi i f(\lambda_j^-) = 0$. We have therefore shown that $B(t)$ is in the domain of V_i. But

$$(1 - V_i)B(t) = -2iR(i)B(t) = -i \frac{\bar{H}(-i, t)}{k(t)} f(t).$$

We claim that these last functions are dense in $L_2(E)$. We will prove this fact by showing that the only $L_2(E)$ function which is orthogonal to all of these functions is equal to zero almost everywhere.

Suppose that

$$\left(\frac{\bar{H}(-i, t)}{k(t)} f(t), g(t)\right) = 0$$

for all $f(t)$ of our form. Then it would follow in particular that

$$\int_E \frac{\bar{H}(-i, t)}{k(t)} h(t)\bar{g}(t) \prod_j (t - \lambda_j^-) \frac{dt}{t - z} = 0,$$

where $h(t)$ is rational in t and $\operatorname{Im} z \neq 0$. Since $\bar{H}(-i, t)/k(t) \neq 0$ a.e., we must conclude that $g(t) = 0$ a.e.

THEOREM 4.3.

$$[1 + 2iR(i)]w_r^- = \sum_{j=1}^{n} w_j^+ \frac{\mu_i(\lambda_j^+)}{\lambda_j^+ - \lambda_r^-},$$

$$[1 - 2iR(-i)]w_r^+ = \sum_{j=1}^{m} w_j^- \frac{\mu_{-i}(\lambda_j^-)}{\lambda_j^- - \lambda_r^+}.$$

We will not make full use of this result in the present paper, but this theorem plays a role in the calculation of the characteristic operator function of the contractions $1 + 2iR(i)$ and $1 - 2iR(-i)$. Note that these are quasi-unitary extensions of the Cayley transform of L in the sense of M. S. Lifshitz. We have already noted (but do not elaborate here) that the characteristic function is computed in terms of the principal function. See, for example, [2].

We turn now to a proof of the first equation.

Consider the inner product

$$\left(R(-i)\frac{\bar{k}(t)}{t - z}, \frac{\bar{H}(-i, t)}{k(t)} \frac{1}{t - \lambda_j^-}\right) = \frac{1}{2} E(-i, z)\left(\frac{H(i, t)}{k(t)}, \frac{1}{t - z}, \frac{\bar{H}(i, t)}{k(t)} \frac{1}{t - \lambda_j^-}\right)$$

$$= \frac{1}{2} \int \left|\frac{H(i, t)}{k(t)}\right|^2 \frac{dt}{(t - z)(t - \lambda_j^-)} E(-i, z).$$

We know that

$$\frac{1}{2\pi} \int \left|\frac{H(i, t)}{k(t)}\right|^2 \frac{dt}{t - z} + \sum_j \frac{\mu_i(\lambda_j^+)}{\lambda_j^+ - z} = E(i, z)\bar{E}(i, \bar{z}) - 1.$$

Hence,

$$\frac{1}{2\pi} \int \left|\frac{H(i, t)}{k(t)}\right|^2 \frac{dt}{(t - z)(t - w)}$$

$$= \frac{1}{z - w} \frac{1}{2\pi} \int \left|\frac{H(i, t)}{k(t)}\right|^2 \frac{dt}{t - z} - \frac{1}{2\pi} \int \left|\frac{H(i, t)}{k(t)}\right|^2 \frac{dt}{t - w}$$

$$= \frac{1}{z - w}\left[\left(E(i, z)\bar{E}(i, \bar{z}) - 1 - \sum_j \frac{\mu_i(\lambda_j^+)}{\lambda_j^+ - z}\right)\right.$$

$$\left. - \left(E(i, w)\bar{E}(i, \bar{w}) - 1 - \sum_j \frac{\mu_i(\lambda_j^+)}{\lambda_j^+ - w}\right)\right]$$

Now, set (as we have seen we may) $w = \lambda_r^-$ for each $r = 1, \ldots, m$. Then since $E(i, \lambda_r^-) = 0$, we get

$$\frac{1}{2\pi}\int\left|\frac{H(i,t)}{k(t)}\right|^2 \frac{dt}{(t-z)(t-\lambda_r^-)} = -\frac{E(i,z)\bar{E}(i,\bar{z})}{\lambda_r^- - z} + \sum_{j=1}^{n} \frac{\mu_i(\lambda_j^+)}{(\lambda_j^+ - z)(\lambda_j^+ - \lambda_r^-)}.$$

Therefore,

$$\frac{1}{2\pi}\left(\frac{k(t)}{t-z}, R(i)w_r^-\right) = -\frac{1}{2i}\frac{E(i,z)}{\lambda_r^- - z} + \frac{E(-i,z)}{2i}\sum_j \frac{\mu_i(\lambda_j^+)}{(\lambda_j^+ - z)(\lambda_j^+ - \lambda_r^-)}.$$

If we take the limit in this expression as z approaches λ on the real axis from the upper and lower half-planes and subtract, we will get—using the Plemelj formulas—

$$\bar{k}(\lambda)\overline{R(i)w_r^-(\lambda)} = \frac{1}{2i}\frac{H(i,\lambda)}{\lambda - \lambda_r^-} + \frac{H(-i,\lambda)}{2i}\sum_{j=1}^{n} \frac{\mu_i(\lambda_j^+)}{(\lambda_j^+ - \lambda)(\lambda_j^+ - \lambda_r^-)}.$$

If we divide by $\bar{k}(\lambda)$ and then take the complex conjugate of both sides of this equation we get the statement of Theorem 4.3.

Note that we could take $z = \lambda_k^-$ above to get

$$\frac{1}{2\pi}\int\left|\frac{H(i,t)}{k(t)}\right|^2 \frac{dt}{(t-\lambda_k^-)(t-\lambda_r^-)} = \sum_{j=1}^{n} \frac{\mu_i(\lambda_j^+)}{(\lambda_j^+ - \lambda_k^-)(\lambda_j^+ - \lambda_r^-)}.$$

LEMMA 4.1. *The range of* $1 + 2iR(i)$ *restricted to the orthogonal complement of the span of* $\{w_j^-\}$ *is the orthogonal complement of the span of* $\{w_r^+\}$.

It is clear that any vector which is orthogonal to w_j^- is mapped by $1 + 2iR(i)$ into a vector orthogonal to w_r^+. To see this let $(x, w_j^-) = 0$ for $j = 1, \ldots, m$. Then by Theorem 4.1,

$$((1 + 2iR(i))x, w_r^+) = (x, (1 - 2i(R(-i))w_r^+) = 0.$$

Now we will show that $(y, w_r^+) = 0$, $r = 1, 2, \ldots, n$, implies that there exists an x with $(x, w_j^-) = 0$, $j = 1, 2, \ldots, m$, and $y = (1 - 2iR(i))x$.

If $(y, w_r^+) = 0$, then $(1 - 2iR(-i))y$ is orthogonal to all of the vectors w_j^- because $((1 - 2iR(-i))y, w_j^-) = (y, (1 + 2iR(i))w_j^-)$.

Therefore, if we set $x = (1 - 2iR(-i))y$, and note that we have already shown that $R(-i)$ and $(H_1 + i)^{-1}$ agree on vectors orthogonal to w_r^+, $r = 1, 2, \ldots, n$, then

$$x = (1 - 2iR(-i))y = (H_1 - i)(H_1 + i)^{-1}y.$$

But this means that $y = (H_1 + i)(H_1 - i)^{-1}x = (1 + 2iR(i))x$ because $(H_1 - i)^{-1}$ and $R(i)$ agree on vectors orthogonal to all w_j^-.

THEOREM 4.4. *The symmetric operator L constructed in Theorem 4.2 has deficiency indices (m, n).*

We have seen that the initial space of the Cayley transform of L is the orthogonal complement of the span of $\{w_j^-\}$—an m-dimensional subspace—and the preceding lemma shows that the orthocomplement of the range of the Cayley transform of L is n-dimensional—and consists of the span of the vectors $\{w_r^+\}$.

V. THE DEFICIENCY SPACES REEXAMINED

Let $Q(l)$ be the smallest invariant subspace of H_2 which contains the range of $D(l) = (1/\pi) d_l \otimes d_l$, where $d_l = [(l - \bar{l})/i]^{1/2} (H_1 - \bar{l})^{-1} k^*$.

We have already identified the absolutely continuous subspace of H_2 with $L_2(E)$—our base space. Since the theory of the principal function is really a theory of pairs of self-adjoint operators, it is desirable to understand the structure of the deficiency spaces of L from a more invariant point of view—that is, without reference to the singular integral operator form. The following theorem is one of several results along these lines. A more comprehensive study of restrictions of one or another of the two operators H_1 and H_2 to certain privileged subspaces will be presented elsewhere in somewhat greater generality than is feasible here.

THEOREM 5.1. *w_j^- and w_r^+ are, respectively, the projections onto the base space of eigenvectors of $H_2 + D(i)$ acting on $Q(i)$ and $H_2 - D(-i)$ acting on $Q(-i)$. That is, there exist vectors x_j^- in $Q(i)$ and x_r^+ in $Q(-i)$ such that $(H_2 + D(i))x_j^- = \lambda_j^- x_j^-$ and $(H_2 - D(-i))x_r^+ = \lambda_r^+ x_r^+$ and $Px_j^- = w_j^-$ while $Px_r^+ = w_r^+$, $j = 1, \ldots, m$, $r = 1, \ldots, n$.*

This is an intrinsic characterization of the deficiency spaces of L. It is valid in general. Thus if either m or n is infinite then the deficiency spaces of L are the closures of the projections onto the base space of the corresponding invariant subspaces of $H_2 + D(i)$ restricted to $Q(i)$ and of $H_2 - D(-i)$ restricted to $Q(-i)$. Thus we can also treat the case when L has infinite deficiency.

We wish to remark also that the theorem also enables us to treat the case where the coefficients $A(\lambda)$ and $k(\lambda)$ appearing in the singular integral operator are $p \times p$ matrices. The deficiency spaces are still formed by projecting the corresponding eigenspaces of the "same" perturbed operators onto the base space.

Proof. We have seen that

$$E(i,z)\bar{E}(i,\bar{z}) = 1 + \frac{1}{\pi}((H_2 - z)^{-1} d_i, d_i) = 1 + \frac{1}{\pi}\int \frac{d\mu_i(t)}{t-z}.$$

Hence H_2 restricted to $Q(i)$ can be represented by multiplication by λ on $L_2(d\mu_i)$. Call this multiplication operator M_+.

We write $L_2(d\mu_i)$ as an orthogonal direct sum. For x in $L_2(d\mu_i)$

$$x = \{x(\lambda), x_1, x_2, \ldots, x_n\} \quad \text{and} \quad \|x\|^2 = \int |x(\lambda)|^2\, d\lambda + \sum_j \mu_i(j)|x_j|^2.$$

Since

$$P(H_1 - \bar{l})^{-1}k^* = \frac{1}{2}\frac{\bar{H}(l, \lambda)}{k(\lambda)}$$

and

$$\frac{1}{\pi}\|(H_1 - \bar{l})^{-1}k^*\|^2 = \frac{1}{2\pi}\int \left|\frac{\bar{H}(l, \lambda)}{k(\lambda)}\right|^2 d\lambda + \frac{1}{\pi}\sum_j \mu_i(j),$$

we can easily see that d_i has the representation

$$d_i = \left\{\frac{1}{\sqrt{2}}\frac{\bar{H}(i, \lambda)}{\bar{k}(\lambda)}, 1, 1, \ldots, 1\right\}.$$

Accordingly,

$$(M_+ + D(i))x = \left\{\lambda x(\lambda) + \frac{\bar{H}(i, \lambda)}{k(\lambda)}\mathscr{A}(x), \lambda_1{}^+ x_1 + \mathscr{A}(x), \ldots, \lambda_n{}^+ \mathscr{A}(x)\right\},$$

where

$$\mathscr{A}(x) = \frac{1}{2\pi}\int x(\lambda)\frac{H(i, \lambda)}{\bar{k}(\lambda)}\, d\lambda + \frac{1}{\pi}\sum_j \mu_i(j)x_j.$$

Now take $x = x_r{}^-$, where we define

$$x_r{}^- = \left\{\frac{\bar{H}(i, \lambda)}{k(\lambda)}\frac{1}{\lambda - \lambda_r{}^-}, \frac{1}{\lambda_1{}^+ - \lambda_r{}^-}, \frac{1}{\lambda_2{}^+ - \lambda_r{}^-}, \ldots, \frac{1}{\lambda_n{}^+ - \lambda_r{}^-}\right\}.$$

Then,

$$\mathscr{A}(x_r{}^-) = \frac{1}{2\pi}\int \left|\frac{H(i, \lambda)}{k(\lambda)}\right|^2 \frac{d\lambda}{\lambda - \lambda_r{}^-} + \frac{1}{\pi}\sum_j \frac{\mu_i(j)}{\lambda_j{}^+ - \lambda_r{}^-}$$

and

$$\frac{1}{2\pi}\int\left|\frac{H(i,\lambda)}{k(\lambda)}\right|^2\frac{d\lambda}{\lambda-z}+\frac{1}{\pi}\sum_j\frac{\mu_i(j)}{\lambda_j{}^+-z}=E(i,z)\bar{E}(i,\bar{z})-1.$$

Since $E(i,\lambda_r{}^-)=0$, we have $\mathscr{A}(x_r{}^-)=-1$. Thus

$$\begin{aligned}(M_+ + D(i))x_r{}^- &= \left\{\lambda\frac{\bar{H}(i,\lambda)}{k(\lambda)}\frac{1}{\lambda-\lambda_r{}^-}-\frac{\bar{H}(i,\lambda)}{k(\lambda)},\frac{\lambda_1{}^+}{\lambda_1{}^+-\lambda_r{}^-}-1,\ldots,\frac{\lambda_n{}^+}{\lambda_n{}^+-\lambda_r{}^-}-1\right\}\\ &= \lambda_r{}^-x_r{}^-.\end{aligned}$$

We turn now to an analysis of the singular space of H_2 on the smallest invariant subspace of this operator that contains the range of $D(-i)$, that is, on $Q(-i)$.

We proceed by first determining the spectral measure of H_2 restricted to $Q(-i)$.

Since

$$E(l,z)\bar{E}(l,\bar{z})=1+\frac{1}{\pi}\left(\frac{l-\bar{l}}{i}\right)^{1/2}k(H_1-l)^{-1}(H_2-z)^{-1}(H_1-\bar{l})^{-1}k^*,$$

we have

$$\begin{aligned}(E(l,z)\bar{E}(l,\bar{z}))^{-1} &= 1-\frac{1}{\pi}\left(\frac{l-\bar{l}}{i}\right)^{1/2}k(H_1-l)^{-1}(H_2+D(\bar{l})-z)^{-1}(H_1-\bar{l})^{-1}k^*.\end{aligned}$$

Since $E(l,z)^{-1}=\bar{E}(\bar{l},\bar{z})$, we get

$$\begin{aligned}k(H_1-\bar{l})^{-1}(H_2-z)^{-1}(H_1-l)^{-1}k^* &= k(H_1-l)^{-1}(H_2+D(\bar{l})-z)^{-1}(H_1-\bar{l})^{-1}k^*.\end{aligned}$$

Take $l=i$. Then we get

$$((H_2-z)^{-1}d_{-i},d_{-i})=((H_2+D(i)-z)^{-1}d_i,d_i)=\int\frac{d\mu_{-i}(t)}{t-z}.$$

Accordingly, the spectral measure of H_2 restricted to $Q(-i)$ is $d\mu_{-i}$, and we can represent H_2 on $Q(-i)$ as multiplication by λ on $L_2(d\mu_{-i})$. Call this operator M_-.

We represent a vector x in $L_2(d\mu_{-i})$ in the form

$$x=\{x(\lambda),x_1,x_2,\ldots,x_m\},$$

with

$$\|x\|^2 = \int |x(\lambda)|^2 \, d\lambda + \sum \mu_{-i}(j)|x_j|^2.$$

Then

$$d_{-i} = \left\{ \frac{1}{\sqrt{2}} \frac{\bar{H}(-i, \lambda)}{k(\lambda)}, 1, \ldots, 1 \right\}.$$

We now form $(M_- - D(-i))x_r^+$, where

$$x_r^+ = \left\{ \frac{\bar{H}(-i, \lambda)}{k(\lambda)} \frac{1}{\lambda - \lambda_r^+}, \frac{1}{\lambda_1^- - \lambda_r^+}, \frac{1}{\lambda_2^- - \lambda_r^+}, \ldots, \frac{1}{\lambda_m^- - \lambda_r^+} \right\},$$

and make a calculation similar to the foregoing one, using the fact that $E(-i, \lambda_r^+) = 0$, to get $(M_- - D(-i)x_r^+ = \lambda_r^+ x_r^+$.

We will improve on these results in Section VII, where we express N_l in terms of the projection onto the singular spectrum of the wave operator.

Before doing this we construct a core for the symmetric operator L.

VI. The Domain of L and Its Extensions

THEOREM 6.1. *Let $f(t)$ be any rational function with simple poles at some finite set z_r of nonreal points $r = 1, 2, \ldots, N$. Let $f(\lambda_j^-) = 0$ and let $f(t)$ vanish at infinity. Consider the set of all functions of the form $[\bar{H}(-i, t)/k(t)]f(t)$. This set of functions is dense in $L_2(E)$ and the formal singular operator restricted to this domain is a symmetric operator with closure L.*

Proof. The function $f(t)$ has a partial fraction expansion in the form $f(t) = \sum_r \beta_r/(t - z_r)$. Form $\bar{k}(t) \sum_r \beta_r E^{-1}(i, z_r)/(t - z_t) = B(t)$. We have $(B(t), w_j^-(t)) = 0$, as was shown in the proof of Theorem 4.2. Therefore $B(t)$ is in the domain of $V = (L + i)/(L - i)$. In fact $(1 - V)B(t) = -i,(\bar{H}(-i, t)/k(t)) \sum_r \beta_r/(t - z_r)$. The functions on the right-hand side of this equation are in fact dense in $L_2(E)$, as we have seen in the proof of Theorem 4.2.

Now

$$L = \frac{1}{i} \frac{1 + V}{1 - V} \quad \text{and} \quad (L - i)(1 - V)|_{D_V} = 2iI|_{D_V}.$$

This is really the defining equation for L. If we show that for y in D_V $(\mathscr{L} - i)(1 - V)y = 2iy$ we will have shown that $\mathscr{L} = L$.

What we will actually do is show this equality for y in a certain set in D_V. We will then be able to assert that the closure of $\mathscr{L}$ restricted to a dense set actually coincides with L.

Consider the vector y_r in D_V defined by

$$y_r = \frac{\bar{k}(t)}{t - z_r} - \sum_{j=1}^{N} w_j^- \left(\frac{k(\sigma)}{\sigma - z_r}, w_j^-\right),$$

$$(1 - V)y_r = -2iR(i)\frac{\bar{k}(t)}{t - z_r} - 2iR(i)\sum_j w_j^- \left(\frac{k(\sigma)}{\sigma - z_r}, w_j^-\right).$$

However, $R(i)w_j^-$ is in the linear span of the functions w_s^+ by Theorem 4.3, and therefore, by Theorem 2.1 together with the relation

$$(\mathscr{L} - i)R(i)[\bar{k}(t)/(t - z_r)] = \frac{\bar{k}(t)}{(t - z_r)}$$

which was established in [8] (and which follows very easily from manipulations like those in Section II) we get

$$(\mathscr{L} - i)(1 - V)y_r = -2i\frac{\bar{k}(t)}{t - z_r}.$$

Now let $a_r = \beta_r E(i, z_r)^{-1}$. We have $\sum a_r y_r = \sum a_r \bar{k}(t)/(t - z_r)$ because $B(t)$ is orthogonal to all the vectors w_j^-, $j = 1, \ldots, m$. But $(1 - V)\sum_r a_r[\bar{k}(t)/(t - z_r)] = -i[\bar{H}(-i, t)/k(t)]f(t)$. Furthermore, we have just established that $(\mathscr{L} - i)(1 - V)\sum_r a_r y_r = -2i\sum_r a_r y_r$, and therefore $\mathscr{L}$ coincides with L on the dense set of vectors $-i[\bar{H}(-i, t)/(k(t)]f(t)$. This shows that this set of functions is a core for the symmetric operator L, and the theorem is proved.

Remark. When the functions $A(t)$ and $k(t)$ are smooth with the exception of a finite number of isolated square integrable power singularities it is possible to show that $\mathscr{L}$ restricted to a domain of smooth functions which vanish in neighborhoods of all the points λ_j^- and that λ_r^+ is a symmetric operator with closure L. Note that the rational function $f(t)$ above vanishes at some of these points (the λ_j^-) and that the function $\bar{H}(-i, t)/k(t)$ vanishes at the other points (the λ_r^+). The formulation of the theorem enables us to treat the general case when the coefficients are not smooth but m and n are finite.

VII. Remarks

We go on now to somewhat more general considerations.

THEOREM 7.1. *$N_l = -(l - \bar{l})^{-2}PP_s(l)P$ where $P_s(l)$ is the projection onto the singular subspace of $H_2 + D(l)$, and P is the projection onto the absolutely continuous spectrum of H_2. Thus $1 - T^*(l)T(l) = PP_s(l)P$.*

We have actually proved this result already in case m and n are finite. The present proof is valid for any values of m and n, and holds also (with only notational changes) in the interesting case where $A(\lambda)$ and $k(\lambda)$ are operator valued. $\int \operatorname{trace} k^*(\lambda)k(\lambda)\,d\lambda < \infty$ and $A(\lambda)$ is self-adjoint.

Let $W_-(l) = \text{s-lim}_{t\to-\infty} \exp it(H_2 + D(l))\exp - itH_2P$ so that $W_-(l)$ is the wave operator corresponding to the perturbation $H_2 \to H_2 + D(l)$. The initial space of $W_-(l)$ is the range of P and the final space of $W_-(l)$ is the absolutely continuous part of $H_2 + D(l)$.

Let $f_l(\lambda) = (A(\lambda) - \bar{l} + k^*(\lambda)k(\lambda))(A(\lambda) - l + k^+(\lambda)k(\lambda))^{-1}$. This unimodular function (unitary operator) should be understood as the symbol in the sense of [12] of $T(l)$, i.e., in the notation of [12] we have $f_l(\lambda) = S_-(M; T(l))$. (In this notation $S_+(M; T(l))f_l(\lambda)_l^{-1} = S(l, \lambda)$, the scattering operator corresponding to the perturbation problem.)

Lemma 7.1

$$PW_-(\bar{l})f_l(\lambda) = T(l) = 1 + (l - \bar{l})R(l).$$

We will take $l = i$ and employ an integral representation for the compression of the wave operator to the range of P that we have proved elsewhere (see [5, Theorem 3.1], and [3, p. 68] for the case of operator valued coefficients:

$$PW_-g(\lambda) = g(\lambda) - \frac{1}{2}\frac{\bar{H}(-i,\lambda)}{k(\lambda)}\int \frac{H(-i,t)}{\bar{k}(t)}\frac{1}{\det_+(t)}\frac{g(t)}{t - (\lambda - i0)}\,dt,$$

where $\det_+(t) = (D_+(t + i0))^{-1} = (E(i, \lambda + i0)\bar{E}(i, \lambda - i0))^{-1}$ is the reciprocal perturbation determinant.

We also note the easily proved equation

$$\frac{A(t) + i + |k(t)|^2}{A(t) - i + |k(t)|^2}H(-i,t) = H(i,t)\det_+(t + i0),$$

which follows from the fact that $D_+(z) = E(i,z)\bar{E}(i,\bar{z})$ and that $E(i,z)$ satisfies a Riemann–Hilbert barrier relation involving the coefficients $A(t)$ and $k(t)$.

Then we get

$$(PW_-f)g(\lambda) = \frac{A(\lambda) + i + |k(\lambda)|^2}{A(\lambda) - i + |k(\lambda)|^2}g(\lambda) - \frac{1}{2\pi}\int \frac{\bar{H}(-i,\lambda)}{k(\lambda)}\frac{H(i,t)}{\bar{k}(t)}g(t)\frac{dt}{t - (\lambda - i0)}.$$

However, since

$$\frac{1}{2\pi}\int \frac{H(i,t)}{(t - w)(t - z)}\,dt = \frac{i}{w - z}(E(i,w) - E(i,z)),$$

we get

$$(PW_- f)\frac{\bar{k}(\lambda)}{\lambda - z} = \frac{i}{\lambda - z}\frac{\bar{H}(-i, \lambda)}{k(\lambda)}E(i, z) + \frac{\bar{k}(\lambda)}{\lambda - z}.$$

But we have seen in the preceding sections that

$$R(i)\frac{\bar{k}(\lambda)}{\lambda - z} = \frac{1}{2}\frac{\bar{H}(-i, \lambda)}{k(\lambda)}\frac{E(i, z)}{\lambda - z}.$$

Since sums of the form $\sum a_r \bar{k}(\lambda)/(\lambda - z_r)$ are dense in $L_2(E)$, the result follows. Now $W_l = 1 + (l - \bar{l})(H_1 - l)^{-1}$, and $PW_l P = T(l)$; thus

$$N_l = \frac{P(1 - W_l^*)(W_l - 1)P}{(l - \bar{l})^2} = \frac{(1 - T^*(l))(T(l) - 1)}{(l - \bar{l})^2}.$$

Since $W_l^* W_l = 1$, we get $N_l = (T^*(l)T(l) - 1)/(l - \bar{l})^2$. Accordingly, $N_{\bar{l}} = (T(l)T^*(l) - 1)/(\bar{l} - l)^2$.

By the lemma,

$$N_{\bar{l}} = (\bar{l} - l)^{-2}(PW_-(\bar{l})f_l(\lambda)\bar{f}_l(\lambda)W_-^*(\bar{l})P - 1) = (\bar{l} - l)^{-2}(PP_{\mathrm{a}}(\bar{l})P - 1)$$
$$= -(\bar{l} - l)^{-2}PP_{\mathrm{s}}(\bar{l})P.$$

The basic commutator identity between H_1 and H_2 can be rewritten in another way. Let $W = (H_1 - \bar{l})^{-1}(H_1 - l)^{-1}$. Then by elementary algebra we get

$$WH_2 = \left(H_2 + \frac{(l - \bar{l})^2}{\pi}(H_1 - l)^{-1}C(H_1 - \bar{l})^{-1}\right)W.$$

Thus if $D(l) = ((l - \bar{l})^2/\pi)(H_1 - l)^{-1}C(H_1 - \bar{l})^{-1}$ we have a (one-dimensional) intertwining triplet $\{W; H_2, H_2 + D\}$ in the sense of [5]. It is known that such a triplet may be characterized up to unitary equivalence by a principal function $G(\lambda, e^{i\theta})$ defined on the direct product of the line and the unit circle.

Further, the transformation properties of these principal functions are such that we can assert that (see [4, 5])

$$G\left(\lambda, \frac{v - \bar{l}}{v - l}\right) = g(v, \lambda).$$

On the other hand, there is a principal function $g'(\lambda, e^{i\theta})$ associated with the intertwining triplet $\{W \cdot f, H_2, H_2 + D(l)\}$, and a theorem in [5] states the following rule for obtaining it.

THEOREM. *Let $\delta(\lambda)$ be the spectral displacement function corresponding* *to the perturbation* $H_2 \to H_2 + D$. *Then* $g'(\lambda, \gamma) = \chi(f(\lambda)S(\lambda),\ f(\lambda))(\gamma) + \chi(\Delta^c)(\lambda)\delta(\lambda)$ *where* $S(\lambda) = e^{-2\pi i\delta(\lambda)}$ *is the scattering operator and* $\chi(f(\lambda)S(\lambda), f(\lambda))(\gamma)$ *denotes the characteristic function, evaluated at* γ, *of the positive arc on the unit circle which extends from* $f(\lambda)S(\lambda)$ *to* $f(\lambda)$. *The function* $f(\lambda)$ *is a unimodular function with support in a Borel set* Δ *of the real line, and* $\chi(\Delta^c)(\lambda)$ *is the value at* λ *of the characteristic function of* Δ^c.

From this result we see immediately that $g'(\lambda, e^{i\theta}) = G(\lambda, e^{i\theta}) = G_l(\lambda, e^{i\theta})$, where $G_l(\lambda, e^{i\theta})$ is the characteristic function of the unit circle arc with endpoints

$$\frac{A(\lambda) - \bar{l} - |k(\lambda)|^2}{A(\lambda) - l - |k(\lambda)|^2} \quad \text{and} \quad \frac{A(\lambda) - \bar{l} + |k(\lambda)|^2}{A(\lambda) - l + |k(\lambda)|^2}.$$

Because the principal function is a complete unitary invariant for an intertwining triplet we can use the cited results to conclude that $(H_1 - \bar{l})/(H - l)$ is unitarily equivalent to the minimal (power) dilation of $W_- f$.

Indeed we have the following scheme:

Let Q_l be the smallest invariant subspace of H_2 which contains the range of $\mathscr{D}(l)$. Let $H_L = H_S(H_2 + \mathscr{D}(l))$, $H_R = H_S(H_2)$, and let P_R denote the projection onto H_R. Set

$$\tilde{H} = \cdots \oplus H_L \oplus H_L \oplus Q_l \oplus H_R \oplus H_R \oplus \cdots.$$

Define $\tilde{W}$ on $\tilde{H}$ for

$$\tilde{x} = [\ldots, x_{-2}, x_{-1}, x_0, x_1, x_2, \ldots]$$

by setting

$$\tilde{W}\tilde{x} = [\ldots, x_{-3}, x_{-2}, x_{-1} + Wx_0, P_R x_0, x_1, \ldots],$$

where $W = W_- f$. Define $\tilde{H}_2$ by setting

$$\tilde{H}_2\tilde{x} = [\ldots, (H_2 + D)x_{-2}, (H_2 + D)x_{-1}, H_2 x_0, H_2 x_1, H_2 x_2, \ldots]$$

and, finally, set

$$\tilde{D}\tilde{x} = [\ldots, 0, 0, Dx_0, 0, 0, \ldots].$$

Then the cited results enable us to assert that $(H_1 - \bar{l})/(H_1 - l)$ is unitarily equivalent to $\tilde{W}$, and H_2 is unitarily equivalent to H_2.

Thus the space on which H_1 and H_2 act is really a disguised copy of the minimal dilation space of the wave operator W.

Now we can prove the following fact.

THEOREM. $(H_1 - \bar{l})/(H_1 - l)$ *is the minimal unitary dilation of* $T(l)$.

For the proof we first observe an order relation between the projection P

onto the absolutely continuous part of H_2 and the projection P_K onto (the Kato space of the perturbation) Q_l.

We claim $PP_K = P$.

This follows at once if we observe that

$$P(H_2 - z)^{-1}(H_1 - l)^{-1}K^* = (H_2 - z)^{-1}R(l)K^* = \frac{H(l, \lambda)}{\bar{K}(\lambda)} \frac{1}{\lambda - z},$$

(letting z vary). The vectors on the right are dense in $L_2(E)$ while the Kato space is generated by the vectors $(H_2 - z)^{-1}(H_1 - l)^{-1}K^*$.

Now we can prove the theorem. Since $(H_1 - \bar{l})/(H_1 - l)$ is the minimal unitary dilation of $W_- f$ we have

$$P_K\left(\frac{H_1 - \bar{l}}{H_1 - l}\right)^n = (W_- f)^n, \qquad n = 0, 1, 2, 3, \ldots,$$

and thus (since the initial space of W_- is the range of P)

$$PP_K\left(\frac{H_1 - \bar{l}}{H_1 - l}\right)^n = P(W_- f)^n = (PW_- f)^N.$$

Hence

$$P\left(\frac{H_1 - \bar{l}}{H_1 - l}\right)^n = [1 + (l - \bar{l})R(l)]^n = T(l)^n.$$

Now we show $(H_1 - \bar{l})/(H_1 - l)$ is a minimal unitary dilation of $T(l)$.

If there were a subspace U of $\tilde{H} \ominus L_2(E)$ invariant for $(H_1 - \bar{l})/(H_1 - l)$ then $((H_2 - z)^{-1}k^*, [1 + (l - \bar{l})(H_1 - l)^{-1}]f) = 0$ for f in U. But $((H_2 - z)^{-1}k^*, f) = 0$; hence we have $((H_2 - z)^{-1}k^*, (H_1 - l)^{-1}f) = E(l, z)((H_2 - z)^{-1}(H_1 - l)^{-1}k^*, f) = 0$. But this implies that f is in the orthocomplement of the Kato space, i.e., $U \subset \tilde{H} \ominus Q_l$. But since $(H_1 - \bar{l})/(H_1 - l)$ is the minimal unitary dilation of $W_- f$, no such invariant subspace can exist.

In the Introduction to this paper we defined an operator $\hat{H}_1$ from the Naimark dilation of the generalized resolution of the identity $F(\lambda)$. We wish to show that the operator H_1 with which we have been occupied ever since is really $\hat{H}_1$, and thus the construction which we have made for the symmetric operator L actually does follow the procedure stated in the Introduction.

THEOREM. $\hat{H}_1 = H_1$.

Let $H_1 = \int \lambda \, dE_\lambda$. We have

$$R(l) = \int \frac{dF_\lambda}{\lambda - l}$$

and by the preceding theorem $PE_\lambda = F_\lambda$. We claim that E_λ is the Naimark minimal orthogonal dilation of the semispectral family F_λ.

If this were not true, there would be a reducing subspace of $\tilde{H} \ominus L_2(E)$ for $(H_1 - \bar{l})/(H_1 - l)$, which we have seen above is not possible.

We can remark that it follows now that $(Lx, y) = \int \lambda \, d(F_\lambda x, y)$ and $\|Lx\|^2 = \int \lambda^2 \, d(F_\lambda x, x)$ for x in the domain of L and $y \in L_2(E)$.

Thus F_λ is a spectral function belonging to L in the sense of M. A. Naimark, and $R(l)$ is a generalized resolvant for L.

The determining function for the contraction $T(l)$ as a solution of a Riemann–Hilbert problem, and index $S(l, \lambda)$.

Let $\bar{\lambda}$ be such that the basic inequality in the proof of Lemma 2.1 holds. We showed that $h_i(\lambda)$ has oscillation one at these points. The basic inequality characterizes the occurrence of singular spectra while the oscillation condition is only necessary and is not sufficient. Nevertheless, when T_l is Fredholm the finite number of points $\bar{\lambda}$ falls into two categories: $h_i(\lambda)$ has left (right) Lebesgue value one at $\bar{\lambda}$ with density one while $h_i(\lambda)$ has right (left) Lebesgue value zero at $\bar{\lambda}$ with density one. If we assign the number plus one to points of the first kind, and the number minus one to points of the second kind then we can define index $S(l, \lambda)$ to be the sum of these integers. This does describe the number of times the essential spectrum of T_l (the unit circle) winds around zero. Of course this integer is also equal to the index of T_l.

The determining function $E(l, z)$ is closely related to another determining function $\varphi(w, z)$. Without elaborating on this relationship here (but see [8, 12]) we wish to indicate briefly that this $\varphi(w, z)$ function can be defined directly in terms of the contraction operator $T(l)$.

If M denotes the operator of multiplication by λ on $L_2(E)$, then the results of [3] imply that

$$\operatorname*{s-lim}_{t \to \pm\infty} \exp(itM) T_l \exp(-itM) = s_\pm(H_2; T_l) = s_\pm(M; T_l)$$

exist.

Furthermore, these symbols commute with M, and are expressed in the direct integral diagonalizing space $(L_2(E))$ for M as $s_\pm(M; T_l) = \int \oplus \, s_\pm(M; T_l)(\lambda) \, d\lambda$, where

$$s_\pm(M; T_l)(\lambda) = \frac{A(\lambda) - \bar{l} \mp |k(\lambda)|^2}{A(\lambda) - l \mp |k(\lambda)|^2}.$$

Thus

$$S(l, \lambda) = s_+(M; T_l)(\lambda) s_-(M; T_l)(\lambda)^{-1}.$$

The real interval $[A(\lambda) - |k(\lambda)|^2, A(\lambda) + |k(\lambda)|^2]$ maps under the linear fractional transformation $t = (e^{i\theta}\bar{l} - l)/(e^{i\theta} - 1)$ to the arc of the unit circle which extends from $S(l, \lambda)f_l(\lambda)$ to $f_l(\lambda)$, and if $G_l(\lambda, e^{i\theta})$, for fixed λ in $\sigma(M)$, denotes the characteristic function of this arc, we have[3]

(a) $\det(s_+(M; T)(\lambda) - z)(s_-(M; T)(\lambda) - z)^{-1} = \exp \int_{|\tau|=1} G_l(\lambda, \tau) \dfrac{d\tau}{\tau - z}$,

(b) $\int G_l(\lambda, e^{i\theta})\, d\theta = 2\pi h_l(\lambda)$.

$G_l(\lambda, e^{i\theta})$ is in fact uniquely determined by (a) and (b).

With

$$d_l = \left[\frac{l - \bar{l}}{i}\right]^{1/2} (H_1 - \bar{l})^{-1} k^*$$

we define

$$\varphi_l(w, z) = 1 + \frac{1}{\pi} d_l \tilde{W}(\tilde{W} - z)(\tilde{H}_2 - w)^{-1} d_l^*,$$

where $\tilde{W}$ is the minimal unitary dilation of $P(H_1 - \bar{l})(H_1 - l)^{-1}$.

We have seen that $\tilde{W}\tilde{H}_2 = (\tilde{H}_2 + (1/\pi) d_l \otimes d_l)\tilde{W}$, and results of [5] show that

$$\det \varphi_l(w, z) = \exp\left(\frac{1}{2\pi i}\right) \int_{-\infty}^{\infty} \int_{|\tau|=1} G_l(\lambda, \tau) \frac{d\tau}{\tau - z} \frac{d\lambda}{\lambda - w}.$$

Furthermore, a little algebra shows that, if $z = \gamma - \bar{l}/\gamma - l$, then

$$\varphi_l(w, z) = \varphi_l\left(w, \frac{\gamma - \bar{l}}{\gamma - l}\right) = E^{-1}(\gamma, w)E(l, w).$$

THEOREM. *For $|z| > 1$, we have*

$$\varphi_l(w, z) = 1 + \frac{1}{\pi} d_{\bar{l}}(T_l - z)^{-1}(M - w)^{-1} d_l.$$

The proof follows from a lemma in [5]. See also the proof of Theorem (3.1) in [3] and Theorem (3.1) in [1]. Furthermore, the following Riemann–Hilbert relation is satisfied:

$$(s_+(M; T(l) - z)(s_-(M; T(l) - z)^{-1} \varphi_l(\lambda + i0, z) = \varphi_l(\lambda - i0, z) \qquad \text{for} \quad |z| \neq 1.$$

[3] The use of determinants here indicates the form of these relations when $A(\lambda)$ and $k(\lambda)$ are matrix or operator valued.

In fact, for $|z|$ large enough, $\varphi_l(w, z)$ is the unique solution (with limit 1 as $w \to \infty$) having boundary values in $\mathscr{L}_p$, $p > 1$. (See [5, Theorem (4.1)] for the proof and the definition of $\mathscr{L}_p$.) Thus our procedure amounts to a way of taking the unique solution (which is defined for $|z| > 1$ by analytic continuation) and extending it into the circle $|z| < 1$. It is remarkable that our dilation procedure accomplishes this automatically.

VIII. Symmetric Toeplitz Operators

Symmetric Toeplitz and Wiener–Hopf operators can be treated by the methods of the previous chapters. However, we will not give a full exposition here. Instead we restrict ourselves to a very brief treatment of the construction of the deficiency spaces for the Toeplitz case. (The unitarily equivalent Wiener–Hopf situation will be discussed in detail elsewhere.)

Let $\mathscr{L} = P_+ W$, where $W(\theta)$ is real and periodic of period 2π and is in $L_2(0, \pi)$, and P_+ denotes the projection of $L_2(0, \pi)$ onto the Hardy space $L_+{}^2$.

Define $L_n = P_+ W_n$ by

$$W_n(\theta) = \begin{cases} W(\theta) & \text{if } \ |W(\theta)| \leqslant n \\ 0 & \text{otherwise.} \end{cases}$$

Then L_n is a bounded self-adjoint operator on $L_+{}^2$, and we study the "limit" of the sequence $\{L_n\}$ in the sense of Section I.

We will not develop the machinery here which is necessary in order to discuss the relationship which exists between symmetric Toeplitz operators and scattering problems.

Instead we will proceed in the most direct fashion to show how the method of Section I gives the deficiency spaces.

We use for this purpose a known result due to Calderon *et al.* [13] and Krein [14], which states:

Theorem.

$$(L_n - w)^{-1}(1 - \bar{\beta}e^{i\theta})^{-1}, (1 - \bar{z}e^{i\theta})^{-1})_{L_+{}^2} = \frac{S_n(w, z)\bar{S}_n(\bar{w}, \beta)}{1 - \bar{\beta}z},$$

where

$$S_n(w, z) = \exp\left(-\frac{1}{4\pi}\int \frac{e^{i\theta} + z}{e^{i\theta} - z} \ln(W_n(\theta) - w)\, d\theta\right).$$

It is quite simple to demonstrate the following facts (see [10] or [15]):

(α) $w\text{-}\lim_{n\to\infty}(L_n - w)^{-1} = R(w)$ exists;

(β) There exists a generalized resolution of the identity $F(t)$ such that

$$(R(w)f, f)_{L_+{}^2} = \int \frac{d(F(t)f, f)}{t - w} \qquad \text{for} \quad f \in L_+{}^2.$$

Let $E(t)$ be the minimal Naimark dilation of $F(t)$ and form $H_1 = \int t\, dE_t$. Then $P(H_1 - w)^{-1} = R(w)$.

We proceed with a direct construction of the operator

$$N(w) \equiv P(H_1 - w)^{-1}(H_1 - \overline{w})^{-1}P - R(w)R(\overline{w}).$$

We can take the limit in n in (α) to get

$$((H_1 - w)^{-1}(H_1 - \overline{w})^{-1}(1 - \overline{\beta}e^{i\theta})^{-1}, (1 - \overline{z}e^{i\theta}))_{L_+{}^2}$$
$$= \frac{1}{1 - \overline{\beta}z} \frac{S(w, z)\overline{S}(\overline{w}, \beta) - S(\overline{w}, z)\overline{S}(w, \beta)}{w - \overline{w}} \tag{1}$$

where

$$S(w, z) = \exp\left(-\frac{1}{4\pi} \int \frac{e^{i\theta} + z}{e^{i\theta} - z} \log(W(\theta) - w)\, d\theta \right).$$

(α) also gives us

$$R(w)(1 - \overline{\beta}e^{i\theta})^{-1} = \frac{S(w, e^{i\theta})\overline{S}(\overline{w}, \beta)}{1 - \overline{\beta}e^{i\theta}}. \tag{2}$$

Thus,

$$\|R(w)(1 - \beta e^{i\theta})^{-1}\|_{L_+{}^2}^2 = \frac{1}{2\pi} \int_0^{2\pi} \left| \frac{S(w, e^{i\theta})\overline{S}(\overline{w}, \beta)}{1 - \overline{\beta}e^{i\theta}} \right|^2 d\theta. \tag{3}$$

Hence

$$(N(w)(1 - ze^{i\theta})^{-1}, (1 - \overline{z}e^{i\theta})^{-1})$$
$$= \frac{1}{1 - |z|^2} \frac{S(w, z)\overline{S}(\overline{w}, z) - S(\overline{w}, z)\overline{S}(w, z)}{w - \overline{w}}$$
$$- \frac{1}{2\pi} \int_0^{2\pi} \left| \frac{S(w, e^{i\theta})}{1 - \overline{z}e^{i\theta}} \right|^2 d\theta\, \overline{S}(\overline{w}, z)S(\overline{w}, z). \tag{4}$$

Now let

$$\Omega_w(z) = \begin{cases} S(w, z)S(\overline{w}, z)^{-1} & \text{if} \quad \operatorname{Im} w > 0 \\ -S(w, z)S(\overline{w}, z)^{-1} & \text{if} \quad \operatorname{Im} w < 0. \end{cases}$$

It is easy to see that there exists a positive measure ν_w such that

$$\operatorname{Im} \Omega_w(z) = \frac{1}{2\pi} \int_0^{2\pi} \frac{1 - |z|^2}{|e^{i\theta} - z|^2} d\nu_w(\theta).$$

Further, as before, the Radon–Nikodym derivative ν_w' of ν_w is easy to compute. We simply observe that

$$|S(w, e^{i\theta})|^2 = \frac{S(w, e^{i\theta})S(\bar{w}, e^{i\theta})^{-1} - \bar{S}(w, e^{i\theta})\bar{S}(w, e^{i\theta})\bar{S}(\bar{w}, e^{i\theta})^{-1}}{w - \bar{w}}.$$

Since $\operatorname{Im} \Omega_w(e^{i\theta}) = \nu_w'(\theta)$, we get, finally, that

$$(N(w)(1 - \bar{z}e^{i\theta})^{-1}, (1 - \bar{z}e^{i\theta})^{-1})S(\bar{w}, z)^{-1}S^{-1}(\bar{w}, z) = \frac{1}{2\pi}\frac{1}{\operatorname{Im} w}\int \frac{d\nu_w^{(s)}(\theta)}{|e^{i\theta} - z|^2} \tag{5}$$

where $\nu_w^{(s)}$ is the singular part of the measure ν_w.

Suppose $\nu_w^{(s)}$ is purely atomic with mass $\nu^{(s)}(w)$ at the points $\{e^{i\theta_j(w)}\}_{j=1}^{n(w)}$. Then

$$(N(w)(1 - \bar{z}e^{i\theta})^{-1}, (1 - \bar{z}e^{i\theta})^{-1})_{L_+^2} = \frac{1}{2\pi}\sum_{j=1}^{n(w)} \frac{\nu_j^{(s)}(w)}{|\operatorname{Im} w|}\frac{S(\bar{w}, z)}{e^{i\theta_j(w)} - z}\frac{\bar{S}(\bar{w}, z)}{e^{-i\theta_j(w)} - \bar{z}}.$$

But it can be seen as before that

(a) $\dfrac{S(w, e^{i\theta})}{e^{i\theta_j(w)} - e^{i\theta}} \in L_+^2,$

(b) $\lim_{\gamma \to e^{i\theta_j(w)}} S(\bar{w}, \gamma) = 0.$

Thus, since

$$\int_0^{2\pi} \frac{S(\bar{w}, e^{i\theta})}{e^{i\theta} - \gamma}\frac{1}{1 - ze^{-i\theta}}\,d\theta = \frac{1}{i}\int_{|z|=1} \frac{S(\bar{w}, \tau)}{(\tau - \gamma)(\tau - z)}\,d\tau$$
$$= \frac{2\pi}{z - \gamma}[S(\bar{w}, z) - S(\bar{w}, \gamma)]$$

for $|w| < 1$, $|z| < 1$, we have

$$\frac{S(\bar{w}, z)}{z - e^{i\theta_j(w)}} = \frac{1}{2\pi}\int_0^{2\pi} \frac{S(\bar{w}, e^{i\theta})}{e^{i\theta} - e^{i\theta_j}}\left[\frac{1}{1 - \bar{z}e^{i\theta}}\right]^* d\theta.$$

Since the functions $1/(1 - ze^{i\theta})$ span L_+^2, we can assert that

$$N(w) = \frac{1}{2\pi}\sum_{j=1}^{n(w)} \frac{\nu_j^{s}(w)}{|\operatorname{Im} w|}\frac{S(\bar{w}, e^{i\theta})}{e^{i\theta} - e^{i\theta_j(w)}} \otimes \frac{S(\bar{w}, e^{i\theta})}{e^{i\theta} - e^{i\theta_j(w)}}.$$

We have sketched the proof of the following result:

THEOREM. *$T(w) \equiv 1 + (w - \bar{w})R(w)$ restricted to the orthocomplement of* *the range of* $N(w)$ *is the Cayley transform of a symmetric operator* L. *The deficiency indices of* L *are* $(n(w), n(\overline{w}))$, *and the corresponding deficiency spaces are the ranges (respectively) of* $N(w)$ *and* $N(\overline{w})$.

We will not provide further information here; but a core like that previously constructed can be found, and a perturbation problem (this time a unitary perturbation problem) can be constructed as well. Again the defect spaces are the projections to the base space of the singular spaces of the perturbed operator.

The main fact is that the phase shift which controls this perturbation problem is $\delta_w(e^{i\theta}) = (1/\pi)\arg(W(e^{i\theta}) - \overline{w})$, $\operatorname{Im} w \geqslant 0$, with $0 \leqslant \delta_w(e^{i\theta}) \leqslant 1$. There is a positive measure $\mu_w(\tau)$ on the circle and a constant $\beta(w)$ such that the perturbation determinant has the form

$$\exp \int_{|\tau|=1} \delta_w(\tau) \frac{d\tau}{\tau - z} = 1 + \beta(w) \int_{|\tau|=1} \tau \frac{d\mu_w(\tau)}{\tau - z}.$$

The unperturbed unitary operator U (which plays the role which H_2 had in the previous sections) is taken to be multiplication by $e^{i\theta}$ on $L_2(d\mu_w)$, and the entire perturbation problem is determined from the above equation.

The measure $\mu_w(\cdot)$ is equivalent to the measure $v_w(\cdot)$; in fact, we have for any Borel set Δ on the circle $(1/2\pi)v_w(\Delta) = \sin \frac{1}{2}\int_0^{2\pi} \delta_w(e^{i\theta})\, d\theta\, \mu_w(\Delta)$.

Again index corresponds to eigenvalues $e^{i\theta}$ where the phase shift has oscillation one. $W(e^{i\theta}) - w/W(e^{i\theta}) - \overline{w}$ wraps around the circle as θ varies.

The phase shift determines a *principal function on the torus* which is the complete unitary invariant of a pair or unitary operators U and Ω with $U^*\Omega U\Omega^* - 1$ having one-dimensional range. Ω plays the role here that $H_1 - \overline{l}/H_1 - l$ played in the previous sections of this paper, and we see in fact that the Toeplitz and singular integral operators both arise from a study of certain compressions of Ω to privileged invariant subspaces of U. In fact the study of the case where $|k(\lambda)|^2/1 + \lambda^2$ is integrable for the singular integral operators is best treated in the context of the two unitary situation.

We close by describing the Wiener–Hopf case. Let $\gamma z = z - i/z + i$ if $\operatorname{Im} z \geqslant 0$. For any function g defined on the unit circle, T, set $Vg(x) = g(\gamma x)$ and define $\Sigma g(x) = \sqrt{2Vg(x)/x + i}$. It is well known (see Devinatz, 16) that Σ is a unitary map from $L^2(T)$ onto $L^2(R)$ which maps $L_+{}^2(T)$ onto $H^2(R)$, the Hardy space of the line. The usual Wiener–Hopf operator acts on $H^2(R)$ in the following way: we let $P_+{}'$ denote the projection from $L^2(R)$ to $H^2(R)$ and we set $L(f)g = fg$ for f in L^∞ and g in $H^2(R)$; then we define the Wiener–Hopf operator by putting $W_f = P_+{}'L(f)|_{H^2(R)}$. W_f is unitarily equivalent to the Toeplitz operator $T_{\tilde{f}} = P_+ V^{-1} f|_{L_+{}^2(T)}$; indeed, $\Sigma T_{\tilde{f}} \Sigma^{-1} = W_f$.

Since it is clear that

$$\frac{1}{2\pi} \int_0^{2\pi} |V^{-1} f(e^{i\theta})|^2\, d\theta = \frac{1}{\pi} \int_0^{\infty} |f(x)|^2 \frac{dx}{1 + x^2}$$

we can see how to define a symmetric Wiener–Hopf operator with real symbol f satisfying $\int_0^\infty |f(x)|^2\, dx/1 + x^2 < \infty$. Our procedure for defining a symmetric limiting operator by constructing a sequence of truncated symbols is clearly equivalent to taking the symmetric Toeplitz operator T_f discussed at the beginning of this section and then defining the corresponding Wiener–Hopf operator to be $\Sigma T_f \Sigma^{-1}$.

In the Fourier transformed representation this corresponds to treating integral operators of the form $\int_0^\infty \hat{f}(t - s)\hat{g}(s)\, ds$ for g in $H^2(R)$, where $\hat{f}$ is the distribution Fourier transform of a real f satisfying $\int_0^\infty |f(x)|^2\, dx/1 + x^2 < \infty$.

All of our theorems on the deficiency spaces of the corresponding unbounded symmetric Toeplitz now carry over in an obvious way to the unbounded Wiener–Hopf operator.

References

1. J. D. Pincus, Commutators and systems of singular integral equations, I, *Acta Math.* **121** (1968), 219–249.
2. R. W. Carey and J. D. Pincus, An invariant for certain operator algebras, *Proc. Nat. Acad. Sci. U.S.A.* **71** (1974) 1952-1956.
3. R. W. Carey and J. D. Pincus, Commutators, symbols, and determining functions, *J. Functional Analysis* **19** (1975) 50–80.
4. R. W. Carey and J. D. Pincus, Mosaic, principal functions and mean motion in Von Neumann Algebras, *Acta Math* **138** (1977), 153–218.
5. R. W. Carey and J. D. Pincus, The structure of intertwining isometries, *Indiana Univ. Math.* **22** (1973) 679–703.
6. I. B. Simonenko, Some general questions on the theory of the Riemann boundary problem, *Izv. Akad. Nauk SSSR* **32** (1968), 1138–1146.
7. R. G. Douglas, Local Toeplitz operators, *J. London Math. Soc.*, to appear.
8. J. D. Pincus, Some applications of operator valued analytic functions of two complex variables, Linear operators and approximation, *ISNM* **20** (1972), 68–79.
9. J. D. Pincus, Symmetric singular integral operators, *Indiana Univ. Math J.* **23** (1973), 537–556.
10. J. D. Pincus and J. Rovnyak, A spectral theory for some unbounded self-adjoint singular integral operators, *Amer. J. Math.* **91** (1969), 619–636.
11. A. Garsia, "Topics in Almost Everywhere Convergence," Markham, Chicago, 1970.
12. R. W. Carey and J. D. Pincus, Construction of seminormal operators with prescribed mosaic, *Indiana Univ. Math. J.* **23** (1974), 1155–1165.
13. A. Calderon, F. Spitzer, and H. Widom, Inversion of Toeplitz matrices, *Illinois J. Math.* **3** (1959), 490–498.
14. M. G. Krein, Integral equations on the half line with kernels depending on the difference of the arguments, *Amer. Math. Soc. Transl., Ser. 2* **22**(1959), 163–288.
15. J. Rovnyak, On the theory of unbounded Toeplitz operators, *Pacific J. Math.* **31** (1969), 481–496.
16. A. Devinatz, On Wiener–Hopf operators, "Functional Analysis," pp. 81–118, Thompson, 1967.

TOPICS IN FUNCTIONAL ANALYSIS
ADVANCES IN MATHEMATICS SUPPLEMENTARY STUDIES, VOL. 3

Change of Variables Formulas with Cayley Inner Functions

MARVIN ROSENBLUM AND JAMES ROVNYAK[†]

Department of Mathematics
University of Virginia
Charlottesville, Virginia

THIS PAPER IS DEDICATED TO M. G. KREIN ON HIS SEVENTIETH BIRTHDAY, AS AN EXPRESSION OF ADMIRATION AND APPRECIATION OF HIS WORK

We develop a calculus for substitution operations in definite and singular integrals of functions on a circle or line, and give applications to distribution function formulas, inversion formulas for singular integrals, and orthogonal expansions.

1. INTRODUCTION

Integration by substitution in elementary calculus is based on the formula

$$\int_a^b (f \circ \xi)(t)\xi'(t)\,dt = \int_c^d f(t)\,dt, \tag{1.1}$$

where f is a given function on $[c, d]$ and ξ is a suitable function that maps $[a, b]$ on $[c, d]$. More sophisticated integration by substitution results can be found, for example, in Hewitt and Stromberg [12, pp. 342–346]. We call such results and (1.1) composition formulas. We shall be concerned in this paper with composition formulas with Cayley inner functions.

A Cayley inner function is an analytic function ξ defined off the real axis, such that $\xi(z) = \xi(z^*)^*$ for all nonreal z, $\operatorname{Im}\xi(z) > 0$ for $y > 0$, and $\xi(x) = \xi(x + i0) = \xi(x - i0)$ is real a.e. Suppose that Δ is a real measurable set such that neither Δ nor its complement in the real line is a Lebesgue null set. Let (c, d) be a real interval which is not empty and not the whole real line. Then (see Theorem 3.15) a Cayley inner function ξ may be chosen such that $\xi(x) \in (c, d)$ for almost all $x \in \Delta$ and $\xi(x) \notin (c, d)$ for almost all $x \in R\backslash\Delta$. Typical composition formulas that we prove in this situation are

$$\int_\Delta \frac{\beta - \alpha^*}{(t - \beta)(t - \alpha^*)}(f \circ \xi)(t)\,dt = \int_c^d \frac{\xi(\beta) - \xi(\alpha)^*}{(t - \xi(\beta))(t - \xi(\alpha)^*)} f(t)\,dt \tag{1.2}$$

[†] This research was supported by NSF Grant MPS 75–04594.

ISBN 0-12-287150-2

and

$$\mathrm{PV}\int_{\Delta}\frac{(f\circ\xi)(t)}{t-\alpha^*}\frac{dt}{t-x}=\frac{\xi(x)-\xi(\alpha)^*}{x-\alpha^*}\,\mathrm{PV}\int_c^d\frac{f(t)}{t-\xi(\alpha)^*}\frac{dt}{t-\xi(x)},\tag{1.3}$$

where f is a given function on (c, d), α, β are any nonreal numbers, and (1.3) holds a.e. on the real line with PV denoting a principal value integral. Under the assumption

$$\lim_{y\to\infty}\xi(iy)/(iy)=\beta>0$$

we obtain additional formulas such as

$$\int_{\Delta}(f\circ\xi)(t)\,dt=\beta^{-1}\int_c^d f(t)\,dt\tag{1.4}$$

and, a.e. on the real line,

$$\mathrm{PV}\int_{\Delta}\frac{(f\circ\xi)(t)}{t-x}dt=\mathrm{PV}\int_c^d\frac{f(t)}{t-\xi(x)}dt.\tag{1.5}$$

We think of formulas of the type (1.2)–(1.5) as special change of variables formulas analogous to (1.1). Such formulas were introduced in [21]. They have proved helpful in reducing problems in analysis on a union of intervals to analogous problems on a single interval (see Rosenblum and Rovnyak [20, 21] and Chandler [7, 8]).

Sections 2 and 3 of this paper generalize the composition formulas used in previous work. These sections give a self-contained treatment of the basic formulas. They are partially expository. Composition formulas for singular integrals are derived in Section 4.

Sections 5–7 are devoted to applications. The first application, in Section 5, is to distribution function formulas. We generalize a result of Loomis [15] by calculating the distribution function for

$$\xi(x)=\mathrm{PV}\int_{-\infty}^{\infty}\frac{d\mu(t)}{t-x},$$

where μ is a finite singular measure. The distribution function for

$$\eta(x)=\mathrm{PV}\int_{\Delta}\frac{dt}{t-x},$$

where Δ is a real measurable set, is calculated separately for x in Δ and for x in $R\backslash\Delta$, and a result of Stein and Weiss [24] is obtained as a byproduct.

In Section 6 we extend a subclass of Ahiezer's inversion formulas for singular integrals [1] so that they are valid over arbitrary real measurable sets instead of bounded open sets. In the case of bounded sets our transforms also map polynomials to polynomials. In one sense our transforms are not as general as Ahiezer's, because Ahiezer's transforms in general induce partial isometries, and ours always induce unitary operators.

One way in which to apply our composition formulas is to map special function theories from a single interval to a union of intervals or some more general set. In Section 7 we illustrate this by deriving generalized Chebychev expansions for functions defined on a real measurable set Δ such that neither Δ nor $R\backslash\Delta$ is a Lebesgue null set. The discussion uncovers some new elementary facts about abstract shift operators on a Hilbert space. These consist of connections between the real and imaginary parts of a shift operator and the Chebychev polynomials (see Theorem 7.2). In Theorem 7.3 we determine the general form of a shift operator S whose real part $X = \operatorname{Re} S$ is given in canonical form. These results are used as aids in calculating generalized Chebychev expansions. The choice of Chebychev polynomials is an arbitrary one. The method is quite general and could be applied to other special functions.

2. Composition Formulas for Harmonic Functions

By D, Γ, D^c we mean, respectively, the sets $|z| < 1$, $|z| = 1$, $|z| > 1$ in the complex plane C. Let σ denote normalized Lebesgue measure on Γ, and for any measurable subset δ of Γ let $L^p(\delta)$, $0 < p \leqslant \infty$, denote the standard complex Lebesgue spaces with respect to the restriction of σ to δ. If $0 < p \leqslant \infty$, then by $H^p(D)$ and $h^p(D)$ we mean, respectively, the usual Hardy spaces of holomorphic and real harmonic functions on D (see Duren [9, p. 2]). In addition, if $1 \leqslant p \leqslant \infty$, we introduce the class $\mathbf{h}^p_{ac}(D)$ of complex harmonic functions $F(z)$ on D which have a Poisson representation

$$F(z) = \int_\Gamma \frac{1 - |z|^2}{|e^{it} - z|^2} f(e^{it})\, d\sigma(e^{it}), \qquad z \in D, \tag{2.1}$$

where $f \in L^p(\Gamma)$. If $F \in \mathbf{h}^p_{ac}(D)$ and $f \in L^p(\Gamma)$ are connected by (2.1), then F has a nontangential boundary function given by $F(e^{it}) = f(e^{it})$ a.e. on Γ. If $1 < p \leqslant \infty$, then $\mathbf{h}^p_{ac}(D) = h^p(D) + ih^p(D)$. However, $\mathbf{h}^1_{ac}(D)$ is a proper subset of $h^1(D) + ih^1(D)$ (see Duren [9, Chapter 1]).

Throughout this section we assume that $B \in H^\infty(D)$, $B \not\equiv 0$, and $|B(z)| < 1$ for $z \in D$. By $B(e^{it})$ we mean the nontangential boundary function formed from values of B inside D. We emphasize this because we extend the domain of B to $D \cup D^c$ by requiring that

$$B(z)B(1/z^*)^* = 1 \tag{2.2}$$

at all points z in $D \cup D^c$ except where one factor is 0, in which case the other factor is defined to be ∞. With this definition we may form a boundary function $B^c(e^{it})$ by taking nontangential limits of values of B in D^c. We have $B^c(e^{it}) = B(e^{it})$ if $|B(e^{it})| = 1$, and $B^c(e^{it}) \neq B(e^{it})$ if $|B(e^{it})| < 1$, assuming e^{it} is excluded from the σ-null set where one or the other boundary function fails to exist.

2.1 LEMMA. (i) *For any measurable subset Δ of Γ, we have*

$$\sigma\{e^{it}: e^{it} \in \Gamma,\, B(e^{it}) \in \Delta\} \leqslant \int_\Delta \frac{1 - |B(0)|^2}{|e^{it} - B(0)|^2}\, d\sigma(e^{it}).$$

(ii) *If γ is a measurable subset of Γ such that $\sigma(\gamma) = 0$, then $\sigma\{e^{it}: e^{it} \in \Gamma,$ $B(e^{it}) \in \gamma\} = 0$.*

Proof. (i) It is sufficient to prove this when Δ is a Borel set. In fact, we need only prove (i) for closed sets Δ, because the class of Borel sets Δ for which the conclusion is true is closed under all unions of expanding sequences and intersections of contracting sequences. Assume then that Δ is closed. If f is the real part of a polynomial, then

$$\int_\Gamma (f \circ B)(e^{it})\, d\sigma(e^{it}) = \int_\Gamma \frac{1 - |B(0)|^2}{|e^{it} - B(0)|^2} f(e^{it})\, d\sigma(e^{it}). \tag{2.3}$$

Let g be any continuous function on Γ such that $0 \leqslant g \leqslant 1$ on Γ and $g = 1$ on Δ. Let $\delta = \{e^{it}: e^{it} \in \Gamma,\, B(e^{it}) \in \Delta\}$. A routine approximation argument using (2.3) yields the inequality in the relation

$$\sigma(\delta) = \int_\delta (g \circ B)(e^{it})\, d\sigma(e^{it}) \leqslant \int_\Gamma \frac{1 - |B(0)|^2}{|e^{it} - B(0)|^2} g(e^{it})\, d\sigma(e^{it}).$$

Since Δ is closed, the conclusion follows by the arbitrariness of g.

(ii) This follows from (i).

If $F \in \mathbf{h}^1_{ac}(D)$, the composite function $F \circ B$ is harmonic on D. If $F(e^{it})$ denotes the boundary function of F, then by Lemma 2.1(ii) we may define $(F \circ B)(e^{it})$ a.e. on Γ by

$$(F \circ B)(e^{it}) = F(B(e^{it})). \tag{2.4}$$

The notation suggests that $(F \circ B)(e^{it})$ is the boundary function for $F \circ B$ on D. It turns out that this and more is true.

2.2 THEOREM. *If $F \in \mathbf{h}^1_{ac}(D)$, then $F \circ B \in \mathbf{h}^1_{ac}(D)$, and*

$$(F \circ B)(z) = \int_\Gamma \frac{1 - |z|^2}{|e^{it} - z|^2} (F \circ B)(e^{it})\, d\sigma(e^{it}), \qquad z \in D. \tag{2.5}$$

In particular, $(F \circ B)(e^{it})$ is the boundary function of $F \circ B$ on D, and $(F \circ B)(e^{it})$ belongs to $L^1(\Gamma)$.

Proof. Using the Poisson representation for $H^1(D)$ (see Duren [9, p. 34]) we easily verify the conclusions of the theorem if $F(z) = z^n$ or $F(z) = z^{*n}$ in D for some $n = 0, 1, 2, \ldots$. Therefore the result follows if F is the real part of a polynomial. Consider an arbitrary $F \in \mathbf{h}^1_{\mathrm{ac}}(D)$. We can assume that F is real valued, since otherwise we need only treat the real and imaginary parts of F separately. Represent F in the form (2.1) where $f \in L^1(\Gamma)$. Select a sequence $\{f_n\}_1^\infty$ such that each f_n is the real part of a polynomial and $f_n \to f$ a.e. on Γ and in $L^1(\Gamma)$ norm. By the special case of the theorem noted above, we have

$$(f_n \circ B)(z) = \int_\Gamma \frac{1 - |z|^2}{|e^{it} - z|^2} (f_n \circ B)(e^{it})\, d\sigma(e^{it}), \qquad z \in D \tag{2.6}$$

for each $n = 1, 2, 3, \ldots$. For $m, n = 1, 2, 3, \ldots$ and $0 < r < 1$,

$$\begin{aligned}
&\int_\Gamma |(f_m \circ B)(re^{it}) - (f_n \circ B)(re^{it})|\, d\sigma(e^{it}) \\
&\quad \leqslant \int_\Gamma \int_\Gamma \frac{1 - |B(re^{i\theta})|^2}{|e^{it} - B(re^{i\theta})|^2} |f_m(e^{it}) - f_n(e^{it})|\, d\sigma(e^{it})\, d\sigma(e^{i\theta}) \\
&\quad = \int_\Gamma \frac{1 - |B(0)|^2}{|e^{it} - B(0)|^2} |f_m(e^{it}) - f_n(e^{it})|\, d\sigma(e^{it}) \\
&\quad \leqslant \frac{1 + |B(0)|}{1 - |B(0)|} \int_\Gamma |f_m - f_n|\, d\sigma.
\end{aligned}$$

Letting $r \uparrow 1$ and applying Fatou's lemma, we see that the sequence $\{(f_n \circ B)(e^{it})\}_1^\infty$ is Cauchy in $L^1(\Gamma)$. We assert that

$$\lim_{n \to \infty} (f_n \circ B)(e^{it}) = (F \circ B)(e^{it}) \tag{2.7}$$

a.e. on Γ. First note that $f_n(z) \to F(z)$, $z \in D$. For since f_n is the real part of a polynomial, we have

$$f_n(z) = \int_\Gamma \frac{1 - |z|^2}{|e^{it} - z|^2} f_n(e^{it})\, d\sigma(e^{it}), \qquad z \in D, \tag{2.8}$$

for each $n = 1, 2, 3, \ldots$. Comparing (2.1) and (2.8), we see that $f_n(z) \to F(z)$, $z \in D$, since $f_n \to f$ in $L^1(\Gamma)$ norm. Thus (2.7) holds if e^{it} is a point where $|B(e^{it})| < 1$. By Lemma 2.1 and the fact that $f_n \to f$ a.e. on Γ, (2.7) holds a.e. on the set of points $e^{it} \in \Gamma$ such that $|B(e^{it})| = 1$. Therefore (2.7) holds a.e.

on Γ. In view of (2.7), (2.5) follows from (2.6) on passage to the limit. The result follows.

2.3 THEOREM. *For any* $F \in \mathbf{h}^1_{\mathrm{ac}}(D)$,

$$\int_\Gamma \frac{1 - zw^*}{(e^{it} - z)(e^{-it} - w^*)} (F \circ B)(e^{it})\, d\sigma(e^{it})$$

$$= \int_\Gamma \frac{1 - B(z)B(w)^*}{(e^{it} - B(z))(e^{-it} - B(w)^*)} F(e^{it})\, d\sigma(e^{it}) \tag{2.9}$$

with absolute convergence of the integrals for any $z, w \in D \cup D^c$. *The expression on the right of* (2.9) *is interpreted by continuity when* z *and/or* w *approach poles of* B *in* D^c.

Proof. It is sufficient to prove (2.9) for real valued F. The function

$$G(z) = \int_\Gamma \frac{e^{it} + z}{e^{it} - z} (F \circ B)(e^{it})\, d\sigma(e^{it}) - \int_\Gamma \frac{e^{it} + B(z)}{e^{it} - B(z)} F(e^{it})\, d\sigma(e^{it})$$

is holomorphic in D, and $\operatorname{Re} G(z) = (F \circ B)(z) - (F \circ B)(z) = 0$ in D by Theorem 2.2. Therefore G is a constant ic where c is real. We obtain (2.9) for $z, w \in D$ from the relation $G(z) + G(w)^* = 0$, $z, w \in D$. We obtain (2.9) for $z \in D$, $w \in D^c$ from $G(z) = G(1/w^*)$, $z \in D$, $w \in D^c$, and similarly for $z \in D^c$, $w \in D$. Finally, (2.9) for $z, w \in D^c$ follows from $G(1/z) + G(1/w)^* = 0$, $z, w \in D^c$.

We make use of a generalized version of Littlewoods's subordination theorem.

2.4 THEOREM. *Let* G *be a subharmonic function defined in* D. *If*

$$\max\{|B(z)| : |z| \leqslant r\} \leqslant \rho, \tag{2.10}$$

where $r, \rho \in (0, 1)$, *then*

$$\int_\Gamma \frac{1 - |z|^2}{|e^{it} - z|^2} (G \circ B)(re^{it})\, d\sigma(e^{it}) \leqslant \int_\Gamma \frac{\rho^2 - |B(rz)|^2}{|\rho e^{it} - B(rz)|^2} G(\rho e^{it})\, d\sigma(e^{it}) \tag{2.11}$$

for all $z \in D$.

Proof. We may assume $G \not\equiv -\infty$. Let F be the least harmonic majorant of G on $\rho D = \{z : |z| < \rho\}$. Then $G \leqslant F$ on ρD and $G = F$ a.e. on $\rho\Gamma = \{z : |z| = \rho\}$ with respect to normalized Lebesgue measure. By (2.10), $G \circ B \leqslant$

$F \circ B$ on rD. Therefore, if $z \in D$, we have

$$\int_\Gamma \frac{1-|z|^2}{|e^{it}-z|^2} (G \circ B)(re^{it})\, d\sigma(e^{it}) \leqslant \int_\Gamma \frac{1-|z|^2}{|e^{it}-z|^2} F(\rho(\rho^{-1}B(re^{it})))\, d\sigma(e^{it})$$

$$= \int_\Gamma \frac{\rho^2-|B(rz)|^2}{|\rho e^{it}-B(rz)|^2} F(\rho e^{it})\, d\sigma(e^{it}),$$

where the equality follows by Theorem 2.3. The result follows.

Littlewood's theorem is obtained by specializing Theorem 2.4 to the case where $B(0) = 0$ and $z = 0$. See Duren [9, p. 10] and Ryff [23].

2.5 Corollary. *Assume $1 \leqslant p < \infty$. If $F \in \mathbf{h}^p_{\mathrm{ac}}(D)$, then $F \circ B \in \mathbf{h}^p_{\mathrm{ac}}(D)$ and*

$$\int_\Gamma \frac{1-|z|^2}{|e^{it}-z|^2} |(F \circ B)(e^{it})|^p\, d\sigma(e^{it}) \leqslant \int_\Gamma \frac{1-|B(z)|^2}{|e^{it}-B(z)|^2} |F(e^{it})|^p\, d\sigma(e^{it}) \quad (2.12)$$

for all $z \in D$.

Proof. Assume $F \in \mathbf{h}^p_{\mathrm{ac}}(D)$. Then $G = |F|^p$ is subharmonic in D, so by Theorem 2.4

$$\int_\Gamma \frac{1-|z|^2}{|e^{it}-z|^2} |(F \circ B)(re^{it})|^p\, d\sigma(e^{it}) \leqslant \int_\Gamma \frac{\rho^2-|B(rz)|^2}{|\rho e^{it}-B(rz)|^2} |F(\rho e^{it})|^p\, d\sigma(e^{it}) \quad (2.13)$$

for all $z \in D$, where $r, \rho \in (0, 1)$ are chosen so that (2.10) holds. Letting $\rho \uparrow 1$, we obtain

$$\int_\Gamma \frac{1-|z|^2}{|e^{it}-z|^2} |(F \circ B)(re^{it})|^p\, d\sigma(e^{it}) \leqslant \int_\Gamma \frac{1-|B(rz)|^2}{|e^{it}-B(rz)|^2} |F(e^{it})|^p\, d\sigma(e^{it}) \quad (2.14)$$

since $F(re^{it}) \to F(e^{it})$ in $L^p(\Gamma)$ norm (see [22, p. 238]). Now let $r \uparrow 1$ in (2.14) and use Fatou's lemma and Theorem 2.2 to deduce (2.12).

We map the results obtained thus far to a half-plane situation. Let Π, R, Π^c denote, respectively, the sets $y > 0$, $y = 0$, $y < 0$ in C, where $z = x + iy$ with x, y real. Unless otherwise specified, we use Lebesgue measure on R. If Δ is a measurable subset of R, by $L^p(\Delta)$, $0 < p \leqslant \infty$, we understand the usual Lebesgue spaces. Let $\mathscr{P}$ denote the class of functions holomorphic and having positive imaginary part in Π. We define $\mathbf{h}^p_{\mathrm{ac}}(\Pi)$, $1 \leqslant p \leqslant \infty$, to be the space of complex harmonic functions F on Π having a Poisson representation

$$F(z) = \frac{1}{\pi} \int_{-\infty}^{\infty} \frac{\operatorname{Im} z}{|t-z|^2} f(t)\, dt, \qquad z \in \Pi, \quad (2.15)$$

where $f(t)/(1 + t^2)^{1/p} \in L^p(R)$ (when $p = \infty$, $f \in L^\infty(R)$). Given $F \in \mathbf{h}^p_{\mathrm{ac}}(\Pi)$ we may recapture the function f appearing in (2.15) as the nontangential boundary function of $F: F(x) = f(x)$ a.e. on R. There is a one-to-one correspondence between $\mathbf{h}^p_{\mathrm{ac}}(D)$ and $\mathbf{h}^p_{\mathrm{ac}}(\Pi)$ determined as follows: $F_0 \in \mathbf{h}^p_{\mathrm{ac}}(D)$ and $F \in \mathbf{h}^p_{\mathrm{ac}}(\Pi)$ correspond if

$$F = F_0 \circ \lambda \qquad \text{on} \quad \Pi \tag{2.16}$$

where

$$\lambda(z) = (z - i)/(z + i) \tag{2.17}$$

denotes the special Cayley transform that maps Π on D, i on 0, and ∞ on 1. If F_0 and f_0 are connected by (2.1), F and f by (2.15), then furthermore we have $f = f_0 \circ \lambda$ a.e. on R.

If $\Phi \in \mathscr{P}$, then

$$B = \lambda \circ \Phi \circ \lambda^{-1} \tag{2.18}$$

belongs to $H^\infty(D)$ and satisfies $|B| < 1$ on D. If $\Phi \equiv i$ on Π, then $B \equiv 0$ on D, a case that we previously excluded for convenience but which we now admit for completeness. Given $\Phi \in \mathscr{P}$, by $\Phi(x)$ we mean the nontangential boundary function formed from values of Φ in Π. As in the disk case, we call attention to the asymmetry in this convention. For, given $\Phi \in \mathscr{P}$, we define Φ on all of $\Pi \cup \Pi^c$ so that

$$\Phi(z^*)^* = \Phi(z) \tag{2.19}$$

for all z in $\Pi \cup \Pi^c$. The nontangential boundary function $\Phi^c(x)$ formed from values of Φ in Π^c is in general not the same as $\Phi(x)$.

It is well known that disk results map in a straightforward way to half-plane results. See Aronszajn and Donoghue [4] for details of such mappings in general. What we need for the sequel is given in the next two results.

2.6 THEOREM. *Assume $\Phi \in \mathscr{P}$.*

(i) *If Δ is a Lebesgue null subset of R, then so is the set $\{x : x \in R, \Phi(x) \in \Delta\}$. In particular, if $F \in \mathbf{h}^1_{\mathrm{ac}}(\Pi)$ and $F(x)$ is its a.e. defined boundary function, we may define $(F \circ \Phi)(x) = F(\Phi(x))$ a.e. on R.*

(ii) *If $F \in \mathbf{h}^1_{\mathrm{ac}}(\Pi)$, then $F \circ \Phi \in \mathbf{h}^1_{\mathrm{ac}}(\Pi)$, and*

$$(F \circ \Phi)(z) = \frac{1}{\pi} \int_{-\infty}^{\infty} \frac{\operatorname{Im} z}{|t - z|^2} (F \circ \Phi)(t)\, dt, \qquad z \in \Pi. \tag{2.20}$$

In particular, $(F \circ \Phi)(t)$ is the boundary function of $F \circ \Phi$ on Π, and $(F \circ \Phi)(t)/(1 + t^2) \in L^1(R)$.

(iii) *Assume* $1 \leqslant p < \infty$. *If* $F \in \mathbf{h}^p_{\mathrm{ac}}(\Pi)$, *then* $F \circ \Phi \in \mathbf{h}^p_{\mathrm{ac}}(\Pi)$, *and*

$$\int_{-\infty}^{\infty} \frac{\operatorname{Im} z}{|t-z|^2} |(F \circ \Phi)(t)|^p \, dt \leqslant \int_{-\infty}^{\infty} \frac{\operatorname{Im} \Phi(z)}{|t-\Phi(z)|^2} |F(t)|^p \, dt, \qquad z \in \Pi. \quad (2.21)$$

Proof. We obtain (i) from Lemma 2.1(ii), (ii) from Theorem 2.2, and (iii) from Corollary 2.5 using the mapping λ given by (2.17). For each part of the theorem, the case in which $\Phi \equiv i$ on Π is easily checked directly.

2.7 THEOREM. *Assume* $\Phi \in \mathscr{P}$. *For any* $F \in \mathbf{h}^1_{\mathrm{ac}}(\Pi)$, *we have*

$$\int_{-\infty}^{\infty} \frac{z - w^*}{(t-z)(t-w^*)} (F \circ \Phi)(t) \, dt = \int_{-\infty}^{\infty} \frac{\Phi(z) - \Phi(w)^*}{(t-\Phi(z))(t-\Phi(w)^*)} F(t) \, dt \quad (2.22)$$

for all $z, w \in \Pi \cup \Pi^c$, *with absolute convergence of the integrals.*

Proof. This follows from Theorem 2.3 except in the case where $\Phi \equiv i$ on Π and $\Phi \equiv -i$ on Π^c, in which case we check (2.22) directly.

More is true under additional assumptions.

2.8 THEOREM. *Assume* $\Phi \in \mathscr{P}$ *and*

$$\beta = \lim_{y \to \infty} \Phi(iy)/(iy) > 0. \quad (2.23)$$

Let $1 \leqslant p < \infty$. *Assume that* $F \in \mathbf{h}^p_{\mathrm{ac}}(\Pi)$, *and, moreover, that the boundary function* $F(x)$ *belongs to* $L^p(R)$.

(i) *We have*

$$\int_{-\infty}^{\infty} |(F \circ \Phi)(t)|^p \, dt \leqslant \beta^{-1} \int_{-\infty}^{\infty} |F(t)|^p \, dt \quad (2.24)$$

and

$$\int_{-\infty}^{\infty} (t-z)^{-1} (F \circ \Phi)(t) \, dt = \int_{-\infty}^{\infty} (t - \Phi(z))^{-1} F(t) \, dt \quad (2.25)$$

for all $z \in \Pi \cup \Pi^c$.

(ii) *In the case* $p = 1$,

$$\int_{-\infty}^{\infty} (F \circ \Phi)(t) \, dt = \beta^{-1} \int_{-\infty}^{\infty} F(t) \, dt. \quad (2.26)$$

Proof. In Theorem 2.6 take $z = iy$, $y > 0$, and write $\Phi(iy) = R(y) + iI(y)$, where R and I are real valued. Then we obtain

$$\begin{aligned} \int_{-\infty}^{\infty} \frac{y^2}{t^2 + y^2} |(F \circ \Phi)(t)|^p \, dt &\leqslant \int_{-\infty}^{\infty} \frac{yI(y)}{(t - R(y))^2 + I(y)^2} |F(t)|^p \, dt \\ &= \int_{-\infty}^{\infty} \frac{I(y)/y}{((t - R(y))/y)^2 + (I(y)/y)^2} |F(t)|^p \, dt. \end{aligned}$$

Let $y \to \infty$. We obtain (2.24) in view of (2.23) and the dominated convergence theorem.

To derive (2.25), write (2.22) in the form

$$\int_{-\infty}^{\infty}\left(\frac{1}{t-z}-\frac{1}{t-w^*}\right)(F\circ\Phi)(t)\,dt=\int_{-\infty}^{\infty}\left(\frac{1}{t-\Phi(z)}-\frac{1}{t-\Phi(w)^*}\right)F(t)\,dt.$$

Then set $w = iv$, where $v > 0$, and let $v \to \infty$.

We obtain (2.26) from (2.25) by setting $z = iy$, multiplying both sides by $-iy$, and letting $y \to \infty$.

3. Composition with Inner and Cayley Inner Functions

Recall that an inner function on D is a function $B \in H^\infty(D)$ such that $|B(e^{it})| = 1$ a.e. on Γ. For any inner function B, write $B = B_b B_s$ where B_b is a Blaschke product and B_s is a singular inner function.

3.1 Definition. The *degree* $\deg B$ of an inner function B is defined to be

(i) 0 if B is a constant of absolute value 1,

(ii) the positive integer n if B_b has exactly n zeros counting multiplicities and B_s is a constant,

(iii) ∞ if B_b has infinitely many zeros or B_s is not a constant.

3.2 Theorem. *If A and B are inner functions, then*

(i) $\deg(AB) = \deg A + \deg B$

(ii) $\deg(A \circ B) = \deg A \deg B$ *provided that the right side is not* $0 \cdot \infty$ *or* $\infty \cdot 0$.

Proof. We leave this to the reader.

3.3 Definition. By a *Cayley inner function* on D we mean a function Ψ which is holomorphic and has a nonnegative imaginary part on D such that the boundary function $\Psi(e^{it})$ of Ψ is real a.e. on Γ.

It is easy to see that Ψ is Cayley inner on D if and only if $\lambda \circ \Psi = B$ is inner on D, where λ is given by (2.17). Moreover if B is any inner function, $B \not\equiv 1$, there exists a Cayley inner function Ψ such that $\lambda \circ \Psi = B$ on D. We define the degree of Ψ, $\deg \Psi$, to be equal to the degree of $B = \lambda \circ \Psi$.

3.4 Theorem. *Assume Ψ is holomorphic on D. The following assertions are equivalent:*

(i) *Ψ is a Cayley inner function on D,*
(ii) *Ψ has a representation*

$$\Psi(z) = \alpha + i \int_{\Gamma} \frac{e^{it} + z}{e^{it} - z} d\mu(e^{it}), \quad z \in D, \tag{3.1}$$

where α is a real constant and μ is a singular measure on Γ,
(iii) *Ψ has an exponential representation*

$$\Psi(z) = \exp\left(\alpha + i\pi \int_{\Delta} \frac{e^{it} + z}{e^{it} - z} d\sigma(e^{it})\right), \quad z \in D, \tag{3.2}$$

where α is a real constant and Δ is a measurable subset of Γ.

Proof. If we assume (i) we obtain (ii) by the Herglotz representation theorem and Fatou's theorem (see Duren [9, p. 4]).

Assume (ii). By considering the Herglotz representation for $\log \Psi$, where $0 \leqslant \operatorname{Im} \log \Psi \leqslant \pi$, we obtain

$$\log \Psi(z) = \alpha + i\pi \int_{\Gamma} \frac{e^{it} + z}{e^{it} - z} f(e^{it})\, d\sigma(e^{it}), \quad z \in D,$$

where $0 \leqslant f \leqslant 1$ a.e. on Γ. Since μ is singular, $\Psi(e^{it})$ is real a.e., and hence $f(e^{it}) = \lim_{r\uparrow 1} \pi^{-1} \operatorname{Im} \log \Psi(re^{it})$ takes the values 0 and 1 a.e. on Γ. It follows that $f = \chi_{\Delta}$ a.e. for some measurable set Δ on Γ. Thus (iii) holds.

The verification that (iii) implies (i) is straightforward, and so the result follows.

Exponential representations are studied in detail by Aronszajn and Donoghue [4].

We borrow a notion that proved useful in the spectral theory of self-adjoint Toeplitz operators due to Ismagilov [13] and Rosenblum [18]. We define the *index* $I(\Delta)$ of a measurable set Δ on Γ to be (i) 0 if Δ is the empty set $\varnothing$ or the full circle Γ modulo a Lebesgue null set, (ii) the positive integer r if, modulo a Lebesgue null set, Δ is the union of r disjoint closed proper arcs, and (iii) ∞ if neither (i) nor (ii) hold. An analytical characterization of index is given in Rosenblum [18]. We recast this characterization in a form suitable to our purpose.

3.5 THEOREM. *Let Ψ be a Cayley inner function on D represented in the form* (3.1) *and* (3.2), *and let $B = \lambda \circ \Psi$, where λ is given by* (2.17). *Let μ be a nonnegative integer or ∞. The following are equivalent:*

(i) $\deg \Psi = n$,
(ii) $\deg B = n$,

(iii) $\dim(H^2(D) \ominus BH^2(D)) = n$,
(iv) *if* $B(z) = \sum_0^\infty b_j z^j$ *in* D, *then*

$$\sum_1^\infty j|b_j|^2 = n,$$

or what is the same thing

$$\frac{1}{\pi} \iint_D |B'(z)|^2 \, dx \, dy = n,$$

(v) $\dim L^2(\mu) = n$, *where* μ *is the singular measure in* (3.1),
(vi) $I(\Delta) = n$, *where* Δ *is the measurable set in* (3.2).

Proof. The equivalence of (i), (ii), and (iii) follows from the definition of degree and well-known facts. See, for example, Kriete [14]. The equivalence of (iii) and (iv) follows from a computation which shows that if P denotes the orthogonal projection mapping $H^2(D)$ on $H^2(D) \ominus BH^2(D)$, then

$$\operatorname{tr}(P) = \sum_0^\infty \langle Pz^n, z^n \rangle = \sum_1^\infty j|b_j|^2 = (1/\pi) \iint_D |B'(z)|^2 \, dx \, dy.$$

We omit the details. We show that (iii) and (v) are equivalent by exhibiting a natural isometry V mapping $H^2(D) \ominus BH^2(D)$ onto $L^2(\mu)$. In fact, we have

$$\left\langle P \frac{1}{1 - u^* z}, \frac{1}{1 - v^* z} \right\rangle = \frac{1 - B(v)B(u)^*}{1 - vu^*}$$

$$= (1 - B(v))(1 - B(u)^*) \frac{1}{2i} \frac{\Psi(v) - \Psi(u)^*}{1 - vu^*} = \int_\Gamma \frac{1 - B(v)}{1 - ve^{-it}} \frac{1 - B(u)^*}{1 - u^* e^{it}} \, d\mu(e^{it})$$

for all $u, v \in D$. It follows that there is a unique isometry V mapping $H^2(D) \ominus BH^2(D)$ onto $L^2(\mu)$ such that

$$V : P \frac{1}{1 - w^* z} \to \frac{1 - B(w)^*}{1 - w^* e^{it}}$$

for each $w \in D$. The equivalence of (v) and (vi) follows from Rosenblum [18].

We study next mapping properties of an inner function B on the circle Γ. First we introduce notation for arcs on Γ. Let t_1, t_2 be distinct points in $[0, 2\pi]$. If $t_1 < t_2$, set

$$\Gamma(t_1, t_2) = \{e^{it} : t_1 < t < t_2\}.$$

If $t_1 > t_2$, set

$$\Gamma(t_1, t_2) = \{e^{it} : t_1 < t < t_2 + 2\pi\}.$$

Notice that $\Gamma(0, 2\pi) = \Gamma \backslash \{1\}$, $\Gamma(2\pi, 0)$ is empty, and in general Γ is the disjoint union of $\Gamma(t_1, t_2)$ and $\Gamma(t_2, t_1)$ and the endpoints of these arcs. For our purpose, isolated points are negligible, and no confusion will arise when we write, for example, $\Gamma(0, 2\pi) = \Gamma$ and otherwise ignore endpoints.

3.6 DEFINITION. Let B be a nonconstant inner function, Δ a measurable subset of Γ, and $\Gamma(t_1, t_2)$ an arc on Γ. We say that B *maps* Δ *on* $\Gamma(t_1, t_2)$ if

(i) $B(e^{it}) \in \Gamma(t_1, t_2)$ for almost all $e^{it} \in \Delta$, and
(ii) $B(e^{it}) \in \Gamma(t_2, t_1)$ for almost all $e^{it} \in \Gamma \backslash \Delta$.

Obviously, 3.6(i) and 3.6(ii) are unaffected if Δ is changed by a null set. If B maps Δ on $\Gamma(t_1, t_2)$, then B maps $\Gamma \backslash \Delta$ on $\Gamma(t_2, t_1)$, and conversely. Every nonconstant inner function maps Γ on Γ. On the other hand, if $0 < \sigma(\Delta) < 1$ and $0 < \sigma(\Gamma(t_1, t_2)) < 1$, the class of nonconstant inner functions that map Δ on $\Gamma(t_1, t_2)$ is highly restricted.

3.7 THEOREM. *Assume that Δ is a measurable subset of Γ and $\Gamma(t_1, t_2)$ is an arc such that $0 < \sigma(\Delta) < 1$ and $0 < \sigma(\Gamma(t_1, t_2)) < 1$. Let ϕ be a linear fractional transformation that maps Π on D and the interval $(-\infty, 0)$ on $\Gamma(t_1, t_2)$. Then the general form of a nonconstant inner function B that maps Δ on $\Gamma(t_1, t_2)$ is given by*

$$B = \phi \circ \Psi \tag{3.3}$$

where

$$\Psi(z) = \exp\left(\alpha + i\pi \int_{\Delta} \frac{e^{it} + z}{e^{it} - z} d\sigma(e^{it})\right), \qquad z \in D \tag{3.4}$$

for some real constant α. For any such B, $\deg B = I(\Delta)$.

Proof. Assume that B is a nonconstant inner function that maps Δ on $\Gamma(t_1, t_2)$. Then $\phi^{-1} \circ B$ is holomorphic and has positive imaginary part on D. Our assumptions imply that $(\phi^{-1} \circ B)(e^{it})$ is negative a.e. on Δ and positive a.e. on $\Gamma \backslash \Delta$. It follows that $\Psi = \phi^{-1} \circ B$ is a Cayley inner function on D having the exponential representation (3.4) for some real constant α. By construction (3.3) holds. The steps in this argument can be reversed to show that if B is given by (3.3) where Ψ has the form (3.4), then B is a nonconstant inner function that maps Δ on $\Gamma(t_1, t_2)$. We obtain $\deg B = I(\Delta)$ by Theorem 3.5.

3.8 EXAMPLE. We give the general form of a nonconstant inner function B that maps a union $\Delta = \bigcup_1^n \Gamma(r_j, s_j)$ of arcs on the single arc $\Gamma(0, \pi)$. For definiteness assume that $0 \leqslant r_1 < s_1 < r_2 < s_2 < \cdots < r_n < s_n < 2\pi$. Easy modifications will handle other cases. The transformation λ defined by (2.17) maps Π on D and $(-\infty, 0)$ on $\Gamma(0, \pi)$. If $0 \leqslant r < s < 2\pi$, by a direct calculation we obtain

$$\exp\left(i\pi \int_{\Gamma(r,s)} \frac{e^{it} + z}{e^{it} - z} \, d\sigma(e^{it}) \right) = e^{i(r-s)/2}(e^{is} - z)/(e^{ir} - z)$$

for all $z \in D$. Hence by Theorem 3.7, the general form for B is $B = \lambda \circ \Psi$, where

$$\Psi(z) = \rho \prod_{1}^{n} e^{i(r_j - s_j)/2}(e^{is_j} - z)/(e^{ir_j} - z), \qquad z \in D,$$

where ρ is any positive real constant. We note that B is a Blaschke product, since it is a rational function.

Theorem 2.3 and Corollary 2.5 take on particularly strong forms when B is inner. As usual B is defined in $D \cup D^c$ by (2.2) with the possibility of some poles in D^c. By the discussion just following (2.2), in the case of an inner function B the two boundary functions formed from values of B in D and D^c coincide.

3.9 Theorem. *Let B be a nonconstant inner function that maps a measurable subset Δ of Γ on an arc $\Gamma(t_1, t_2)$. Then for each $f(e^{it}) \in L^1(\Gamma(t_1, t_2))$, $(f \circ B)(e^{it}) \in L^1(\Delta)$, and*

$$\int_{\Delta} \frac{1 - zw^*}{(e^{it} - z)(e^{-it} - w^*)} (f \circ B)(e^{it}) \, d\sigma(e^{it})$$
$$= \int_{\Gamma(t_1,t_2)} \frac{1 - B(z)B(w)^*}{(e^{it} - B(z))(e^{-it} - B(w)^*)} f(e^{it}) \, d\sigma(e^{it}) \tag{3.5}$$

for all $w, z \in D \cup D^c$, with interpretations by continuity of the integral on the right at the poles of B.

Proof. In Theorem 2.3 choose F to be the $\mathbf{h}^1_{\mathrm{ac}}(D)$ function whose boundary function is

$$F(e^{it}) = \begin{cases} f(e^{it}) & \text{a.e. on} \quad \Gamma(t_1, t_2), \\ 0 & \text{on} \quad \Gamma(t_2, t_1). \end{cases}$$

By our assumptions on B,

$$(F \circ B)(e^{it}) = \begin{cases} (f \circ B)(e^{it}) & \text{a.e. on} \quad \Delta \\ 0 & \text{a.e. on} \quad \Gamma \backslash \Delta. \end{cases}$$

The result follows.

A consequence is that in the case of inner functions, equality holds in the inequality of Corollary 2.5.

3.10 COROLLARY. *Let B be a nonconstant inner function that maps a measurable subset Δ of Γ on an arc $\Gamma(t_1, t_2)$. Assume $1 \leqslant p < \infty$. If $f(e^{it}) \in L^p(\Gamma(t_1, t_2))$, then $(f \circ B)(e^{it}) \in L^p(\Delta)$, and*

$$\int_\Delta \frac{1 - |z|^2}{|e^{it} - z|^2} |(f \circ B)(e^{it})|^p \, d\sigma(e^{it})$$

$$= \int_{\Gamma(t_1, t_2)} \frac{1 - |B(z)|^2}{|e^{it} - B(z)|^2} |f(e^{it})|^p \, d\sigma(e^{it}) \tag{3.6}$$

for all $z \in D$.

Proof. Apply Theorem 3.9 to the function $|f(e^{it})|^p$.

We map these results to the real line R.

3.11 DEFINITION. By a *Cayley inner function on Π* we mean a function ξ in $\mathscr{P}$ whose boundary function $\xi(x) = \xi(x + i0)$ is real a.e. We always extend the domain of ξ to $\Pi \cup \Pi^c$ by requiring that $\xi(z^*)^* = \xi(z)$ for all z in $\Pi \cup \Pi^c$.

For any Cayley inner function ξ on Π, we have $\xi(x) = \xi(x + i0) = \xi(x - i0)$ a.e. on R. There is a one-to-one correspondence between the class of Cayley inner functions on Π and the class of inner functions $B \not\equiv 1$ on D, determined by $\xi = \lambda \circ B \circ \lambda^{-1}$, where λ is given by (2.17). We define $\deg \xi = \deg B$ in this case.

If a and b are real numbers or $\pm\infty$, we define the *arc* (a, b) to be (i) the usual set $\{x : x \text{ real}, a < x < b\}$ if $a \leqslant b$, and (ii) the set $(-\infty, b) \cup (a, \infty)$ if $a > b$. Thus if we ignore endpoints, then in general (a, b) and (b, a) partition R into two disjoint arcs.

The *index* $I(\Delta)$ of a measurable subset Δ of R is defined to be the index of the image $\lambda(\Delta)$ of Δ under the mapping (2.17). Thus, intuitively, $I(\Delta)$ is the number of arcs, not the number of intervals, in Δ. For example, $\Delta = (-\infty, -1) \cup (1, \infty)$ has index 1 because Δ is the single arc $(1, -1)$.

3.12 THEOREM. *Assume ξ is holomorphic in Π. The following statements are equivalent:*

(i) *ξ is a Cayley inner function on Π,*
(ii) *ξ has a representation in the form*

$$\xi(z) = \alpha + \beta z + \int_{-\infty}^{\infty} \left(\frac{1}{t - z} - \frac{t}{1 + t^2} \right) d\mu(t), \qquad z \in \Pi \cup \Pi^c, \tag{3.7}$$

where $\alpha, \beta \in R$, $\beta \geqslant 0$, *and* μ *is a positive singular measure on* R *such that*

$$\int_{-\infty}^{\infty} (1 + t^2)^{-1} \, d\mu(t) < \infty, \tag{3.8}$$

(iii) ξ *has a representation in the form*

$$\xi(z) = \exp\left(\alpha + \int_{\Delta} \left(\frac{1}{t - z} - \frac{t}{1 + t^2}\right) dt\right), \qquad z \in \Pi \cup \Pi^c, \tag{3.9}$$

where $\alpha \in R$ *and* Δ *is a measurable subset of* R.

Proof. This is the half-plane version of Theorem 3.4.

3.13 THEOREM. *Assume that* ξ *is a Cayley inner function on* Π *and* n *is a nonnegative integer or* ∞. *The following statements are equivalent*:

(i) $\deg \xi = n$,
(ii) $\dim \mathscr{L}(\xi) = n$, *where* $\mathscr{L}(\xi)$ *is the Hilbert space of holomorphic functions on* $\Pi \cup \Pi^c$ *with reproducing kernel function*

$$L(w, z) = [\xi(z) - \xi(w)^*]/(z - w^*), \qquad w, z \in \Pi \cup \Pi^c, \tag{3.10}$$

(iii) $\operatorname{sgn} \beta + \dim L^2(\mu) = n$, *where* β *and* μ *are as in Theorem* 3.12(ii) *and* $\operatorname{sgn} \beta = 1$ *or* 0 *according as* $\beta > 0$ *or* $\beta = 0$,
(iv) $I(\Delta) = n$, *where* Δ *is as in Theorem* 3.12(iii).

Proof. The equivalence of (i), (iii), and (iv) follows from Theorem 3.5. The equivalence of (ii) and (iii) is proved by using the identity

$$L(w, z) = \beta + \int_{-\infty}^{\infty} \frac{d\mu(t)}{(t - z)(t - w^*)}, \qquad w, z \in \Pi \cup \Pi^c$$

to set up an isometry mapping $\mathscr{L}(\xi)$ onto the direct sum of $L^2(\mu)$ and a space that has dimension 0 or 1 according as $\beta = 0$ or $\beta > 0$.

3.14 DEFINITION. Let ξ be a nonconstant Cayley inner function on Π, Δ a measurable subset of R, and (a, b) an arc on R. We say that ξ maps Δ on (a, b) if

(i) $\xi(x) \in (a, b)$ for almost all x in Δ, and
(ii) $\xi(x) \in (b, a)$ for almost all x in $R \setminus \Delta$.

Notice that ξ maps Δ on (a, b) if and only if ξ maps $R \setminus \Delta$ on (b, a). Every nonconstant Cayley inner function ξ on Π maps $\Delta = R$ on $(a, b) = R$.

3.15 THEOREM. *Let Δ be a measurable subset of R such that neither Δ nor $R\backslash\Delta$ is a Lebesgue null set. Let ϕ be a linear fractional transformation that maps Π onto Π and $(-\infty, 0)$ onto some arc (a, b). Then the general form of a nonconstant Cayley inner function on Π that maps Δ on (a, b) is given by $\xi = \phi \circ \Phi$, where*

$$\Phi(z) = \exp\left(\alpha + \int_\Delta \left(\frac{1}{t - z} - \frac{t}{1 + t^2}\right) dt\right), \qquad z \in \Pi, \qquad (3.11)$$

for some real constant α. We have $\deg \xi = I(\Delta)$.

Proof. This is the half-plane version of Theorem 3.7. A direct proof for everything but the final assertion is given in [21].

An example given in [21] illustrates this theorem. Let $\Delta = \bigcup_1^n (c_j, d_j)$, where $-\infty < c_1 < d_1 < c_2 < d_2 < \cdots < c_n < d_n < \infty$, and let $-\infty < a < b < \infty$. Then

$$\xi(z) = \left(b - a\prod_1^n [(d_j - z)/(c_j - z)]\right) \Big/ \left(1 - \prod_1^n [(d_j - z)/(c_j - z)]\right),$$

where $z \in \Pi \cup \Pi^c$, is a Cayley inner function on Π that maps Δ on (a, b).

We now formulate the main substitution formulas for Cayley inner functions on Π.

3.16 THEOREM. *Assume that ξ is a nonconstant Cayley inner function on Π that maps a real measurable set Δ on an arc (a, b). If $f(x)/(1 + x^2) \in L^1(a, b)$, then $(f \circ \xi)(x)/(1 + x^2) \in L^1(\Delta)$, and*

$$\int_\Delta \frac{z - w^*}{(t - z)(t - w^*)} (f \circ \xi)(t)\, dt = \int_{(a,b)} \frac{\xi(z) - \xi(w)^*}{(t - \xi(z))(t - \xi(w)^*)} f(t)\, dt \qquad (3.12)$$

for all $w, z \in \Pi \cup \Pi^c$.

Proof. This follows from Theorem 3.9.

3.17 THEOREM. *Let ξ be a nonconstant Cayley inner function on Π that maps a real measurable set Δ on an arc (a, b). Assume that ξ satisfies* (2.23). *If $f \in L^p(a, b)$ for some p, $1 \leqslant p < \infty$, then $f \circ \xi \in L^p(\Delta)$, and we have*

$$\int_\Delta \frac{(f \circ \xi)(t)}{t - z}\, dt = \int_{(a,b)} \frac{f(t)}{t - \xi(z)}\, dt, \qquad z \in \Pi \cup \Pi^c, \qquad (3.13)$$

and

$$\int_\Delta |(f \circ \xi)(t)|^p\, dt = \beta^{-1} \int_{(a,b)} |f(t)|^p\, dt. \qquad (3.14)$$

In the case $p = 1$, we also have

$$\int_{\Delta} (f \circ \xi)(t)\, dt = \beta^{-1} \int_{(a,b)} f(t)\, dt. \tag{3.15}$$

Proof. We deduce (3.13) and (3.15) from Theorem 2.8 in the same way that Theorem 3.9 was obtained from Theorem 2.3. Then (3.14) follows on replacing f by $|f|^p$ in (3.15).

3.18 EXAMPLE. Assume that Δ is a measurable subset of R having finite measure $m(\Delta)$. Let $-\infty < a < b < \infty$. By Theorem 3.15,

$$\xi(z) = \left[a \exp\left(\int_{\Delta} dt/(t-z)\right) - b\right] \Big/ \left[\exp\left(\int_{\Delta} dt/(t-z)\right) - 1\right], \qquad z \in \Pi \cup \Pi^c,$$

is a Cayley inner function that maps Δ on (a, b). Condition (2.23) is satisfied, and

$$\beta = \lim_{y \to \infty} \xi(iy)/(iy) = (b-a)/m(\Delta).$$

Thus, for example, if $f \in L^1(a, b)$, then $f \circ \xi \in L^1(\Delta)$, and

$$\int_{\Delta} (f \circ \xi)(t)\, dt = \frac{m(\Delta)}{b-a} \int_a^b f(t)\, dt$$

by Theorem 3.17.

4. COMPOSITION FORMULAS INVOLVING SINGULAR INTEGRALS

Let Δ be a measurable subset of the real line. Let f be a complex valued function on Δ such that $f(t)/(1 + |t|) \in L^1(\Delta)$. Then the Hilbert transform (relative to Δ) of f is defined a.e. on R by

$$\tilde{f}(x) = \mathrm{PV} \frac{1}{\pi} \int_{\Delta} \frac{f(t)}{t-x}\, dt, \tag{4.1}$$

where PV indicates that the integral is taken in the principal value sense. If we set

$$F(z) = \frac{1}{2\pi i} \int_{\Delta} \frac{f(t)}{t-z}\, dt, \qquad z \in \Pi \cup \Pi^c, \tag{4.2}$$

then

$$F(x + i0) - F(x - i0) = \begin{cases} f(x) & \text{a.e. on } \Delta \\ 0 & \text{a.e. on } R \backslash \Delta, \end{cases} \tag{4.3}$$

$$F(x + i0) + F(x - i0) = -i\tilde{f}(x) \qquad \text{a.e. on } R. \tag{4.4}$$

4.1 THEOREM. *Let ξ be a nonconstant Cayley inner function on Π that maps Δ on some arc (a, b). If $f(t)/(1+t^2) \in L^1(a, b)$, then $(f \circ \xi)(t)/(1+t^2) \in L^1(\Delta)$, and for any $w \in \Pi \cup \Pi^c$,*

$$\mathrm{PV}\frac{1}{\pi}\int_{\Delta}\frac{(f\circ\xi)(t)}{t-w^*}\frac{dt}{t-x}=\frac{\xi(x)-\xi(w)^*}{x-w^*}\mathrm{PV}\frac{1}{\pi}\int_{(a,b)}\frac{f(t)}{t-\xi(w)^*}\frac{dt}{t-\xi(x)} \tag{4.5}$$

for almost all real x. If $f(t)/(1+|t|) \in L^1(a, b)$, then in addition

$$\mathrm{PV}\frac{1}{\pi}\int_{\Delta}\frac{\xi(t)-\xi(x)}{t-x}\frac{(f\circ\xi)(t)}{t-w^*}\,dt=\frac{\xi(x)-\xi(w)^*}{x-w^*}\frac{1}{\pi}\int_{(a,b)}\frac{f(t)}{t-\xi(w)^*}\,dt \tag{4.6}$$

for almost all real x.

Proof. The first assertion follows from Theorem 3.16. It guarantees the existence of all the integrals that we need. Set

$$F(z)=\frac{1}{2\pi i}\int_{(a,b)}\frac{f(t)}{t-\xi(w)^*}\frac{dt}{t-z}$$

$$G(z)=\frac{1}{2\pi i}\int_{\Delta}\frac{(f\circ\xi)(t)}{t-w^*}\frac{dt}{t-z}$$

for $z \in \Pi \cup \Pi^c$. By Theorem 3.16, we have

$$G(z)=\frac{\xi(z)-\xi(w)^*}{z-w^*}(F\circ\xi)(z), \qquad z\in\Pi\cup\Pi^c. \tag{4.7}$$

Now $F|\Pi$ is a linear combination of four functions in $\mathscr{P}$. Therefore, with the aid of Theorem 2.6 we can prove that

$$(F\circ\xi)(x+i0)=F_+(\xi(x)) \qquad \text{a.e. on} \quad R,$$

where $F_+(x) = F(x + i0)$ a.e. on R. Similarly,

$$(F\circ\xi)(x-i0)=F_-(\xi(x)) \qquad \text{a.e. on} \quad R,$$

where $F_-(x) = F(x - i0)$ a.e. on R. Thus using (4.4), (4.7) yields (4.5).
Using (4.5) we see that the left side of (4.6) is equal a.e. to

$$\frac{\xi(x)-\xi(w)^*}{x-w^*}\left[\mathrm{PV}\frac{1}{\pi}\int_{(a,b)}\frac{tf(t)}{t-\xi(w)^*}\frac{dt}{t-\xi(x)}\right.$$
$$\left.-\,\xi(x)\,\mathrm{PV}\frac{1}{\pi}\int_{(a,b)}\frac{f(t)}{t-\xi(w)^*}\frac{dt}{t-\xi(x)}\right],$$

which simplifies to the right side of (4.6).

4.2 THEOREM. *Let ξ be a nonconstant Cayley inner function on Π that maps Δ on an arc (a, b). Assume that ξ satisfies (2.23). If $f \in L^p(a, b)$ for some p, $1 \leqslant p < \infty$, then $f \circ \xi \in L^p(\Delta)$, and we have*

$$\mathrm{PV}\frac{1}{\pi}\int_\Delta \frac{(f \circ \xi)(t)}{t - x}\,dt = \mathrm{PV}\frac{1}{\pi}\int_{(a,b)} \frac{f(t)}{t - \xi(x)}\,dt \tag{4.8}$$

a.e. on R.

Proof. If $f \in L^p(a, b)$, then $f \circ \xi \in L^p(\Delta)$ by Theorem 3.17. Set

$$F(z) = \frac{1}{2\pi i}\int_\Delta \frac{(f \circ \xi)(t)}{t - z}\,dt,$$

$$G(z) = \frac{1}{2\pi i}\int_\Delta \frac{f(t)}{t - z}\,dt$$

for $z \in \Pi \cup \Pi^c$. By Theorem 3.17, $F(z) = (G \circ \xi)(z)$ for all $z \in \Pi \cup \Pi^c$. Therefore, an argument similar to one used in the proof of Theorem 4.1 yields (4.8).

5. APPLICATIONS. SOME DISTRIBUTION FUNCTION RESULTS

Let m denote Lebesgue measure on the real line R. If f is a real measurable function on a measurable set S in R, then the distribution function of f on S is given by

$$\lambda \to m\{t : t \in S \text{ and } |f(t)| > \lambda\}. \tag{5.1}$$

We derive properties of this function and examine results of Loomis [15] and Stein and Weiss [24]. It is convenient to introduce the set function

$$\delta \to m\{t : t \in S \text{ and } f(t) \in \delta\},$$

where δ ranges over the class of measurable subsets of R.

5.1 THEOREM. *If B is a nonconstant inner function on D, then for any measurable subset δ of Γ,*

$$\sigma\{e^{it} : e^{it} \in \Gamma \text{ and } B(e^{it}) \in \delta\} = \int_\delta \frac{1 - |B(0)|^2}{|1 - B(0)e^{-it}|^2}\,d\sigma(e^{it}).$$

Proof. In Theorem 3.9, choose $\Delta = \Gamma$, f a suitable characteristic function, and set $w = z = 0$.

Throughout the rest of this section ξ denotes a nonconstant Cayley inner function on Π.

5.2 THEOREM. *Assume ξ satisfies* (2.23), *and let $\beta > 0$ be the constant appearing in* (2.23). *Then for any measurable subset δ of R,*

$$m\{t : t \in R \text{ and } \xi(t) \in \delta\} = m(\delta)/\beta. \tag{5.2}$$

Proof. It is sufficient to prove this when $m(\delta) < \infty$. In this case (5.2) follows from (3.15) in Theorem 3.17, where in Theorem 3.17 we choose $\Delta = (a, b) = R$ and $f = \chi_\delta$, the characteristic function of δ.

5.3 THEOREM. *Suppose that for some real constant a,*

$$\lim_{y \to \infty} (-iy)\xi(a + iy) = \beta^{-1}, \tag{5.3}$$

where $0 < \beta < \infty$. Then for any measurable subset δ of R, we have

$$m\{t : t \in R \text{ and } \xi(t) \in \delta\} = \beta^{-1} \int_\delta t^{-2}\, dt.$$

Proof. The function $\xi_1(z) = -1/\xi(a + z)$, $z \in \Pi$, is Cayley inner on Π. Let $\delta_1 = \{t : t \in R \text{ and } -1/t \in \delta\}$. Then

$$m\{t : t \in R \text{ and } \xi(t) \in \delta\} = m\{t : t \in R \text{ and } \xi_1(t) \in \delta_1\}.$$

But ξ_1 satisfies (2.23), and the constants β of (2.23) and (5.3) coincide. Therefore, by Theorem 5.2,

$$\begin{aligned} m\{t : t \in R \text{ and } \xi_1(t) \in \delta_1\} &= \beta^{-1} m(\delta_1) \\ &= \beta^{-1} \int_{-\infty}^{\infty} \chi_\delta(-1/t)\, dt = \beta^{-1} \int_\delta t^{-2}\, dt. \end{aligned}$$

The result follows.

5.4 EXAMPLE. Theorem 5.3 yields a quick proof of a distribution function result used by Loomis [15] and subsequent authors. Let

$$\xi(x) = \sum_1^n d_j/(a_j - x), \qquad x \in R,$$

where $d_j > 0$ and $a_j \in R$ for each $j = 1, \ldots, n$. Then for each $\lambda > 0$,

$$m\{t : t \in R \text{ and } |\xi(t)| > \lambda\} = 2\lambda^{-1} \sum_1^n d_j.$$

This follows from Theorem 5.3 on choosing $\xi(z) = \sum_1^n d_j/(a_j - z)$ for $z \in \Pi \cup \Pi^c$ and $\delta = \{t : |t| > \lambda\}$.

More generally, we have

5.5 COROLLARY. *Let*

$$\xi(x) = \mathrm{PV} \int_{-\infty}^{\infty} d\mu(t)/(t-x) \qquad \text{a.e. on} \quad R, \tag{5.4}$$

where μ is a finite positive singular measure on R. Then for any real measurable set δ,

$$m\{t : t \in R \text{ and } \xi(t) \in \delta\} = \mu(R) \int_\delta t^{-2}\, dt.$$

In particular, for any $\lambda > 0$,

$$m\{t : t \in R \text{ and } |\xi(t)| > \lambda\} = 2\lambda^{-1}\mu(R).$$

Proof. The function $\xi(z) = \int_{-\infty}^{\infty} d\mu(t)/(t-z)$, $z \in \Pi \cup \Pi^c$, satisfies the assumptions in Theorem 5.3 and has boundary function (5.4) a.e. (see, for example, Butzer and Nessel [6, p. 307]). We find that $\beta^{-1} = \mu(R)$, and so the result follows.

5.6 THEOREM. *Let*

$$\eta(x) = \mathrm{PV} \int_\Delta dt/(t-x) \qquad \text{a.e. on} \quad R,$$

where Δ is a real measurable set with $m(\Delta) < \infty$. The following formulas hold for any real measurable set δ:

$$m\{t : t \in \Delta \text{ and } \eta(t) \in \delta\} = \frac{m(\Delta)}{4} \int_\delta \operatorname{sech}^2(t/2)\, dt, \tag{5.5}$$

$$m\{t : t \in R \backslash \Delta \text{ and } \eta(t) \in \delta\} = \frac{m(\Delta)}{4} \int_\delta \operatorname{csch}^2(t/2)\, dt, \tag{5.6}$$

$$m\{t : t \in R \text{ and } \eta(t) \in \delta\} = m(\Delta) \int_\delta \frac{\cosh t}{\sinh^2 t}\, dt. \tag{5.7}$$

In particular, for any $\lambda > 0$,

$$m\{t : t \in \Delta \text{ and } |\eta(t)| > \lambda\} = 2m(\Delta)/(e^\lambda + 1), \tag{5.8}$$

$$m\{t : t \in R \backslash \Delta \text{ and } |\eta(t)| > \lambda\} = 2m(\Delta)/(e^\lambda - 1), \tag{5.9}$$

$$m\{t : t \in R \text{ and } |\eta(t)| > \lambda\} = 2m(\Delta)/\sinh \lambda. \tag{5.10}$$

Formula (5.10) is a result of Stein and Weiss [24].

Proof. By Theorem 3.15 the function

$$\xi(z) = \left[1 + \exp\left(\int_\Delta dt/(t - z)\right)\right] \Big/ \left[1 - \exp\left(\int_\Delta dt/(t - z)\right)\right], \qquad z \in \Pi \cup \Pi^c,$$

is a Cayley inner function on Δ that maps Δ on $(-1, 1)$. Condition (2.23) is satisfied with $\beta = 2/m(\Delta)$. Setting $\Phi(z) = \int_\Delta dt/(t - z)$, we find that

$$\Phi(x \pm i0) = \eta(x) \pm \pi i \chi_\Delta(x) \qquad \text{a.e. on} \quad R,$$

and therefore

$$\xi(x) = \begin{cases} (1 - e^{\eta(x)})/(1 + e^{\eta(x)}) & \text{a.e. on} \quad \Delta \\ (1 + e^{\eta(x)})/(1 - e^{\eta(x)}) & \text{a.e. on} \quad R \backslash \Delta. \end{cases}$$

Fix δ and set

$$\delta_1 = \left\{ t : t \in (-1, 1) \text{ and } \log \frac{1 - t}{1 + t} \in \delta \right\}.$$

Then

$$m\{t : t \in \Delta \text{ and } \eta(t) \in \delta\} = m\{t : t \in R \text{ and } \xi(t) \in \delta_1\},$$

and by Theorem 5.2

$$\begin{aligned} m\{t : t \in R \text{ and } \xi(t) \in \delta_1\} &= \tfrac{1}{2} m(\Delta) m(\delta_1) \\ &= \tfrac{1}{2} m(\Delta) \int_{-1}^{1} \chi_\delta\left(\log \frac{1 - t}{1 + t}\right) dt \\ &= \frac{m(\Delta)}{4} \int_\delta \operatorname{sech}^2(t/2)\, dt. \end{aligned}$$

Therefore (5.5) follows.

Similarly, setting

$$\delta_2 = \left\{ t : t \in (1, -1) \text{ and } \log \frac{t - 1}{t + 1} \in \delta \right\}$$

we find that

$$\begin{aligned} m\{t : t \in R \backslash \Delta \text{ and } \eta(t) \in \delta\} &= m\{t : t \in R \text{ and } \xi(t) \in \delta_2\} \\ &= \tfrac{1}{2} m(\Delta) m(\delta_2) \\ &= \tfrac{1}{2} m(\Delta) \int_{|t| > 1} \chi_\delta\left(\log \frac{t - 1}{t + 1}\right) dt \\ &= \frac{m(\Delta)}{4} \int_\delta \operatorname{csch}^2(t/2)\, dt, \end{aligned}$$

which proves (5.6). We get (5.7) by combining (5.5) and (5.6). The relations (5.8), (5.9), and (5.10) follow on specializing to the case $\delta = \{t : |t| > \lambda\}$.

6. Applications. Ahiezer Tranforms

Let ξ and η be nonconstant Cayley inner functions on Π. Assume that ξ maps Δ on (a, b) and η maps E on (c, d), where Δ, E are real measurable sets and (a, b), (c, d) are arcs on R. We fix positive measurable weight functions p and q on (a, b) and (c, d) respectively. By $L_p^2(a, b)$ we mean the weighted Lebesgue space with norm

$$\|f\|_p = \left(\int_{(a,b)} |f(t)|^2 p(t)\, dt \right)^{1/2}$$

and similarly for $L_q^2(c, d)$ and other weighted Lebesgue spaces.

We give conditions under which every bounded linear operator

$$T : L_p^2(a, b) \to L_q^2(c, d) \tag{6.1}$$

induces in a natural way a bounded linear operator

$$\hat{T} : L^2_{p \circ \xi}(\Delta) \to L^2_{q \circ \eta}(E). \tag{6.2}$$

We shall apply our results to obtain, in particular, a class of transforms studied by Ahiezer [1].

By $\mathscr{L}(\xi)$ and $\mathscr{L}(\eta)$ we mean the spaces associated with ξ and η as in Theorem 3.13(ii).

6.1 Theorem. *Assume that p satisfies*

$$\int_{(a,b)} \frac{p(t)}{1 + t^2}\, dt < \infty,$$

hence also

$$\int_{\Delta} \frac{(p \circ \xi)(t)}{1 + t^2}\, dt < \infty. \tag{6.3}$$

Then there is an onto isometry

$$P : \mathscr{L}(\xi) \otimes L_p^2(a, b) \to L^2_{p \circ \xi}(\Delta) \tag{6.4}$$

such that

$$P : \frac{\xi(z) - \xi(w)^*}{z - w^*} \otimes f(t) \to \frac{\xi(t) - \xi(w)^*}{t - w^*} (f \circ \xi)(t) \tag{6.5}$$

for each $f \in L_p^2(a, b)$ *and* $w \in \Pi \cup \Pi^c$.

Proof. By Theorem 3.16, we have

$$\int_{\Delta} \frac{\xi(t) - \xi(w)^*}{t - w^*} (f \circ \xi)(t) \left[\frac{\xi(t) - \xi(z)^*}{t - z^*} (g \circ \xi)(t) \right]^* (p \circ \xi)(t)\, dt$$

$$= \frac{\xi(z) - \xi(w)^*}{z - w^*} \int_{(a,b)} f(t) g(t)^* p(t)\, dt$$

for all $f, g \in L_p{}^2(a, b)$ and all $w, z \in \Pi \cup \Pi^c$. Hence there exists an isometry (6.4) satisfying (6.5). To see that the range of P is all of $L^2_{p \circ \xi}(\Delta)$, in (6.5) choose $f(t) = 1/(t - \xi(w)^*)$, where $w \in \Pi \cup \Pi^c$. We see that the range of P contains every function $1/(t - w^*)$, and so the result follows.

Given an operator (6.1), we shall construct (6.2) so that it is unitarily equivalent to

$$J \otimes T : \mathscr{L}(\xi) \otimes L_p{}^2(a, b) \to \mathscr{L}(\eta) \otimes L_q{}^2(c, d), \tag{6.6}$$

where $J : \mathscr{L}(\xi) \to \mathscr{L}(\eta)$ is given by

$$J : \frac{\xi(z) - \xi(w)^*}{z - w^*} \to \frac{\eta(z) - \eta(w)^*}{z - w^*}, \qquad w \in \Pi \cup \Pi^c, \tag{6.7}$$

on a fundamental set. A condition must be added to insure the existence of J as a bounded operator.

6.2 LEMMA. *For J to exist as a bounded operator on $\mathscr{L}(\xi)$ to $\mathscr{L}(\eta)$ it is necessary and sufficient that there exist an $M > 0$ such that*

$$\operatorname{Im} \eta(z) \leqslant M^2 \operatorname{Im} \xi(z), \qquad z \in \Pi. \tag{6.8}$$

Proof. A routine argument shows that J exists as an operator bounded by M if and only if the function $[\Phi(z) - \Phi(w)^*]/(z - w^*)$, $z, w \in \Pi \cup \Pi^c$, where $\Phi(z) = M^2 \xi(z) - \eta(z)$, is positive definite (see Aronszajn [3]). This is equivalent to (6.8) (see de Branges [5, Theorem 5]).

6.3 THEOREM. *Assume that p satisfies (6.3) and q satisfies the analogous condition. Assume also that ξ and η satisfy (6.8). Then every bounded linear operator (6.1) induces a unique bounded linear operator (6.2) such that if $T : f(t) \to g(t)$, then*

$$\hat{T} : \frac{\xi(t) - \xi(w)^*}{t - w^*} (f \circ \xi)(t) \to \frac{\eta(t) - \eta(w)^*}{t - w^*} (g \circ \eta)(t) \tag{6.9}$$

for each $w \in \Pi \cup \Pi^c$. The operator $\hat{T}$ is unitarily equivalent to the operator (6.6) where J satisfies (6.7).

Proof. By Lemma 5.2, J exists as a bounded operator, and hence so does $J \otimes T$. By Theorem 6.1 we have an isometry P determined by (6.4) and (6.5) and an analogous isometry

$$Q: \mathscr{L}(\eta) \otimes L_q^2(c, d) \to L_{q \circ \eta}(E).$$

A direct verification shows that the operator $\hat{T} = Q(J \otimes T)P^{-1}$ has the required properties.

We now apply this result in a special situation. Ahiezer [1] has given conditions for the validity of the formulas

$$g(x) = \mathrm{PV}\frac{1}{\pi}\int_{\Delta}\frac{f(t)}{t-x}\,p(t)\,dt, \tag{6.10}$$

$$f(x) = -\mathrm{PV}\frac{1}{\pi}\int_{\Delta}\frac{g(t)}{t-x}\,q(t)\,dt, \tag{6.11}$$

and

$$\int_{\Delta}|f(t)|^2 p(t)\,dt = \int_{\Delta}|g(t)|^2 q(t)\,dt, \tag{6.12}$$

where Δ is a bounded real set that consists of a finite or infinite union of intervals and p and q are nonnegative weight functions integrable over Δ and such that $p(x)g(x) = 1$ on Δ (see also Mihlin [16]). A remarkable feature of the correspondence in (6.10) and (6.11) is that f is a polynomial of exact degree n if and only if g has the same property.

We shall derive conditions for the validity of (6.10)–(6.12) for certain p, q and more general measurable real sets Δ. Our treatment is different from that of Ahiezer. Under our assumptions the correspondence yields a unitary mapping. Ahiezer is concerned with a more general situation which yields partial isometries.

6.4 THEOREM. *Let Δ be a measurable subset of R such that neither Δ nor $R \backslash \Delta$ is a Lebesgue null set. Let ξ be a nonconstant Cayley inner function on Π that maps Δ on $(0, \infty)$. Let*

$$p(t) = \xi(t)^{-1/2} \quad \text{and} \quad q(t) = \xi(t)^{1/2}$$

a.e. on Δ. Then for each $f \in L_p^2(\Delta)$, the integral

$$(Af)(x) = \mathrm{PV}\frac{1}{\pi}\int_{\Delta}\frac{f(t)}{t-x}\,p(t)\,dt \tag{6.13}$$

exists a.e. on Δ and defines a function in $L_q^2(\Delta)$. The operator A so defined is *a unitary mapping from* $L_p^{\,2}(\Delta)$ *onto* $L_q^{\,2}(\Delta)$, *with inverse given by*

$$(A^{-1}g)(x) = -\mathrm{PV}\frac{1}{\pi}\int_\Delta \frac{g(t)}{t-x}\,q(t)\,dt \tag{6.14}$$

a.e. on Δ *for each* $g \in L_q^{\,2}(\Delta)$.

Proof. In the special case where $\Delta = (0, \infty)$, $\xi(z) = z$, $p_0(x) = x^{-1/2}$, $q_0(x) = x^{1/2}$ the theorem is proved by making a simple change of variables in well known formulas for Hilbert transforms. See Ahiezer and Glazman [2, pp. 114–115]. Denote the unitary operator so obtained by

$$A_0 : L^2_{p_0}(0, \infty) \to L^2_{q_0}(0, \infty).$$

We use this special case in deriving the general case.

Let us first show that (6.13) exists a.e. on Δ for any $f \in L_p^{\,2}(\Delta)$. Let $f(x) = (x - i)f_1(x)$. Then

$$f_1(x)p(x) = [f(x)\xi(x)^{-1/4}][(x - i)\xi(x)^{1/4}]^{-1} \in L^{6/5}(\Delta) \cap L^1(\Delta), \tag{6.15}$$

because the first factor is in $L^2(\Delta)$ and the second is in $L^3(\Delta) \cap L^2(\Delta)$ by Theorem 3.16. The existence of (6.13) is now proved by showing that the integral is equal to

$$(Af)(x) = (x - i)\,\mathrm{PV}\frac{1}{\pi}\int_\Delta \frac{f_1(t)}{t-x}\,p(t)\,dt + \frac{1}{\pi}\int_\Delta f_1(t)p(t)\,dt \tag{6.16}$$

a.e. on Δ. We do not yet know that $Af \in L_q^{\,2}(\Delta)$.

Next we evaluate (6.13) on a dense set. Suppose $A_0 : f_0(x) \to g_0(x)$ for some $f_0 \in L^2_{p_0}(0, \infty)$, $g_0 \in L^2_{q_0}(0, \infty)$, and let $w \in \Pi \cup \Pi^c$. Then applying Theorem 4.1 with $f(x) = (x - \xi(w)^*)f_0(x)/x^{1/2}$, we obtain

$$A : \frac{\xi(x) - \xi(w)^*}{x - w^*}(f_0 \circ \xi)(x) \to \frac{\xi(x) - \xi(w)^*}{x - w^*}(g_0 \circ \xi)(x). \tag{6.17}$$

Now let $\hat{A}_0 : L_p^{\,2}(\Delta) \to L_q^{\,2}(\Delta)$ be the operator whose existence is guaranteed by Theorem 6.3. Since A_0 is unitary, by Theorem 6.3 so is $\hat{A}_0$. By (6.17), $\hat{A}_0$ and A coincide on a dense subset of $L_p^{\,2}(\Delta)$. Consider any $f \in L_p^{\,2}(\Delta)$. Let $\{f_n\}_1^\infty$ be a sequence in the dense set in $L_p^{\,2}(\Delta)$ on which A and $\hat{A}_0$ coincide such that $\|f - f_n\|_p \to 0$. Using (6.15) and (6.16) and well-known properties of Hilbert transforms, we see that there exists a subsequence $\{f_{n(k)}\}_{k=1}^\infty$ such that $(Af_{n(k)})(x) \to (Af)(x)$ a.e. on Δ. Then by Fatou's lemma,

$$\begin{aligned}\int_\Delta |(Af)(t)|^2 q(t)\,dt &\leq \liminf_{k\to\infty} \int_\Delta |(Af_{n(k)})(t)|^2 q(t)\,dt \\ &= \liminf_{k\to\infty} \|\hat{A}_0 f_{n(k)}\|_q^2 = \|f\|_p^2.\end{aligned}$$

Therefore A is a bounded operator from $L_p^2(\Delta)$ to $L_q^2(\Delta)$, and so $A = \hat{A}_0$ is unitary.

A similar argument yields (6.14).

Ahiezer transforms are known to map polynomials to polynomials. We show that the operator A in Theorem 6.4 has this property if Δ is a bounded set.

6.5 THEOREM. *In the situation of Theorem* 6.4 *assume that Δ is a bounded set. Then the spaces $L_p^2(\Delta)$ and $L_q^2(\Delta)$ contain all polynomials, and A and A^{-1} map polynomials of exact degree n into polynomials of exact degree n.*

Proof. To show that polynomials belong to the two spaces, we need only show that $\xi^{\pm 1/2} \in L^1(\Delta)$. By Theorem 3.15,

$$\xi(z) = -C \exp\left(-\int_\Delta dt/(t-z)\right), \qquad z \in \Pi \cup \Pi^c$$

for some $C > 0$. By the same theorem, the function

$$\eta(z) = \left[1 + \exp\left(\int_\Delta dt/(t-z)\right)\right] \Big/ \left[1 - \exp\left(\int_\Delta dt/(t-z)\right)\right], \qquad z \in \Pi \cup \Pi^c,$$

is Cayley inner on Π and maps Δ on $(-1, 1)$. Thus

$$\xi(z) = C[1 + \eta(z)]/[1 - \eta(z)], \qquad z \in \Pi \cup \Pi^c.$$

Since η satisfies (2.23) we may apply Theorem 3.17 to obtain

$$\begin{aligned}\int_\Delta \xi(t)^{\pm 1/2}\, dt &= C^{\pm 1/2} \int_\Delta \left(\frac{1 + \eta(t)}{1 - \eta(t)}\right)^{\pm 1/2} dt \\ &= \tfrac{1}{2} m(\Delta) C^{\pm 1/2} \int_{-1}^{1} \left(\frac{1+t}{1-t}\right)^{\pm 1/2} dt \\ &< \infty.\end{aligned}$$

Therefore our spaces contain all polynomials.

Now we have

$$\begin{aligned}(A1)(x) &= \mathrm{PV} \frac{1}{\pi} \int_\Delta \frac{1}{t-x}\, p(t)\, dt \\ &= C^{-1/2}\, \mathrm{PV} \frac{1}{\pi} \int_\Delta \frac{1}{t-x} \left(\frac{1 - \eta(t)}{1 + \eta(t)}\right)^{1/2} dt \\ &= C^{-1/2}\, \mathrm{PV} \frac{1}{\pi} \int_{-1}^{1} \frac{1}{t - \eta(x)} \left(\frac{1-t}{1+t}\right)^{1/2} dt \\ &= -C^{-1/2}\end{aligned}$$

a.e. on Δ, where the third equality follows by Theorem 4.2 and the fourth equality follows by Erdélyi *et al.* [11, p. 248]. Since

$$\mathrm{PV}\frac{1}{\pi}\int_\Delta \frac{tf(t)}{t-x}\,p(t)\,dt$$

$$= x\,\mathrm{PV}\frac{1}{\pi}\int_\Delta \frac{f(t)}{t-x}\,p(t)\,dt + \frac{1}{\pi}\int_\Delta f(t)p(t)\,dt$$

a.e. on Δ for any $f \in L_p^2(\Delta)$, the conclusion of the theorem for A, and hence for A^{-1}, follows by induction.

7. Applications. Generalized Chebychev Expansions

Suppose ξ is a Cayley inner function on Π that maps a real measurable set Δ on the arc (a, b). Let $\{\phi_n\}_0^\infty$ be an orthogonal system of functions in $L^2(a, b)$. By Theorem 3.16,

$$\int_\Delta \frac{\xi(t)-\xi(\alpha)^*}{t-\alpha^*}(\phi_j \circ \xi)(t)\left[\frac{\xi(t)-\xi(\beta)^*}{t-\beta^*}(\phi_k \circ \xi)(t)\right]^* dt$$

$$= \frac{\xi(\beta)-\xi(\alpha)^*}{\beta-\alpha^*}\int_\Delta \phi_j(t)\phi_k(t)^*\,dt \tag{7.1}$$

for all $\alpha, \beta \in \Pi \cup \Pi^c$. Thus we obtain an associated orthogonal system of functions in $L^2(\Delta)$. In this section we shall study this situation in some detail in the special case where $\{\phi_n\}_0^\infty$ is a system of weighted Chebychev polynomials. Our notation for the Chebychev polynomials is that of Erdélyi *et al.* [10].

The Chebychev polynomials $\{T_n\}_0^\infty$ and $\{U_n\}_0^\infty$ are defined by the formal expansions

$$\frac{1-xt}{1-2xt+t^2} = \sum_0^\infty T_n(x)t^n$$

and

$$\frac{1}{1-2xt+t^2} = \sum_0^\infty U_n(x)t^n.$$

They satisfy $T_n(\cos\theta) = \cos(n\theta)$ and $U_n(\cos\theta) = \sin((n+1)\theta)/\sin\theta$ for each $n = 0, 1, 2, \ldots$ and θ real. The orthogonality relations for the Chebychev polynomials are given by

$$\int_{-1}^{1} U_j(t)U_k(t)(1-t^2)^{1/2}\,dt = \begin{cases} 0, & j \neq k \\ \pi/2, & j = k, \end{cases} \tag{7.2}$$

and

$$\int_{-1}^{1} T_j(t)T_k(t)(1 - t^2)^{-1/2}\, dt = \begin{cases} 0, & j \neq k \\ \pi, & j = k = 0 \\ \pi/2, & j = k \neq 0. \end{cases} \tag{7.3}$$

Each f in $L^2(-1, 1)$ can be developed in orthogonal expansions, convergent in the norm of $L^2(-1, 1)$, of the form

$$f(t) = \sum_0^\infty a_n(1 - t^2)^{1/4} U_n(t) = \sum_0^\infty b_n(1 - t^2)^{-1/4} T_n(t). \tag{7.4}$$

The coefficients in these expansions satisfy

$$\int_{-1}^{1} |f(t)|^2\, dt = (\pi/2) \sum_0^\infty |a_n|^2 = \pi|b_0|^2 + (\pi/2) \sum_1^\infty |b_n|^2. \tag{7.5}$$

Throughout this section Δ denotes a real measurable set such that neither Δ nor $R\backslash\Delta$ is a Lebesgue null set. We assume that ξ is a nonconstant Cayley inner function that maps Δ on $(-1, 1)$. Then $|\xi| < 1$ a.e. on Δ. We define $\mathscr{E}_\pm$ to be the closed linear span in $L^2(\Delta)$ of all functions of the form

$$(1 - \xi(t)^2)^{\pm 1/4} \frac{\xi(t) - \xi(\alpha)^*}{t - \alpha^*}, \qquad \alpha \in \Pi \cup \Pi^c. \tag{7.6}$$

We note that functions of the form (7.6) belong to $L^2(\Delta)$ by Theorem 3.16. By Theorem 3.13, $\dim \mathscr{E}_\pm = I(\Delta)$, where $I(\Delta)$ is the index of Δ.

We begin by deriving an expansion for functions in $L^2(\Delta)$ analogous to (7.4)–(7.5).

7.1 THEOREM. *Each f in $L^2(\Delta)$ may be developed in orthogonal expansions*

$$f(t) = \sum_0^\infty e_n^+(t)(U_n \circ \xi)(t) = \sum_0^\infty e_n^-(t)(T_n \circ \xi)(t), \tag{7.7}$$

where $e_n^\pm \in \mathscr{E}_\pm$ for each $n = 0, 1, 2, \ldots$ and the series converge in the metric of $L^2(\Delta)$. We have

$$\|f\|^2 = \sum_0^\infty \|e_n^+\|^2 = \|e_0^-\|^2 + \frac{1}{2} \sum_1^\infty \|e_n^-\|^2 \tag{7.8}$$

where $\|\cdot\|$ denotes the norm in $L^2(\Delta)$.

Proof. We assert that $\{(U_n \circ \xi)\mathscr{E}_+\}_0^\infty$ and $\{(T_n \circ \xi)\mathscr{E}_-\}_0^\infty$ are orthogonal systems of sets in $L^2(\Delta)$, and

$$\|(U_n \circ \xi)e\| = \|e\|, \qquad e \in \mathscr{E}_+, \quad n \geqslant 0, \tag{7.9}$$

$$\|(T_n \circ \xi)e\| = \begin{cases} \|e\|, & e \in \mathscr{E}_-, \quad n = 0 \\ \tfrac{1}{2}\|e\|, & e \in \mathscr{E}_-, \quad n > 0. \end{cases} \tag{7.10}$$

In fact, the orthogonality relations and (7.9) and (7.10) follow from (7.1) for finite linear combinations of functions of the form (7.6), and in general by approximation.

To prove the theorem, we now need only show that there is no nonzero element f of $L^2(\Delta)$ which is orthogonal to either of the subspaces $\bigvee_0^\infty (U_n \circ \xi)\mathscr{E}_+$ or $\bigvee_0^\infty (T_n \circ \xi)\mathscr{E}_-$. It is sufficient to show that whenever

$$\int_\Delta (1 + |t|)^{-1} |g(t)|\, dt < \infty$$

and

$$\int_\Delta \frac{\xi(t) - \xi(\alpha)^*}{t - \alpha^*} \xi(t)^n g(t)\, dt = 0, \qquad \alpha \in \Pi \cup \Pi^c, \quad n \geqslant 0,$$

then $g = 0$ a.e. on Δ. For such g we have

$$\int_\Delta \frac{\xi(t)^{n+1} g(t)}{t - \alpha^*} dt = \xi(\alpha)^{*n} \int_\Delta \frac{g(t)}{t - \alpha^*} dt \tag{7.11}$$

for all $\alpha \in \Pi \cup \Pi^c$ and all $n \geqslant 0$. Recall that $|\xi(t)| < 1$ a.e. on Δ and $|\xi(t)| > 1$ a.e. on $R \backslash \Delta$. In particular, the inequality $|\xi(\alpha)| > 1$ holds on some nonempty open subsets of Π and Π^c. Letting $n \to \infty$ in (7.11), we see that

$$\int_\Delta \frac{g(t)}{t - \alpha^*} dt = 0,$$

first for all α such that $|\xi(\alpha)| > 1$, and then for all α in $\Pi \cup \Pi^c$ by analytic continuation. Hence $g = 0$ a.e. on Δ, and the theorem follows.

We explore some questions related to the expansions in Theorem 7.1. One question is how to calculate such expansions. For the sake of conciseness we concentrate on the expansion associated with the polynomials $\{U_n\}_0^\infty$. It is convenient to digress briefly and establish a connection between the Chebychev polynomials and abstract shift operators on a Hilbert space.

Let $\mathscr{H}$ be a complex Hilbert space. By a *shift operator* on $\mathscr{H}$ we mean an isometric operator S mapping $\mathscr{H}$ into itself such that $\bigcap_0^\infty S^j\mathscr{H} = (0)$, or what is the same thing, $\lim_{n\to\infty} S^{*n}f = 0$ for every f in $\mathscr{H}$. We define the *multiplicity* of S to be the minimum dimension of a subspace $\mathscr{E}$ of $\mathscr{H}$ such that $\mathscr{H} = \bigvee_0^\infty S^j\mathscr{E}$. The minimum is attained with $\mathscr{E} = \ker S^* = \mathscr{H} \ominus S\mathscr{H}$, and for this choice of $\mathscr{E}$ we have $S^j\mathscr{E} \perp S^k\mathscr{E}$ whenever $j \neq k$, $j, k = 0, 1, 2, \ldots,$ and

$$\mathscr{H} = \sum_0^\infty \oplus S^n\mathscr{E}.$$

Also with the special choice $\mathscr{E} = \ker S^*$, the orthogonal projection of $\mathscr{H}$ on $\mathscr{E}$ is given by $P_0 = I - SS^*$. Then the orthogonal projection of $\mathscr{H}$ on $S^n\mathscr{E}$ is given by $P_n = S^n P_0 S^{*n}$ for any $n = 0, 1, 2, \ldots$, and we have

$$f = \sum_0^\infty S^n P_0 S^{*n} f \tag{7.12}$$

in the sense of norm convergence for each f in $\mathscr{H}$. (See Radjavi and Rosenthal [17] and Rosenblum and Rovnyak [19]; further references to shift operators are cited in these works.)

The connection between shift operators and Chebychev polynomials seems to have escaped previous notice.

7.2 THEOREM. *Let S be a shift operator on a Hilbert space $\mathscr{H}$. Write $S = X + iY$ where $X = \operatorname{Re} S$ and $Y = \operatorname{Im} S$. Let $\mathscr{E} = \ker S^*$ and $P_0 = I - SS^*$. Then for each $e \in \mathscr{E}$,*

$$S^n e = U_n(X)e \quad \text{and} \quad iYS^n e = T_{n+1}(X)e \tag{7.13}$$

for all $n = 0, 1, 2, \ldots$. Furthermore, for each f in $\mathscr{H}$ we have

$$f = \sum_0^\infty U_n(X) P_0 U_n(X) f \tag{7.14}$$

in the sense of norm convergence.

Proof. Since $U_0(x) = 1$ and $U_1(x) = 2x$, we have $S^0 e = e = U_0(X)e$ and $S^1 e = (S + S^*)e = 2Xe = U_1(X)e$. Suppose we know that $S^j e = U_j(X)e$ for all $j \leqslant n + 1$ for some $n = 0, 1, 2, \ldots$. Since $U_{n+2}(x) = 2xU_{n+1}(x) - U_n(x)$, we have

$$\begin{aligned} S^{n+2}e &= (S + S^*)S^{n+1}e - S^n e \\ &= [2XU_{n+1}(X) - U_n(X)]e \\ &= U_{n+2}(X)e. \end{aligned}$$

Hence the first relation in (7.13) follows by induction.

Since $T_1(x) = x$, we have $iYS^0 e = \frac{1}{2}(S - S^*)e = \frac{1}{2}(S + S^*)e = Xe = T_1(X)e$. But for any $n \geqslant 1$, $T_{n+1}(x) = \frac{1}{2}[U_{n+1}(x) - U_{n-1}(x)]$, so by what we proved above,

$$\begin{aligned} iYS^n e &= \tfrac{1}{2}(S + S^*)S^n e \\ &= \tfrac{1}{2}[U_{n+1}(X) - U_{n-1}(X)]e \\ &= T_{n+1}(X)e, \end{aligned}$$

which is the second relation in (7.13).

By the first relation in (7.13) we see that $S^n P_0 = U_n(X) P_0$, and hence $P_0 S^{*n} = P_0 U_n(X)$. Thus (7.14) follows from (7.12).

We construct a model for a shift operator S in which $X = \operatorname{Re} S$ is in canonical form. It is known (see Rosenblum [18]) that X has uniform absolutely continuous spectrum on $[-1, 1]$ with multiplicity equal to the multiplicity of S. Let us assume that $\mathscr{H}$ is separable. Denote by $\mathscr{C}$ any conveniently chosen Hilbert space whose dimension is equal to the multiplicity of S. Then the simplest model for $X = \operatorname{Re} S$ is the operator

$$X_0 : f(x) \to x f(x) \tag{7.15}$$

in the Lebesgue space $L_{\mathscr{C}}{}^2(-1, 1)$ of weakly measurable $\mathscr{C}$-valued functions $f(x)$ on $(-1, 1)$ such that

$$\|f\|^2 = \int_{-1}^{1} |f(t)|_{\mathscr{C}}^2 \, dt < \infty,$$

where $|\cdot|_{\mathscr{C}}$ denotes the norm in $\mathscr{C}$. Therefore, what we seek is the form of a shift operator S_0 on $L_{\mathscr{C}}{}^2(-1, 1)$ satisfying $X_0 = \operatorname{Re} S_0$.

7.3 THEOREM. *The general form of a shift operator* S_0 *on* $L_{\mathscr{C}}{}^2(-1, 1)$ *for which* $\operatorname{Re} S_0 = X_0$ *has the form* (7.15) *is given by*

$$S_0 : f(x) \to x f(x) - \mathrm{PV} \frac{1}{\pi} \int_{-1}^{1} \frac{(1 - x^2)^{1/4} (1 - t^2)^{1/4}}{t - x} C(x)^* C(t) f(t) \, dt, \tag{7.16}$$

where C *is a weakly measurable function on* $(-1, 1)$ *whose values are unitary operators on* $\mathscr{C}$, *and the principal value integral is interpreted weakly by forming inner products with arbitrary vectors in* $\mathscr{C}$. *If* S_0 *is given by* (7.16), *then the subspace* $\mathscr{E}_0 = \ker S_0{}^*$ *is the set of functions of the form*

$$e(x) = (1 - x^2)^{1/4} C(x)^* a \tag{7.17}$$

where $a \in \mathscr{C}$. *For any* $e \in \mathscr{E}_0$, *we have*

$$S_0 : U_n(x) e(x) \to U_{n+1}(x) e(x) \tag{7.18}$$

for all $n = 0, 1, 2, \ldots$.

Proof. Suppose first that C is a function as described in the theorem. By (7.2) there is a unique shift operator S_0 on $L_{\mathscr{C}}{}^2(-1, 1)$ such that (7.18) holds for each e of the form (7.17). Using the identities (see Erdélyi *et al.* [10])

$$U_{n+1}(x) = x U_n(x) + T_{n+1}(x) \tag{7.19}$$

$$T_{n+1}(x) = -\mathrm{PV} \frac{1}{\pi} \int_{-1}^{1} \frac{(1 - t^2)^{1/2}}{t - x} U_n(t) \, dt, \qquad -1 < x < 1, \tag{7.20}$$

we obtain

$$U_{n+1}(x) = xU_n(x) - \mathrm{PV}\frac{1}{\pi}\int_{-1}^{1} \frac{(1-t^2)^{1/2}}{t-x} U_n(t)\,dt, \qquad -1 < x < 1, \tag{7.21}$$

for each $n = 0, 1, 2, \ldots$. It follows that S_0 is given by (7.16) on the special functions appearing in (7.18). By linearity and continuity, S_0 is given by (7.16) for all functions in $L_{\mathscr{C}}^2(-1, 1)$. A direct calculation shows that $\mathrm{Re}\, S_0 = X_0$, where X_0 is given by (7.15).

Conversely, let S_0 be any shift operator on $L_{\mathscr{C}}^2(-1, 1)$ such that $\mathrm{Re}\, S_0 = X_0$ is given by (7.15). Define S_1 to be the shift operator on $L_{\mathscr{C}}^2(-1, 1)$ given by (7.16) with $C \equiv 1$, the identity operator on $\mathscr{C}$.

Since $\mathrm{Re}\, S_1 = X_0 = \mathrm{Re}\, S_0$, S_0 and S_1 have the same multiplicity, and so they are unitarily equivalent. Let W be a unitary operator on $L_{\mathscr{C}}^2(-1, 1)$ such that $S_0 = W^* S_1 W$. Then $S_0^* = W^* S_1^* W$, and so $WX_0 = X_0 W$. Therefore W has the form $W: f(x) \to C(x)f(x)$ where C is a function as in the theorem. By the definition of S_1, we obtain (7.16) for S_0. This completes the proof.

As an example, consider the case where $\mathscr{C}$ is the one-dimensional space of complex numbers with the absolute value norm. In this case f, C, and e are scalar valued functions, and any shift operator S_0 of the form (7.16) has multiplicity 1. Shifts of higher multiplicity can be obtained by replacing (7.15) by a more general multiplication operator as in (7.22) below.

7.4 THEOREM. *Let Δ be a real measurable set such that neither Δ nor $R \backslash \Delta$ is a Lebesgue null set, and let ξ be a nonconstant Cayley inner function that maps Δ on $(-1, 1)$. Let C denote any complex valued measurable function on Δ such that $|C(x)| = 1$ a.e. on Δ. Define operators X and S on $L^2(\Delta)$ by*

$$X: f(x) \to \xi(x)f(x) \tag{7.22}$$

and

$$S: f(x) \to \xi(x)f(x) - \mathrm{PV}\frac{1}{\pi}\int_{\Delta} \frac{(1-\xi(x)^2)^{1/4}(1-\xi(t)^2)^{1/4}}{t-x} C(x)^* C(t) f(t)\,dt. \tag{7.23}$$

Then S is a shift operator on $L^2(\Delta)$, and $\mathrm{Re}\, S = X$. The multiplicity of S is equal to the index of Δ. The kernel $\mathscr{E}_+$ of S^* is the closed span in $L^2(\Delta)$ of all functions of the form

$$e(x) = (1-\xi(x)^2)^{1/4} \frac{\xi(x) - \xi(w)^*}{x - w^*} C(x)^* \tag{7.24}$$

where $w \in \Pi \cup \Pi^c$. *The projection* $P_0 = I - SS^*$ *of* $L^2(\Delta)$ *onto* $\mathscr{E}_+$ *is given by*

$$P_0 : f(x) \to \mathrm{PV}\frac{2}{\pi}\int_\Delta \frac{\xi(t) - \xi(x)}{t - x}(1 - \xi(x)^2)^{1/4}(1 - \xi(t)^2)^{1/4} f(t)\,dt \quad (7.25)$$

for any $f \in L^2(\Delta)$. *Finally, for any* $e \in \mathscr{E}_+$, *we have*

$$S : (U_n \circ \xi)(x)e(x) \to (U_{n+1} \circ \xi)(x)e(x) \quad (7.26)$$

for all $n = 0, 1, 2, \ldots$.

Proof. It follows from Theorem 7.1 that there exists a unique shift operator S on $L^2(\Delta)$ such that $\ker S^* = \mathscr{E}_+$ and (7.26) holds. By (7.21) and Theorem 4.1, we have

$$\frac{\xi(x) - \xi(w)^*}{x - w^*}(U_{n+1} \circ \xi)(x) = \xi(x)\frac{\xi(x) - \xi(w)^*}{x - w^*}(U_n \circ \xi)(x)$$
$$- \mathrm{PV}\frac{1}{\pi}\int_\Delta \frac{\xi(t) - \xi(w)^*}{t - w^*}\frac{(1 - \xi(t)^2)^{1/2}}{t - x}(U_n \circ \xi)(t)\,dt$$

a.e. on Δ for any nonreal number w and $n = 0, 1, 2, \ldots$. It follows that S is given by (7.23) on the special functions appearing in (7.26), and hence on all functions in $L^2(\Delta)$ by linearity and continuity. A direct calculation shows that the real part of S is given by (7.22). Using Theorem 3.13, we see that $\dim \mathscr{E}_+ = I(\Delta)$, and therefore the multiplicity of S is equal to the index of Δ. It remains to prove (7.25). We have

$$P_0 = I - SS^* = (S + S^*)S + S^*(S + S^*) - (S + S^*)^2$$

or

$$P_0 = 2XS + 2S^*X - 4X^2.$$

We obtain (7.25) from this using (7.22) and (7.23).

In principle, Theorem 7.4 permits the calculation of the coefficients $e_0^+, e_1^+, e_2^+, \ldots$ in the first expansion of (7.7) in Theorem 7.1. For by Theorem 7.2 we have

$$e_n^+ = P_0 U_n(X)f, \qquad n = 0, 1, 2, \ldots, \quad (7.27)$$

and this may be calculated using (7.25). We conclude with an example of one such expansion in which the coefficients can be calculated in closed form. The example is

$$\frac{(1 - \xi(x)^2)^{1/4}}{x - w} = \sum_{n=0}^{\infty}(-2)\frac{\xi(x) - \xi(w)}{x - w}(1 - \xi(x)^2)^{1/4}$$
$$\times \{\xi(w) - i[1 - \xi(w)^2]^{1/2}\}^{n+1}(U_n \circ \xi)(x), \quad (7.28)$$

where w is any number in $\Pi \cup \Pi^c$. It is sufficient to treat the case where $w \in \Pi$, since then we can obtain the other case by taking complex conjugates. We use that branch of $(1 - z^2)^{1/2}$ in the upper half-plane which is positive on the positive imaginary axis. To show that (7.28) is a special instance of the first expansion of (7.7) in Theorem 7.1, by (7.27) and (7.25) it is enough to show that a.e. on Δ

$$\mathrm{PV}\frac{1}{\pi}\int_\Delta \frac{\xi(t) - \xi(x)}{t - x}\frac{(1 - \xi(t)^2)^{1/2}(U_n \circ \xi)(t)}{t - w}\,dt$$
$$= -\frac{\xi(x) - \xi(w)}{x - w}\{\xi(w) - i[1 - \xi(w)^2]^{1/2}\}^{n+1}$$

for any $n = 0, 1, 2, \ldots$. By Theorem 4.1, formula (4.6), this will follow if we can show that

$$\frac{1}{\pi}\int_{-1}^{1} \frac{(1 - t^2)^{1/2}U_n(t)}{t - w}\,dt = -[w - i(1 - w^2)^{1/2}]^{n+1}, \qquad w \in \Pi. \tag{7.29}$$

We use a method given by Tricomi [25, p. 181]. The function $F(z) = -[z - i(1 - z^2)^{1/2}]^{n+1}$ belongs to $H^2(\Pi)$, and its boundary function satisfies

$$\operatorname{Im} F(x) = \begin{cases} (1 - x^2)^{1/2}U_n(x), & |x| < 1 \\ 0, & |x| > 1. \end{cases}$$

Therefore (7.29) follows from Cauchy's theorem in the form

$$\frac{1}{\pi}\int_{-\infty}^{\infty} \frac{\operatorname{Im} F(t)}{t - z}\,dt = F(z), \qquad z \in \Pi.$$

This completes the proof of (7.28).

Added in Proof. After completing this paper, the authors discovered that in the special case of rational functions, composition formulas of the type proved in Theorems 3.16 and 3.17 were first proved by the great English mathematician, G. Boole, "On the comparison of transcendents, with certain applications to the theory of definite integrals", *Phil. Trans.* **CXLVII** (1857), 745–803. In this fascinating paper, Boole develops a calculus which is equivalent to the calculus of residues. Boole discovered this calculus independently of Cauchy. Additional references to relevant nineteenth century literature, including a note by Cayley, can be found in J. W. L. Glaisher, "Note on certain theorems in definite integration", *Messenger Math.* **8** (1879), 63–74. More recently, applications of Boole's ideas have been given in ergodic theory and probability theory. See, e.g., R. L. Adler and B. Weiss, "The ergodic infinite measure preserving transformation of Boole", *Israel J. Math.* **16** (1973), 263–278, G. Letac, "Which functions preserve Cauchy laws?", *Proc. Amer. Math. Soc.* **67** (1977), 277–286, and J. H. Neuwirth, "Ergodicity of some mappings of the circle and the line", preprint. We are indebted to J. V. Ryff and our colleagues N. F. G. Martin and L. Pitt for telling us about the connection between Boole's work and ergodic theory and for supplying the Glaisher reference. We also thank

D. Sarason for calling our attention to the paper of O. D. Cereteli, "Certain properties of inner functions", *Sakharth. SSR Mecn. Akad. Moambe* **82**(2) (1976), 313–316 [in Russian, with Georgian and English summaries; *M R* 55, #3260, which derives the Stein–Weiss distribution function formula by a method similar to the method that we use. Cereteli discusses this result and related results in Chapter I (which he wrote) of the survey article by B. V. Khvedelidze, "The method of Cauchy-type integrals in the discontinuous boundary-value problems in the theory of holomorphic functions of a complex variable", *J. Soviet Math.* **7** (1977), 309–415.

REFERENCES

1. N. I. AHIEZER, On some inversion formulas for singular integrals, *Izv. Akad. Nauk SSSR, Ser. Mat.* **9** (1945), 275–290; *MR* **7**, 439.
2. N. I. AHIEZER AND I. M. GLAZMAN, "Theory of Linear Operators in Hilbert Space," Vol. I, Ungar, New York, 1961.
3. N. ARONSAJN, Theory of reproducing kernels, *Trans. Amer. Math. Soc.* **68** (1950), 337–404; *MR* **14**, 479.
4. N. ARONSZAJN AND W. F. DONOGHUE, On exponential representations of analytic functions in the upper halfplane with positive imaginary part, *J. Analyse Math.* **5** (1956–1957), 321–388.
5. L. DE BRANGES, "Hilbert Spaces of Entire Functions," Prentice–Hall, Englewood Cliffs, New Jersey, 1968.
6. P. L. BUTZER AND R. J. NESSEL, "Fourier Analysis and Approximation," Vol. 1, Birkhäuser–Verlag, Basel, and Academic Press, New York, 1971.
7. J. D. CHANDLER, JR., Analysis on unions of intervals, Ph.D. Thesis, Univ. of Virginia, 1976.
8. J. D. CHANDLER, JR., A generalization of a theorem of S. N. Bernstein, *Proc. Amer. Math. Soc.* **63** (1977), 95–100.
9. P. L. DUREN, "Theory of H^p Spaces," Academic Press, New York, 1970; *MR* **42**, #3552.
10. A. ERDÉLYI *et al.*, Bateman manuscript project, *in* "Higher Transcendental Functions," Vol. II, McGraw–Hill, New York, 1953.
11. A. ERDÉLYI *et al.*, Bateman manuscript project, *in* "Tables of Integral Transforms," Vol. II, McGraw–Hill, New York, 1954.
12. E. HEWITT AND K. STROMBERG, "Real and Abstract Analysis," Springer–Verlag, Berlin and New York, 1965.
13. R. S. ISMAGILOV, The spectrum of Toeplitz matrices, *Dokl. Akad. Nauk SSSR* **149** (1963), 769–772; *Soviet Math. Dokl.* **4** (1963), 462–465; *MR* **26**, #4190.
14. T. L. KRIETE, A generalized Paley–Wiener theorem, J. *Math. Anal. Appl.* **36** (1971), 529–555; *MR* **44**, #5473.
15. L. H. LOOMIS, A note on the Hilbert transform, *Bull. Amer. Math. Soc.* **52** (1946), 1082–1086; *MR* **8**, #377.
16. L. G. MIHLIN, Singular integral equations, *Uspehi Mat. Nauk* (*N.S.*) **3**, No. 3(25) (1948), 29–112; English transl. *Amer. Math. Soc. Ser. 1* **10** (1962), 84–198; *MR* **10**, #305.
17. H. RADJAVI AND P. ROSENTHAL, "Invariant Subspaces," Springer–Verlag, Berlin and New York, 1973.
18. M. ROSENBLUM, A concrete spectral theory for selfadjoint Toeplitz operators, *Amer. J. J. Math.* **87** (1965), 709–718. *MR* **31** #6127.
19. M. ROSENBLUM AND J. ROVNYAK, The factorization problem for nonnegative operator valued functions, *Bull. Amer. Math. Soc.* **77** (1971), 287–318; *MR* **42**, #8315.
20. M. ROSENBLUM AND J. ROVNYAK, Restrictions of analytic functions, II, *Proc. Amer. Math. Soc.* **51** (1975), 335–343; *MR* **53**, #3764b.

21. M. Rosenblum and J. Rovnyak, Cayley inner functions and best approximation, *J. Approximation Theory* **17** (1976), 241–253.
22. W. Rudin, "Real and Complex Analysis," McGraw–Hill, New York, 1966.
23. J. V. Ryff, Subordinate H^p functions, *Duke Math. J.* **33** (1966), 347–354; *MR* **33**, #289.
24. E. M. Stein and G. Weiss, An extension of a theorem of Marcinkiewicz and some of its applications, *J. Math. Mech.* **8** (1959), 263–284; *MR* **21**, #5888.
25. F. Tricomi, "Integral Equations," Wiley (Interscience), New York, 1957.

AMS (MOS) subject classifications: primary 30A76; secondary 42A40, 45E05, 33A65, 28A65.

TOPICS IN FUNCTIONAL ANALYSIS
ADVANCES IN MATHEMATICS SUPPLEMENTARY STUDIES, VOL. 3

The Complex-Wave Representation of the Free Boson Field

I. E. SEGAL†

Department of Mathematics
Massachusetts Institute of Technology
Cambridge, Massachusetts

DEDICATED TO MARK G. KREIN ON THE OCCASION OF HIS SEVENTIETH BIRTHDAY

Associated with any given complex Hilbert space **H**, there is a corresponding free boson field $\Phi = (\mathbf{K}, W, \Gamma, v)$ consisting of a complex Hilbert space **K**, a Weyl system W, a representation Γ of the unitary group $U(\mathbf{H})$ on **H** into $U(\mathbf{K})$, and a unit vector v in **K** which is invariant under Γ and cyclic for W. The representation for φ which is in many ways mathematically the simplest is one in which **K** is a Hilbert space of entire antiholomorphic functions on **H**.

Originally introduced (Segal, 1962) as the L_2-completion of the antiholomorphic polynomials on **H**, the elements of **K** are here shown to be identifiable with functions well defined at every point of **H**, which are antiholomorphic and satisfy a simple L_2-boundedness condition. As the representation which achieves a variety of diagonalization for the creation operators, it is called the *complex-wave* representation, and is compared here with the *real-wave* (or functional integration) representation, which diagonalizes the hermitian field operators. This representation is given a standardized form with simple Fourier–Wiener transformation properties, and the intertwining operator between the two representations is given explicitly.

1. INTRODUCTION

The concept of a free quantum field of particles satisfying Bose–Einstein statistics is commonly introduced in a highly structured manner involving space–time, invariant wave equations thereon, etc.; but in algebraic essence, all that is involved is a given complex Hilbert space **H**. The concrete structure of this space may vary from application to application, but many of the central results of the theory are independent of this special structure.

This article treats the abstract free quantum field cited, called for brevity the *free boson field*, and especially its relation to a Hilbert space of entire antiholomorphic functions. It thus provides an application of ideas which

† Research supported in part by NSF Grant MPS 72–05098.

ISBN 0–12–287150–2

in many ways go back to M. G. Krein's intensely fruitful and pioneering work on this subject and related ones in modern functional analysis.

Mathematically, the free boson field[1] over a given complex Hilbert space $\mathbf{H}$, to be denoted as $\Phi(\mathbf{H})$, may be defined as a quadruple $(\mathbf{K}, W, \Gamma, v)$ consisting of:

(1) a complex Hilbert space $\mathbf{K}$;

(2) a Weyl system W on $\mathbf{H}$ with values in $U(\mathbf{K})$; i.e. a (strongly, as throughout this article) continuous mapping $z \to W(z)$ from $\mathbf{H}$ to the unitary operators on $\mathbf{K}$ satisfying the Weyl relations:

$$W(z)W(z') = \exp((i/2)\operatorname{Im}\langle z, z'\rangle)W(z + z')$$

for arbitrary z and z' in $\mathbf{H}$;

(3) a continuous representation Γ from $U(\mathbf{H})$ into $U(\mathbf{K})$ satisfying the relation

$$\Gamma(U)W(z)\Gamma(U)^{-1} = W(Uz)$$

for arbitrary $U \in U(\mathbf{H})$ and $z \in \mathbf{H}$;

(4) a unit vector v in $\mathbf{K}$ having the properties that $\Gamma(U)v = v$ for all $U \in U(\mathbf{H})$, and that the $W(z)v$, $z \in \mathbf{H}$, span $\mathbf{K}$ topologically. (For brevity, we cite the latter condition as cyclicity of v for W.)

The constraints (1)–(4) do not uniquely determine the free boson field, but serve to do so with the addition of

(5) Γ is "positive" in the sense that if A is any nonnegative self-adjoint operator in $\mathbf{H}$, then $d\Gamma(A)$ is likewise nonnegative, where for any self-adjoint A in $\mathbf{H}$ $d\Gamma(A)$ denotes the self-adjoint generator of the one-parameter unitary group $[U(e^{itA}): t \in R^1]$ provided by Stone's theorem.

In the present article we shall take conditions (1)–(5) as the definition of the free boson field. This axiomatic characterization must, of course, be supplemented by suitable existence considerations, which will be supplied by the concrete representations of $\Phi(\mathbf{H})$ given below.

Historically, the free boson field was introduced by Dirac in 1926 in purely formal analogy with the Heisenberg quantization of a single-particle system; the Hilbert spaces $\mathbf{H}$ and $\mathbf{K}$ are totally implicit in this treatment, the representation Γ does not occur at all, and the Weyl system W appears only in the form of an infinite sequence of putative operators: p_1, q_1; p_2, q_2; . . . , satisfying the Heisenberg commutative relations, whose existence

[1] More precisely, this is a concrete free boson field; *the* (abstract) free boson field is the unitary equivalence class of all concrete ones.

Dirac merely hypothesized. An explicit construction showing formal existence of this early version of the free boson field was first given by Fock in 1932. A construction which assumes and exploits the Hilbertian character of the so-called *single-particle space* **H** was first given in Cook's thesis (1953). That the Fock–Cook representation satisfies conditions (1)–(5) was clear except for the Weyl relations, which follow from its unitary equivalence with the functional integration representation given by Segal (1956).

In a general scientific way, the Fock–Cook representation expresses the particle properties of the quantum field, by providing explicit diagonalizations for the so-called *occupation numbers* $d\Gamma(P)$, where P ranges over a maximal commuting set of projections on **H**. These are the properties typically most directly observable in high-energy experiments. The functional integration representation expresses the wave properties, by virtue of its explicit diagonalization of the values of the field at different points of space, at a fixed time. The wave properties are conceptually fundamental, and technically the functional integration representation has been the basis for progress in constructive quantum field theory during the past decade.

However, in each of the two classic representations—particle and wave—for the free boson field, some of the important operators have highly complicated actions. On the other hand, in the complex-wave (or antiholomorphic) representation, virtually all of the important operators have a remarkably simple formal appearance, despite the greater complexity of the physical interpretation of the representation.

The present work gives an explicit treatment of the complex-wave representation from first principles, and shows in particular that its pointwise properties are greatly superior to those of the wave representation described above, which we distinguish by the prefix "real", in view of the hermitian character of the operators diagonalized by it, in contrast to the nonhermitian character of the operators pseudodiagonalized by the complex-wave representation. The results promise to be useful in group-theoretic investigations, in particular by facilitating application of the reproducing kernel technique exemplified by Rossi and Vergne (1976), used in forthcoming work extending to the infinite-dimensional symplectic group Sp(**H**) the harmonic representation well known in the finite-dimensional case. The relations between the two wave representations are treated explicitly employing normalizations which yield notably simple expressions for intertwining operators and relevant transforms. It is probable that similar developments are valid in the case of the free fermion field.

Although primarily concerned here with the case of an infinite-dimensional Hilbert space **H**, the case when **H** is finite-dimensional is naturally included. There is a flexibility in the presentation of the finite-dimensional analog of the free boson field which is lacking in the infinite-dimensional case, due to the applicability of the Stone–von Neumann theorem in the former case but not the latter. As a consequence, although in principle it must be possible to compare the representations for Heisenberg systems in spaces of holomorphic functions given by Bargmann (1961) and Satake (1971) with aspects of the restriction of the present work to the finite-dimensional case, it would be tedious as well as of doubtful utility to do so, the more so because of their use of representation spaces of holomorphic rather than antiholomorphic functions. This use of antiholomorphic functions is indicated by a combination of positivity and covariance constraints, of a simple mathematical and essential physical nature. While complex conjugation maps the one class of functions into the other, it is represented by an antiunitary operator, which reverses positivity. With a given conjugation in the Hilbert space **H**—as in the case $\mathbf{H} = C^n$ with its usual linear and inner product structures—or by the consideration of holomorphic functions on the dual of **H**, rather than antiholomorphic ones on **H**, the treatment may be effected in terms of holomorphic functions alone, but at the cost of diminished covariance and/or increased circumlocution.

It may also be of interest to indicate the relation with heuristic developments due to theoretical physicists. In the case of wave functions in one dimension, the harmonic oscillator hamiltonian $n = (1/2)(p^2 + q^2 - 1)$, the position operator q, and the "creation" operator $c = 2^{-1/2}(p + iq)$ all have simple spectra in the sense that they admit a cyclic vector. According to the general ideas of quantum mechanics, any one of them may be used as a complete state labeling operator. In Fock (1928) the use of c was originated in this connection, and it was noted that this led to the consideration of holomorphic wave functions; but no Hilbert space of such functions was formulated, and the only specific domain of definition cited was the unit disk. The more comprehensive treatment of Dirac (1949) remains formal, and the case of a quantum field is reduced to the one-dimensional case via a representation as a direct product. In Dirac's work there is again no Hilbert space **K**, no representation Γ of $U(\mathbf{H})$ (the analysis being dependent on a choice of basis in **H**), and only formally defined field operators. In brief, not only mathematical questions of existence, uniqueness, and clear formulation, but also, in part essential formal features are not found in the cited work.

The complex wave representation was formulated and treated in our 1960 lectures at the Summer Seminar in Applied Mathematics, Boulder, Colorado (Segal, 1963). A more detailed account of some of its features is included in Segal (1962). The present work relates it more closely to classical complex analysis, and more explicitly to the real-wave representation, while including an exposition of its general background and developing a formalism adapted to further group-theoretic developments; its basic results date to an earlier period.

2. The Free Boson Field

We treat here the existence and uniqueness of the free boson field, and related questions, essentially from first principles. We begin with the question of uniqueness. It is interesting, in this connection, that constraint (5) in the preceding section is equivalent to the apparently much weaker condition that its conclusion holds for just one nontrivial operator A, as follows from

Theorem 1. *Let* **H** *be a given complex Hilbert space, and let* W *be a Weyl system over* **H** *with representation space* **K** *and cyclic vector* v *in* **K**.

Suppose there exists a nonnegative self-adjoint operator A *in* **H** *which annihilates no nonzero vector, and a one-parameter unitary group* Γ' *on* K *with the properties*:

(a) $\Gamma'(t)W(z)\Gamma'(-t) = W(e^{itA}z)$ *for all* $t \in R^1$ *and* $z \in \mathbf{H}$;
(b) $\Gamma'(t)v = v$ *for all* $t \in R^1$;
(c) $\Gamma'(t) = e^{itH}$, *where* H *is self-adjoint and nonnegative.*

Then there exists a unique representation Γ *of* $U(\mathbf{H})$ *into* $U(\mathbf{K})$ *which extends* Γ' *in the sense that* $\Gamma(e^{itA}) = \Gamma'(t)$ *for all* $t \in R^1$, *and such that* $(\mathbf{K}, W, \Gamma, v)$ *satisfies conditions* (1)–(5).

Proof. Let $H = d\Gamma(A)$, and set

$$f(u) = \langle e^{-uH}W(z)v, W(z)v\rangle,$$

where $u = s + it$ with $s > 0$, and z is arbitrary in **H**. Then f is bounded and continuous in the half-plane $s \geqslant 0$, and holomorphic in the interior $s > 0$; we denote the totality of all such functions as **B**.

By virtue of the Weyl relations, the boundary values $f(it)$ may be expressed as

$$f(it) = \exp((i/2)\,\mathrm{Im}\langle z_t, z\rangle)\langle W(z_t - z)v, v\rangle, \qquad z_t = e^{-itA}z.$$

Now setting

$$g(u) = \exp(-\langle e^{-uA}z, z\rangle/2),$$

then $g \in \mathbf{B}$ also. Accordingly, $fg \in \mathbf{B}$ and fg has the boundary values

$$(fg)(it) = \langle W(z_t - z)v, v\rangle \exp(-\mathrm{Re}\langle z_t, z\rangle/2).$$

Replacing z by $-z$, it follows that $\langle W(-z_t + z)v, v\rangle \exp(-\mathrm{Re}\langle z_t, z\rangle/2)$ is also the boundary value function of an element of **B**. But this function is the complex conjugate of the function $(fg)(it)$. Accordingly, fg must be a constant, which may be evaluated as $\exp(-\langle z, z\rangle/2$, by setting $t = 0$, yielding the equation

$$\langle W(z_t - z)v, v\rangle = \exp(-\|z_t - z\|^2/4).$$

By virtue of the triviality of the null space of A, the $z_t - z$ are dense in **H** as z and t vary, as follows from the spectral theorem; and it follows in turn by continuity that

$$\langle W(z)v, v\rangle = \exp(-\|z\|^2/4)$$

for all $z \in \mathbf{H}$.

This result implies that the inner product $\langle W(Uz)v, W(Uz')z\rangle$, where z and z' are arbitrary in **H**, is independent of $U \in U(\mathbf{H})$; and this implies in turn that the linear transformation T_0:

$$\sum_i a_i W(z_i)v \to \sum_i a_i W(Uz_i)v$$

is welldefined and isometric, on the domain **D** of all finite linear combinations of the $W(z)v$, $z \in \mathbf{H}$. Since **D** is dense in **K**, T_0 extends uniquely to a unitary operator on all of **K**, which we denote as $\Gamma(U)$; and by construction, $\Gamma(U)W(z)\Gamma(U)^{-1} = W(Uz)$ and $\Gamma(U)v = v$ for all $U \in U(\mathbf{H})$. That Γ is a representation of $U(\mathbf{H})$ follows from its intertwining relation with W, noting also that $\Gamma(U)$ is the unique unitary operator on **K** which transforms $W(z)$ into $W(Uz)$ and leaves v invariant, by virtue of the cyclicity of v. In order to show that Γ is continuous from $U(\mathbf{H})$ into $U(\mathbf{K})$, it suffices to show that $\Gamma(U)W(z)v$ is a continuous function of U, for any fixed $z \in \mathbf{H}$, at the element $U = I$ of $U(\mathbf{H})$. To this end it suffices in turn to show that

$$\langle \Gamma(U)W(z)v, W(z)v\rangle \to 1$$

as $U \to I$; but by the Weyl relations,

$$\langle \Gamma(U)W(z), W(z)v\rangle = \exp((i/2)\operatorname{Im}\langle Uz, z\rangle)\exp(-\|Uz - z\|^2/4),$$

which is continuous in U.

To show the positivity of the representation Γ, suppose now that A is any nonnegative self-adjoint operator in **H**; it suffices to show that $\Gamma(e^{itA})$ is a positive-frequency function of t, i.e., that $\int\langle\Gamma(e^{itA})w, w'\rangle g(t)\,dt = 0$ for functions $g \in L_1(R^1)$ whose Fourier transforms vanish on the negative half-axis, and for arbitrary vectors w and w' in **K**. To this end it suffices in turn to show that $\langle\Gamma(e^{itA})W(z)v, W(z')v)\rangle$ is a positive-frequency function of t for arbitrary z and z' in **H**. But by the earlier evaluation of $\langle W(z)v, v\rangle$ and the Weyl relations, this function is $\exp(-\|z\|^2/4 - \|z'\|^2/4 + \langle z_{-t}, z'\rangle/4)$; and since $\langle z_{-t}, z'\rangle$ is a positive-frequency function of t, so also is $\exp(\langle z_{-t}, z'\rangle)$.

Finally, the unicity of the representation Γ follows from the cyclicity of v under W, which implies that any unitary operator which commutes with all $W(z)$ and leaves v invariant must be the identity.

COROLLARY 1.1. *Any two systems* $(\mathbf{K}, W, \Gamma, v)$ *satisfying conditions* (1)–(5) *are unitarily equivalent.*

Proof. Given a second such system, denoted by primes, there exists a unique unitary transformation T from $\mathbf{K}$ onto $\mathbf{K}'$ which carries $W(z)v$ into $W'(z)v'$, for all $z \in \mathbf{H}$, by an argument employed in the preceding proof. The respective Weyl systems are then unitarily equivalent via T, and the definition of $\Gamma(U)$ as the unique unitary on $\mathbf{K}$ which transforms $W(z)$ into $W(Uz)$ and leaves v invariant shows that it is transformed by T into $\Gamma'(U)$.

We turn now to the question of existence, which will be treated in terms of the theory of integration in Hilbert space. To fix the notation and summarize the parts of the theory required here, let $\mathbf{H}'$ be a given real Hilbert space and let g denote the weak probability distribution (or generalized random process) on $\mathbf{H}'$ known as the centered isotropic normal distribution. This means that if $x_1, \ldots, x_n$ are the coordinates of any point $x \in \mathbf{H}$ relative to an arbitrary orthonormal basis, then the joint probability distribution of $x_1, \ldots, x_n$ has the element

$$dg = \sigma^{-n}(2\pi)^{-n/2} \exp(-(x_1{}^2 + \cdots + x_n{}^2)/2\sigma^2)\, dx_1 \cdots dx_n,$$

where σ is a positive constant whose square is called the *variance*. An integration theory can be developed in the infinite-dimensional case by considering g to determine a positive linear functional E (the *expectation*) defined on the algebra $\mathbf{A}(\mathbf{H}')$ of all functions f on $\mathbf{H}'$ of the form $f(x) = F(Px)$, where P is the projection of $\mathbf{H}'$ onto a linear subspace $\mathbf{M}'$ of finite dimension m, F is a bounded complex-valued Baire function on $\mathbf{M}'$, and

$$E(f) = \int_{\mathbf{M}'} F(x)\, dg(x).$$

The elements of $\mathbf{A}(\mathbf{H}')$ are called *bounded tame functions*; the term *tame function* is used when the boundedness restriction is removed; and the term *square-integrable tame function* for those tame functions f such that $E(|f|^2) < \infty$ ($E(|f|^2)$ being defined by substitution in the foregoing equation).

Alternatively, g may be regarded in an obvious manner as a probability measure defined on the ring $\mathbf{R}$ of all subsets of $\mathbf{H}'$ whose characteristic

[2] We shall occasionally refer to such a vector as a *functional* to distinguish it from a function on $\mathbf{H}'$ with which it may be related; thus, if f is a tame function, $\theta(f)$ is a tame functional.

functions are tame; it is then countably additive on the subring $\mathbf{R}_{\mathbf{M}'}$ consisting of those subsets whose characteristic functions are *based on* $\mathbf{M}'$, which is defined as meaning that they have the form given above; but is not countably additive on $\mathbf{H}'$ as a whole, when $\mathbf{H}'$ is infinite dimensional. As a result, the Riesz–Fischer theorem does not hold in its usual form, and the Hilbert space completion, denoted $L_2(\mathbf{H}', g)$, of $\mathbf{A}(\mathbf{H}')$ with respect to the inner product $\langle f, f' \rangle = E(f\overline{f'})$ can not be identified in the usual way with an algebra of functions on $\mathbf{H}'$ modulo an ideal of null functions. The canonical homomorphism of $\mathbf{A}(\mathbf{H}')$ into $L_2(\mathbf{H}, g)$ will be denoted as θ; thus for any vector $f \in \mathbf{A}(\mathbf{H}')$, $\theta(f)$ is the corresponding residue class modulo the subspace of $\mathbf{A}(\mathbf{H}')$ consisting of vectors of zero norm, as a vector[2] in the completion $L_2(\mathbf{H}', g)$.

Now if $\mathbf{H}$ is a given complex Hilbert space, it has also the structure of a real Hilbert space, with inner product equal to the real part of the complex inner product given in $\mathbf{H}$. In this way, g may be well defined on $\mathbf{H}$, and the space $L_2(\mathbf{H}, g)$ is correspondingly such. In the case of a real space $\mathbf{H}'$, the space $L_2(\mathbf{H}', g)$ may be defined in terms of polynomials, as well as tame functions, and the completion of the algebra $\mathbf{P}'$ of all polynomials on $\mathbf{H}'$, with respect to the same inner product

$$\langle \theta(f), \theta(f') \rangle = \int_{\mathbf{M}'} f(x)\overline{f'(x)}\, dg(x).$$

Here $\mathbf{M}'$ is any finite-dimensional subspace of $\mathbf{H}'$ on which the polynomials f and f' are based; a polynomial on $\mathbf{H}'$ being defined as a function of the form $f(x) = p(\langle x, e_1 \rangle, \ldots, \langle x, e_n \rangle)$, where p is a polynomial function on R^n and the e_j are arbitrary vectors, finite in number in $\mathbf{H}'$. It is no essential loss of generality, and we shall henceforth always assume, that the e_j are orthonormal in this representation. The possibility of defining $L_2(\mathbf{H}', g)$ as the completion of the polynomials may be regarded as a generalization of the completeness of the Hermite functions in $L_2(R^1)$.

This means that if $\mathbf{H}$ is a complex Hilbert space, then $L_2(\mathbf{H}, g)$ consists of the completion of the algebra $\mathbf{P}'(\mathbf{H})$ of the functions of the form

$$f(x) = p(\mathrm{Re}\langle x, e_1 \rangle, \ldots, \mathrm{Re}\langle x, e_n \rangle);$$

such a function we call a *real-analytic polynomial*. But in addition to the algebra $\mathbf{P}'$, there are two other simple unitarily invariant algebras of polynomials: the *complex-analytic*, defined as those of the form

$$f(x) = p(\langle x, e_1 \rangle, \ldots, \langle x, e_n \rangle),$$

where p is a polynomial function on C^n; and the *complex-antianalytic*, i.e., the complex conjugates of those just indicated. We denote the totality of complex-antianalytic polynomials on **H** as **P(H)**; the representation space **K** for the complex-wave representation will consist of the closure of **P** in $L_2(\mathbf{H}, g)$.

We round out the presentation of this representation by first defining a system $(\mathbf{K}', W', \Gamma', v')$ which fails to represent the free boson field only in that condition (5) is violated, and that v' is not cyclic for the $W'(z)$. To begin with, let $\mathbf{K}' = L_2(\mathbf{H}, g)$. Next, for any vector $z \in \mathbf{H}$, we define an operator $W_0'(z)$ on the subspace $\theta(\mathbf{S})$ of $\mathbf{K}'$, where **S** denotes the totality of all square-integrable tame functions on **H**, as follows. For any function f on **H**, let f_z denote the function given by the equation

$$f_z(u) = f(u - \sigma z)\exp(\langle z, u\rangle/2\sigma - \langle z, z\rangle/4).$$

If f is tame, so also is f_z; and if $\theta(f) = 0$, $\theta(f_z) = 0$ also, since straightforward computation shows that $\|\theta(f)\|^2 = \|\theta(f_z)\|^2$ for arbitrary tame functions f and vectors z in **H**. The mapping $\theta(f) \to \theta(f_z)$ is therefore a well-defined isometry of $\theta(\mathbf{S})$ into itself, and defines an operator $W_0'(z)$ on $\theta(\mathbf{S})$.

It is straightforward to verify that

$$W_0'(z)W_0'(z') = \exp((i/2)\operatorname{Im}\langle z, z'\rangle)W_0'(z + z')$$

for arbitrary z and z' in **H**. Setting $z' = -z$, it follows that $W_0'(z)$ is invertible, and so extends to a unique unitary transformation, to be denoted as $W'(z)$, on all of $\mathbf{K}'$. By continuity, the Weyl relations remain valid for W. It is not difficult to show in addition that the mapping $z \to W(z)$ is continuous from **H** to $U(\mathbf{K}')$, and it follows that W is a Weyl system on **H** with representation space $\mathbf{K}'$.

We next define a representation Γ_0' of $U(\mathbf{H})$ on $\theta(\mathbf{A})$ (where $\mathbf{A} = \mathbf{A}(\mathbf{H})$) as follows: For arbitrary $f \in \mathbf{A}$ and $U \in U(\mathbf{H})$, $\Gamma_0'(U)$ sends $\theta(f)$ into $\theta(f^U)$, where $f^U(z) = f(U^{-1}z)$. Since it is readily verified that $f^U \in \mathbf{A}$ when $f \in \mathbf{A}$ and that the map $f \to f^U$ preserves the expectation functional E, $\Gamma_0'(U)$ is well defined and is isometric from $\theta(\mathbf{A})$ into itself. It is straightforward to verify also that $\Gamma_0'(UU') = \Gamma_0'(U)\Gamma_0'(U')$ for arbitrary U and U' in $U(\mathbf{H})$; i.e., Γ_0' is a representation of $U(\mathbf{H})$ on $\theta(\mathbf{A})$, it being obvious that $\Gamma_0'(e) = I$. In particular, $\Gamma_0'(U)$ is invertible for all $U \in U(\mathbf{H})$, and so extends uniquely to a unitary operator $\Gamma'(U)$ defined on all of **K**. It follows that Γ' is a representation of $U(\mathbf{H})$ into $U(\mathbf{K})$; and it is not difficult to verify that Γ' is continuous from $U(\mathbf{H})$ into $U(\mathbf{K}')$ by checking strong continuity on the spanning subset of $\mathbf{K}'$ consisting of the finite products of real-linear functionals.

Now let v' denote the functional identically 1 on **H**, or more exactly $\theta(1)$; then $\Gamma'(U)v' = v'$ for all $U \in U(\mathbf{H})$; and the required intertwining relation between Γ' and W' is readily verified on the dense subset $\theta(\mathbf{S})$ and thence, by continuity, holds on all of $\mathbf{K}'$. Thus the quadruple $\varphi' = (\mathbf{K}', W', \Gamma', v')$ satisfies all of the conditions on the free boson field over **H** except the positivity condition (5), and the cyclicity for v'. We call Φ' the *regular boson field over* **H** in analogy with the term *regular representation* for groups.

The complex-wave representation may now be defined as that given by restriction to the subspace spanned by the complex-antianalytic polynomials.

THEOREM 2. *Let* $\Phi' = (\mathbf{K}', W', \Gamma', v')$ *denote the regular boson field over the given complex Hilbert space* **H**. *Let* **K** *denote the cyclic subspace under the action of* W' *generated by* v'. *Then*:

(a) **K** *is spanned by the complex-antianalytic polynomials on* **H**;

(b) *Setting* $W(z) = W'(z)|\mathbf{K}$ *and* $\Gamma(U) = \Gamma'(U)|\mathbf{K}$ *for all* $z \in \mathbf{H}$ *and* $U \in U(\mathbf{H})$, *and* $v = v'$, *then the quadruple* $\Phi = (\mathbf{K}, W, \Gamma, v)$ *is a representation of the free boson field over* **H**.

Proof. By direct computation $W(z)v = \theta(f)$, where f is the tame function on **H**, $f(u) = \exp(\langle z, u\rangle/2\sigma - \langle z, z\rangle/4)$. Thus the functionals $\theta(b_z)$ are all in the subspace spanned by the $W(z)v$, where $b_z(u) = \exp(\langle z, u\rangle)$, $z \in \mathbf{H}$. Since $b_{z+z'} = b_z b_{z'}$, the set of all finite linear combinations of the b_z forms a ring, which contains also, for any $\epsilon > 0$ and finite ordered orthonormal set of vectors $e_1, \ldots, e_m$ and positive integers $n_1, \ldots, n_m$ the vector $c_\epsilon = \prod_{j=1}^m [(\exp(\epsilon\langle e_j, u\rangle) - 1)/\epsilon]^{n_j}$; and it follows by dominated convergence that $\theta(c_\epsilon) \to \theta(\prod_{j=1}^m \langle e_j, \cdot\rangle^{n_j})$, showing that the image under θ of any complex antianalytic polynomial is in **K**. Conversely, the application of dominated convergence to the power series expansion of b_z in terms of $\langle z, u\rangle$ shows that $W(z)v$ is in the closure of $\theta(\mathbf{P})$.

Thus, v is a cyclic vector for W, and to conclude that Φ is the free boson field over **H** it is only necessary to show that the representation Γ is positive (condition 5). It suffices to show that the "number of particles" operator $d\Gamma(I)$ has a nonnegative spectrum. In fact, $\Gamma(e^{it})$ sends the monomial $\theta(w)$, where $w(u) = \langle e_1, u\rangle^{n_1}, \ldots, \langle e_m, u\rangle^{n_m}$, into $\exp(it(n_1 + \cdots + n_m))\theta(w)$, showing that $d\Gamma(I)$ has a spectrum consisting of the nonnegative integers (and, incidentally, is diagonalized by the totality of the $\theta(w)$ relative to a fixed orthonormal basis).

We now adapt the definition of Nachbin (1969) to present purposes.

DEFINITION. An *extended homogeneous polynomial* of degree n on **H** is a function F on **H** to C which has the form $F(z) = \tilde{F}(z, z, \ldots, z)$, where $\tilde{F}$ is an n-linear functional on **H** to C with the property that

$$|\tilde{F}(z_1, \ldots, z_n)| \leqslant C \prod_{j=1}^{n} \|z_j\|$$

for some constant C. An *entire function* on $\mathbf{H}$ is a function F of the form $F(z) = \sum_{n=0}^{\infty} F_n(z)$, where F_n is an extended homogeneous polynomial of degree n, and the series is convergent for all $z \in \mathbf{H}$. An *antientire* function is one which is the complex conjugate of an entire function.

THEOREM 3. *If φ is any vector in $\mathbf{K}$, then*

(1) *the function F on $\mathbf{H}$ given by the equation $F(z) = \langle \varphi, \theta(\exp(\langle z/2\sigma^2, \cdot \rangle)) \rangle$ is an antientire function on $\mathbf{H}$;*

(2) $\sup_{\mathbf{M}} \int_{\mathbf{M}} |F(z)|^2 \, dg(z) = \|\varphi\|^2$, *the supremum being taken over the set of all finite-dimensional subspaces $\mathbf{M}$ of $\mathbf{H}$;*

(3) *if $\varphi = \theta(f)$, f being a tame antientire function, then $F(z) = f(z)$, $z \in \mathbf{H}$.*

Conversely, if F is an antientire function on $\mathbf{H}$ such that

$$\sup_{\mathbf{M}} \int_{\mathbf{M}} |F(z)|^2 \, dg(z) < \infty,$$

then there exists a unique vector $\varphi \in \mathbf{K}$ such that the foregoing holds.

Proof. We use the term *antimonomial* for a vector in $\mathbf{K}$ (or, on occasion, a function on $\mathbf{H}$, the context indicating which) which is a constant multiple of one of the form $\theta(w)$ (respectively, w), where w is as above. Note to begin with that any two antimonomials of different (total) degrees $n_1 + \cdots + n_m$ are orthogonal in $\mathbf{K}$, being proper vectors of $d\Gamma(I)$ with different proper values. Using the notation b_z' for the function $\exp(\langle z/2\sigma^2, \cdot \rangle)$,

$$b_z'(u) = \sum_{j=0}^{\infty} (j!)^{-1} \langle z, u \rangle^j (2\sigma^2)^{-j}$$

so that

$$\theta(b_z') = \sum_{j=0}^{\infty} (j!)^{-1} \theta(\langle z, \cdot \rangle^j)(2\sigma^2)^{-j},$$

the series being convergent in $\mathbf{K}$ and the terms mutually orthogonal. It follows that $F(z) = \sum_{j=0}^{\infty} (j!)^{-1} \langle \varphi, \langle z, \cdot \rangle^j \rangle (2\sigma^2)^{-j}$.

On the other hand, each summand in the latter series is an extended homogeneous polynomial, since $\langle \varphi, \langle z, \cdot \rangle^j \rangle$ is obtained by setting $z_1 = z_2 = \cdots = z_j$ in the multilinear function on $\mathbf{H}$, $\varphi \to \langle \varphi, \theta(q) \rangle$, where $q(u) = \langle z_1, u \rangle \cdots \langle z_j, u \rangle$; and by the Schwarz and Hölder inequalities,

$$|\langle \varphi, \theta(q) \rangle| \leqslant \|\varphi\| \prod_{i=1}^{j} \|\theta(\langle z_i, \cdot \rangle)\|_{2j},$$

where $\|\cdot\|_p$ indicates the norm in L_p; but $\|\theta(\langle z, \cdot \rangle)\|_p = c_p \|z\|$ for some finite constant c_p. Thus F is antientire.

Observe next that the antimonomials $\psi_{\mathbf{n}}$, where $\mathbf{n}$ stands for the multi-index $n_1, n_2, \ldots, = 2^{-n/2}(n!)^{-1/2}\sigma^{-n}\theta(w)$, w being as earlier and n being $\sum_j n_j$, form an orthonormal basis in $\mathbf{K}$, relative to any given orthonormal basis $e_1, e_2, \ldots$ in $\mathbf{H}$. Orthogonality is a consequence of the vanishing of the integrals $\int_C |z|^s z^t \, dg(z)$ when s and t are nonnegative integers with $t \neq 0$, as follows from rotational invariance; normalization follows by evaluation of the integral when $t = 0$. It follows that for any vector $\varphi \in \mathbf{K}$, $\|\varphi\|^2 = \sum_{\mathbf{n}} |a_n|^2$ where $a_{\mathbf{n}} = \langle \varphi, \psi_{\mathbf{n}} \rangle$.

Now let φ be arbitrary in $\mathbf{K}$, and let $\{p_n\}$ be a sequence in $\mathbf{P}$ such that $\theta(p_n) = \varphi$ in $L_2(\mathbf{H}, g)$. Then for any positive integers m and n there exists a finite-dimensional subspace $\mathbf{L}$ of $\mathbf{H}$ on which p_m and p_n may be based, so that

$$\|p_m - p_n\|^2 = \int_{\mathbf{L}} |p_m(u) - p_n(u)|^2 \, dg(u).$$

On the other hand, for any element $f \in \mathbf{P}$, and any finite-dimensional subspace $\mathbf{M}$ of $\mathbf{H}$, which is spanned by basis vectors in the given collection,

$$\int_{\mathbf{M}} |F(u)|^2 \, dg(u) = \sum_{\mathbf{n} \in \mathbf{n}(\mathbf{M})} |a_{\mathbf{n}}|^2,$$

where $\mathbf{n}(\mathbf{M})$ denotes the set of all multi-indices $\mathbf{n}$ such that $n_j = 0$ when e_j is not in $\mathbf{M}$. It follows—taking note of the arbitrariness of the given orthonormal set in $\mathbf{H}$—that the norm of $f \,|\, \mathbf{M}$ in $L_2(\mathbf{M}, g)$ is an increasing function of $\mathbf{M}$ and converges (trivially) as $\mathbf{M} \to \mathbf{H}$ to the norm of $\theta(f)$ in $L_2(\mathbf{H}, g)$. It follows in turn that the sequence of restrictions $\{p_n | \mathbf{L}\}$ is convergent for fixed $\mathbf{L}$ to an element of $L_2(\mathbf{H}, g)$, which is in fact a square-integrable tame functional based on $\mathbf{L}$. We denote this vector as $\varphi_{\mathbf{L}}$; it is independent of the Cauchy sequence $\{p_n\}$ and depends only on φ.

Moreover, the antientire function on $\mathbf{L}$ corresponding to φ is $\langle \theta(\varphi_{\mathbf{L}}), \theta(b_z')\rangle$, which is easily seen to be identical with the restriction to $\mathbf{L}$ of the antientire function F on $\mathbf{H}$ which corresponds to φ. The θ involved here is of course relative to the subspace $\mathbf{L}$ rather than $\mathbf{H}$, but an approximation argument parallel to that just indicated shows that this dependence is immaterial. Now denoting the cited restriction as $F_{\mathbf{L}}$, it follows that

$$\|\varphi_L\|^2 = \|F_{\mathbf{L}}\|^2 = \sum_{\mathbf{n} \in \mathbf{n}(\mathbf{L})} |a_{\mathbf{n}}|^2,$$

where conventional usage is now followed in the deletion of θ when a countably additive measure is involved. Thus $\|F_{\mathbf{L}}\|^2$ is bounded by and, in fact as $\mathbf{L} \to \mathbf{H}$, converges to the constant $\sum_{\mathbf{n}} |a_n|^2 = \|\varphi\|^2$.

Now suppose conversely that F is a given antientire function on $\mathbf{H}$ for which $\|F \,|\, \mathbf{M}\|_{L_2(\mathbf{M}, g)}$ remains bounded as $\mathbf{M}$ ranges over the set of all finite-dimensional subspaces of $\mathbf{H}$. Let $f^{\mathbf{M}}$ be the function on $\mathbf{H}$ given by the equation

$$f^{\mathbf{M}}(u) = \sum_{\mathbf{n} \in \mathbf{n}(\mathbf{M})} a_{\mathbf{n}} \Psi_{\mathbf{n}}(u),$$

where $\Psi_{\mathbf{n}}(u) = 2^{-n/2}(n!)^{-1/2}\sigma^{-n}\langle e_1, u\rangle^{n_1} \cdots \langle e_m, u\rangle^{n_m}$ and $a_n = \langle F|\mathbf{M}, \psi_{\mathbf{n}}\rangle$. Then $f_{\mathbf{M}}$ is a square-integrable tame function on $\mathbf{H}$ which is antianalytic and based on $\mathbf{M}$. Evidently

$$\|f^{\mathbf{M}} - f^{\mathbf{L}}\|^2 = \sum_{\mathbf{n} \in \mathbf{n}(\mathbf{M}) \Delta \mathbf{n}(\mathbf{L})} |a_{\mathbf{n}}|^2,$$

where Δ denotes the symmetric difference. It follows that as $\mathbf{M} \to \mathbf{H}$, $\{\theta(f^{\mathbf{M}})\}$ is convergent to a vector $\varphi \in \mathbf{K}$. By the same argument as in the preceding paragraph, the antientire function $z \to \langle \varphi, \theta(b_z)\rangle$ corresponding to φ is identical with the original antientire function F. Unicity of φ is evident from the fact that

$$F(u) = \sum_{\mathbf{n}} a_{\mathbf{n}} 2^{-n/2}(n!)^{-1/2}\sigma^{-n}\langle e_1, u\rangle^{n_1} \cdots \langle e_m, u\rangle^{n_m},$$

as follows by a limiting argument from the finite-dimensional case, the series being convergent by virtue of the Schwarz inequality.

It remains only to show (3). This reduces to the relation: $f(z) = \int f(u) \exp(\langle u, z/2\sigma^2\rangle)\, dg(u)$. The inequality $|f(z)| \leqslant \exp(\langle z, z\rangle/4\sigma^2)\|f\|$ which results from the application of the Schwarz inequality to the power series expansion for f implies that it suffices to establish this relation for the spanning set of functions of the form $f(u) = \exp(\langle u, w\rangle)$, $w \in \mathbf{H}$. In this case the relation follows by evaluation of an elementary Gaussian integral.

It follows that the Weyl system operators $W(z)$ and the representation Γ of $U(\mathbf{H})$ for the free boson field may be defined in a pointwise fashion, and not merely as Hilbert space limits, in accordance with

COROLLARY 3.1. *The linear boson field $\Phi(\mathbf{H})$ over $\mathbf{H}$ may be represented as follows:*

(1) *$\mathbf{K}$ is the Hilbert space of all antientire functions F on $\mathbf{H}$ for which the norm*

$$\|F\| = \sup_{\mathbf{M}} \|F|\mathbf{M}\|_{L_2(\mathbf{M}, g)}$$

is finite, the supremum being taken over all finite-dimensional subspaces $\mathbf{M}$ of $\mathbf{H}$, with the inner product

$$\langle F, F'\rangle = \lim_{\mathbf{M} \to \mathbf{H}} \langle F|\mathbf{M}, F'|\mathbf{M}\rangle_{L_2(\mathbf{M}, g)}.$$

(2) *For any $z \in \mathbf{H}$, $W(z)$ is the operator*

$$F(u) \to F(u - \sigma z)\exp(\langle z, u\rangle/2\sigma - \langle z, z\rangle/4), \qquad F \in \mathbf{K}.$$

(3) *For any* $T \in U(\mathbf{H})$, $\Gamma(T)$ *is the operator*

$$F(u) \to F(T^{-1}u), \qquad F \in \mathbf{K}.$$

(4) v *is the function identically* 1 *on* $\mathbf{H}$.

Proof. This involves straightforward limiting arguments of the type already employed and the formalities are consequently omitted.

It is useful to take explicit note of certain facts obtained in the course of the proof of Theorem 3.

COROLLARY 3.2. *If* $e_1, e_2, \ldots$ *is any orthonormal basis for* $\mathbf{H}$, *then an orthonormal basis for the preceding space* $\mathbf{K}$ *is provided by the polynomials*

$$P(z) = 2^{-n/2}(n_1!n_2!\cdots)^{-1/2}\sigma^{-n}\langle e_1, z\rangle^{n_1}\langle e_2, z\rangle^{n_2}\cdots,$$

as the n_i *range independently over the nonnegative integers subject to the constraint that* $n = \sum_i n_i$ *be finite.*

The corresponding expansion for any vector $F \in \mathbf{K}$ *is convergent pointwise, and*

$$|F(z)| \leqslant \|F\| \exp(\langle z, z\rangle/4\sigma^2), \qquad z \in \mathbf{H}.$$

Remark. It would be interesting to have an effective general criterion on a given weak probability distribution m on $\mathbf{H}$ under which the closure in the Hilbert space $L_2(\mathbf{H}, m)$ of the complex-analytic polynomials can be identified with a class of entire functions on $\mathbf{H}$, in a fashion analogous to that here established for the isotropic normal distribution. The class of such distributions m probably includes many non-Gaussian ones which must be treated by approximation methods rather than explicit evaluations.

The subclass of those which are quasi-invariant under vector translations in $\mathbf{H}$ would provide a class of Weyl systems which may be useful in quantum field theory. Another interesting subclass is that admitting a temporal invariance group whose natural action on $\tilde{L}_2(\mathbf{H}, m)$ has a positive generator. The question of whether there exist such m other than mixtures of normal distributions appears physically apposite.

We now rationalize the term *complex-wave representation* by making explicit the sense in which this representation pseudodiagonalizes the creation operators; the basic result in Segal (1962) is sharpened by determination of maximal domains and actions thereon. We first recall the following

DEFINITION. For any representation $\Phi = (\mathbf{K}, W, \Gamma, v)$ of the free boson field over the given Hilbert space $\mathbf{H}$ and a vector $z \in \mathbf{H}$, the *creation operator* *for the vector* z, to be denoted as $C(z)$, is defined as the operator $2^{-1/2}(dW(z) - i\,dW(iz))$, where $dW(z)$ denotes the self-adjoint generator of the one-parameter group $[W(tz): t \in R^1]$. The *annihilation operator for the vector* z, to be denoted as $C^*(z)$, is defined as the operator

$$2^{-1/2}(dW(z) + i\,dW(iz)).$$

COROLLARY 3.3. *The operators* $C(z)$ *and* $C^*(z)$ *are closed, densely defined, and mutually adjoint. In the complex-wave representation,* $C(z)$ *has domain consisting of all* $F \in \mathbf{K}$ *such that* $\langle z, \cdot \rangle F(\cdot) \in \mathbf{K}$, *and sends any such function into* $a^{-1}\langle z, \cdot\rangle F(\cdot)$, *where* $a = i\sigma\sqrt{2}$; $C^*(z)$ *has domain consisting of all* $F \in \mathbf{K}$ *such that* $\partial_z F \in \mathbf{K}$, *where* $(\partial_z F)(u) = \lim_{\epsilon \to 0} \epsilon^{-1}(F(u + \epsilon z) - F(u))$, *and sends any such function into* $-a\partial_z F$.

Proof. That $C(z)$ and $C^*(z)$ are closed, densely defined, and mutually adjoint is valid for an arbitrary Weyl system, by familiar smoothing arguments, and we omit the proof. Since convergence in the Hilbert space $\mathbf{K}$ implies pointwise convergence of the corresponding antientire functions on $\mathbf{H}$, the actions of $C(z)$ and $C^*(z)$ may be computed by taking pointwise limits. This is readily done and yields the cited results.

It remains to show that the domains of $C(z)$ and $C^*(z)$ include all vectors to which the corresponding pointwise-defined operators are applicable, as operators in $\mathbf{K}$. To this end let $\mathbf{D}$ denote the algebraic direct sum of the subspaces $\mathbf{K}_n$ of $\mathbf{K}$ consisting of the extended homogeneous polynomials of degree n, and let F denote any element of $\mathbf{K}$ such that $\langle z, \cdot\rangle F(\cdot)$ is again in $\mathbf{K}$. It is easily seen that multiplication by $\langle z, \cdot\rangle$ is bounded on $\mathbf{K}_n$ and maps it into $\mathbf{K}_{n+1}$; consequently, $\|\langle z, \cdot\rangle F(.)\|^2 = \sum_n \|\langle z, \cdot\rangle F_n(\cdot)\|^2$, where F_n denotes the component of F in $\mathbf{K}_n$. On the other hand, it is readily verified that $\mathbf{K}_n$ is in the domains of the $dW(z)$ for all z, and hence in the domains of all the $C(z)$ and $C^*(z)$. Setting $F^N = \sum_{n<N} F_n$, it follows that $F^N \to F$ and that $C(z)F^N \to \langle z, \cdot\rangle F(.)$ and it follows in turn, since $C(z)$ is closed, that F is in the domain of $C(z)$. The same argument applies to $C^*(z)$ with $n + 1$ replaced by $n - 1$.

3. THE PARTIAL FOURIER–WIENER TRANSFORM

In conventional Fourier analysis on R^n, there have been a variety of normalizations employed for the constants a and b in the Fourier transform: $f \to \hat{f}$, $\hat{f}(x) = a\int \exp(ib\langle x, y\rangle)f(y)\,dy$. The values $a = 1$, $b = \pm 2\pi$ are the only ones for which the transform is both unitary and multiplicative on convolutions; but other values are used quite commonly, without substantially affecting the applicability of the transform; all the various transforms are equivalent via scale transformations combined with constant factors. This is no longer the case for infinite-dimensional Fourier analysis, for which a direct change of scale is inadmissible, not being absolutely continuous. In addition, in the analysis associated with a complex Hilbert space **H**, the question of invariance with respect to $U(\mathbf{H})$, rather than multiplicativity or invariance under translations (both of which are lost on transition to Gaussian from Lebesgue measure, as is essential) becomes of major importance.

For this reason, as well as for the natural normalization of the analog to the Fourier transform in the infinite-dimensional case and for the better understanding of the relation of the real- and complex-wave representations, it is useful to treat the partial Fourier–Wiener transform in $L_2(\mathbf{H}', g)$, $\mathbf{H}'$ being a given real Hilbert space, and its transformation properties under $U(\mathbf{H})$, **H** being the complexification of $\mathbf{H}'$. We recall that the full Fourier–Wiener transform for functionals on Hilbert space may be regarded as a canonical form of the transform of the same name introduced by Cameron (1945) and Cameron and Martin (1945) for functionals on Wiener space, and may be characterized (Segal, 1956) as the unique unitary operator **F** on $L_2(\mathbf{H}', g)$ which acts as follows on polynomials f on $\mathbf{H}'$: $\theta(f) \to \theta(\tilde{f})$, where

$$\tilde{f}(u) = \int_{\mathbf{H}'} f(2^{1/2}x + iu)\, dg(x).$$

The partial transform $\mathbf{F}(\mathbf{M}')$ with respect to a given closed linear subspace $\mathbf{M}'$ of $\mathbf{H}'$ is correspondingly defined as the unique unitary operator which acts as follows on polynomials: $\theta(f) \to \theta(f')$, where

$$f'(u) = \int_{\mathbf{M}'} f(2^{1/2}x + i(\mathbf{M}')u)\, dg(x),$$

where $i(\mathbf{M}') = iP + (I - P)$, P being the projection of $\mathbf{H}'$ onto $\mathbf{M}'$; alternatively, $i(\mathbf{M}')$ may be defined on **H**, rather than only on $\mathbf{H}'$, as the unique unitary operator which multiplies elements of the complexification **M** of $\mathbf{M}'$ by i, and leaves fixed elements of the orthocomplement to **M**.

It is readily verified that if f is a polynomial then f' is well defined and is likewise a polynomial. Its extendability to a unitary operator follows from the clarification of its relation to the Wiener transform given by

LEMMA 4.1. *Let $\mathbf{H}'$ be a given real Hilbert space, $\mathbf{M}'$ a given closed linear subspace, and $\mathbf{N}'$ its orthocomplement. Let Λ denote the unitary transformation from $L_2(\mathbf{M}', g) \otimes L_2(\mathbf{N}', g)$ onto $L_2(\mathbf{H}', g)$ which carries $\theta(f) \otimes \theta(h)$ into $\theta(p)$ for any polynomials f and h on $\mathbf{M}'$ and $\mathbf{N}'$, respectively, where $p(u) = f(Pu)h((I - P)u)$, P being the projection of $\mathbf{H}'$ onto $\mathbf{M}'$. Then*

$$\Lambda^{-1}\mathbf{F}(\mathbf{M}')\Lambda = \mathbf{F}_{\mathbf{M}'} \otimes I,$$

where $\mathbf{F}_{\mathbf{M}'}$ *denotes the (full) Wiener transform on* $L_2(\mathbf{M}', g)$ *and* I *denotes the identity on* $L_2(\mathbf{N}', g)$.

Proof. It is not difficult to verify the existence of the cited unitary transformation Λ, and its unicity; formal details will be omitted. To complete the proof, it suffices to show that

$$\mathbf{F}(\mathbf{M}')\Lambda(\theta(f) \otimes \theta(h)) = \Lambda(\mathbf{F}_{\mathbf{M}'} \otimes I)\theta(p),$$

where f, h, and p are as above. For it then follows that $\mathbf{F}(\mathbf{M}')$ as defined preserves the inner products of any two product functions such as p, the totality of which has image under θ which spans $L_2(\mathbf{H}', g)$, which implies the existence and unicity of a unitary extension of $\mathbf{F}(\mathbf{M}')$ as defined on polynomials. The bounded operators $\mathbf{F}(\mathbf{M})\Lambda$ and $\Lambda(\mathbf{F}_{\mathbf{M}'} \otimes I)$ are thereby shown to agree on a dense set, and hence to be equal, completing the proof.

On the other hand, setting $Q = I - P$, $\mathbf{F}(\mathbf{M}')$ transforms $\theta(p)$, where $p(u) = f(Pu)h(Qu)$, into $\theta(p)$, where

$$p(u) = \int f(P(2^{1/2}x + i(\mathbf{M}')u))h(Q(2^{1/2}x + i(\mathbf{M}')u))\,dg(u);$$

writing $x = x_1 + x_2$ and $u = u_1 + u_2$, with x_1, $u_1 \in \mathbf{M}'$ and $x_2, u_2 \in \mathbf{N}'$, and recalling that $dg(u) = dg(u_1)\,dg(u_2)$ in the sense already indicated, it follows that

$$p(u) = \int f(2^{1/2}x_1 + iu_1)h(2^{1/2}x_2 + u_2)\,dg(u_1)\,dg(u_2),$$

noting that $h(Q(2^{1/2}x + i(\mathbf{M}'))) = h(x_2 + u_2)$. It results that

$$p(u) = \tilde{f}(u_1)h(u_2),$$

implying agreement of the operators in question on the spanning set of direct product functionals.

Remark. By a change of variable, the Fourier–Wiener transform may also be represented as an integral operator: $\theta(f) \to \theta(\tilde{f})$, where

$$\tilde{f}(u) = \exp(\langle u, u\rangle/4\sigma^2, \int_{\mathbf{H}'} \exp(i\langle u, x\rangle/2^{1/2}\sigma^2)f(\sqrt{2}x)\,dg(x).$$

This expression is notably simple in that the kernel is identical with that of the conventional Fourier transform, apart from a fixed factor, despite the use of Gaussian measure. On the other hand, the simplicity is in part specious, because the integral on the right side does not define a tame function, even when f is such, without the fixed posterior factor; nor can the image under θ of the integral otherwise be given a nontrivial meaning as a measurable functional.

4. Standardization of the Real-Wave Representation

Let $\mathbf{H}$ be a given complex Hilbert space, and let $\mathbf{H}'$ be a given determining real subspace, by which we mean that every vector $z \in \mathbf{H}$ can be uniquely expressed in the form $z = x + iy$ with x and y in $\mathbf{H}'$, $\langle x, y\rangle$ being real. We wish next to represent the free boson field over $\mathbf{H}$ in $L_2(\mathbf{H}', g)$ in a canonical way.

To begin with, we propose to have $v = 1$, the unit functional on $\mathbf{H}'$. Next we wish to constrain Γ as follows:

(a) If U is a real element of $U(\mathbf{H})$ (i.e., one leaving $\mathbf{H}'$ invariant), then the action of $\Gamma(U)$ on the tame functional $\theta(f)$ is to carry it into $\theta(f^U)$, where $f^U(x) = f(U^{-1}x)$. (This means that the orthogonal group $O(\mathbf{H}')$ acts on functions on $\mathbf{H}'$ in formally the usual way.)

(b) If $\mathbf{M}'$ is any real subspace of $\mathbf{H}'$, with complexification $\mathbf{M} = \mathbf{M}' + i\mathbf{M}'$, and if $i(\mathbf{M})$ denotes the element of $U(\mathbf{H})$ which is multiplication by i on $\mathbf{M}$ and the identity on the orthocomplement of $\mathbf{M}$, then $\Gamma(i(\mathbf{M}))$ should be the partial Wiener transform $\mathbf{F}(\mathbf{M}')$ with respect to $\mathbf{M}'$.

Finally, the Weyl system W is to be constrained as follows: $W(iy)$ is to act essentially by multiplication by the fixed tame function $\exp(ib\langle u, y\rangle)$, for some suitable constant b; to be quite precise, $W(iy)$ is to be the unique unitary operator which sends $\theta(f)$ into $\theta(\exp(ib\langle\cdot, y\rangle)f(\cdot))$, for any polynomial f on $\mathbf{H}'$.

Theorem 4. *There exists a unique Weyl system W and representation Γ of $U(\mathbf{H})$ which satisfy the foregoing constraints and are such that $(L_2(\mathbf{H}', g), W, \Gamma, v)$ represents the free boson field over* $\mathbf{H}$.

W is given specifically by the following action on tame functionals: $W(z)\theta(f) = \theta(f')$, where

$$f'(u) = f(u - \sigma\sqrt{2}z)\exp(\langle z, u\rangle/\sigma\sqrt{2} - \langle z, x\rangle/2).$$

Proof. That there exists a unitary operator $W(z)$ on $L_2(\mathbf{H}', g)$, acting in the indicated manner on tame functionals; and that the resulting map $z \to W(z)$ satisfies the Weyl relations and is continuous follows by a variation of the proof in the case of the complex-wave representation. That v is a cyclic vector for the $W(z)$ follows from its cyclicity for the $W(iy)$ alone, $y \in \mathbf{H}'$, which in turn results from the observation that the $\theta(\exp(i\langle y, \cdot\rangle))$ span the tame functionals, and hence all of $L_2(\mathbf{H}', g)$.

To show the existence of the representation Γ, it suffices to show that the functional $\varphi(z) = \langle W(z)v, v\rangle$ is unitarily invariant, i.e., $\varphi(Uz) = \varphi(z)$ for all $z \in \mathbf{H}$. For the mapping $\sum_i a_i W(z_i)v \to \sum_i a_i W(Uz_i)v$ is then isometric from a dense subset of $L_2(\mathbf{H}', g)$ to itself, and so extends uniquely to a unitary operator $\Gamma(U)$ defined on all of $L_2(H', g)$. It then follows, by arguments similar to ones used earlier, that Γ is a representation of $U(\mathbf{H})$, and it is obvious that $\Gamma(U)v = v$ for all $U \in U(\mathbf{H})$. Straightforward computation shows in fact $\langle W(z)v, v\rangle = \exp(-\langle z, z\rangle/4)$; and from this form continuity follows as in the proof of Theorem 1.

To show that $\Gamma(U)$ has the stated form when U is real, note that the operator S_0 on the tame functionals which carries $\theta(f)$ into $\theta(f^U)$ is isometric and invertible, and therefore extends to a unique unitary operator S on $L_2(\mathbf{H}', g)$; and that by the orthogonal invariance of the expression for W, $SW(z)S^{-1} = W(Uz)$, $z \in \mathbf{H}$. This means that S transforms the $W(z)$ in identical fashion to $\Gamma(U)$, and since both S and $\Gamma(U)$ leave v fixed and v is cyclic for W, $S = \Gamma(U)$.

To show that $\Gamma(i_{\mathbf{M}})$ has the indicated form, note

LEMMA 4.2. *Let f be an entire tame function on $\mathbf{H}$ of order <2. Then $\mathbf{F}(\mathbf{M}')\theta(f)$ is a tame functional, and has the form $\theta(\tilde{f})$ with*

$$\tilde{f}(u) = \int_{\mathbf{M}'} f(2^{1/2}x + i(\mathbf{M}')u)\, dg(x).$$

Proof. Let $f_n(z)$ denote the nth partial sum in the expansion of $f(z)$ into homogeneous polynomials; then $|f_n(z)| \leqslant c\exp(c'\|z\|^{2-\epsilon})$ for some $\epsilon > 0$. Since $\theta(f_n) \to \theta(f)$ in $L_2(\mathbf{H}, g)$, $\mathbf{F}_{\mathbf{M}}\theta(f_n) \to \mathbf{F}_{\mathbf{M}}\theta(f)$. Now if f is based on $\mathbf{N}'$, then so also are the f_n; the expression for the partial Wiener transform shows that so also are the $\tilde{f}_n$, and it follows that $\mathbf{F}_{\mathbf{M}'}\theta(f)$ is also based on $\mathbf{N}'$. On the functions involved here, therefore, convergence in $L_2(\mathbf{H}', g)$ is equivalent to convergence in $L_2(\mathbf{N}', g)$.

On the other hand,

$$\tilde{f}(u) = \int_{\mathbf{M}'} f(2^{1/2}x + i(\mathbf{M})u)\, dg(u) = \int_{P_{N'}\mathbf{M}'} f(2^{1/2}x + i(\mathbf{M})u)\, dg(x),$$

since as a function of $x \in \mathbf{M}'$ $f(2^{1/2}x + iu)$ depends only on $P_{N'}x$, where $P_{N'}$ denotes the projection of x onto $\mathbf{N}'$; the same is true with f replaced by f_n, and applying dominated convergence in the finite-dimensional context of $P_{N'}\mathbf{M}'$ it follows that

$$\tilde{f}_n(u) = \int_{P_{N'}\mathbf{M}'} f_n(2^{1/2}x + i(\mathbf{M})u)\, dg(x) \to \int_{P_{N'}\mathbf{M}'} f(2^{1/2}x + i(\mathbf{M})u)\, dg(x) = \tilde{f}(u),$$

for all $u \in \mathbf{H}'$. Since all functions involved are based on $\mathbf{N}'$, the Riesz–Fischer theorem applies, and shows that $\theta(\tilde{f}) = \mathbf{F}(\mathbf{M}')\theta(f)$.

Now resuming the proof of Theorem 4, it follows from the Weyl relations that it suffices to show that

$$\Gamma(i(\mathbf{M}))W(x)\Gamma(i(\mathbf{M}))^{-1} = W(i(\mathbf{M})x)$$

for all $x \in \mathbf{H}'$, or equivalently that

$$\mathbf{F}(\mathbf{M}')W(x)\theta(f) = W(i(\mathbf{M})x)\mathbf{F}(\mathbf{M}')\theta(f)$$

for all functions f on $\mathbf{H}'$ of the form $f(u) = \exp(i\langle t, u\rangle)$ for a fixed vector $t \in \mathbf{H}'$. By the lemma connecting partial with full Wiener transforms, it suffices to consider the case in which $\mathbf{M}' = \mathbf{H}'$. We note first that direct computation based on the preceding lemma establishes

LEMMA 4.3. $\mathbf{F}$ *sends* $\theta(\exp(i\langle t, \cdot\rangle))$ *into* $\exp(-a^2\langle t, t\rangle/2)\theta(\exp(-\langle t, \cdot\rangle))$.

Now $\mathbf{F}W(x)\theta(\exp(i\langle t, \cdot\rangle))$ can be evaluated by straightforward substitution as $\exp(-a^2\langle t, t\rangle/2)\theta(\exp(i\langle x, \cdot\rangle/2 - \langle t, \cdot\rangle))$, and the same is true of $W(ix)\mathbf{F}\theta(\exp(i\langle t, \cdot\rangle))$, completing the proof that all constraints are satisfied.

Consider now the unicity question. Since $W(ix)$ is given for $x \in \mathbf{H}'$, modulo the constant b, and since $\mathbf{F}W(ix)\mathbf{F}^{-1} = W(-x)$, $W(x)$ is similarly determined. On the other hand, the satisfaction of the Weyl relations implies that $W(x)$ and $W(iy)$ satisfy the following relations: $W(iy)W(x) = \exp(i\langle x, y\rangle)W(x)W(iy)$. Explicit computation shows that this relation is satisfied if and only if $b = -a^{-1}$. Since $W(z) = W(x)W(iy)\exp((i/2)\langle x, y\rangle)$, the Weyl system is then completely determined.

To show the unicity of Γ, it evidently suffices to show that $U(\mathbf{H})$ is generated by the real unitaries together with the $i(\mathbf{M})$. In fact, it is generated by the $i(\mathbf{M})$ alone. For since the totality of unitaries on $\mathbf{H}$ which are the identity on a cofinite-dimensional subspace form a strongly dense subset of $U(\mathbf{H})$, it suffices in turn to treat the case in which $\mathbf{H}$ is two dimensional, which can be achieved by several methods. The least computational (and one applicable to $U(n)$ for any finite n) is to observe that the closure of the group generated by the $i(\mathbf{M})$ is nonabelian and invariant and hence must be all of $U(n)$, which is simple modulo its center, as a topological group.

Remark. Some qualitatively weaker constraints, derivable in fact from the preceding proof, will serve equally to standardize the real-wave representation, but our purpose in the sequel is to compare similar conventional forms for the real- and complex-wave representations, rather than to characterize them abstractly.

5. THE INTERTWINING OPERATOR BETWEEN THE COMPLEX AND REAL REPRESENTATIONS

Let $\mathbf{H}'$ denote a real Hilbert space, $\mathbf{H}$ its complexification, and let σ and σ' denote arbitrary real nonvanishing parameters; let g denote the isonormal distribution of variance σ^2 on $\mathbf{H}$ and g' the isonormal distribution of variance σ'^2 on $\mathbf{H}'$. For arbitrary $z \in \mathbf{H}$, let $W_c(z, \sigma)$ denote the operator on functions over $\mathbf{H}$:

$$W_c(z, \sigma): f(u) \to f(u - \sigma z) \exp(\langle z, u \rangle / 2\sigma - \langle z, z \rangle / 4, \qquad u \in \mathbf{H},$$

and let $\overline{W}_c(z, \sigma)$ denote the operator on the subspace $\mathbf{K}$ of $L_2(\mathbf{H}, g)$ described in Theorem 3, obtained by restriction of $W_c(z, \sigma)$. For arbitrary $z \in \mathbf{H}$, let $W_r(z, \sigma)$ denote the operator on functions over $\mathbf{H}'$:

$$W_r(z, \sigma): f(u) \to f(u - \sigma\sqrt{2}x) \exp(\langle z, u \rangle / \sigma\sqrt{2} - \langle z, x \rangle / 2), \qquad u \in \mathbf{H}',$$

and let $\overline{W}_r(z, \sigma)$ denote the unitary operator on the space $\mathbf{K}' = L_2(\mathbf{H}', g')$ which carries any tame square-integrable functional $\theta(f)$ into $\theta(W_r(z, \sigma)f)$. By Theorem 1, for any given σ and σ' there exists a unique unitary transformation $T_{\sigma,\sigma'}$ from $\mathbf{K}$ onto $\mathbf{K}'$ which carries $\overline{W}_c(z, \sigma)$ into $\overline{W}_r(z, \sigma')$ and such that $1_{\mathbf{H}} \to \theta(1_{\mathbf{H}'})$. Simple expressions for the T and T^{-1} on sufficiently regular functionals are readily obtained as follows.

Since the intertwining operators between the $\overline{W}_c(z, \sigma)$ for different values of σ are simple changes of scale, as are those between the $\overline{W}_r(z, \sigma)$, it will suffice to treat $T_{\sigma,\sigma'}$ for two specific values of σ. Inspection indicates that the choice $\sigma' = \sigma\sqrt{2}$ should be convenient and we simply take $\sigma = 2^{-1/2}$, $\sigma' = 1$, and denote the corresponding operator simply as T.

It follows that T is uniquely determined by unitarity together with the constraint that

$$T: \exp(\langle z, \cdot \rangle) \to \exp(-\langle z, \bar{z} \rangle / 2) \theta(\exp(\langle z, \cdot \rangle)),$$

where $\bar{z} = x - iy$ if $z = x + iy$ with $x, y \in \mathbf{H}'$. More comprehensively, we have

THEOREM 5. *The (unitary) intertwining operator T between the complex- and real-wave representations (specifically, between $\overline{W}_c(\cdot, 2^{-1/2})$ and $\overline{W}_r(\cdot, 1)$) has the following action on tame antientire functions F on $\mathbf{H}$ of order <2: $F \to \theta(f)$, where*

$$f(u) = \int_{\mathbf{H}} \exp(\langle z, u \rangle - \langle z, \bar{z} \rangle / 2) F(z) \, dg(z).$$

Its inverse carries any tame functional $\theta(f) \in L_2(\mathbf{H}', g')$ into the antientire function F on $\mathbf{H}$ given by the equation

$$F(z) = \exp(-\langle \bar{z}, z \rangle / 2) \int_{\mathbf{H}'} \exp(\langle u, z \rangle) f(u) \, dg'(u);$$

more generally, for arbitrary $\varphi \in L_2(\mathbf{H}', g')$, $T^{-1}\varphi = F$, where

$$F(z) = \exp(-\langle \bar{z}, z \rangle / 2) \langle \theta(f), \theta(\exp(\langle z, \cdot \rangle)) \rangle.$$

Proof. In case $F(z) = \exp(\langle w, z\rangle)$ for some $w \in \mathbf{H}$, the reproducing kernel property of this function shows that

$$\int_{\mathbf{H}} \exp(\langle z, u\rangle - \langle z, \bar{z}\rangle/2)F(z)\,dg(z) = (TF)(u). \tag{*}$$

It follows from the unitarity of T and the dominated convergence argument employed in the first part of the proof of Theorem 2 that equation (*) is valid if F is any antianalytic polynomial. It follows in turn, by dominated convergence, as in the proof of Lemma 4.2, that the same is true for an arbitrary tame antientire function of order <2.

To establish the form given for the inverse on tame functionals, an elementary Gaussian integration shows its validity when f has the form $f(u) = \exp(\langle w, u\rangle)$ for some fixed vector $w \in \mathbf{H}$. It then follows for arbitrary tame functionals from their approximability in $L_2(\mathbf{H}', g')$ by linear combinations of the $\theta(\exp(\langle w, \cdot\rangle))$, together with the continuity of $\int_{\mathbf{H}'} \exp(\langle u, z\rangle) f(u)\,dg'(u)$ as a function of $\theta(f) \in L_2(\mathbf{H}', g)$, for square-integrable functionals based on a fixed finite-dimensional subspace. This integral representation is equivalent to the equation

$$F(z) = \exp(\langle \bar{z}, z\rangle/2)\langle \varphi, \theta(\exp(\langle z, \cdot\rangle))\rangle,$$

where $\varphi = \theta(f)$, whose validity for general $\varphi \in L_2(\mathbf{H}', g')$ follows by the unitarity of T and the density of the tame functionals, together with the continuity of $F(z)$ as a function of F, for fixed z.

Remark 1. It is interesting to note that as a result of the standardizations and choice of respective variance parameters the projective mapping corresponding to T (i.e., from $L_2(\mathbf{H}, g)$ modulo equivalence via constant factors to $L_2(\mathbf{H}', g')$ similarly projectified) is simple restriction, from $\mathbf{H}$ to $\mathbf{H}'$, on the vectors of the form $\exp(\langle z, \cdot\rangle)$; these are the so-called *coherent states* in physical applications.

Remark 2. The intertwining operators between the particle and wave representations are of a different nature from those between the two wave representations, since the (Fock–Cook) representation space of the particle representation is a space of tensors of varying rank, rather than a function space. In the case of the real-wave representation, it is given in Segal (1956). In the case of the complex-wave representation it is simpler, and readily derived from the cited work.

Remark 3. In the case when $\mathbf{H}$ is finite dimensional, a version of the operator T utilizing $L_2(\mathbf{H}')$ with respect to euclidean rather than Gaussian measure and holomorphic rather than antiholomorphic functions has been given by Bargmann (1961); cf. Bargmann (1962) and Satake (1971).

Added in Proof (27 July 1978). I am indebted to P. Kree for bringing to my attention his independent work inclusive of a result similar to Theorem 3. See Lecture Notes in Mathematics, No. 474, Springer-Verlag, Berlin and New York, 1975, pp. 16–47, and references given therein.

References

V. Bargmann, On a Hilbert space of analytic functions and an associated integral transform, Part I, *Commun. Pure Appl. Math.* (1961), 187–214.

V. Bargmann, Acknowledgment, Proc. Nat. Acad. Sci. U.S.A. **48** (1962), 2204.

R. H. Cameron, Some examples of Fourier–Wiener transforms of analytic functionals, *Duke Maths. J.* **12** (1945), 485–488.

R. H. Cameron and W. T. Martin, Fourier–Wiener transforms of analytic functionals, *Duke Math. J.* **12** (1945), 489–507.

J. M. Cook, The mathematics of second quantization, *Trans. Amer. Math. Soc.* **74** (1953), 222–245.

P. A. M. Dirac, La seconde quantification, *Ann. Inst. H. Poincaré* **11** (1949), 15–47.

V. Fock, Verallgemeinerung und Lösung der Diracschen statistischen Gleichung, *Z. Physik.* **49** (1928), 339–357.

L. Nachbin, "Topology on Spaces of Holomorphic Mappings," Springer–Verlag, Berlin and New York, 1969.

H. Rossi and M. Vergne, Analytic continuation of the holomorphic discrete series, *Acta Math.* **136** (1976), 1–59.

I. Satake, Factors of automorphy and Fock representations, *Advances in Math.* **7** (1971), 83–110.

I. E. Segal, Tensor algebras over Hilbert spaces, I, *Trans. Amer. Math. Soc.* **81** (1956), 106–134.

I. E. Segal, Mathematical characterization of the physical vacuum for a linear Bose–Einstein field, *Illinois J. Math.* **6** (1962), 500–523.

I. E. Segal, Mathematical problems of relativistic physics, *Proc. Summer Seminar on Appl. Math., Boulder Colorado, 1960.* Amer. Math. Soc., Providence, Rhode Island, 1963.

TOPICS IN FUNCTIONAL ANALYSIS
ADVANCES IN MATHEMATICS SUPPLEMENTARY STUDIES, VOL. 3

Families of Pseudodifferential Operators

HAROLD WIDOM[†]

Natural Science Division
University of California, Santa Cruz
Santa Cruz, California 95064

DEDICATED TO M. G. KREIN ON THE OCCASION OF HIS SEVENTIETH BIRTHDAY

I. INTRODUCTION

It has long been suspected that asymptotic expansions such as the Minakshisundaram–Pleijel expansion for the heat operator [10] and the continuous analogue [7] of the strong Szegö limit theorem for Toeplitz determinants [16] are two parts of the same picture, that there ought to be one theory which gives both. Such a theory is developed here.

Minakshisundaram and Pleijel showed that for the Laplace–Beltrami operator on a compact n-dimensional Riemannian manifold, with eigenvalues λ_i ($\to -\infty$) and suitably normalized eigenfunctions $\varphi_i(x)$, one has an asympotic expansion

$$\sum_i e^{\lambda_i t}|\varphi_i(x)|^2 \sim (4\pi t)^{-n/2}(u_0(x) + u_1(x)t + u_2(x)t^2 + \cdots)$$

as $t \to 0$. In particular,

$$\sum_i e^{\lambda_i t} \sim (4\pi t)^{-n/2}(a_0 + a_1 t + a_2 t^2 + \cdots).$$

The explicit determination of the u_i and a_i has been the subject of considerable study in the last decade and the first four of each have been determined. We mention only that

$$u_0 = 1, \qquad u_1 = (1/6) \times \text{scalar curvature} \tag{1.1}$$

and that a_i is gotten from $u_i(x)$ by integrating with respect to the Riemann volume element. Similar expansions exist for all strongly elliptic differential operators on compact manifolds [4, 14].

The continuous analogue of the Szegö limit theorem says that under certain conditions on a function k defined on the real line, if T_α denotes convolution

[†] Supported by a grant from the National Science Foundation.

ISBN 0-12-599102-9

by $\alpha k(\alpha x)$ on $L_2(0, 1)$ then as $\alpha \to \infty$

$$\log \det(I + T_\alpha) = \alpha F(0) + \int_0^\infty xF(x)F(-x)\,dx + o(1), \tag{1.2}$$

where

$$F(x) = (1/2\pi) \int_{-\infty}^{\infty} e^{-ix\xi} \log\left[1 + \int_{-\infty}^{\infty} e^{iy\xi} k(y)\,dy\right] d\xi.$$

Now these two results have the following in common. In both cases one has a family of pseudodifferential operators depending on a parameter (in the first case the family is simply t times the Laplacian) and what is determined is the asymptotic behavior of the trace of a function of the operator (in the first case the exponential, in the second case the logarithm). We remind the reader that a pseudodifferential operator on R^n is an operator S from $C_0{}^\infty(R^n)$ to $C^\infty(R^n)$ of the form

$$Sf(x) = \int \hat{f}(\xi)\, e^{i\xi \cdot x} \sigma(x, \xi)\, d\xi, \tag{1.3}$$

where

$$\hat{f}(\xi) = (2\pi)^{-n} \int f(x) e^{-i\xi \cdot x} dx$$

and the "symbol" $\sigma(x, \xi)$ satisfies certain conditions; a pseudodifferential operator on a C^∞ manifold M is an operator from $C_0{}^\infty(M)$ to $C^\infty(M)$ which locally is of the above form. (All this is made precise in the next section.)

Let us see how the parameters appear in the symbols of the two families of operators described above or, more precisely, simplified versions of them. Consider first, instead of t times the Laplacian on a manifold, t times the Laplacian on R^n. Since the Laplacian has symbol $-|\xi|^2$, the Laplacian times t has symbol

$$-t|\xi|^2 = -|t^{1/2}\xi|^2.$$

Second, consider convolution by $\alpha k(\alpha x)$ on $L_2(-\infty, \infty)$ rather than $L_2(0, 1)$. Its symbol is

$$\int_{-\infty}^{\infty} k(x) e^{-ix\xi/\alpha}\, dx.$$

Thus in both cases the appearance of ξ in the symbol (x appears in neither symbol because the operators have constant coefficients) is as ξ/α with α large; in the first case $\alpha = t^{-1/2}$.

This, then, is the general context: a family of pseudodifferential operators on R^n depending on a distinguished parameter α such that, roughly speaking, in the symbol of the operators ξ and α appear as ξ/α; for families on manifolds the operators, transferred locally to R^n, must be of this form. There may be other parameters floating around but their role is to be harmless; nevertheless, permitting their existence turns out to be very useful. The parameter α may run through any subset of $[-1, \infty)$. If the subset is $\{1\}$ only we have the case of a single pseudodifferential operator.

The main goal will be to determine, in case α is unbounded, asymptotic expansions for traces of functions of these operators. If $\sigma(x, \xi)$ is sufficiently small for $|x|$, $|\xi|$ large, the operator (1.3) will extend to a trace class operator on $L_2(R^n)$ with trace

$$(2\pi)^{-n} \iint \sigma(x, \xi)\, dx\, d\xi.$$

Consequently, for the family of operators S_α given by

$$S_\alpha f(x) = \int \hat{f}(\xi) e^{i\xi \cdot x} \sigma(x, \xi/\alpha)\, d\xi \tag{1.4}$$

(we call $\sigma(x, \xi)$ the symbol of this family) the trace of S_α equals

$$(\alpha/2\pi)^n \iint \sigma(x, \xi)\, dx\, d\xi. \tag{1.5}$$

It is well known how to determine the symbol of a product of pseudodifferential operators on R^n. See [12], for example. From products one gets inverses and from inverses one gets, via Cauchy's theorem, analytic functions. Thus everything is quite straightforward in R^n. In particular it is no problem to get as many terms as desired of the asymptotic expansion of

$$\log \det(I + T_\alpha),$$

where now T_α is the integral operator on $L_2(R^n)$ with kernel

$$\alpha^n k(x, \alpha(x - y))$$

and k is a sufficiently well-behaved function of two variables.

The real difficulty comes in the extension to manifolds; it is not even clear what the symbol of a pseudodifferential operator should be. Everyone is familiar with the principal symbol, which is a function on the cotangent bundle of the manifold, but it is the complete symbol we need if we are to get asymptotic expansions. The principal symbol can give nothing beyond the first term. Shih [15] and Petersen [13] independently showed how one could think of the symbol as defined on an object obtained from jet bundles and found formulas for the symbol of a product; Petersen showed also how the introduction of a connection on the manifold allowed one to think of the symbol as a formal asymptotic expansion on the cotangent bundle itself and (very important) that every such expansion was the symbol of a pseudodifferential operator. However, these results do not help us much for a couple of reasons: the multiplication of symbols was not sufficiently explicit to give us answers (and it is answers that we want), and pseudodifferential operators of order $-\infty$ had symbol zero whereas we want to compute the traces of pseudodifferential operators which may easily have order $-\infty$—and the trace should, by analogy with (1.5), be the integral of the symbol over its domain. Nevertheless, some of the ideas of this paper are drawn from Petersen's work.

The bulk of our work will be the development of a symbolic calculus for families of pseudodifferential operators of the type described above. Since a single pseudodifferential operator is a special case, there is some novelty here even for a single operator. Here is the idea behind our definition of the symbol. Consider the operator S of the form (1.3). Then, formally at least,

$$\sigma(x_0, \xi) = Se^{i\xi \cdot (x - x_0)}\big|_{x = x_0}.$$

The function $\xi \cdot (x - x_0)$ vanishes at $x = x_0$ and is linear in x with gradient ξ. This suggests that for an operator on a manifold M, with $v \in T^*(M)$, the cotangent bundle of M, one should define

$$\sigma(v) = Se^{il(v,x)}\big|_{x = \pi(v)} \tag{1.6}$$

(π being the usual projection $T^*(M) \to M$), where l vanishes at $x = \pi(v)$, its differential at $\pi(v)$ equals v, and l is in some sense linear at $\pi(v)$. To make sense of linearity take a connection on the manifold and consider its associated covariant derivative operation (which takes any tensor field on M and sends it into a tensor field of covariance degree one larger—formulas will be written down later). Then we say a function on M is linear at x (with respect to the particular connection chosen) if for each $k \geqslant 2$ its kth covariant derivative at x, symmetrized, is equal to zero. With this definition a function $l \in C^\infty(T^*(M) \times M)$ with the required properties can be found, and σ defined by (1.6) is called the symbol of S. In the case of our pseudodifferential families the symbol is defined by

$$\sigma(v) = Se^{i\alpha l(v,x)}\big|_{x = \pi(v)}$$

and so the symbol is a function on the cartesian product of $T^*(M)$ with the parameter space. For a single operator the symbol is a function on $T^*(M)$. (It always depends, of course, on the particular connection chosen.)

Actually one ought to be a little careful and define the symbol to be an equivalence class rather than a function, the reason being that there is nonuniqueness (apart even from the choice of connection) in the definition. This nonuniqueness occurs in two places. First, the function l is by no means unique. Second, if the manifold is not compact S cannot be applied to $e^{il(v,x)}$ and so one must first multiply by a C^∞ function with compact support which is identically 1 in a neighborhood of $\pi(v)$; this introduces another nonuniqueness. Therefore one introduces an equivalence relation on $C^\infty(T^*(M))$ and the symbol should be considered an equivalence class. (We shall be consistently careless in this regard.) In the case of a single pseudodifferential operator a symbol is equivalent to zero if it is very small at ∞ (and an operator of order $-\infty$ has symbol equivalent to zero), but for families the symbol must also be very small everywhere for α large (so that a pseudodifferential family of order $-\infty$ will normally not have symbol equivalent to zero).

The symbol of a product will turn out to be an infinite series (a power series in α^{-1}) each term of which is bilinear differential operator applied to the symbols of the factors. From this it will be seen that the symbol of an analytic function of the family is given by a similar series. The coefficients of the operators arising in this series involve the covariant derivatives of $l(v, x)$ evaluated at $x = \pi(v)$, and these in turn can be expressed in terms of the torsion and curvature tensors of the connection and their covariant derivatives. Although the computations become unpleasant as one goes far out in the series, the first couple of terms are not at all bad, and in particular (1.1) will be very easy to obtain. It will also be clear how to program a machine to write any $u_i(x)$ in terms of the curvature tensor and its covariant derivatives. Moreover, one can compute as many terms as desired of the manifold version of (1.2) in terms of the symbol of the given family.

We should also mention that one gets information about the general form of the asymptotic expansion for e^{-tA}, where A is an arbitrary strongly elliptic operator. See, for example, [4, Theorem 1.6.1] and [2a, p. 303; 2b, p. 279]. In the latter the results of Seeley [14] and a sophisticated argument were used to show that if A is of order 2 each $u_i(x)$ is (in any given coordinate system) a polynominal in the coefficients of A and their derivatives times the determinant of the coefficients of the quadratic terms of the symbol of A raised to the power $-N - \frac{1}{2}$, with N a sufficiently large integer. This will be an immediate consequence of the present work.

The continuous Szegö theorem (1.2) concerns an interval, which is a manifold with boundary. Since only manifolds without boundary are considered here this does not drop out directly. This matter will be taken up in a forthcoming paper [17], where it will be seen how to apply the theory to obtain (in principle) a complete asymptotic expansion for quite general families of convolution-like operators.

The author, an amateur at many of the things contained in this paper, was helped greatly by talks with colleagues; special thanks go to R. Narasimhan and J. Guckenheimer.

II. Pseudodifferential Families on R^n

A pseudodifferential operator of order ω on R^n is an operator of the form (1.3) where σ is a complex-valued C^∞ function on $R^n \times R^n$ with the following property: For any multi-indices j, k and any compact set K in R^n

$$\frac{\partial^j}{\partial x^j}\frac{\partial^k}{\partial \xi^k}\sigma(x, \xi) = O((1 + |\xi|)^{\omega - |k|}) \tag{2.1}$$

uniformly for $x \in K$. (We shall not concern ourselves with the more general classes of pseudodifferential operators which have been investigated, for example, in [5].)

A pseudodifferential family or order ω is a set of pseudodifferential operators depending on a distinguished parameter α which runs over a subset of $[1, \infty)$ and possibly other parameters as well such that if S denotes an operator of the family (its dependence on the parameters is not explicitly indicated in the notation), then

$$Sf(x) = \int \hat{f}(\xi)e^{i\xi \cdot x}\sigma(x, \xi/\alpha)\,d\xi, \tag{2.2}$$

where σ, a function of x, ξ, and the parameters (the dependence of σ on the parameters is not explicitly indicated in the notation), satisfies (2.1) uniformly for all parameter values. We call σ the symbol of the family.

A pseudodifferential family will be denoted typically by $\{S\}$. Clearly, the pseudodifferential families of a given order and parameter set form a linear space.

Proposition 2.1. *Let $\{S\}$ and $\{T\}$ be pseudodifferential families (with the same parameter set), let $\varphi \in C_0^\infty(R^n)$, and denote multiplication by φ by M_φ. Then $\{SM_\varphi T\}$ is a pseudodifferential family with order the sum of the orders of $\{S\}$ and $\{T\}$.*

Proof. Let $\{S\}$, $\{T\}$ have symbols σ, τ, respectively. We may incorporate M_φ into T (by replacing $\tau(x, \xi)$ by $\varphi(x)\tau(x, \xi)$) and so take $\varphi \equiv 1$ and τ of compact support in x. Then

$$STf(x) = \int \hat{f}(\eta)e^{i\eta \cdot x}\rho(x, \eta/\alpha)\,d\eta,$$

where

$$\rho(x, \eta/\alpha) = (2\pi)^{-n} \iint \sigma(x, \xi/\alpha)\tau(y, \eta/\alpha)e^{i(\xi - \eta)\cdot(x - y)}\,dy\,d\xi.$$

Thus

$$\rho(x, \eta) = (\alpha/2\pi)^n \iint \sigma(x, \xi)\tau(y, \eta)e^{i\alpha(\xi - \eta)\cdot(x - y)}\,dy\,d\xi, \tag{2.3}$$

which equals

$$\alpha^n \int \sigma(x, \xi)\hat{\tau}(\alpha(\xi - \eta), \eta)e^{i\alpha(\xi - \eta)\cdot x}\, d\xi.$$

Here $\hat{\tau}$ denotes the Fourier transform of τ with respect to its first variable. Since $\tau(x, \eta)$ vanishes for x outside a compact set and each derivative with respect to x is at most a constant times

$$(1 + |\eta|)^{\omega(T)} \qquad (\omega(T) = \text{order of } \{T\}),$$

one sees that for any $N > 0$ $|\hat{\tau}(\xi, \eta)|$ is at most a constant times

$$(1 + |\xi|)^{-N}(1 + |\eta|)^{\omega(T)}.$$

Combining this with the given bounds for σ shows that for x in any compact set $\rho(x, \eta)$ has absolute value at most a constant times

$$\alpha^n \int \frac{(1 + |\xi|)^{\omega(S)}(1 + |\eta|)^{\omega(T)}}{(1 + \alpha|\xi - \eta|)^N}\, d\xi, \tag{2.4}$$

and it is routine to verify that this is at most a constant times $(1 + |\eta|)^{\omega(S) + \omega(T)}$ if N is chosen large enough.

Thus we have the right sort of bound on ρ but we must also consider its derivatives with respect to x and η. First, what is the effect of an x differentiation of some order in (2.3)? It results in some x differentiations of $\sigma(x, \xi)$, which does not change the form of the bounds, and the introduction of factors of the form $\alpha(\xi_i - \eta_i)$, which also causes no difficulty because the exponent N in (2.4) was arbitrary.

What of $\partial^k/\partial\eta^k$ applied to (2.3), with some multi-index k? This results in a sum of terms each involving

$$\frac{\partial^p}{\partial\eta^p}\tau(y, \eta) \quad \text{and} \quad \frac{\partial^q}{\partial\eta^q} e^{i\alpha(\xi - \eta)(x - y)}$$

with $p + q = k$. The first results in a lowering of $\omega(T)$ to $\omega(T) - |p|$. Since

$$\frac{\partial^q}{\partial\eta^q} \quad \text{and} \quad \frac{\partial^q}{\partial\xi^q}$$

have the same effect on $e^{i\alpha(\xi - \eta)\cdot(x - y)}$ (except for sign), integration by parts shows that our $\partial^q/\partial\eta^q$ can be effected by applying $\partial^q/\partial\xi^q$ to $\sigma(x, \xi)$, and this results in a lowering of $\omega(S)$ to $\omega(S) - |q|$. Thus we find that

$$\left|\frac{\partial^k}{\partial\eta^k}\rho(x, \eta)\right|$$

is on compact sets at most a constant times

$$(1 + |\eta|)^{\omega(S) + \omega(T) - |k|}.$$

Combination of the preceding two arguments gives the same estimate for each

$$\frac{\partial^j}{\partial x^j}\frac{\partial^k}{\partial \eta^k}\rho(x, \eta)$$

and the proposition is established.

The next two lemmas are prefatory to proving the invariance of pseudodifferential families under local coordinate transformations.

LEMMA 2.2. *Suppose $\sigma(x, \xi, y)$ (which may depend also on α and other parameters) belongs to $C^\infty(R^n \times R^n \times R^n)$ for each parameter value, has compact (x, y)-support, and for arbitrary multi-indices j, k, l*

$$\frac{\partial^j}{\partial x^j}\frac{\partial^k}{\partial \xi^k}\frac{\partial^l}{\partial y^l}\sigma(x, \xi, y) = O((1 + |\xi|)^{\omega - |k|})$$

uniformly in x, y, and all parameter values. Then the operators from $C_0{}^\infty(R^n)$ to $C^\infty(R^n)$ defined by

$$Sf(x) = (2\pi)^{-n}\iint e^{i\xi\cdot(x-y)}\sigma(x, \xi/\alpha, y)f(y)\,dy\,d\xi$$

comprise a pseudodifferential family of order ω.

Proof. The lemma is proved by straightforward computation. (See [11, p. 155] for the case of a single operator.)

LEMMA 2.3. *Suppose $k(x, y)$ (which may depend also on α and other parameters) satisfies for each N, j, k*

$$\frac{\partial^j}{\partial x^j}\frac{\partial^k}{\partial y^k}k(x, y) = ((1 + \alpha|x - y|)^{-N}).$$

Then the set $\{S\}$ defined by

$$Sf(x) = \alpha^n \int k(x, y)f(y)\,dy$$

is a pseudodifferential family of order $-\infty$ (i.e., of order ω for all ω).

Proof. The operator S is given by (1.3) with

$$\sigma(x, \xi) = (2\pi)^{-n}\int k(x, x + (y/\alpha))e^{i\xi\cdot y}\,dy$$

and the conclusion follows by straightforward estimation.

Next we consider the effect of a change of coordinates. The statement is slightly awkward since we consider a set of coordinate changes rather than one; this will be very useful later. Recall that a subset $\mathscr{X}$ of $C^\infty(U)$ (U is an open subset of R^n here) is called bounded if each partial derivative has on each compact subset of U a bound which holds uniformly for all elements of $\mathscr{X}$. A subset $\mathscr{X}$ of $C^\infty(U, R^n)$ is called bounded if each of its component functions is a bounded subset of $C^\infty(U)$.

Suppose now that χ is a diffeomorphism from U into R^n and that φ and ψ both belong to $C_0^\infty(U)$. If S is a function from $C_0^\infty(R^n)$ to $C^\infty(R^n)$ then T defined by

$$\begin{aligned} (Tf(\chi(x))) &= \varphi(x)(S\psi \cdot (f \circ \chi))(x), \qquad && x \in U, \\ (Tf)(y) &= 0, && y \notin \chi(U), \end{aligned}$$

is also a function from $C_0^\infty(R^n)$ to $C^\infty(R^n)$ (in fact, to $C_0^\infty(R^n)$). Note that although $f \circ \chi$ is only defined on U, $\psi \cdot (f \circ \chi)$ has an obvious definition as a function in $C^\infty(R^n)$.

Now suppose that $\{S\}$ is a pseudodifferential family. Then $\{T\}$ is a set of operators with the same parameter set as $\{S\}$. However, we shall allow χ to run through a set $\mathscr{X}$ of diffeomorphisms and so T involves not only the original parameters of S but a new one, namely χ. The content of the following proposition is that under a certain assumption on $\mathscr{X}$, $\{T\}$ is a pseudodifferential family with respect to the enlarged parameter set.

PROPOSITION 2.4. *Let $\mathscr{X}$ be a bounded subset of $C^\infty(U, R^n)$ such that each $\chi \in \mathscr{X}$ is a diffeomorphism and such that on compact subsets of U the determinants $|d\chi|$ are bounded away from zero uniformly for $\chi \in \mathscr{X}$. Then if $\{S\}$ is a pseudodifferential family of order ω so is $\{T\}$.*

Proof. [This contains nothing essentially new beyond what goes into Kuranishi's proof of invariance of pseudodifferential operators under coordinate change. (See the proof of [12, Theorem 6], for example.) We give the details, though, because they are not completely straightforward.] Write $\lambda = \chi^{-1}$, which maps $\chi(U)$ onto U. We have

$$(S\psi \cdot f)(x) = (2\pi)^{-n} \iint e^{i\xi \cdot (x-y)} \sigma(x, \xi/\alpha)\psi(y) f(y)\, dy\, d\xi, \tag{2.5}$$

so for $x \in \mathscr{X}(U)$

$$\begin{aligned} (S\psi \cdot (f \circ \chi))(\lambda(x)) &= (2\pi)^{-n} \iint e^{i\xi \cdot (\lambda(x) - y)} \\ &\qquad \times \sigma(\lambda(x), \xi/\alpha)\psi(y) f(\chi(y))\, dy\, d\xi \\ &= (2\pi)^{-n} \iint_{\chi(U)} e^{i\xi \cdot (\lambda(x) - \lambda(y))} \sigma(\lambda(x), \xi/\alpha)\psi(\lambda(y)) \\ &\qquad \times |d\chi(y)| f(y)\, dy\, d\xi. \end{aligned}$$

Thus for $x \in \chi(U)$

$$Tf(x) = \varphi(\lambda(x))(2\pi)^{-n} \iint_{\chi(U)} e^{i\xi \cdot (\lambda(x) - \lambda(y))} \sigma(\lambda(x), \xi/\alpha) \times \psi(\lambda(y))|d\chi(y)| f(y)\, dy\, d\xi \tag{2.6}$$

and for $x \notin \chi(U)$ we have $Tf(x) = 0$.

Now let $u \in C^\infty(R^n)$ be equal to 1 for $|x| \leqslant \epsilon$ and 0 for $|x| \geqslant 2\epsilon$. The number ϵ will be chosen later. In (2.5) the integrand may be given an extra factor

$$1 = [1 - u(x - y)] + u(x - y).$$

Then S may be correspondingly written as a sum of two operators S' and S'' and consequently also T as the sum of T' and T''.

First, T' is the integral operator with kernel $k(x, y)$, equal to zero when $x \notin \chi(U)$ or $y \notin \chi(U)$ and otherwise equal to

$$(2\pi)^{-n}\varphi(\lambda(x))\psi(\lambda(y))[1 - u(\lambda(x) - \lambda(y))] \int e^{i\xi \cdot (\lambda(x) - \lambda(y))} \sigma(\lambda(x), \xi/\alpha)\, d\xi. \tag{2.7}$$

Replacing ξ by $\alpha\xi$ and integrating by parts shows that this equals (except for sign)

$$\left(\frac{\alpha}{2\pi}\right)^n \varphi(\lambda(x))\psi(\lambda(y))(i\alpha)^{-|M|} \frac{1 - u(\lambda(x) - \lambda(y))}{[\lambda(x) - \lambda(y)]^M} \times \int e^{i\alpha\xi \cdot (\lambda(x) - \lambda(y))} \frac{\partial^M}{\partial \xi^M} \sigma(\lambda(x), \xi)\, d\xi. \tag{2.8}$$

(Here M is a multi-index and $[\lambda(x) - \lambda(y)]^M$ means

$$\prod_{i=1}^n [\lambda(x) - \lambda(y)]_i^{M_i}.$$

The reader may be troubled by the possible divergence of the integral in (2.7) (but not in (2.8) if $|M|$ is large enough). However, one can insert a factor $e^{-\delta|\xi|^2}$ into the integral in (2.6) (or, rather, the corresponding integral for T'), perform the partial integrations, and then let $\delta \to 0$. This gives (2.8).) If $|M|$ is large enough the last integral is bounded uniformly for all x, y, χ. Since the determinants $|d\chi|$ are uniformly bounded away from zero on compact subsets of U (we have in mind the union of the supports of φ and ψ) we have if $\varphi(\lambda(x))\psi(\lambda(y)) \neq 0$,

$$\max_{1 \leqslant i \leqslant n} |[\lambda(x) - \lambda(y)]_i| \geqslant c|x - y|$$

for some constant $c > 0$ independent of χ. It follows that given a positive integer N one can find a multi-index M with $|M| = N$ and

$$|[\lambda(x) - \lambda(y)]^N| \geqslant c^N|x - y|^N.$$

Since $k(x, y) = 0$ for $|x - y|$ sufficiently small we have established, for each N, the bound

$$k(x, y) = O((1 + \alpha|x - y|)^{-N}). \tag{2.9}$$

Since, once again, $|d\chi|$ are bounded away from zero on compact sets, each partial derivative of

$$\varphi(\lambda(x))\psi(\lambda(y))$$

is uniformly bounded. Using this fact it is easy to check that there is a bound like (2.9) for each partial derivative of k. Therefore, by Lemma 2.3, T' is a pseudodifferential family of order $-\infty$.

To take care of T'' note first that if x belongs to any compact subset of U (we have in mind the support of φ) and if $|x - y|$ is less than the distance from the support of φ to the complement of U then the entire line segment from x to y belongs to U and so for any χ

$$\begin{aligned}\chi(x) - \chi(y) &= \int_0^1 \frac{d}{ds}\chi(sx + (1 - s)y)\,ds \\ &= \left(\int_0^1 d\chi(sx + (1 - s)y)\,ds\right)(x - y) \\ &= (H_\chi(x, y))(x - y),\end{aligned}$$

say. Clearly, if x and y are close enough $H_\chi(x, y)$ will be a nonsingular linear transformation of R^n. More precisely, there is an $\epsilon > 0$ such that if x belongs to the support of φ and $|x - y| \leqslant \epsilon$ then $H_\chi(x, y)$ is nonsingular for all $\chi \in \mathscr{X}$. This is the ϵ we use in the choice of u.

Replacing x, y by $\lambda(x)$, $\lambda(y)$ shows that if $\lambda(x)$ belongs to the support of φ then

$$x - y = H_\chi(\lambda(x), \lambda(y)(\lambda(x) - \lambda(y)) \tag{2.10}$$

if $|\lambda(x) - \lambda(y)| \leqslant \epsilon$. Now for $x \in \chi(U)$

$$\begin{aligned}T''f(x) = \varphi(\lambda(x))(2\pi)^{-n} &\iint e^{i\xi\cdot(\lambda(x)-\lambda(y))}\sigma(\lambda(x), \xi/\alpha) \\ &\times u[\lambda(x) - \lambda(y)]\psi(\lambda(y))|d\chi(y)|f(y)\,dy\,d\xi\end{aligned}$$

and the integrand vanishes unless $|\lambda(x) - \lambda(y)| \leqslant \epsilon$, in which case (2.10) holds with invertible $H_\chi(x, y)$. Thus we may replace ξ by

$$H_\chi(x, y)^t\xi$$

(t = transpose) in the last integral; we obtain

$$T''f(x) = \varphi(\lambda(x))(2\pi)^{-n} \iint e^{i\xi \cdot (x-y)} \sigma(\lambda(x), H_\chi(\lambda(x), \lambda(y))^t \xi/\alpha) \times u[\lambda(x) - \lambda(y)] \psi(\lambda(y)) |d\chi(y)| \, |H_\chi(x, y)| f(y) \, dy \, d\xi.$$

Hence T'' is of the form of S of Lemma 2.2 with $\sigma(x, \xi, y)$ given by

$$\varphi(\lambda(x))\psi(\lambda(y))\sigma(\lambda(x), H_\chi(\lambda(x), \lambda(y))^t\xi)u[\lambda(x) - \lambda(y)] |d\chi(y)| \, |H_\chi(\lambda(x), \lambda(y))|$$

and the conditions of that lemma are satisfied uniformly for $\chi \in \mathscr{X}$. This completes the proof of the proposition.

Next comes the main localization result. We use the following notational convention: If S is an operator (or one of a family of operators) then $S_y f(x, y)$ denotes the result of thinking of x as fixed, applying S to the function $f(x, \cdot)$, and evaluating the result at y.

Proposition 2.5. *Let U be an open set in R^n, $\mathscr{H}$ a bounded subset of $C^\infty(R^n, R^n)$ such that*

$$\{|dh| : h \in \mathscr{H}\}$$

is bounded away from zero on compact subsets of U, and $\mathscr{F}$ a bounded subset of $C^\infty(R^n \times R^n)$ such that each $f(x, \cdot)$ is supported in a fixed compact subset K of U. Suppose further that

$$\frac{\partial^l}{\partial y^l} f(x, y)\big|_{y=x} = 0 \text{ for } |l| \leqslant m - 1 \tag{2.11}$$

for all x in a neighborhood of a compact set C. Then for any pseudodifferential family $\{S\}$ of order ω and any multi-indices j and k

$$\frac{\partial^j}{\partial x^j} \frac{\partial^k}{\partial \xi^k} \{S_y f(x, y) e^{i\alpha\xi \cdot (h(y) - h(x))}|_{y=x}\} = O(\alpha^{-m}(1 + |\xi|)^{\omega - |k| - m}) \tag{2.12}$$

uniformly for $h \in \mathscr{H}$, $f \in \mathscr{F}$, $x \in C$.

Proof. We shall reduce things, in stages, to the point where one need only check a simple special use.

Suppose the desired estimates have been established in the special case $k = 0$. The $\partial^k/\partial\xi^k$ on the left of (2.12) introduces a factor $(i\alpha)^{|k|}$ and replaces the set $\mathscr{F}$ by

$$\left\{ f(x, y) \prod_{i=1}^{n} [h_i(y) - h_i(x)]^{k_i} : f \in \mathscr{F}, h \in \mathscr{H} \right\},$$

and this set satisfies the hypotheses on $\mathscr{F}$ but with m replaced by $m + |k| + 1$.

Thus the desired estimates for general k follow. We may therefore assume $k = 0$.

It follows from the assumption on $\mathscr{H}$ that if for some $\epsilon > 0$ x and y are within ϵ of each other and of K then $h(x) = h(y)$ for some h implies $x = y$. If one covers K by finitely many precompact open sets of diameter less than ϵ and then takes a partition of 1 (for K) subordinate to this covering one sees that it is no loss of generality to assume that U is precompact, that all h are homeomorphisms on U, and that the $|dh|$ are bounded away from zero uniformly for $h \in \mathscr{H}$.

Clearly, we need consider only the two special cases: (i) $C \subset U$ and (ii) $C \cap \bar{U} = \varnothing$. We shall show that case (ii) follows from case (i). Assume therefore that the assertions hold in case (i) and that we are in case (ii). The left side of (2.12) does not change on C if f is multiplied by a function of x equal to 1 on a neighborhood of C and 0 on U. Thus it may be assumed that $f(x, y) = 0$ for $x \in U$, so (2.11) holds with m arbitrary.

Let V be a precompact neighborhood of C which is disjoint from $\bar{U}$ and find $\rho_u, \rho_v \in C^\infty$ such that ρ_u (resp. ρ_v) equals 1 (resp. 0) on U and 0 (resp. 1) on V. Since $\bar{U}$ is precompact, all $h(U)$ lie in a fixed compact set H. Let λ be an affine transformation of determinant 1 taking $\bar{V}$ to the complement of H. Having these, and given any $h \in \mathscr{H}$, define

$$h' = \rho_u h + \rho_v \lambda.$$

These comprise a new family $\mathscr{H}'$ which has the properties of $\mathscr{H}$ but with U replaced by $U \cup V$. Since $C \subset U \cup V$ we are in case (i) and so we have, uniformly for $x \in C$ and m arbitrary,

$$\frac{\partial^j}{\partial x^j} \{S_y f(x, y) e^{i\alpha\xi \cdot (h'(y) - h'(x))}|_{y=x}\} = O(\alpha^{-m}(1 + |\xi|)^{\omega - m}).$$

What is desired is a similar estimate for the left side with each h' replaced by h. Replacement of $h'(y)$ by $h(y)$ has no effect since $h' = h$ on the support of $f(x, \cdot)$. Replacement of $h'(x)$ by $h(x)$ introduces, after differentiation, factors of the form

$$\frac{\partial^j}{\partial x^j} e^{i\alpha\xi \cdot (h(x) - h'(x))} = O(\alpha^{|j|}|\xi|^{|j|}),$$

and since m was arbitrarily large in the last estimate this also has no effect.

Thus we may assume $C \subset U$. If $\varphi, \psi \in C_0^\infty(U)$ are equal to 1 on neighborhoods of C and K, respectively, then the left side of (2.12), with $k = 0$, is equal on C to

$$\frac{\partial^j}{\partial x^j} \{\varphi(x) S_y \psi(y) f(x, y) e^{i\alpha\xi \cdot (h(y) - h(x))}|_{y=x}\}.$$

If now we write, for $x, y \in U$,

$$x' = h(x), \qquad y' = h(y), \qquad f(x, y) = f'(h(x), h(y))$$

and set $f'(x', y') = 0$ otherwise, then

$$\varphi(x) S_y \psi(y) f(x, y) e^{i\alpha\xi \cdot (h(y) - h(x))} = T_{y'} f'(x', y') e^{i\alpha\xi \cdot (y' - x')},$$

where $\{T\}$ is the pseudodifferential family obtained from $\{S\}$ as in Proposition 2.4 with

$$\mathscr{X} = \{h | U : h \in \mathscr{H}\}.$$

This shows that it may be assumed to begin with that $\mathscr{H}$ consists of the identity function only, i.e., that what is to be estimated is

$$\frac{\partial^j}{\partial x^j} \{S_y f(x, y) e^{i\alpha\xi \cdot (x - y)}|_{y = x}\}$$

for $x \in C$. If S is given by (2.2) then, with the circumflex denoting the Fourier transform with respect to the second variable,

$$S_y f(x, y) e^{i\alpha\xi \cdot (x - y)} = \int \hat{f}(x, \eta) e^{i\eta \cdot y} \sigma(y, \xi + \alpha^{-1}\eta)\, d\eta, \tag{2.13}$$

and at $y = x$ this is

$$\int \hat{f}(x, \eta) e^{i\eta \cdot x} \sigma(x, \xi + \alpha^{-1}\eta)\, d\eta.$$

By Taylor's theorem

$$\sigma(x, \xi + \alpha^{-1}\eta) = \sum_{|l| \leqslant m - 1} \frac{(\alpha^{-1}\eta)^l}{l!} \frac{\partial^l}{\partial \xi^l} \sigma(x, \xi) \tag{2.14}$$

$$+ \sum_{|l| = m} \frac{m}{l!} (\alpha^{-1}\eta)^l \int_0^1 u^{m-1} \frac{\partial^l}{\partial \xi^l} \sigma(x, \xi + (1 - u)\alpha^{-1}\eta)\, du.$$

Since

$$\int \hat{f}(x, \eta) e^{i\eta \cdot x} \eta^l\, d\eta = (2\pi)^n i^{-|l|} \frac{\partial^l}{\partial y^l} f(x, y)|_{y = x},$$

the first sum on the right side of (2.14) contributes zero to the right side of (2.13) for all $x \in C$ and the same holds after applying any $\partial^j/\partial x^j$. It remains to consider $\partial^j/\partial x^j$ of the contribution of the second sum on the right side of (2.14). Since each

$$\frac{\partial^j}{\partial x^j} f(x, \eta) e^{i\eta \cdot x} \eta^l = O((1 + |\eta|)^{-N}$$

for all N, uniformly for $f \in \mathscr{F}$ and $x \in C$, the contribution of each term of the second sum is $O(\alpha^{-m})$ times

$$\max_{0 \leqslant u \leqslant 1} \int (1 + |\eta|)^{-N}(1 + |\xi + (1 - u)\alpha^{-1}\eta|^{\omega - m})\, d\eta,$$

and if N is large enough this is $O((1 + |\xi|)^{\omega - m})$. This completes the proof.

COROLLARY 2.6. *Let U and $\mathscr{H}$ be as in Proposition 2.5, $\mathscr{F}$ a bounded set of functions in $C^\infty(R^n)$ all supported in a compact subset K of U, and C a compact set disjoint from K. Then for any pseudodifferential family $\{S\}$, any multi-indices j and k, and any $N > 0$*

$$\frac{\partial^j}{\partial x^j}\frac{\partial^k}{\partial \xi^k} e^{-i\alpha\xi \cdot h(x)}(Sfe^{i\alpha\xi \cdot h})(x) = O(\alpha^{-N}(1 + |\xi|)^{-N})$$

uniformly for $x \in C$.

Proof. Find $\varphi \in C_0^\infty(R^n)$ which vanishes on a neighborhood of K and equals one on a neighborhood of C, and apply the proposition to the family

$$\{\varphi(x)f(y): f \in \mathscr{F}\}.$$

III. PSEUDODIFFERENTIAL FAMILIES ON MANIFOLDS

Throughout, M will denote a paracompact n-dimensional C^∞ manifold. We define a pseudodifferential family of order ω on M to be a family $\{S\}$ of operators from $C_0^\infty(M)$ to $C^\infty(M)$ which, when restricted to any coordinate neighborhood of M and transferred to R^n in the obvious way, becomes a pseudodifferential family in R^n of order ω. More precisely, for any diffeomorphism χ from an open set $U \subset M$ to R^n and for any $\varphi, \psi \in C_0^\infty(M)$ it is required that the set $\{T\}$ defined by

$$\begin{aligned} (Tf)(\chi(x)) &= \varphi(x)(S\psi \cdot (f \circ \chi))(x), && x \in U \\ Tf(y) &= 0, && y \in \chi(U) \end{aligned}$$

be a pseudodifferential family of order ω on R^n. It is important to keep in mind that U may be disconnected.

Proposition 2.4 tells us that this definition is at least as general as the original one if $M = R^n$. It is also immediate that the main localization result, Proposition 2.5, carries over to the general case if the following modifications are made: U is an open set in M, $\mathscr{H}$ is a subset of $C^\infty(M, R^n)$, and $\mathscr{F}$ is a subset of $C^\infty(M \times M)$. Moreover, $x \in M$ and $\xi = (\xi_1, \ldots, \xi_n) \in R^n$. We shall refer to this generalization also as "Proposition 2.5".

IV. The Symbol

Given any pseudodifferential family, its symbol will be defined to be a function on the cartesian product of $T^*(M)$ with the parameter set inherent in the family. This parameter set is in turn the cartesian product of the range of the distinguished parameter α, a subset of $[1, \infty)$, and whatever parameter set the other parameters run over. These parameter sets will not be mentioned explicitly, and any bounds involved in definitions and theorems are tacitly assumed to be uniform with respect to the parameters.

Local coordinates $x_1, \ldots, x_n$ on an open set U in M induce a set of dual local coordinates

$$(x_1, \ldots, x_m, \xi_1, \ldots, \xi_n)$$

in $T^*(U)$: for $v \in T^*(U)$ the x_i are the coordinates of $\pi(v)$ and

$$v = \sum \xi_i \, dx_i(\pi(v)).$$

We shall generally write dual coordinates as (x, ξ) and think of x as a point of M.

Definition. (a) $\sum_\omega$ *consists of those functions* σ *in* $C^\infty(T^*(M))$ *which satisfy, on compact subsets of coordinate neighborhoods in* M *and for all multi-indices* j *and* k,

$$\frac{\partial^j}{\partial x^j} \frac{\partial^k}{\partial \xi^k} \sigma = O((1 + |\xi|)^{\omega - |k|}).$$

(b) $\sum$ *is the union of all* $\sum_\omega$. (c) Z (*the set of null symbols*) *consists of those* σ *such that on compact subsets, for all multi-indices* j *and* k, *and all* $N > 0$,

$$\frac{\partial^j}{\partial x^j} \frac{\partial^k}{\partial \xi^k} \sigma = O(\alpha^{-N}(1 + |\xi|)^{-N}).$$

Although the true symbol space should be $\sum/Z$ we shall use $\sum$ instead and write $\sigma \equiv \tau$ to mean $\sigma - \tau \in Z$. Note also that as a consequence of our tacit assumption σ is really a mapping from the parameter set to $C^\infty(T^*(M))$ such that the estimates in (a) or (b) are uniform on the range of σ.

Proposition 4.1. *Suppose* M *is covered by a family of open sets* U *for each of which there is given a* $\sigma_U \in \sum_\omega(U)$ *such that*

$$\sigma_{U_1} - \sigma_{U_2} \in Z(U_1 \cap U_2)$$

if $U_1 \cap U_2 \neq \varnothing$. *Then there is a* $\sigma \in \sum_\omega(M)$ *such that each*

$$\sigma - \sigma_U \in Z(U),$$

and any two such σ *differ by an element of* $Z(M)$.

Proof. Take a partition of unity $\{\varphi_V\}$ subordinate to a locally finite refinement $\{V\}$ of $\{U\}$ and set

$$\sigma = \sum \varphi_V \sigma_V$$

where if $V \subset U$ one defines $\sigma_V = \sigma_U | V$.

DEFINITION. *If $\sigma_m \in \sum$ $(m = 1, 2, \ldots)$ we say that σ_m converges to σ if for each compact subset K of a coordinate neighborhood, for any multi-indices j, k, and any $N > 0$, there is an $\bar{m}$ such that for each $m > \bar{m}$*

$$\frac{\partial^j}{\partial x^j} \frac{\partial^k}{\partial \xi^k} (\sigma_m - \sigma) = O(\alpha^{-N}(1 + |\xi|^{-N})) \tag{4.1}$$

uniformly for $x \in K$.

Note that $\sigma \in \sum$ necessarily, and that if each $\sigma_m \in \sum_\omega$ then $\sigma \in \sum_\omega$ also. Moreover, if σ_m converges to σ and $\tau_m \equiv \sigma_m$ then also τ_m converges to σ.

One says, of course, that an infinite series converges if its sequence of partial sums does.

In order to prove convergence of a sequence $\{\sigma_m\} \subset \sum(M)$ it suffices to prove that each point has a neighborhood U such that the sequence $\{\sigma_m | U\}$ converges in $\sum(U)$. This follows from Proposition 4.1. Another useful fact is that sometimes one need only check (4.1) for the special case $j = k = 0$.

PROPOSITION 4.2. *If $\sigma, \sigma_m \in \sum_\omega$ and if for each compact subset K of a coordinate neighborhood and each $N > 0$ there is an $\bar{m}$ such that $m > \bar{m}$ implies*

$$\sigma_m - \sigma = O(\alpha^{-N}(1 + |\xi|)^{-N})$$

uniformly for $x \in K$, then σ_m converges to σ.

Proof. In the following c will denote various constants. It is a trivial fact that for functions of one variable on an interval, with $\| \ \| = \max | \ |$,

$$\|f'\|^2 \leqslant c\|f\|(\|f'\| + \|f''\|),$$

and so by induction for any p

$$\|f^{(n)}\|^{2^n} \leqslant c\|f\| \prod_{k=1}^{n} \|f^{(k)} + f^{(k+1)}\|^{2^{k-1}}.$$

It follows that for boxes of any fixed dimensions in $R^n \times R^n$ and any multi-indices j, k

$$\left\| \frac{\partial^j}{\partial x^j} \frac{\partial^k}{\partial \xi^k} f \right\| \leqslant c\|f\|^{\epsilon} \sum \prod \left\| \frac{\partial^p}{\partial x^p} \frac{\partial^q}{\partial \xi^q} f \right\|^{\epsilon_{p,q}}$$

where each product on the right is taken over a finite set of pairs of multi-indices and

$$\epsilon, \epsilon_{p,q} > 0, \qquad \epsilon + \sum \epsilon_{p,q} = 1.$$

We apply this inequality to the functions $f = \sigma_m - \sigma$ on the cartesian product of K with unit cubes in R^n and find

$$\left\| \frac{\partial^j}{\partial x^j} \frac{\partial^k}{\partial \xi^k} (\sigma_m - \sigma) \right\| \leqslant c \|\sigma_m - \sigma\|^{\epsilon} \sum \prod \left\| \frac{\partial^p}{\partial x^p} \frac{\partial^q}{\partial \xi^q} (\sigma_m - \sigma) \right\|^{\epsilon_{p,q}}.$$

Since $\sigma, \sigma_m \in \sum_\omega$ and $\sum \epsilon_{p,q} \leqslant 1$ the product in the right is, uniformly in m,

$$O((1 + |\xi|)^{\max(\omega, 0)}).$$

However, our assumption implies that the first factor on the right is, for $m > \bar{m}$,

$$O(\alpha^{-\epsilon N}(1 + |\xi|)^{-\epsilon N}),$$

and since N is arbitrarily large the result follows.

Proposition 4.3. *Suppose $\sigma_m \in \sum_{\omega_m}$ with $\omega_m \downarrow -\infty$. Then $\sum_{m=0}^{\infty} \alpha^{-m} \sigma_m$ converges to an element of $\sum_{\omega_0}$.*

Proof. It may be assumed that all σ_m vanish for x outside some compact subset of a coordinate neighborhood U; as usual (x, ξ) denote dual coordinates in $T^*(U)$. Define

$$M_m = 2^m \left\{ 1 + \max \left| \frac{\partial^j}{\partial x^j} \frac{\partial^k}{\partial \xi^k} \sigma_m (1 + |\xi|)^{-\omega_m} \right| \right\}$$

where the max is taken over all (x, ξ) and over all multi-indices j, k with $|j| \leqslant m$, $|k| \leqslant m$. Let ρ be any function in $C^\infty[0, \infty)$ satisfying

$$0 \leqslant \rho(u) \leqslant 1, \qquad \rho(u) \begin{cases} = 0 & \text{for} \quad 0 \leqslant u \leqslant 1, \\ = 1 & \text{for} \quad u \geqslant 2. \end{cases}$$

Finally, set

$$\sigma = \sum_{m=0}^{\infty} \alpha^{-m} \rho(M_m^{-1}(\alpha^2 + |\xi|^2)^{1/2}) \sigma_m. \tag{4.2}$$

Since the mth term vanishes unless

$$M_m \leqslant (\alpha^2 + |\xi|^2)^{1/2}$$

and $M_m \to \infty$ we have a finite sum in the neighborhood of each $v \in T^*(M)$.

Thus $\sigma \in C^{\infty}(T^*(M))$. Moreover, since

$$\rho(M_m^{-1}(\alpha^2 + |\xi|^2)^{1/2}) = 1$$

unless

$$(\alpha^2 + |\xi^2|)^{1/2} \leqslant 2M_m,$$

each

$$\sigma_m \equiv \rho(M_m^{-1}(\alpha^2 + |\xi^2|)^{1/2})\sigma_m.$$

Thus it suffices to show that the series in (4.2) converges to σ in the sense we have been using. But it is easy to check that each

$$\frac{\partial^k}{\partial \xi^k} \rho(M_m^{-1}(\alpha^2 + |\xi|^2)^{1/2})$$

is bounded (uniformly in α, ξ, and m), from which it follows first that each partial sum in (4.2) belongs to $\sum_{\omega_0}$, and second that

$$\sum_{m \geqslant \max(|j|,|k|,N)} \alpha^{-m} \left| \frac{\partial^j}{\partial x^j} \frac{\partial^k}{\partial \xi^k} \rho(M_m^{-1}(\alpha^2 + |\xi|^2)^{1/2})\sigma_m \right|$$

is at most a constant times

$$\sum_{m \geqslant N} \alpha^{-m} 2^{-m} (\alpha^2 + |\xi^2|)^{1/2} (1 + |\xi|)^{\omega_m} \leqslant 2^{-N+1} \alpha^{-N+1} (1 + |\xi|)^{\omega_N + 1}.$$

This completes the proof.

As mentioned in the Introduction we define the symbol of a pseudodifferential family only after introducing a connection on the manifold. Such a connection always exists and is determined by the Christoffel symbols Γ^i_{jk} which must satisfy certain relations under coordinate change [9, §9.2]. The covariant derivative ∇ is defined as follows (on covariant tensor fields): If $t = t_{k_1 \cdots k_m}$ (in local coordinates $x_1 \cdots x_n$) is an mth-order covariant tensor then ∇t is the $(m+1)$st-order covariant tensor given by

$$(\nabla t)_{k_1 \cdots k_m k} = \frac{\partial t_{k_1 \cdots k_m}}{\partial x_k} - \sum_{\nu=1}^{m} \Gamma^i_{k k_\nu} t_{k_1 \cdots k_{\nu-1} i k_{\nu+1} \cdots k_m}. \tag{4.3}$$

Note that if t is scalar then $\nabla t = dt$. If g and h are covariant tensors of orders p, q, respectively, then one has the product formula

$$(\nabla gh)_{i_1 \cdots i_p j_1 \cdots j_q k} = (\nabla g)_{i_1 \cdots i_p k} h_{j_1 \cdots j_q} + g_{i_1 \cdots i_p} (\nabla h)_{j_1 \cdots j_q k},$$

and similarly for higher order covariant derivatives. It follows that if S denotes symmetrization with respect to all indices, and

$$g \odot h = S(gh)$$

then

$$S\nabla^k gh = \sum_{i=0}^{k} \binom{k}{i} (S\nabla^i g) \odot (S\nabla^{k-i} h).$$

For a scalar function f define

$$\partial^k f = S\nabla^k f.$$

The last identity yields

$$\partial^k f_1 \cdots f_r = \sum_{i_1 + \cdots + i_r = k} \frac{k!}{i_1! \cdots i_r!} \partial^{i_1} f_1 \odot \cdots \odot \partial^{i_r} f_r. \tag{4.4}$$

Observe that in terms of local coordinates $x_1, \ldots, x_n$ one has

$$(\partial^k f)_{i_1 \cdots i_k} = \frac{\partial^k f}{\partial x_{i_1} \cdots \partial x_{i_k}} + \sum_{|j| < k} \lambda_{j i_1 \cdots i_k} \frac{\partial^j f}{\partial x^j}, \tag{4.5}$$

where the λ's are polynomials in the Γ^i_{jk} and their derivatives. This follows easily from (4.3).

PROPOSITION 4.4. *There exists a real-valued function $l \in C^\infty(T^*(M) \times M)$ such that*

(i) *$l(v, y)$ is linear in v on each fiber $\pi^{-1}(x)$, $x \in M$,*
(ii) *for each $v \in T^*(M)$*

$$\partial^k l(v, y)\big|_{y = \pi(v)} = \begin{cases} 0, & k \neq 1 \\ v, & k = 1. \end{cases}$$

Proof. Suppose first that each point of M has a neighborhood U such that an l_U can be found for $T^*(U)$. Then if $\{V\}$ is a locally finite refinement of $\{U\}$, φ_V a partition of unity subordinate to $\{V\}$, and for each V

$$\psi_V \in C_0^\infty(V), \qquad \psi_V = 1 \quad \text{on} \quad \operatorname{supp} \varphi_V,$$

then one takes

$$l(v, y) = \sum\nolimits_V \begin{cases} \psi_V(\pi(v)) l_V(v, y) \varphi_V(y) & \pi(v) \in V \\ 0 & \pi(v) \notin V \end{cases}$$

(where if $V \subset U$ one sets $l_V = l_U$).

That one can find l locally follows from the form (4.5) of ∂^k and the following fact which follows from the standard proof [11, §1.54] of Borel's theorem on formal power series: Let V be an open set in R^n and suppose that for each multi-index j there is given a function $f_j \in C^\infty(V)$; then for each open set U

with compact closure contained in V one can find a function $g \in C^\infty(U \times U)$ such that for each k

$$\frac{\partial^j}{\partial y^j} g(x, y)\big|_{y=x} = f_j(x). \tag{4.6}$$

To construct l in U, where there are local coordinates $x_1, \ldots, x_n$, it suffices to construct functions $l_i(x, y)$ satisfying

$$\partial^k l_i(x, y)\big|_{y=x} = \begin{cases} 0, & k \neq 1 \\ dx_i(x), & k = 1, \end{cases} \tag{4.7}$$

for then one defines

$$l(\textstyle\sum \xi_i \, dx_i, y) = \sum \xi_i l_i(y).$$

In view of (4.5) the conditions (4.7) are of the form (4.6) for suitable recursively defined f_j and so the l_i exist.

We shall now define the symbol of a pseudodifferential family $\{S\}$, given a particular connection on M. If in terms of local coordinates $x_1, \ldots, x_n$ one sets

$$l_i(x, y) = l(dx_i(x), y)$$

and thinks of $\{l_i(x, y)\}$ as being, for each x, an element of $C^\infty(M, R^n)$, then the determinant

$$|d_y\{l_i(x, y)\}|$$

is nonzero when $y = x$. It follows that each point has a coordinate neighborhood U with compact closure such that the corresponding determinants

$$|d_y\{l_i(x, y)\}|$$

are uniformly bounded away from zero for $x, y \in U$. For each such U, let $\varphi_U \in C_0^\infty(M)$ be identically 1 on a neighborhood of $\bar{U}$ and define

$$\sigma_U(v) = S_y \varphi_U(y) e^{i\alpha l(v,y)}\big|_{y=\pi(v)}, \qquad v \in T^*(U).$$

It follows from Proposition 2.5 (with $m = 0$ and $\mathscr{H}$ the set of all $\{l_i(x, \cdot)\}$, $x \in U$) that $\sigma_U \in \sum_\omega(U)$ if $\{S\}$ has order ω and from the same proposition (with m arbitrarily large) that different choices of φ_U give rise to σ_U differing by an element of $Z(U)$ and that the set $\{\sigma_U\}$ satisfies the hypothesis of Proposition 4.1. Thus there is defined an element σ of $\sum_\omega(M)$ and we call this the symbol of M. Of course there are infinitely many such σ but any two differ by an element of $Z(M)$. We shall see later that, modulo Z, σ is also independent of l (but not, of course, independent of the connection).

The first job will be to determine, in terms of σ,

$$S_y f(y) e^{i\alpha\xi \cdot (h(y) - h(x))}\big|_{y=x},$$

where $h \in C^\infty(M, R^n)$ satisfies $|dh| \neq 0$ on the support of f. This is the content of Lemma 4.7. The entire symbolic calculus will be an easy consequence of this. We proceed in stages.

For each $x \in M$ and $\tau \in C^\infty(T^*(M))$ the restriction of τ to $\pi^{-1}(x)$ has a kth-order derivative which is, for each $v \in \pi^{-1}(x)$, a k-linear functional on $\pi^{-1}(x)$, or, equivalently, an element of the dual space of the k-fold tensor product of T_x^*. This dual space is just the k-fold tensor product of T_x. Thus for each v the kth-order derivative $D^k\tau$, evaluated at v, is a (symmetric) contravariant tensor of order k. Since for any $f \in C^\infty(M)$, the symmetrized kth-order covariant derivative $\partial^k f$ is a covariant tensor of order k, the (contracted) product

$$\partial^k f(x)\, D^k \tau(v), \qquad x = \pi(v),$$

makes sense and is indeed an element of $C^\infty(T^*(M))$. We shall write this simply as $\partial^k f\, D^k\tau$ if there is no question of the names of the variables; if there is a question subscripts may be used. Note that if $\tau \in \sum_\omega$ then $\partial^k f\, D^k \tau \in \sum_{\omega - k}$.

As usual dual coordinates systems will be used without comment, and $l_i(x, y)$ always means $l(dx_i(x), y)$ in terms of local coordinates $x_1, \ldots, x_n$.

LEMMA 4.5. *For any indices $j_1, \ldots, j_k$ and any $\tau \in C^\infty(T^*(M))$*

$$\partial_y{}^k l_{j_1}(x, y) \cdots l_{j_k}(x, y)\, D^k \tau(y, \xi)\big|_{y=x} = k!\, \frac{\partial^k}{\partial \xi_{j_1} \cdots \partial \xi_{j_k}} \tau(x, \xi).$$

Proof. From its definition, $D^k\tau$ evaluated at

$$dx_{j_1} \otimes \cdots \otimes dx_{j_k}$$

is equal to

$$\frac{\partial^k \tau}{\partial \xi_{j_1} \cdots \partial \xi_{j_k}}.$$

But one of the characteristic properties of l implies

$$d_y l_i(x, y)\big|_{y=x} = dx_i.$$

Combining this with (4.4) and the rest of (4.7) gives

$$\partial_y{}^k l_{j_1}(x, y) \cdots l_{j_k}(x, y)\big|_{y=x} = k!\, dx_{j_1} \odot \cdots \odot dx_{j_k}$$

and the result follows.

LEMMA 4.6. *If $\mathscr{F}$ is a bounded family of functions $f \in C^\infty(M)$ all supported in a fixed compact set then for any pseudodifferential family $\{S\}$ with symbol σ*

$$S_y f(y) e^{i\alpha l(v,y)}\big|_{y=\pi(v)} \equiv \sum_{m=0}^{\infty} \frac{(i\alpha)^{-m}}{m!} \partial^m f(\pi(v))\, D^m \sigma(v).$$

Remarks. The left side is a function in $C^\infty(T^*(M))$ for each parameter value associated with $\{S\}$ and for each $f \in \mathscr{F}$. Thus we are dealing with a new parameter set which is the cartesian product of the original one with $\mathscr{F}$; it is to this enlarged parameter set that the sum on the right and the symbol "$\equiv$" refer. The conclusion of the lemma therefore implies uniformity with respect to $\mathscr{F}$ as well as the other parameters.

Proof. Use of a partition of unity and Proposition 2.5 allow us to assume that the common support of the functions of $\mathscr{F}$ is contained in one of the coordinate neighborhoods U used in the definition of σ and to concern ourselves only with $T^*(U)$. In terms of the functions $l_i(x, y)$ we can write for $x \in U$ and each M

$$f(y) = \sum_{\substack{m<M \\ i_1 \cdots i_m}} c_{i_1 \cdots i_m}(x) l_{i_1}(x, y) \cdots l_{i_m}(x, y) + g(x, y)$$

where for $f \in \mathscr{F}$ the $c_{i_1 \cdots i_m}$ form a bounded set in $C^\infty(U)$ and g a bounded set in $C^\infty(U \times M)$, all functions of which vanish for x outside some compact subset of U and satisfy

$$\frac{\partial^k}{\partial y^k} g(x, y)\big|_{y=x} = 0 \qquad \text{for} \quad |k| \leqslant M - 1. \tag{4.8}$$

(For each x the $l_i(x, \cdot)$ form a local coordinate system in U and the c's are the Taylor coefficients of f.) Thus if $\varphi_U \in C_0{}^\infty(M)$ equals 1 on U then in terms of dual coordinates (x, ξ) for v we have

$$\begin{aligned} S_y f(y) e^{i\alpha l(v,y)}\big|_{y=\pi(v)} &= S_y f(y)\varphi(y) \exp(i\alpha \textstyle\sum \xi_i l_i(x, y))\big|_{y=x} \\ &= \displaystyle\sum_{\substack{m<M \\ i_1 \cdots i_m}} c_{i_1 \cdots i_m}(x) S_y \varphi_U(y) l_{i_1}(x, y) \cdots l_{i_m}(x, y) \exp(i\alpha \textstyle\sum \xi_i l_i(x, y))\big|_{y=x} \\ &\quad + S_y \varphi_U(y) g(x, y) \exp(i\alpha \textstyle\sum \xi_i l_i(x, y))\big|_{y=x}. \end{aligned} \tag{4.9}$$

Now

$$\sigma(x, \xi) = S_y \varphi_U(y) \exp(i\alpha \textstyle\sum \xi_i l_i(x, y))\big|_{y=x}$$

and so

$$\frac{\partial^m \sigma}{\partial \xi_{i_1} \cdots \partial \xi_{i_m}} = (i\alpha)^m S_y \varphi_U(y) l_{i_1}(x, y) \cdots l_{i_m}(x, y) \exp(i\alpha \textstyle\sum \xi_i l_i(x, y))|_{y=x}.$$

Hence by Lemma 4.5 the sum on the right side of (4.9) equals

$$\sum_{\substack{m<M \\ i_1 \cdots i_m}} \frac{(i\alpha)^{-m}}{m!} c_{i_1 \cdots i_m}(x)\, \partial^m l_{i_1}(x, y) \cdots l_{i_m}(x, y)\, D^m \sigma(y, \xi)|_{y=x}$$

$$= \sum_{\substack{m<M \\ i_1 \cdots i_m}} \frac{(i\alpha)^{-m}}{m!} \partial^m f(x)\, D^m \sigma(x, \xi)$$

by (4.8). Also from (4.8), this time with Proposition 2.5, we find that on compact subsets of U

$$\frac{\partial^j}{\partial x^j} \frac{\partial^k}{\partial \xi^k} S_y \varphi_U(y) g(x, y) \exp(i\alpha \textstyle\sum \xi_i l_i(x, y))|_{y=x} = O(\alpha^{-M}(1 + |\xi|^{\omega - |k| - M})$$

and the lemma is proved.

Next comes the main lemma.

Lemma 4.7. *Let $\mathscr{F}$ be a bounded set of functions $f \in C^\infty(M)$ all supported in the fixed compact set K and $\mathscr{H}$ a bounded set of functions $h \in C^\infty(M, R^n)$ such that the determinants $|dh|$ are uniformly bounded away from zero on K. Then for a pseudodifferential family $\{S\}$ with symbol σ*

$$S_y f(y) e^{i\alpha\xi \cdot (h(y) - h(x))}|_{y=x}$$

$$= \sum_{\substack{k \geq 0 \\ m_1, \ldots, m_k \geq 2}} \frac{(i\alpha)^{k - \Sigma_0^k m_i}}{k! m_0! m_1! \cdots m_k!} \partial^{m_0} f(x) \odot \partial^{m_1} \xi \cdot h(x) \odot \cdots \odot \partial^{m_k} \xi \cdot h(x)$$

$$\times D^{\Sigma_0^k m_i} \sigma(d(\xi \cdot h(x))).$$

Remark. See the remark following the statement of Lemma 4.6. Here $\mathscr{H}$ is also a new parameter set. It is easy to see that the coefficient of $(i\alpha)^{-m}$ in the power series on the right belongs to $\sum_{\omega - m}$ if $\{S\}$ has order ω, so the sum makes sense by Proposition 4.3. Of course all terms corresponding to the same power of α are to be combined; since $m_1, \ldots, m_k \geq 2$, only nonpositive powers occur and there are only finitely many terms involving any given power.

Proof. With $h = \{h_i\}$ $(i = 1, \ldots, n)$ let us write

$$\lambda(x, y) = \{l(dh_i(x), y)\}_{i=1}^n$$

so that $\lambda \in C^\infty(M \times M, R^n)$. Note that

$$d_y\lambda(x, y)|_{y=x} = dh(x).$$

For x lying in a compact set, $u \in [0, 1]$, and $h \in \mathscr{H}$ the functions $k_{u,x}$ given by

$$k_{u,x}(y) = (1 - u)h(y) + u\lambda(x, y)$$

run over a bounded subset of $C^\infty(M, R^n)$. Each $k_{u,x}$ satisfies

$$d_y k_{u,x}(y)|_{y=x} = dh(x)$$

and the determinants of these differentials are uniformly bounded away from zero in K (and so also in a neighborhood of K). Therefore each point of K has a neighborhood U for which the family

$$\mathscr{K} = \{k_{u,x} : 0 \leqslant u \leqslant 1, x \in U, h \in \mathscr{H}\}$$

has the property that

$$\{|d_y k_{u,x}(y)| : k_{u,x} \in \mathscr{K}\}$$

is bounded away from zero. A partition of unity and Proposition 2.5 show that we may assume K itself is a subset of U and that the assertion of the lemma need be proved only in $T^*(U)$.

Taylor's theorem gives

$$e^t = \sum_{k=0}^{N-1} \frac{t^m}{m!} + \frac{t^N}{(N-1)!} \int_0^1 u^{N-1} e^{(1-u)t}\, du.$$

Set

$$t = i\alpha\xi \cdot [h(y) - h(x) - \lambda(x, y)],$$

multiply both sides by $f(y)e^{i\alpha\xi \cdot \lambda(x,y)}$, and apply S_y. What results is

$$S_y f(y) e^{i\alpha\xi \cdot (h(y) - h(x))}$$

$$= \sum_{k=0}^{N-1} \frac{(i\alpha)^k}{k!} S_y f(y) \{\xi \cdot [h(y) - h(x) - \lambda(x, y)]\}^k e^{i\alpha\xi \cdot \lambda(x,y)}$$

$$+ \frac{(i\alpha)^N}{(N-1)!} \int_0^1 u^{N-1} S_y f(y) \{\xi \cdot [h(y) - h(x) - \lambda(x, y)]\}^N e^{i\alpha\xi \cdot k_{u,x}(y)}\, du.$$

Consider first the integral. The Nth power that appears is equal to a homogeneous polynomial of degree N in the ξ_i with coefficients that form a bounded subset of $C^\infty(M \times M)$, each vanishing to order $2N$ at $y = x$ for each x. These coefficients times $f(y)$ therefore comprise a subset of $C^\infty(M \times M)$ satisfying the hypotheses of Proposition 2.5 for the family $\mathscr{F}$ (with C any compact subset of U) and m replaced by $2N$. Moreover, the family $\mathscr{K}$ satisfies the conditions on $\mathscr{H}$. We therefore conclude that the term involving the integral on the right side contributes

$$O(\alpha^{-N}(1 + |\xi|)^{\omega - N})$$

uniformly on compact subsets of U.

It remains to evaluate the terms

$$S_y f(y)\{\xi \cdot [h(y) - h(x) - \lambda(x, y)]\}^k e^{i\alpha\xi \cdot \lambda(x,y)}|_{y=x}. \tag{4.10}$$

Keeping in mind that

$$\xi \cdot \lambda(x, y) = l(\textstyle\sum \xi_i \, dh_i(x), y) = l(d(\xi \cdot h(x)), y)$$

we see from Lemma 4.6 that (4.10) equals

$$\sum_{m=0}^{\infty} \frac{(i\alpha)^{-m}}{m!} \partial_y{}^m f(y)\{\xi \cdot [h(y) - h(x) - \lambda(x, y)]\}_{y=x}^k D^m \sigma(d(\xi \cdot h(x)))$$

(modulo Z with convergence in $\sum$). Now (4.4) gives

$$\partial^m f(y)\{\xi \cdot [h(y) - h(x) - \lambda(x, y)]\}^k$$
$$= \sum_{m_0 + \cdots + m_k = m} \frac{m!}{m_0! m_1! \cdots m_k!} \partial^{m_0} f(y) \odot \partial^{m_1}\{\xi \cdot [h(y) - h(x) - \lambda(x, y)]\} \odot$$
$$\cdots \odot \partial^{m_k}\{\xi \cdot [h(y) - h(x) - \lambda(x, y)]\},$$

and this is to be evaluated at $y = x$. Since

$$[h(y) - h(x) - \lambda(x, y)]_{y=x} = 0,$$
$$d_y[h(y) - h(x) - \lambda(x, y)]_{y=x} = 0,$$

any term with any of $m_1, \ldots, m_k$ equal to 0 or 1 will give zero. Since

$$\partial^m \lambda(x, y)|_{y=x} = 0 \qquad \text{if} \quad m \geqslant 2$$

we have for $m_i \geqslant 2$

$$\partial^{m_i}\{\xi \cdot [h(y) - h(x) - \lambda(x, y)]\}_{y=x} = \partial^{m_i}\xi \cdot h(x).$$

This establishes the formula of the lemma interpreted as follows: For any M there exists an $\bar{M}$ such that the difference between the left side and any partial sum of the right side which includes all powers of α greater than $-\bar{M}$ is

$$O(\alpha^{-M}(1 + |\xi|)^{-M})$$

uniformly on compact subsets of U. Since each term of the sum belongs to $\sum_\omega$, and since the left side belongs to $\sum_\omega$ by Proposition 2.5 (with $m = 0$), the full conclusion of the lemma follows, by Proposition 4.2.

As a first application of Lemma 4.7 let $l'(v, y)$ be another function with the properties of l given by Proposition 4.4, and σ' the symbol associated with it as σ was associated with l. We apply Lemma 4.7 with $f = \varphi_U$ and

$$h(y) = h_z(y) = \{l'(dx_i(z), y)\}, \qquad z \in U, \tag{4.11}$$

and then set $z = x$. All terms of the series vanish on U except the one with $k = 0$, $m_0 = 0$, and we deduce that for each M

$$\sigma' - \sigma = O(\alpha^{-M}(1 + |\xi|)^{-M})$$

on compact subsets of U. Since $\sigma, \sigma' \in \sum_\omega$, we deduce from Proposition 4.2 that $\sigma \equiv \sigma'$. Thus, modulo Z, the symbol depends on the connection only.

We can also see how the symbol changes under change of connection.

THEOREM 4.8. *If σ is the symbol of $\{S\}$ with respect to a certain connection and a second connection is introduced with corresponding function $l'(v, x)$ then $\sigma'(v)$, the symbol of $\{S\}$ with respect to the second connection, is given by*

$$\sum_{m_1, \ldots, m_k \geq 2} \frac{(i\alpha)^{k - \Sigma m_i}}{k! m_1! \cdots m_k!} \partial^{m_1} l'(v, \pi(v)) \odot \cdots \odot \partial^{m_k} l'(v, \pi(v)) D^{\Sigma m_i} \sigma(v). \tag{4.12}$$

Proof. Apply Lemma 4.7 with the functions h given by (4.11).

Before stating the next result there is a remark to be made about series such as (4.12). They are of the form

$$P(\sigma) = \sum_{m=0}^{\infty} \alpha^{-m} P_m(\sigma),$$

where P_m is a differential operator on $C^\infty(T^*(M))$ taking each $\sum_\omega$ into $\sum_{\omega - m}$ and P_0 is the identity. Proposition 4.3 implies that each such series converges in $\sum$. Now the series

$$\sum_{m=0}^{\infty} \alpha^{-m} P_m$$

has the formal two-sided inverse

$$\sum_{m=0}^{\infty} \alpha^{-m} Q_m$$

where the Q_m are defined by

$$Q_0 = \text{identity}, \qquad Q_m = -\sum_{r=1}^{\infty} P_r Q_{m-r}.$$

The Q_m so defined also send each $\sum_\omega$ to $\sum_{\omega-m}$, as may be proved by induction, and so

$$Q(\sigma) = \sum_{m=0}^{\infty} \alpha^{-m} Q_m(\sigma)$$

is a well-defined element of $\sum_\omega$ for each $\sigma \in \sum_\omega$. Moreover,

$$PQ(\sigma) \equiv QP(\sigma) \equiv \sigma$$

for each σ. Thus, modulo Z, P is a one-one mapping of $\sum_\omega$ onto itself.

The next result, critical for all that follows, is a simple consequence of this fact and Theorem 4.8.

THEOREM 4.9. *Given any $\sigma \in \sum$ and any connection on M there is a pseudodifferential family which with respect to that connection has symbol σ.*

Proof. Suppose the result has been established for all σ vanishing outside some $T^*(U)$ with U a coordinate neighborhood. Let $\{U\}$ be a locally finite covering of M by coordinate neighborhoods and $\{\varphi_U\}$ a partition of unity subordinate to this covering. Then if σ is a given element of $\sum$ and for each U $\{S_U\}$ is a pseudodifferential family with symbol $\varphi_U\sigma$, then $\{S\}$ defined by

$$Sf(x) = \sum_U \varphi_U(x)(S_U f)(x)$$

is a pseudodifferential family with symbol σ.

Thus we may assume to begin with that σ is supported in a coordinate neighborhood, and this implies that we may assume $M = R^n$. We shall use Theorem 4.8 where the first connection is our given one and the second is the usual Euclidean connection. Let σ' denote the sum of the series (4.12). If $\{S\}$ has symbol σ with respect to the given connection then by the theorem it will have symbol σ' with respect to the Euclidean connection. In fact the converse is true: For if $\{S\}$ has symbol σ' with respect to the Euclidean connection and σ_0 with respect to the given connection then the theorem tells us that the series with σ replaced by σ_0 also gives σ'. As we have seen, this implies $\sigma_0 \equiv \sigma$, so $\{S\}$ has symbol σ with respect to the original connection, as desired.

What is left to prove then (changing notation) is that given any $\sigma \in \sum(R^n)$ there is a pseudodifferential family on R^n which, with respect to the Euclidean connection, has symbol σ. But of course there is; it is given by formula (2.2).

If $\{S\}$ and $\{T\}$ are pseudodifferential families on M with the same parameter set and $f \in C_0^\infty(M)$ then Proposition 2.1, transferred from R^n to M, tells us that

$$\{SM_f T\}$$

is also a pseudodifferential family. The problem is to express the symbol of this family in terms of σ and τ, the symbols of $\{S\}$ and $\{T\}$, respectively.

To find the symbol, let U be one of the little neighborhoods involved in its definition and let $\varphi_U \in C_0^\infty(M)$ be identically one on a neighborhood of $\bar{U}$. Then modulo $Z(U)$ the symbol is, with $\pi(v) = x \in U$,

$$S_y f(y) T_y \varphi_U(y) e^{i\alpha l(v,y)}\big|_{y=x}$$
$$= S_y f(y) e^{i\alpha l(v,y)} \{T_z \varphi_U(z) e^{i\alpha[l(v,z)-l(v,y)]}\big|_{z=y}\}\big|_{y=x}. \tag{4.13}$$

Now we would like to apply Lemma 4.6 with $\mathscr{F}$ the set of functions

$$f(y) T_z \varphi_U(z) e^{i\alpha[l(v,z)-l(v,y)]}\big|_{z=y}.$$

This family is not bounded. However, if $\{T\}$ has order ω then the above family multiplied by

$$\alpha^{-\omega}(1 + |\xi|^2)^{-\omega/2}$$

is. (Here of course v has dual coordinates (x, ξ).) This shows that Lemma 4.6 is indeed applicable and we find that for $x \in V$ (4.13) is given by

$$\sum_{p=0}^{\infty} \frac{(i\alpha)^{-p}}{p!} \partial^p \{f(y) T_z \varphi_U(z) e^{i\alpha[l(v,z)-l(v,y)]}\big|_{z=y}\}\big|_{y=x} D^p \sigma(v).$$

But from Lemma 4.7, with $\mathscr{H}$ the family of all

$$\{l(dx_i(z), y)\}, \qquad z \in U,$$

we obtain for $y \in U$

$$T_z \varphi_U(z) e^{i\alpha[l(v,z)-l(v,y)]}\big|_{z=y}$$
$$\equiv \sum_{m_1,\ldots,m_k \geq 2} \frac{(i\alpha)^{k-\Sigma m_i}}{k! m_1! \cdots m_k!} \partial^{m_1} l(v, y) \odot \cdots \odot \partial^{m_k} l(v, y) D^{\Sigma m_i} \tau(d_y l(v, y)).$$

(Note that $\partial^{m_0} \varphi_U = 0$ on U for $m_0 > 0$.)

Combining these things shows that $\{SM_f T\}$ has symbol

$$\sum_{m_1,\ldots,m_k \geq 2} \frac{(i\alpha)^{k-\Sigma m_i - p}}{k! m_1! \cdots m_k! p!} \partial^p \{f(y) \partial^{m_1} l(v, y)$$
$$\odot \cdots \odot \partial^{m_k} l(v, y) D^{\Sigma m_i} \tau(d_y l(v, y))\}\big|_{y=x} D^p \sigma(v).$$

Now we want to expand the ∂^p of the curly brackets in this formula. An apparent difficulty is that the product inside the curly brackets is not the product of functions on M but rather the function resulting from the mutual action of covariant and contravariant tensors of the same order, $\sum m_i$.

Actually this is no problem since (4.4) holds for ∂^k applied to an arbitrary product of tensors of whatever types. Now ∂^k denotes ∇^k followed by symmetrization with respect to the k new indices introduced and the $\odot$ denotes symmetrization with respect to these same indices. (The definition of ∇t for t an arbitrary tensor is similar to (4.3). See [9, p. 102], for example.) It is important to keep in mind that

$$\partial^{j+k} \neq \partial^j \partial^k$$

in general. The left side is more symmetric than the right.

The last formula can now be rewritten

$$\sum_{\substack{\bar{p}+p_0+p_1+p_k=p \\ m_1,\ldots,m_k \geq 2}} \frac{(i\alpha)^{k-\Sigma m_i - p}}{k!m_1!\cdots m_k!\bar{p}!p_0!\cdots p_k!}$$
$$\times \{\partial^{\bar{p}} f(y) \odot \partial^{p_1}\partial^{m_1} l(v,y) \odot \cdots \odot \partial^{p_k}\partial^{m_k} l(v,y)$$
$$\odot\, \partial^{p_0} D^{\Sigma m_i} \tau(d_y l(v,y))\}\big|_{y=x} D^p \sigma(v).$$

The last $\odot$ above also implies contraction with respect to the $\sum m_i$ indices arising from the $D^{\Sigma m_i}$ and the various ∂^{m_i}. We shall write

$$\partial^{p_0} D^{\Sigma m_i} \tau(d_y l(v,y))\big|_{y=x} \tag{4.14}$$

simply as

$$\partial^{p_0} D^{\Sigma m_i} \tau(v).$$

This is independent of the choice of l but depends only, as does ∂^p applied to functions, on the connection. The final result therefore takes the following form.

THEOREM 4.10. *If $\{S\}$ has symbol σ and $\{T\}$ has symbol τ then $\{SM_f T\}$ has symbol*

$$\sum_{\substack{\bar{p}+p_0+\cdots+p_k=p \\ m_1,\ldots,m_k \geq 2}} \frac{(i\alpha)^{k-\Sigma m_i - p}}{k!m_1!\cdots m_k!\bar{p}!p_0!\cdots p_k!}$$
$$\times\, \partial^{\bar{p}} f(\pi(v))\, \partial^{p_1}\partial^{m_1} l(v,\pi(v)) \odot \cdots \odot \partial^{p_k}\partial^{m_k} l(v,\pi(v)) \odot \partial^{p_0} D^{\Sigma m_i}\tau(v)\, D^p\sigma(v).$$

COROLLARY 4.11. *If M is compact, $\{S\}$ has symbol σ, and $\{T\}$ has symbol $\{\tau\}$ then $\{ST\}$ has symbol*

$$\sum_{\substack{p_0+p_1+\cdots+p_k=p \\ m_1,\ldots,m_k \geq 2}} \frac{(i\alpha)^{k-\Sigma m_i - p}}{k!m_1!\cdots m_k!p_0!\cdots p_k!}$$
$$\times\, \partial^{p_1}\partial^{m_1} l(v,\pi(v)) \odot \cdots \odot \partial^{p_k}\partial^{m_k} l(v,\pi(v)) \odot \partial^{p_0} D^{\Sigma m_i}\tau(v)\, D^p\sigma(v).$$

In order explicitly to evaluate the terms of this sum one has to know the covariant derivatives

$$\nabla_y^k l(v, y)\big|_{y=\pi(v)},$$

which we shall write simply as $\nabla^k l(v, \pi(v))$. Once we know these then the terms $\partial^p \partial^m l$ are obtained by taking $k = p + m$ and symmetrizing over the first m and last p indices. These covariant derivatives are also needed for the computation of $\partial^n D^m \tau$.

We shall now show that $\nabla^k l(v, \pi(v))$ is linear for v in each $\pi^{-1}(x)$ (this is obvious) with coefficients expressible in terms of the torsion and curvature tensors of the connection and their covariant derivatives. The Ricci identity [8, p. 291] expresses for any tensor t

$$(\nabla^2 t)_{i,j} - (\nabla^2 t)_{j,i}$$

as a sum of terms (with $\pm$ signs) each of which is either t times the curvature tensor followed by contraction on an index or ∇t times the torsion tensor followed by contraction on an index. A consequence of this is that for any k and any permutation α on $\{1, \ldots, k\}$

$$(\nabla^k l)_{i_{\alpha(1)} \cdots i_{\alpha(k)}}$$

equals

$$(\nabla^k l)_{i_1 \cdots i_k}$$

plus a sum of terms each of which is a product of a $(k-1)$st- or lower-order covariant derivative of l and covariant derivatives of the curvature and torsion tensors followed by contraction on certain indices. If one now sums over all permutations α and then sets $y = \pi(v)$, what results is that zero ($k!$ times $\partial^k l(v, \pi(v))$) is equal to

$$k!(\nabla^k l)_{i_1 \cdots i_k}(v, \pi(v))$$

plus a sum of terms each of which is a lower-order covariant derivative evaluated at $(v, \pi(v))$ times a product of covariant derivatives of the curvature and torsion tensors followed by contraction on certain indices. Thus the assertion concerning the general form of $\nabla^k l(v, \pi(v))$ is proved by induction.

We shall carry out this procedure only up to $k = 3$ and under the assumption that the connection is symmetric so the torsion tensor vanishes. In terms of dual coordinates (x, ξ) for v,

$$(\nabla l)_i = \xi_i. \tag{4.15}$$

(This is just the fact $dl(v, \pi(v)) = v$ rewritten.) Since the connection is symmetric

$$(\nabla^2 l)_{j,i} = (\nabla^2 l)_{i,j}.$$

Adding $(\nabla^2 l)_{i,j}$ to both sides and using $\partial^2 l = 0$ give

$$(\nabla^2 l)_{i,j} = 0.$$

Next we use the identity

$$t_{ikj} = t_{ijk} + t_p R^p{}_{ikj}$$

which holds for any covariant vector t_i. (The summation convention is used and R is the curvature tensor.) This and (4.15) give

$$\begin{aligned}(\nabla^3 l)_{ijk} &= (\nabla^3 l)_{ijk},\\ (\nabla^3 l)_{ikj} &= (\nabla^3 l)_{ijk} + \xi_p R^p{}_{ikj},\\ (\nabla^3 l)_{jki} &= (\nabla^3 l)_{jik} + \xi_p R^p{}_{jki}\\ &= (\nabla^3 l)_{ijk} + \xi_p R^p{}_{jki}.\end{aligned}$$

Symmetry with respect to the first two indices shows that the other three permutations of $i\,j\,k$ give the same results. Thus summing these three relations and using $\partial^3 l = 0$ give

$$\begin{aligned}0 &= 3(\nabla^3 l)_{ijk} + \xi_p R^p{}_{ikj} + \xi_p R^p{}_{jki},\\ (\nabla^3 l)_{ijk} &= -\tfrac{1}{3}\xi_p (R^p{}_{ikj} + R^p{}_{jki}).\end{aligned}$$

In particular,

$$(\partial\partial^2 l)_{ijk} = -\tfrac{1}{3}\xi_p (R^p{}_{ikj} + R^p{}_{jki}) \tag{4.16}$$

$$(\partial^2\partial l)_{ijk} = -\tfrac{1}{6}\xi_p (R^p{}_{jki} + R^p{}_{kji}); \tag{4.17}$$

the last equation used the antisymmetry of R with respect to its last two indices.

V. Compact Manifolds

Denote by $\mathscr{P}$ the linear space of pseudodifferential families on M (with a certain parameter set) and by $\mathscr{Z}$ the subspace of those families whose symbols vanish (i.e., belong to Z). (A family $\{S\}$ belongs to $\mathscr{Z}$ if and only if for each N $\{\alpha^N S\}$ is a pseudodifferential family of order $-N$. In particular if the parameter set is $\alpha = \{1\}$, the case of a single operator, $\mathscr{Z}$ is the set of operators of order $-\infty$.) What the symbol does is give a linear isomorphism between

$$\textstyle\sum/Z \quad \text{and} \quad \mathscr{P}/\mathscr{Z}.$$

In case M is compact (this is henceforth assumed) $\mathscr{P}$ is an algebra with $\mathscr{Z}$ a two-sided ideal and Corollary 4.11 tells us how multiplication in $\sum/Z$ is to be defined so that the above isomorphism is an algebra isomorphism. We shall denote this multiplication by $*$ so that the series in the statement of the corollary defines $\sigma * \tau$. This is of the form

$$\sigma * \tau \equiv \sum_{m=0}^{\infty} (i\alpha)^{-m} P_m(\sigma, \tau)$$

(do not confuse these P_m, and the Q_m to come, with the P_m and Q_m of the remark following Theorem 8.4) where P_m is a bilinear differential operator from $C^\infty(T^*(M)) \times C^\infty(T^*(M))$ to $C^\infty(T^*(M))$ with the property that

$$\sigma \in \sum\nolimits_{\omega_1}, \tau \in \sum\nolimits_{\omega_2} \qquad \text{imply} \quad P_m(\sigma, \tau) \in \sum\nolimits_{\omega_1 + \omega_2 - m}, \tag{5.1}$$

The first few P_m are

$$P_0(\sigma, \tau) = \sigma\tau,$$
$$P_1(\sigma, \tau) = \partial\tau\, D\sigma,$$
$$P_2(\sigma, \tau) = \tfrac{1}{2}\partial^2\tau\, D^2\sigma + \tfrac{1}{2}\partial\, \partial^2 l\, D^2\tau\, D\sigma.$$

That this multiplication $*$ is associative is certainly not directly obvious, but follows from our multiplicative isomorphism and the associativity of multiplication in $\mathscr{P}$.

We shall now give a sufficient condition for the invertibility of σ modulo Z. This inverse is not to be confused with the reciprocal, which refers to pointwise multiplication.

Definition. *We call σ elliptic of order ω if it has a reciprocal modulo Z, an element σ^{-1} of $\sum$ such that*

$$\sigma\sigma^{-1} \equiv 1,$$

and $\sigma \in \sum_\omega$, $\sigma^{-1} \in \sum_{-\omega}$.

It is easy to see that σ is elliptic of order ω if and only if there is an inequality

$$a(1 + |\xi|)^\omega \leqslant \sigma \leqslant A(1 + |\xi|)^\omega \tag{5.2}$$

with appropriate positive constants a and A in each coordinate neighborhood for $\alpha + |\xi|$ sufficiently large. The "only if part" is trivial. For the "if" part take any $\rho \in C^\infty[0, \infty)$ such that

$$\rho(t) \begin{cases} = 0 & \text{for} \quad t \leqslant 1 \\ = 1 & \text{for} \quad t \geqslant 2. \end{cases}$$

Then if σ satisfies (5.2) and M is sufficiently large the well-defined function

$$\rho(M(\alpha + |\xi|^2))\sigma^{-1}$$

belongs to $\sum_{-\omega}$ and is a reciprocal of σ modulo Z. (This is of course local; one uses a partition of unity for the global construction.) Note that (5.2) implies that ω, if it exists, is unique.

THEOREM 5.1. *If σ is elliptic then it is invertible modulo Z with inverse given by*

$$\sum_{m=0}^{\infty} (i\alpha)^{-m} Q_m(\sigma),$$

where the operators Q_m are defined recursively by

$$Q_0(\sigma) = \sigma^{-1}, \qquad Q_m(\sigma) = -\sigma^{-1} \sum_{r=1}^{m} P_r(\sigma, Q_{m-r}(\sigma)).$$

Proof. One checks by induction, using (5.1), that with $Q_m(\sigma)$ as defined above

$$Q_m(\sigma) \in \sum_{-\omega-m}.$$

Therefore by Proposition 4.3 the series in the statement of the theorem, which is formally the right inverse of σ, converges in $\sum$ and so represents an actual right inverse of σ modulo Z. Similarly there is a left inverse and the two must be equal.

The first few Q_m are

$$\begin{aligned}
Q_0(\sigma) &= \sigma^{-1},\\
Q_1(\sigma) &= -\sigma^{-1}\,\partial\sigma^{-1}\,D\sigma = \sigma^{-3}\,\partial\sigma\,D\sigma,\\
Q_2(\sigma) &= -\sigma^{-1}\{\partial(\sigma^{-3}\,\partial\sigma\,D\sigma)\,D\sigma + \tfrac{1}{2}\partial^2\,\sigma^{-1}\,D^2\sigma + \tfrac{1}{2}\partial\,\partial^2 l\,D^2\sigma^{-1}\,D\sigma\}.
\end{aligned}$$

Once we can find inverses we can construct an analytic functional calculus via Cauchy's theorem.

Define the *spectrum* of σ to be the set of complex numbers λ such that $\sigma - \lambda$ is not elliptic of order greater than or equal to zero, and the *resolvent set* to be the complement of the spectrum.

The characterization (5.2) of ellipticity implies that the resolvent set is open. The same characterization and Theorem 5.1 imply that for any open set U with compact closure contained in the resolvent set there is a function $R_\lambda(\sigma)$, the resolvent, defined and analytic for $\lambda \in U$, belonging to $\sum_0$ (where now one thinks of the parameter set as augmented by U), and satisfying

$$R_\lambda(\sigma) * (\sigma - \lambda) \equiv (\sigma - \lambda) * R_\lambda(\sigma) \equiv 1. \tag{5.3}$$

On overlapping sets the resolvents agree modulo Z.

We can now define an analytic function of σ in the obvious way, if $\sigma \in \sum_0$. In this case the spectrum is compact. Let f be any function analytic on its spectrum. Then we define

$$f_a(\sigma) \equiv -(1/2\pi i) \int_\Lambda f(\lambda) R_\lambda(\sigma)\, d\lambda,$$

where Λ is the boundary of a sufficiently nice bounded open set containing the spectrum and on the closure of which f is analytic. The resolvent R_λ is defined as above with U an appropriate neighborhood of Λ and $f_a(\sigma)$ is well defined modulo Z. The subscript "a" refers to the fact that f is being applied to σ in the sense of the analytic functional calculus. Just plain $f(\sigma)$ denotes pointwise application of f to σ. We leave to the reader the verification of the facts that for f, g analytic in suitable regions, one has

$$f_a(\sigma) * g_a(\sigma) \equiv (fg)_a(\sigma), \qquad f_a(g_a(\sigma)) \equiv (f \circ g)_a(\sigma). \tag{5.4}$$

From the series of Theorem 5.1 for the inverse one can write down a series for $f_a(\sigma)$ for any f. With the $Q_m(\sigma)$ as defined there one can easily see that for $m \geqslant 1$

$$Q_m(\sigma) = \sum_{k=2}^{2m} \sigma^{-k-1} Q_{m,k}(\sigma),$$

where the $Q_{m,k}$ are operators, each sending $\sum_\omega$ to $\sum_{\omega-m}$, with the property that for any complex number λ

$$Q_{m,k}(\sigma - \lambda) = Q_{m,k}(\sigma).$$

The first few $Q_{m,k}$ are given by

$$\begin{aligned} Q_{1,2}(\sigma) &= \partial\sigma\, D\sigma \\ Q_{2,2}(\sigma) &= \tfrac{1}{2}\partial^2\sigma\, D^2\sigma + \tfrac{1}{2}\partial\, \partial^2 l\, D^2\sigma\, D\sigma \\ Q_{2,3}(\sigma) &= -\partial^2\sigma\, D\sigma\, D\sigma - \partial\sigma\, \partial D\sigma\, D\sigma - \partial\sigma\, \partial\sigma\, D^2\sigma \\ Q_{2,4}(\sigma) &= 3\, \partial\sigma\, \partial\sigma\, D\sigma\, D\sigma. \end{aligned} \tag{5.5}$$

It follows that

$$R_\lambda(\sigma) \equiv (\sigma - \lambda)^{-1} + \sum_{m=1}^{\infty} (i\alpha)^{-m} \sum_{k=2}^{2m} (\sigma - \lambda)^{-k-1} Q_{m,k}(\sigma),$$

where, for the right-hand side, σ has been replaced if necessary by an equivalent element of $\sum$ such that $\sigma - \lambda$ is bounded away from zero for λ in a neighborhood of Λ. Integration therefore gives the basic result

$$f_a(\sigma) \equiv f(\sigma) + \sum_{m=1}^{\infty} (i\alpha)^{-m} \sum_{k=2}^{2m} (-1)^k \frac{f^k(\sigma)}{k!} Q_{m,k}(\sigma). \tag{5.6}$$

Thus to a first order $f_a(\sigma)$ equals $f(\sigma)$. The series gives the complete correction to this approximation.

The functions with which we shall be most concerned are the exponential and the logarithm. The former is applicable to any element of $\sum_0$ but the latter to a restricted class. We shall now extend our definition of $\log_a \sigma$ to denote any element of $\sum_0$ satisfying

$$\exp_a(\log_a \sigma) \equiv \sigma.$$

Such a $\log_a \sigma$ may exist even if there is no single-valued logarithm defined on the spectrum; if there is then the two definitions of $\log_a \sigma$ agree, by the second identity of (5.4).

THEOREM 5.2. *If there exists an element* $\log \sigma \in \sum_0$ *satisfying*

$$\exp \log \sigma \equiv \sigma \tag{5.7}$$

then a $\log_a \sigma$ *exists and is given by*

$$\log \sigma - \sum_{m=1}^{\infty} (i\alpha)^{-m} \sum_{k=2}^{2m} k^{-1} \sigma^{-k} Q_{m,k}(\sigma). \tag{5.8}$$

Proof. By (5.6) for any $\tau \in \sum_0$

$$\exp_a \tau \equiv \exp \tau \left[1 + \sum_{m=1}^{\infty} (i\alpha)^{-m} \sum_{k=2}^{2m} \frac{(-1)^k}{k!} Q_{m,k}(\tau) \right]. \tag{5.9}$$

If there is an analytic logarithm on the spectrum of σ, for example if always

$$|1 - \sigma(v)| \leqslant \tfrac{1}{2}, \tag{5.10}$$

then with $\log \sigma$ denoting any analytic branch the series (5.8) substituted for τ in (5.9) gives an equality. On the left is σ and on the right is σ plus a power series in α with coefficients which are certain differential operators applied to σ. Since this holds for all σ satisfying (5.10) these operators must be identically zero. But then for any $\log \sigma \in \sum_0$ and satisfying (5.7) substitution of (5.8) into (5.9) gives an identity, and this is equivalent to the assertion of the theorem.

Having defined a functional calculus on $\sum$ (more exactly, on $\sum/Z$) one has immediately a functional calculus on $\mathscr{P}$ (more exactly, on $\mathscr{P}/\mathscr{Z}$). If $\{S\}$ is a pseudodifferential family of order zero with symbol σ and f is analytic on the spectrum of σ, or if $f = \log$ and the hypothesis of Theorem 5.2 is satisfied, then $\{f_a(S)\}$ denotes any pseudodifferential family with symbol $f_a(\sigma)$. The relationship between this functional calculus and the usual functional calculus for operators on Hilbert space, which is crucial, will be explained shortly (see Theorem 5.5).

The Hilbert spaces with which we shall be concerned are the spaces H_p. For Euclidean space R^n one defines $H_p(R^n)$ to be the completion of $C_0{}^{\infty}(R^n)$ under the norm given by

$$\|f\|_p^2 = \int |\hat{f}(\xi)|^2(1 + |\xi|^2)^p \, d\xi.$$

For a compact manifold M with a finite partition of unity $\{\varphi_j\}$ subordinate to a system of coordinate neighborhoods $H_p(M)$ is the completion of $C^{\infty}(M)$ under the norm given by

$$\|f\|_p^2 = \sum \|\varphi_j f\|_p^2,$$

where the $\varphi_j f$ on the right side are thought of as transferred to R^n. Of course different choices of $\{\varphi_j\}$ lead to different norms, but these are all equivalent.

THEOREM 5.3. *If $\{S\}$ is a pseudodifferential family of order ω then S sends each H_p into $H_{p-\omega}$ and the operators*

$$S: H_p \to H_{p-\omega}$$

are bounded with norms $O(\alpha^{\max(-\omega,0)})$. If $\omega < -n$ then the operators

$$S: H_p \to H_p$$

are of trace class with trace norms $O(\alpha^{n+\epsilon})$ for any $\epsilon > 0$.

Proof. The first statement follows from Lemma 1 of [14] (which yields boundedness in the case of a single operator) and the estimate

$$\sup_{\eta} \frac{(1 + |\eta/\alpha|)^{\omega}}{(1 + |\eta|)^{\omega}} = O(\alpha^{\max(-\omega,0)}).$$

The second statement follows from the first since the identity

$$i: H_{p-\omega} \to H_p$$

is of trace class for $\omega < -n$. This follows from the fact that i is Hilbert–Schmidt if $\omega < -n/2$. Although this is surely well known we have been unable to find a proof in the literature, so here is one.

It suffices to show that if $\varphi \in C_0{}^{\infty}(R^n)$ then M_{φ}, multiplication by φ, is Hilbert–Schmidt from $H_{p-\omega}(R^n)$ to $H_p(R^n)$. Consider the maps

$$H_{p-\omega} \xrightarrow{\alpha} H_0 \xrightarrow{\beta} H_0 \xrightarrow{\gamma} H_p$$

defined by

$$\widehat{\alpha f}(\xi) = \hat{f}(\xi)(1 + |\xi|^2)^{(p-\omega)/2}$$

$$\widehat{\beta f}(\xi) = (2\pi)^{-n} \int \hat{\varphi}(\xi - \eta)(1 + |\eta|^2)^{(\omega - p)/2}(1 + |\xi|^2)^{p/2}\hat{f}(\eta)\, d\eta$$

$$\widehat{\gamma f}(\xi) = \hat{f}(\xi)(1 + |\xi|^2)^{-p/2}.$$

Then α and γ are isometries and β is Hilbert–Schmidt since $\omega < -n/2$ implies

$$\iint |\hat{\varphi}(\xi - \eta)|^2(1 + |\eta|^2)^{\omega - p}(1 + |\xi|^2)^p\, d\xi\, d\eta < \infty.$$

Thus $M_\varphi = \gamma \circ \beta \circ \alpha$ is Hilbert–Schmidt.

We next show that ellipticity of (the symbol of) $\{S\}$ implies that these operators are uniformly invertible.

THEOREM 5.4. *If $\{S\}$ is a pseudodifferential family with symbol elliptic of order ω then for any p the operators*

$$S : H_p \to H_{p-\omega}$$

are invertible for sufficiently large α and the inverses have norm $O(\alpha^{\max(\omega,0)})$.

Proof. Let τ be an inverse of σ modulo Z as given by Theorem 5.1, and $\{T\}$ a pseudodifferential family with symbol τ. Clearly $\{T\}$ has order $-\omega$, so the operators

$$T : H_{p-\omega} \to H_p$$

are bounded with norm $O(\alpha^{\max(\omega,0)})$. Moreover,

$$\{ST\} \in I + \mathscr{Z},$$

and so

$$ST : H_{p-\omega} \to H_{p-\omega}$$

differs from the identity by an operator of norm $O(\alpha^{-N})$ for each N. Thus ST has, for α large, a uniformly bounded inverse and so S has a right inverse of norm $O(\alpha^{\max(\omega,0)})$. Similarly, S has a left inverse of norm $O(\alpha^{\max(\omega,0)})$ and the result is established.

We can now see how the functional calculus developed in the above paragraphs relates to the usual functional calculus for Hilbert space operators. If S is a bounded operator on the space H_p and f is analytic on the spectrum of S then $f(S)$ denotes, as usual,

$$-(1/2\pi i) \int_\Lambda f(\lambda)(S - \lambda I)^{-1}\, d\lambda$$

over an appropriate Λ surrounding the spectrum of S.

THEOREM 5.5. *Let $\{S\}$ be a pseudodifferential family of order zero with symbol σ. Let U be an open set containing the spectrum of σ and p an arbitrary real number. Then if α is sufficiently large the spectrum of S, thought of as an operator on H_p, is contained in U. Moreover, if f is analytic in U then the operators $f_a(S)$ differ from $f(S)$ by trace class operators with trace norm $O(\alpha^{-N})$ for each N.*

Proof. First let Λ be any compact set disjoint from the spectrum of σ. Then associated with the resolvent $R_\lambda(\sigma)$ $(\lambda \in \Lambda)$ is the pseudodifferential family

$$\{R_\lambda(S)\}$$

with the resolvent as symbol. Formula (5.3) shows that

$$\{R_\lambda(S)(S - \lambda I)\} \in I + \mathscr{L}, \qquad \{(S - \lambda I)R_\lambda(S)\} \in I + \mathscr{L},$$

where Λ is thought of as augmented to the parameter set. Since $\alpha^N \mathscr{L} \subset \mathscr{L}$ for each N, Theorem 5.3 implies that if one thinks of our operators as acting on H_p then

$$R_\lambda(S)(S - \lambda I) - I, \qquad (S - \lambda I)R_\lambda(S) - I$$

have trace norms $O(\alpha^{-N})$, uniformly for $\lambda \in \Lambda$. This implies that $S - \lambda I$ is invertible for α large and

$$(S - \lambda I)^{-1} - R_\lambda(S)$$

has trace norm $O(\alpha^{-N})$ uniformly for $\lambda \in \Lambda$. This gives the first assertion of the theorem since Λ is an arbitrary compact set disjoint from the spectrum of σ and also the second assertion since Λ may consist of contours appropriate to the definitions of f and f_a and one simply multiplies by $f(\lambda)$ and integrates over λ. Since

$$f(S) = -(1/2\pi i) \int_\Lambda f(\lambda)(S - \lambda I)^{-1}\, d\lambda,$$

$$f_a(S) = -(1/2\pi i) \int_\Lambda f(\lambda) R_\lambda(S)\, d\lambda \mod \mathscr{L} \tag{5.11}$$

the result is established. (Lest the reader have any doubt about (5.11), recall that $\{f_a(S)\}$ is a pseudodifferential family with symbol $f_a(\sigma)$. Since $\{R_\lambda(S)\}$ has symbol $R_\lambda(\sigma)$ the symbol of the right side of (5.11) is

$$-(1/2\pi i) \int f(\lambda) R_\lambda(\sigma)\, d\lambda = f_a(\sigma),$$

as desired.)

The following corollary, combined with Theorem 5.2, will allow us to compute determinants.

COROLLARY 5.6. *If $\{S\}$ is a pseudodifferential family of order zero whose symbol σ satisfies the hypothesis of Theorem 5.2 then for any p and any $N > 0$*

$$S - \exp \log_a S$$

is a trace class operator on H_p of trace norm $O(\alpha^{-N})$.

Proof. Apply Theorem 5.5 to the family $\{\log_a S\}$ and the function exp. Since

$$\exp_a \log_a S \equiv S \mod \mathscr{L}$$

the result follows.

The next project will be the asymptotic evaluation of the traces of, or more generally the kernels of, pseudodifferential families of order $< -n$.

Let μ be a measure on M such that with respect to any local coordinate system $x_1, \ldots, x_n$

$$d\mu = \rho\, dx_1 \cdots dx_n, \tag{5.12}$$

where ρ is a positive C^∞ function. We shall call such a μ a positive C^∞ measure. If μ is such a measure and S is a pseudodifferential operator of order $< -n$ then S is of the form

$$Sf(x) = \int k(x, y) f(y)\, d\mu(y),$$

where k is a continuous function on $M \times M$.

This may be seen as follows. For $f, g \in C^\infty(M)$ define

$$\langle f, g \rangle = \int fg\, d\mu.$$

This extends to a bounded bilinear form on $H_p \times H_{-p}$ for any p, and in fact any continuous linear functional on H_p is of the form

$$f \to \langle f, g \rangle$$

for some $g \in H_{-p}$. Now suppose S has order $-n - \delta$. Then S is continuous from $H_{-(n+\delta)/2}$ to $H_{(n+\delta)/2}$ and

$$(\varphi, x) \to \varphi(x)$$

is a continuous function on $H_{(n+\delta)/2} \times M$. Hence we can find $k(x, y)$ such that

$$x \to k(x, \cdot)$$

is continuous from M to $H_{(n+\delta)/2}$ such that for $f \in H_{-(n+\delta)/2}$

$$Sf(x) = \int k(x, y)\, d\mu(y).$$

Since $H_{(n+\delta)/2}$ consists entirely of continuous functions, k is a continuous function of both its variables.

We shall determine the asymptotic form of the kernels k in the case of a pseudodifferential family $\{S\}$. This will be expressed in terms of the symbol of $\{S\}$ and measures on the fibers $\pi^{-1}(x)$ induced by the given measure μ. If (x, ξ) are local dual coordinates for $T^*(U)$ and in U the measure μ is given by (5.12) then we define the measure μ_x on $\pi^{-1}(x)$ by

$$d\mu_x = \rho(x)^{-1}\, d\xi_1 \cdots d\xi_n.$$

This measure is independent of the local coordinates chosen since

$$\Omega = dx_1 \wedge \cdots \wedge dx_n \wedge dx_n \wedge d\xi_1 \wedge \cdots \wedge d\xi_n \tag{5.13}$$

is a $2n$-form on M.

THEOREM 5.7. *If $\{S\}$ is a pseudodifferential family of order less than $-n$ with symbol σ, and k is the kernel of S with respect to a positive C^∞ measure μ, then*

(a) *$k(x, y) = O(\alpha^{-N})$ for each N uniformly for x, y belonging to disjoint compact sets;*
(b) *$k(x, x) = (\alpha/2\pi)^n \int_{\pi^{-1}(x)} \sigma\, d\mu_x + O(\alpha^{-N})$ for each N uniformly in x.*

Remark. Since the symbol depends on the connection while the kernel does not the formula of (b) seems strange. However, look at Theorem 4.8, which tells how the symbol changes under change of connection. Each term of the series corresponding to a negative power of $i\alpha$ (that is, each term with $k > 0$) is a homogenous polynomial of the ξ_i times a linear combination of partial derivatives of σ with respect to the ξ_i of order larger than the degree of the corresponding homogeneous polynomial. Successive integration by parts shows that the integral of this term with respect to $d\xi_1 \cdots d\xi_n$ (and so with respect to $d\mu_x$) is zero. It follows that change of connection, and so change of symbol, can only change the integral in (b) by $O(\alpha^{-N})$ for each N.

Proof. We can write S as a finite sum

$$\sum M_{\varphi_i} S M_{\varphi_i},$$

where $\varphi_i, \psi_i \in C^\infty(M)$ and for each i the union of the supports of φ_i and ψ_i is contained in a (possibly disconnected) coordinate neighborhood U and it suffices to prove the assertions for each family $\{M_{\varphi_i} S M_{\varphi_i}\}$.

In this case $k(x, y)$ vanishes for x or y outside a compact subset of U, so we need only concern ourselves with x, y belonging to a fixed compact subset of U. Moreover, by the preceding remark we may assume the connection on U is the trivial one (all $\Gamma^i_{jk} = 0$) with respect to the given coordinates on U. But then in terms of the coordinates (x, ξ) S is given by

$$Sf(x) = \int \hat{f}(\xi) e^{i\xi \cdot x} \sigma(x, \xi/\alpha)\, d\xi, \qquad f \in C_0^{\infty}(U).$$

This may be written

$$Sf(x) = (2\pi)^{-n} \int \left[\int e^{i\xi \cdot (x-y)} \sigma(x, \xi/\alpha)\, d\xi \right] f(x)\, dx$$

so that the kernel of S with respect to μ is just

$$k(x, y) = (2\pi)^{-n} \int e^{i\xi \cdot (x-y)} \sigma(x, \xi/\alpha)\, d\mu_x(\xi).$$

Now (a) follows by $N + n$ partial integrations and (b) is immediate.

COROLLARY 5.8. *If $\{S\}$ is a pseudodifferential family of order less than $-n$ with symbol σ then the trace of S is given by*

$$\operatorname{tr} S = (\alpha/2\pi)^n \int_{T^*(M)} \sigma\Omega + O(\alpha^{-N})$$

for all N, where Ω is the canonical $2n$-form (5.13).

Proof. The trace of S is

$$\int k(x, x)\, d\mu$$

and the conclusion follows from part (b) of the theorem since (with slight abuse of notation)

$$\Omega = d\mu\, d\mu_x.$$

VI. APPLICATIONS

As in the preceeding section M is a compact manifold of dimension n. The first application is the manifold version of Szegö's theorem (1.2). In fact we derive a complete asymptotic expansion. The operators $Q_{k,m}$ and $2n$-form Ω are as in the preceding section.

THEOREM 6.1. *Let $\{S\}$ be a pseudodifferential family of order less than $-n$ with symbol σ. Assume there is an element*

$$\log(1 + \sigma) \in \textstyle\sum_{-\delta} \qquad (\delta > 0)$$

satisfying

$$\exp \log(1 + \sigma) \equiv 1 + \sigma.$$

Then for α *sufficiently large* $\det(I + S) \neq 0$ *and one has the asymptotic expansion, for* $\alpha \to \infty$,

$$\log\det(I + S) \sim (\alpha/2\pi)^n \left\{ \int_{T^*(M)} \log(1 + \sigma)\Omega \right.$$

$$\left. - \sum_{m=1}^{\infty} (i\alpha)^{-m} \sum_{k=2}^{2m} k^{-1} \int_{T^*(M)} (1 + \sigma)^{-k} Q_{m,k}(\sigma)\Omega \right\}.$$

Remark. The determinant refers to S as a trace class operator on any of the spaces H_p. If $\{S\}$ has order $\omega < -n$, the hypothesis that $\log(1 + \sigma)$ belongs to $\sum_{-\delta}$ for some δ implies that it belongs to $\sum_{\omega}$.

Proof. Corollary 5.6 applied to the family $\{(I + S)\}$ tells us that

$$I + S - \exp\log_{\mathrm{a}}(1 + S)$$

is of trace class with trace norm $O(\alpha^{-N})$ for each N. The hypothesis implies that $1 + \sigma$ is an elliptic symbol of order zero and so by Theorem 5.4 for sufficiently large α the operators $I + S$ on H_p are invertible with uniformly bounded inverses. Thus

$$I - (I + S)^{-1} \exp\log_{\mathrm{a}}(1 + S)$$

is of trace class with trace norm $O(\alpha^{-N})$ for each N and it follows that

$$\det\{(I + S)^{-1} \exp\log_{\mathrm{a}}(1 + S)\} = 1 + O(\alpha^{-N}).$$

This implies, by the multiplicativity of the determinant,

$$\begin{aligned} \log\det(I + S) &= \log\det\exp\log_{\mathrm{a}}(I + S) + O(\alpha^{-N}) \\ &= \operatorname{tr}\log_{\mathrm{a}}(I + S) + O(\alpha^{-N}). \end{aligned}$$

The symbol of $\log_{\mathrm{a}}(I + S)$ is given by Theorem 5.2 and the trace is given in terms of the symbol by Corollary 5.7.

For the next application, to asymptotic expansions of heat kernels, we must extend the functional calculus of the last section to certain pseudo-differential families of positive order. By analogy with differential operators, we call a symbol σ strongly elliptic of order ω if $\sigma \in \sum_{\omega}$ and there is an inequality

$$\mathscr{R}(\sigma + A) \geqslant a(1 + |\xi|)^{\omega}, \tag{6.1}$$

with an appropriate constants A, $a > 0$ in each coordinate neighborhood for sufficiently large $\alpha + |\xi|$. Equivalently, σ differs by an element of Z from a symbol for which the inequality holds for all α, ξ.

We shall be concerned only with the case $\omega > 0$ since otherwise we would be dealing with the situation already treated in the last section. For a strongly elliptic symbol σ we define the exponential

$$\exp_{\mathrm{a}}(-\sigma)$$

as follows. First modify σ by adding to it an element of Z so that the new symbol, still denoted by σ, satisfies (6.1) for all α, ξ. Then in particular the range of σ is contained in a wedge

$$|\arg(\lambda + A)| \leqslant (\pi/2) - \delta, \qquad \delta > 0.$$

Let Λ be the contour (described downward)

$$|\arg(\lambda + 2A)| = \pm(\pi - \delta)/2.$$

We now proceed as before. Since $\sigma - \lambda$ is bounded away from zero for $\lambda \in \Lambda$, the symbol $\sigma - \lambda$ is elliptic of order ω when Λ is thought of as augmenting the parameter space. Hence by Theorem 5.1 there is a resolvent $R_\lambda(\sigma)$ belonging to $\sum_{-\omega}$ and satisfying

$$R_\lambda(\sigma) * (\sigma - \lambda) \equiv (\sigma - \lambda) * R_\lambda(\sigma) \equiv 1.$$

With this resolvent we define

$$\exp_{\mathrm{a}}(-\sigma) = -(1/2\pi i) \int_\Lambda e^{-\lambda} R_\lambda(\sigma)\, d\lambda$$

and, modulo Z, this is independent of the choices made during the construction.

It is obvious that $\exp_{\mathrm{a}}(-\sigma) \in \sum_{-\omega}$, but a little thought shows that in fact

$$\exp_{\mathrm{a}}(-\sigma) \in \textstyle\sum_{-\infty}$$

(that is, $\exp_{\mathrm{a}}(-\sigma) \in \sum_{-N}$ for all N).

Let us now look at Theorem 5.4 in the case of $\omega > 0$ and replace p by $p + \omega$. The theorem says that if $\{S\}$ is elliptic of order ω then

$$S : H_{p+\omega} \to H_p$$

has a bounded inverse

$$S^{-1} : H_p \to H_{p+\omega}$$

for α sufficiently large. But for $\omega > 0$ the identity

$$i : H_{p+\omega} \to H_p$$

is continuous and so we may think of $\{S^{-1}\}$ as being a family of bounded operators on H_p, for α sufficiently large. The various S^{-1} for different p are just restrictions of each other.

If we apply all this to the preceding situation of a pseudodifferential family with strongly elliptic symbol σ we see that, with Λ as before,

$$e^{-S} = -(1/2\pi i) \int_\Lambda e^{-\lambda}(S - \lambda I)^{-1}\, d\lambda$$

is well defined for α sufficiently large. Moreover, the analog of Theorem 5.5 is easily checked in this case: The operators

$$e^{-S} \quad \text{and} \quad \exp_a(-S)$$

differ by trace class operators of trace norm $O(\alpha^{-N})$ for each N. Thus an application of Theorem 5.7 and (5.9), which also holds in this case, give the following.

THEOREM 6.2. *Let $\{S\}$ be pseudodifferential family with strongly elliptic symbol of positive order. Then for any p the operators e^{-S} are trace class on H_p for α sufficiently large, and if μ is any positive C^∞ measure on M the kernel $k(x, y)$ of e^{-S} with respect to μ has the following asymptotic behavior as $\alpha \to \infty$:*

(a) *$k(x, y) = O(\alpha^{-N})$ for each N uniformly for x, y belonging to disjoint compact sets;*

(b)

$$k(x, x) \sim (\alpha/2\pi)^n \int_{\pi^{-1}(x)} e^{-\sigma}\left[1 + \sum_{m=1}^{\infty} (i\alpha)^{-m} \sum_{k=2}^{2m} \frac{1}{k!} Q_{m,k}(\sigma)\right] d\mu_x$$

uniformly in x.

COROLLARY 6.3. *Under the same assumptions on $\{S\}$*

$$\operatorname{tr} e^{-S} \sim (\alpha/2\pi)^n \int_{T^*(M)} e^{-\sigma}\left[1 + \sum_{m=1}^{\infty} (i\alpha)^{-m} \sum_{k=2}^{2m} \frac{1}{k!} Q_{m,k}(\sigma)\right] \Omega.$$

Note that we may write $Q_{m,k}(\sigma)$ rather then $Q_{m,k}(-\sigma)$ since the two are the same, as may be proved by induction on m.

How does all this relate to the heat kernel? Let A be an rth-order differential operator on R^n (with C^∞ coefficients):

$$Af(x) = \sum_{|k| \leqslant r} a_k(x) \frac{\partial^k f}{\partial x^k}. \tag{6.2}$$

Then

$$Af(x) = \int f(\xi) e^{i\xi \cdot x} \sum_{|k| \leqslant r} a_k(x)(i\xi)^k\, d\xi$$

so A is a pseudodifferential operator with symbol (with respect to the trivial connection)

$$\sum_{|k|\leqslant r} a_k(x)(i\xi)^k.$$

However, we also have, for a parameter α,

$$\alpha^{-r}Af(x) = \int f(\xi)e^{i\xi\cdot x} \sum_{|k|\leqslant r} a_k(x)\alpha^{|k|-r}(i\xi/\alpha)^k \, d\xi$$

and the function

$$\sum_{|k|\leqslant r} a_k(x)\alpha^{|k|-r}(i\xi)^k$$

satisfies the conditions (2.1) with $\omega = r$. Hence $\{\alpha^{-r}A\}$ is a pseudodifferential family of order r.

Transferring this to the manifold M shows that if A is any rth-order differential operator on M then $\{\alpha^{-r}A\}$ is a pseudodifferential family of order r. Moreover, it follows immediately from the definition that if A has symbol $\sigma(v)$ with respect to some connection then $\{\alpha^{-r}A\}$ has symbol $\alpha^{-r}\sigma(\alpha v)$.

The operator A is called (this is the standard definition) strongly elliptic of order r (necessarily even) if in local coordinates where A is given by (6.2) there is an inequality

$$\mathscr{R} \sum_{|k|=r} a_k(x)(i\xi)^k \geqslant a|\xi|^r$$

with some $a > 0$ in each compact subset of the coordinate neighborhood. This is equivalent to the assertion that its symbol satisfy

$$\mathscr{R}\sigma \geqslant a|\xi|^r,$$

for $|\xi|$ sufficiently large, in each compact subset of a coordinate neighborhood. This in turn implies that $\alpha^{-r}\sigma(\alpha v)$, the symbol of the family $\{\alpha^{-r}A\}$, is strongly elliptic in our sense.

Thus Theorem 6.2 gives the asymptotic expansion of the heat kernel associated with A [4, 14] in terms of the $Q_{m,k}$. One sets $\alpha = t^{-1/r}$.

THEOREM 6.3. *Let A be a strongly elliptic operator of order r on M with symbol σ and $k(x, y)$ the kernel of e^{-tA} with respect to a positive C^∞ measure μ. Then as $t \to 0+$,*

(a) *$k(x, y) = O(t^N)$ for each N uniformly for x, y belonging to disjoint compact sets;*

(b)

$$k(x, x) \sim (2\pi t^{1/r})^{-n} \int_{\pi^{-1}(x)} \exp(-t\sigma(t^{-1/r}v)) \times \left[1 + \sum_{m=1}^{\infty} i^{-m} t^{m/r} \sum_{k=2}^{2m} \frac{1}{k!} Q_{m,k}(t\sigma(t^{-1/r}v))\right] d\mu_x$$

uniformly in x.

Let us see how this looks when expanded in powers of $t^{1/r}$. After replacing t by α^{-r} in it, a typical integral becomes (in dual coordinates and suppressing x terms)

$$\int \exp(-\alpha^{-r}\sigma(\alpha\xi))\sigma(\alpha\xi)^k Q_{m,k}(\alpha^{-r}\sigma(\alpha\xi))\, d\xi. \tag{6.3}$$

Now $Q_{m,k}$ takes functions of a given parity (even or odd with respect to ξ) into functions of the same or opposite parity depending on whether m is even or odd. It follows that

$$Q_{m,k}(\alpha^{-r}\sigma(\alpha\xi))$$

is a polynomial in α^{-1} and the ξ_i in each term of which the power of α^{-1} and the degree in ξ are of the same or opposite parity depending on whether m is even or odd. In the series expansion in powers of α^{-1} of

$$\exp\left(-\alpha^{-r} \sum_{|k|<r} a_k(x)(\alpha\xi)\right)$$

the powers of α^{-1} in each term and the degrees in ξ have the same parity. It follows that any term in the expansion of (6.3) in powers of α which involves an odd power of α with m even or an even power of α with m odd is a sum of terms of the form

$$\int \xi^l \exp\left\{-\sum_{|k|=r} a_k(x)\xi^k\right\} d\xi,$$

where l is a multi-index with $|l|$ odd. Since r is even this integral vanishes. Thus in the expansion of the integral (6.3) in powers of α^{-1} only even powers occur for m even and only odd powers occur if m is odd, so α^{-m-rk} times this integral has in its expansion only even powers of α. It follows that the expansion (b) takes the form

$$k(x, x) \sim (2\pi t^{1/r})^{-n} \sum_{k=0}^{\infty} t^{2k/r} v_k(x). \tag{6.4}$$

By thinking a little more about what was done above one can easily determine the general nature of the v_k. In terms of local coordinates, with

$$d\mu = \rho\, dx_1 \cdots dx_m$$

each $v_k(x)$ is $\rho(x)^{-1}$ times a sum of terms each of which is a polynomial in the coefficients of A and their derivatives times

$$\int \exp\left(-\sum_{|k|=r} a_k(x)\xi^k\right) q(\xi)\, d\xi,$$

where q is a polynomial in the ξ_i.

In case $r = 2$, where the $a_k(x)$ may be thought of as a matrix $(a_{ij}(x))$, the integral is easily seen to be a polynomial in the a_{ij} times

$$(\det a_{ij})^{-N-1/2},$$

where N is half the degree of q. This is clear if $q = 1$ since then the integral is exactly

$$(\det a_{ij})^{-1/2}\pi^{n/2}.$$

If the assertion is true for a given q then replacing a fixed a_{ij} by $a_{ij} + \gamma$, differentiating with respect to γ, and setting $\gamma = 0$ gives the result for $\xi_i\xi_j q$. This establishes the assertion for homogeneous polynomials of even degree, while for homogeneous polynomials of odd degree the integral is zero. This description of the nature of the $v_k(x)$ was used, and derived differently, in [2a,b].

In case $r = 2$ the more common way of writing (6.4) is

$$k(x, x) \sim (4\pi t)^{-n/2} \sum_{k=0}^{\infty} t^k u_k(x). \tag{6.5}$$

Let us now work through the computations of u_0 and u_1 in case M is a Riemannian manifold and A is the negative of the Laplacian on M. Thus

$$Af = -(\nabla^2 f)_{ij} g^{ij}$$

where (g^{ij}) is the inverse of the metric tensor (g_{ij}) and ∇ is the covariant derivative associated with the metric. Let us first compute the symbol of A relative to the connection associated with the metric [9, §11.2]. We have

$$\nabla^2 e^{il(v,y)} = \{-\nabla l(v, y)\nabla l(v, y) + i\nabla^2 l(v, y)\} e^{il(v,y)}.$$

Since g^{ij} is symmetric and $\partial^2 l(v, y)$ vanishes at $y = \pi(v)$ we deduce that A has symbol

$$\sigma(v) = (dv)_i (dv)_j g^{ij} = \xi_i \xi_j g^{ij}$$

since $v = \sum \xi_i \, dx_i$. Hence also

$$t\sigma(t^{-1/2}v) = \xi_i \xi_j g^{ij}.$$

As the measure μ we use the Riemann volume element

$$d\mu = (\det g_{ij})^{1/2} \, dx_1 \cdots dx_n$$

and Theorem 6.3b gives in this case (keeping in mind that only integral powers of t can arise from the integral)

$$k(x, x) \sim (4\pi^2 t)^{-n/2} (\det g_{ij})^{-1/2} \int \exp(-\xi_i \xi_j g^{ij}) \times \left[1 + \sum_{m=1}^{\infty} (-1)^m t^m \sum_{k=2}^{4m} \frac{1}{k!} Q_{2m,k}(\xi_i \xi_j g^{ij}) \right] d\xi.$$

Thus in the expansion (6.5)

$$u_0(x) = 1,$$

$$u_1(x) = -\pi^{-n/2} (\det g_{ij})^{-1/2} \int \exp(-\xi_i \xi_j g^{ij}) \sum_{k=2}^{4} \frac{1}{k!} Q_{2,k}(\xi_i \xi_j g^{ij}) \, d\xi. \quad (6.6)$$

The $Q_{2,k}$ are listed in (5.5) and we must compute various operators applied to

$$\sigma = \xi_i \xi_j g^{ij}.$$

Since D is just partial differentiation with respect to the ξ_i we have

$$(D\sigma)^i = 2\xi_j g^{ij}, \qquad (D^2 \sigma)^{ij} = 2g^{ij}.$$

Next we must compute $\partial\sigma$ and $\partial^2\sigma$, which, recall, are defined by (4.14). Writing l for $l(v, \cdot)$ we have

$$\sigma(dl) = (dl)_i (dl)_j \, g^{ij}.$$

First, since

$$\nabla^2 l(\pi(v)) = 0$$

and since (g^{ij}) has covariant derivative identically zero [9, p. 119], we obtain

$$\partial\sigma = 0.$$

Second, these same two facts together with

$$(dl)_i (\pi(v)) = \xi_i$$

and (4.4) give

$$(\partial^2 \sigma)_{kl} = 2\xi_i (\partial^2 \, \partial l)_{jkl} \, g^{ij}$$

and by (4.17) this is

$$-\tfrac{1}{3}\xi_i\xi_p(R^p{}_{klj} + R^p{}_{lkj})g^{ij} = \tfrac{1}{3}\xi_i\xi_p(R^p{}_k{}^i{}_l + R^p{}_l{}^i{}_k)$$

with the usual convention for raising exponents having been used.

Hence using (4.16) one finds

$$Q_{2,2}(\sigma) = \tfrac{1}{3}\xi_i\xi_p(R^p{}_k{}^{ik} + R^p{}_l{}^{il}) - \tfrac{2}{3}\xi_l\xi_p(R^p{}_i{}^{li} + R^p{}_j{}^{lj}) = -\tfrac{2}{3}\xi_i\xi_p R^p{}_k{}^{ik}.$$

One also finds

$$Q_{2,3}(\sigma) = \tfrac{4}{3}\xi_i\xi_p\xi_k\xi_l(R^{pkil} + R^{plik}) = 0$$

by the asymmetry of R, and $Q_{2,4} = 0$.

To evaluate the integral in (6.6) we may assume the local coordinates chosen so that at $\pi(v) = x$

$$g^{ij} = g_{ij} = \delta_{ij},$$

and we obtain

$$u_1(x) = \tfrac{1}{3}\pi^{-n/2}\int \exp(-\textstyle\sum \xi_j{}^2) \displaystyle\sum_{i,p,k} \xi_i\xi_p R^p{}_k{}^i{}_k \, d\xi.$$

The contribution of $\sum_{i \neq p}$ is clearly zero so

$$\begin{aligned} u_1(x) &= \tfrac{1}{3}\pi^{-n/2}\int \exp(-\textstyle\sum \xi_j{}^2) \displaystyle\sum_{i,k} \xi_i{}^2 R^i{}_k{}^i{}_k \, d\xi \\ &= \tfrac{1}{6}\sum_{i,k} R^i{}_k{}^i{}_k. \end{aligned}$$

References

1. S. Agmon, "Lectures on Elliptic Boundary Value Problems," Van Nostrand–Reinhold, Princeton, New Jersey, 1965.

2a. M. Atiyah, R. Bott, and V. K. Patodi, On the heat equation and the index theorem, *Invent. Math.* **19** (1973), 279–330.

2b. M. Atiyah, R. Bott, and V. K. Patodi, On the heat equation and the index theorem, *Invent. Math.* **28** (1975), 277–280.

3. J. J. Duistermatt and V. W. Guillemin, The spectrum of positive elliptic operators and periodic bicharacteristics, *Invent. Math.* **29** (1975), 39–79.

4. P. Greiner, An asymptotic expansion for the heat equation, *Arch. Rational Mech. Anal.* **41** (1971), 163–218.

5. L. Hörmander, Pseudo-differential operators and hypoelliptic equations, *Proc. Symp. Pure Math* **10** (1968), 138–183.

6. N. Jacobson, "Lie Algebras," Wiley (Interscience), New York, 1962.

7. M. Kac, Toeplitz matrices, translation kernels, and a related problem in probability theory, *Duke Math. J.* **21** (1954), 501–509.

8. E. Kreyszig, "Introduction to Differential Geometry and Riemannian Geometry," Univ. of Toronto Press, Toronto, 1968.
9. D. Laugwitz, "Differential and Riemannian Geometry," Academic Press, New York, 1965.
10. S. Minakshisundaram and A. Pleijel, Some properties of the eigenfunctions of the Laplace operator on Riemannian manifolds, *Canad. J. Math.* **1** (1949), 242–256.
11. R. Narasimhan, "Analysis on Real and Complex Manifolds," North-Holland Publ., Amsterdam, 1968.
12. L. Nirenberg, Pseudo-differential operators, *Proc. Symp. Pure Math.* **16** (1970), 149–167.
13. B. E. Petersen, On the calculus of symbols for pseudo-differential operators, *Proc. Symp. Pure Math.* **16** (1970), 169–173. [Ph.D. Thesis, M.I.T., 1968.]
14. R. T. Seeley, Complex powers of an elliptic operator, *Proc. Symp. Pure Math.* **10** (1968), 288–307.
15. W. Shih, On the symbol of a pseudo-differential operator, *Bull. Amer. Math. Soc.* **74** (1968), 657–659.
16. G. Szegö, On certain hermitian forms associated with the Fourier series of a positive function, *Commun. Sem. Math. Univ. Lund, Tome Suppl.* (1952) 228–237.
17. H. Widom, Asymptotic expansions of determinants for families of trace class operators, *Indiana U. Math. J.* **27** (1978), 449–478.

A
B
C 8
D 9
E 0
F 1
G 2
H 3
I 4
J 5